M-Theory and Quantum Geometry

NATO Science Series

A Series presenting the results of activities sponsored by the NATO Science Committee. The Series is published by IOS Press and Kluwer Academic Publishers, in conjunction with the NATO Scientific Affairs Division.

A. Life Sciences	IOS Press
B. Physics	Kluwer Academic Publishers
C. Mathematical and Physical Sciences	Kluwer Academic Publishers
D. Behavioural and Social Sciences	Kluwer Academic Publishers
E. Applied Sciences	Kluwer Academic Publishers
F. Computer and Systems Sciences	IOS Press
1. Disarmament Technologies	Kluwer Academic Publishers
2. Environmental Security	Kluwer Academic Publishers
3. High Technology	Kluwer Academic Publishers
4. Science and Technology Policy	IOS Press
5. Computer Networking	IOS Press

NATO-PCO-DATA BASE

The NATO Science Series continues the series of books published formerly in the NATO ASI Series. An electronic index to the NATO ASI Series provides full bibliographical references (with keywords and/or abstracts) to more than 50000 contributions from international scientists published in all sections of the NATO ASI Series.
Access to the NATO-PCO-DATA BASE is possible via CD-ROM "NATO-PCO-DATA BASE" with user-friendly retrieval software in English, French and German (WTV GmbH and DATAWARE Technologies Inc. 1989).

The CD-ROM of the NATO ASI Series can be ordered from: PCO, Overijse, Belgium

Series C: Mathematical and Physical Sciences – Vol. 556

M-Theory and Quantum Geometry

Edited by

Lárus Thorlacius
Science Institute,
University of Iceland

and

Thordur Jonsson
Science Institute,
University of Iceland

Kluwer Academic Publishers

Dordrecht / Boston / London

Published in cooperation with NATO Scientific Affairs Division

Proceedings of the NATO Advanced Study Institute on
Quantum Geometry
Akureyri, Iceland
August 9–20, 1999

A C.I.P. Catalogue record for this book is available from the Library of Congress.

ISBN 0-7923-6474-0

Published by Kluwer Academic Publishers,
P.O. Box 17, 3300 AA Dordrecht, The Netherlands.

Sold and distributed in North, Central and South America
by Kluwer Academic Publishers,
101 Philip Drive, Norwell, MA 02061, U.S.A.

In all other countries, sold and distributed
by Kluwer Academic Publishers,
P.O. Box 322, 3300 AH Dordrecht, The Netherlands.

Printed on acid-free paper

Printed in the Netherlands.

Table of Contents

PREFACE

The fundamental structure of matter and spacetime at the shortest length scales remains an exciting frontier of basic research in theoretical physics. A unifying theme in this area is the quantization of geometrical objects. The majority of lectures at the Advanced Study Institute on Quantum Geometry in Akureyri was on recent advances in superstring theory, which is the leading candidate for a unified description of all known elementary particles and interactions. The geometric concept of one-dimensional extended objects, or strings, has always been at the core of superstring theory but in recent years the focus has shifted to include also higher-dimensional objects, so called D-branes, which play a key role in the non-perturbative dynamics of the theory.

A related development has seen the strong coupling regime of a given string theory identified with the weak coupling regime of what was previously believed to be a different theory, and a web of such "dualities" that interrelates all known superstring theories has emerged. The resulting unified theoretical framework, termed M-theory, has evolved at a rapid pace in recent years.

D-branes have also advanced our understanding of quantum effects in black hole physics, in particular the nature of black hole entropy and thermodynamics. An unexpected correspondence between the near-horizon physics of certain black holes and conformal quantum field theories has also been uncovered. The conformal theories are related to the gauge theory of the strong interaction so this line of development is of phenomenological interest besides offering a promising new approach to some quantum gravity problems.

Some alternative approaches to quantization of gravity were also discussed at the Advanced Study Institute. One of the most successful is "dynamical triangulations" which endeavors to construct the quantum geometry of spacetime using simple geometrical building blocks. This approach includes extensive numerical simulations of systems in various dimensions and has also provided techniques for quantizing geometric objects in a mathematically rigorous fashion.

Many individuals and organizations contributed to the success of the meeting in Akureyri. First of all I would like to thank the ASI students for their participation and enthusiasm for the program, and the lecturers for providing excellent reviews and up to date information on a wide range of rapidly developing subjects. The organizing of the Advanced Study Institute has from beginning to end been shared with Thordur Jonsson, with valuable input from our fellow organizers Paolo DiVecchia and Andy Strominger. I especially wish to thank Gerlinde Xander of the Science Institute at the University of Iceland, for her efforts in preparing for and ensuring the smooth running of the meeting, and also at the later stages of finalizing reports and preparing the proceedings for publication. I also want to thank Ellen Pedersen of Nordita in Copenhagen for her valuable assistance with the accounts.

The manager and staff at Hotel Edda in Akureyri provided an excellent environment for the participants, both inside and outside the lecture hall. The Advanced Study Institute was made possible by generous funding from a number of agencies. Our principal sponsor was the NATO Scientific and Environmental Affairs Division, with additional funding coming from NorFa and Nordita. The Icelandic Ministry of Education and the Town Council of Akureyri sponsored social events for the ASI participants.

Finally, I thank the staff of Kluwer Academic Publishers for their patience and cooperation.

Lárus Thorlacius

LECTURERS

Jan Ambjørn
Niels Bohr Institute, Blegdamsvej 17, 2100 Copenhagen Ø, DANMARK
ambjorn@nbi.dk

Curtis Callan
Dept. of Physics, Princeton University, Princeton, NJ 08544, USA
callan@feynman.princeton.edu

Thordur Jonsson
Science Institute, Univ. of Iceland, Dunhaga 3, IS-107 Reykjavik, ICELAND
thjons@raunvis.hi.is

Jerzy Jurkiewicz
Institute of Physcis, Jagellonian University, Reymonta 4, 30-059 Krakow, POLAND
jurkiewi@thrisc.if.uj.edu.pl

Jan Louis
MLU Halle, FB Physik, Friedemann-Bach-Platz 6, D-06108 Halle, GERMANY
jlouis@hera.physik.uni-halle.de

Yuri Makeenko
Inst. of Theoretical and Experimental Physics, B. Chermushkinskaya 25,
117218 Moscow, RUSSIA
makeenko@itep.ru

Andrew Strominger
Dept. of Physics, Harvard University, Cambridge, MA 02138, USA
strominger@physics.harvard.edu

Leonard Susskind
Department of Physics, Stanford University, Stanford, CA 94305, USA
susskind@sewerrat.stanford.edu

Washington Taylor
Center for Theoretical Physics, M I T, Cambridge, MA 02139, USA
wati@mit.edu

Larus Thorlacius
Science Institute, Univ. of Iceland, Dunhaga 3, IS-107 Reykjavik, ICELAND
lth@raunvis.hi.is

Paolo Di Vecchia
NORDITA, Blegdamsvej 17, DK-2100 Copenhagen, DENMARK
DIVECCHIA@nbivms.nbi.dk

Herman Verlinde
Princeton University, Dept. of Physics, Princeton University, Princeton, NJ 08544, USA
verlinde@princeton.edu

ASI Students

Carlo Angelantonj
Centre de Physique Theorique, Ecole Polytechnique, FR-91128 Palaiseau Cedex, FRANCE
angelant@cpht.polytechnique.fr

Sergei Antropov
Physical Faculty, Dniepropetrovsk State University, Naucnyj Str. 13, Dniepropetrovsk-10, UKRAINE
santrop@usa.net

Peter Austing
Theoretical Physics, 1 Keble Road, Oxford OX1 3NP, UK
pja@thphys.ox.ac.uk

Maximo Banados
Dept. de Fisica Teorica, Univ. de Zaragoza, Cuidad Universitaria, Zaragoza 50009, SPAIN
max@posta.unizar.es

Daniela Bigatti
Phys. Dept., Katholieke Univ. Leuven, Celestijnenleen 200 D, B-3001 Heverlee, BELGIUM
daniela.bigatti@fys.kuleuven.ac.be

Björn Brinne
FYSIKUM, Stockholm University, Box 6730, S-113 85 Stockholm, SWEDEN
brinne@physto.se

Ruth Britto-Pacumio
Dept. of Physics, Harvard University, Cambridge, MA 02138, USA
britto@physics.harvard.edu

Zdzislaw Burda
Institute of Physcis, Jagellonian University, Reymonta 4, 30-059 Krakow, POLAND
burda@agrest.if.uj.edu.pl

Gaspare Carbone
S.I.S.S.A. - I.S.A.S. via Beirut No. 2-4, I-34013 Trieste, ITALY
carbone@sissa.it

Daniel Cartin
Dept. of Physics, Penn State University, State College, PA 16802, USA
cartin@gravity.phys.psu.edu

Andrew H. Chamblin
Pembroke College, Cambridge CB2 1RF, UK
H.A.Chamblin@amtp.cam.ac.uk

Chang Shih Chan
Graduate College, Princeton, NJ 08544, USA
cschan@princeton.edu

Joao Correia
Niels Bohr Institute, Blegdamsvej 17, DK-2100 Copenhagen, DENMARK
correia@nbi.dk

Vladimir Damgov
Bulg. Academy of Sciences, Space Research Institute, JK"Mladost 1A" Bl. 522, Vh. 5, Ap. 98, 1784 Sofia, BULGARIA
vdamgov@bas.bg

Liana David
The University of Edinburgh, Dept. of Mathematics and Statistics, Mayfield Road, Edinburgh EH9 3JZ, UK
liana@maths.ed.ac.uk

Jaques Distler
Phys. Dept., University of Texas, RLM 9-308, MS C1600, Austin, TX 78712 USA
distler@golem.ph.utexas.edu

Olivier Dore
L A P T H, BP 110, F-74940 Annecy-le-Vieux, FRANCE
dore@lapp.in2p3.fr

Brynjolfur Eyjolfsson
Menntaskolinn a Akureyri, Eyrarlandsvegur 28, IS-600 Akureyri, ICELAND
bey@ma.is

Stephen Fairhurst
Penn State University, University Park, PA 16802, USA
fairhurs@phys.psu.edu

Daniel Friedan
Rutgers University, 126 Frelinghuysen Road, Piscataway, NJ 08854 , USA
friedan@physics.rutgers.edu

S. James Gates, Jr.
Dept. of Physics, Univ. of Maryland at College Park,
College Park, MD 207 42-4111, U S A
gates@katherine.physics.umd.edu

Florian Girelli
Village, Montagagne 09240, FRANCE
florian.girelli@usa.net

Ulf Gran
Dept. of Theortical Physics and, Mechanics, Chalmers University of Technology, S-41296 Göteborg, SWEDEN
gran@fy.chalmers.se

Marc Grisaru
Dept. of Physics, Brandeis University, Waltham, MA 02454, USA
marc@binah.cc.brandeis.edu

Vidar Gudmundsson
Science Institute, University of Iceland, Dunhaga 3, IS-107 Reykjavik, ICELAND
vidar@raunvis.hi.is

Holger Günther
MLU Halle FB Physik, Friedemann-Bach-Platz 6, D-06108 Halle, GERMANY
guenther@hera.physik.uni-halle.de

Parvis Haggi
Institute of Theoretical Physics, Box 6730, S-11385 Stockholm, SWEDEN
parviz@physto.se

Troels Harmark
Niels Bohr Institute, Blegdamsvej 17, DK-2100 Copenhagen, DENMARK
harmark@nbi.dk

Carl Herrmann
Centre de Phys. Theorique, CNRS - Luminy Case 907,
F-13288 Marseille Cedex 9, FRANCE
herrmann@cpt.univ-mrs.fr

Sigbjorn Hervik
Dept. of Physics, UiO, Postbox 1048, Blindern, N-0316 Oslo, NORWAY
sigbjorn.hervik@fys.uio.no

Laur Järv
Centre for Particle Theory, Dept. of Math. Sc., Science Labs., South Road, Durham, DH1 3LE, UK
Laur.Jarv@durham.ac.uk

Niels Karlsson
Univ. of Akureyri, Akureyri, ICELAND
niels@unak.is

Justin Khoury
Dept. of Physics, Princeton University, Princeton, NJ 08544, USA
jkhoury@princeton.edu

Marcia E. Knutt
Physics Dept., McGill University, 3600 University St.,
Montreal, QC H3A2T8, CANADA
knutt@physics.mcgill.ca

Kristjan R. Kristjansson
Science Institute, University of Iceland, Dunhaga 3, IS-107 Reykjavik, ICELAND
kristk@hi.is

Morten Krogh
Dept. of Physics, Princeton University, Princeton, NJ 08544, USA
krogh@princeton.edu

Luis Martin Kruczenski
Lindsbergsgatan 7B, S-75 240 Uppsala, SWEDEN
martink@teorfys.uu.se

Teresia Mansson
FYSIKUM, Box 6730, S-113 85 Stockholm, SWEDEN
teresia@physto.se

Dennis Gerard McKeon
Dept. of Applied Mathematics, University of Western Ontario,
London Ont., CANADA N6A 5B7
dgmckeon@julian.uwo.ca

Jeremy Michelson
Jefferson Physical Labs., Harvard University, Cambridge, MA 02138, USA
jeremy@yukawa.harvard.edu

Stefano Monni
Dept. of Applied Mathematics and, Theoretical Physics, University of Cambridge,
Silver Street, Cambridge CB3 9EW, UK
S.Monni@damtp.cam.ac.uk

Eric Novac
Dept. of Physics, Penn State University, University Park, PA 16802, USA
novak@phys.psu.edu

Sergei Odintsov
Tomsk State Pedagogical University, 634041 Tomsk, RUSSIA
odintsov@mail.tomsknet.ru

Aksel Hiorth Orsland
Dept. of Physics, Postbox 1048, Blindern, N-0316 Oslo, NORWAY
a.h.orsland@fys.uio.no

Ossi Pasanen
Helsinki Institute of Physics, P.O. Box 9, FIN-00014 University of Helsinki,
FINLAND
ossi.pasanen@helsinki.fi

Konstantin Savvidy
Dept. of Physics, Princeton University, Princeton, NJ 08544, USA
ksavvidi@princeton.edu

Jan Pieter van der Schaar
Inst. for Theoretical Physics, Nyenbrogh 4, 9747 AG Groningen,
THE NETHERLANDS
schaar@phys.rug.nl

Koenraad Schalm
Inst. for Theoretical Physics, State University of New York,
Stony Brook, NY 11794-3840, USA
konrad@insti.physics.sunysb.edu

Michael Schulz
Dept. of Physics, LeConte Hall, University of California, Berkeley, CA 94720, USA
mschulz@socrates.berkeley.edu

Khalil Shadiev
2-a Abdullo Kahhor Street, Samarkand, REPUBLIC OF UZBEKISTAN, 703024
lyceum@samuni.silk.org

Paul Shocklee
Dept. of Physics, Princeton University, Princeton, NJ 08544, USA
shocklee@princeton.edu

Sergey Solodukhin
Spinoza Institute, University of Utrecht, Leuvenlaan 4, , 3584 CE Utrecht, THE NETHERLANDS
S.Solodukhin@phys.uu.nl

Harald G. Svendsen
10 B 655 Kringsja, Sognsveien 218, N-0864 Oslo, NORWAY
h.g.svendsen@fys.uio.no

Oyvind Tafjord
507 West Drive, Princeton, NJ 08540, USA
otafjord@princeton.edu

Gudmar Thorleifsson
University of Bielefeld, FB Physik, P.O. Box 10 01 31, D-33501 Bielefeld, GERMANY
thorleif@physik.uni-bielefeld.de

Graziano Vernizzi
Physics Department, University of Parma, I-43100 Parma, ITALY
vernizzi@prix7.pr.infn.it

Anastasia Volovich
Dept. of Physics, Harvard University, Oxford Street # 17,
Cambridge, MA 02138, USA
nastya@physics.harvard.edu

John Wheater
Theoretical Physcis, 1 Keble Road, Oxford OX1 3NP, UK
j.wheater1@physics.ox.ac.uk

Maxim Zabzine
FYSIKUM, Stockholm University, Box 6730 S-11385 Stockholm, SWEDEN
zabzin@physto.se

Leopoldo Pando Zayas
Dept. of Physics, University of Michigan, Ann Arbor, MI 48105, USA
lpandoz@umich.edu

D BRANES IN STRING THEORY, I

PAOLO DI VECCHIA
Nordita,
Blegdamsvej 17, 2100 Copenhagen OE
Denmark
email: divecchia@nbivms.nbi.dk

AND

ANTONELLA LICCARDO
Dipartimento di Fisica, Universitá di Napoli,
and I.N.F.N. , Sezione di Napoli
Mostra d'Oltremare, Pad. 19, 80125 Napoli
Italy
email: liccardo@na.infn.it

Abstract. In these lectures we present a detailed description of the origin and of the construction of the boundary state that is now widely used for studying the properties of D branes.

1. Introduction

The existence of Dp-branes in string theories has been an essential ingredient for concluding that the five consistent and perturbatively inequivalent supersymmetric string theories in ten dimensions belong to a unique eleven dimensional theory that is called M-theory. In the framework of string theories their existence was required by T-duality in theories with both open and closed strings [1]. On the other hand classical solutions of the low-energy string effective action coupled to graviton, dilaton and $(p+1)$-form R-R potential were later constructed [2]. Since their tension is proportional to the inverse of the string coupling constant they correspond to new non-

[1]See Ref. [1] and references therein.
[2]See Ref. [2] and references therein.

L. Thorlacius and T. Jonsson (eds.), M-Theory and Quantum Geometry, 1-59.

perturbative states of string theory. At the end of 1995 Polchinski [3] provided strong arguments for identifying this new states with the Dp-branes required by T-duality opening the way to study their properties in string theory. In particular their interaction can be computed through the one-loop open string annulus diagram. On the other hand, since the very early days of string theory it is known that this one-loop open string diagram can be equivalently rewritten as a tree diagram in the closed string theory in which a closed string is generated from the vacuum, propagates for a while and then annihilates again in the vacuum. The state that describes the creation of closed string from the vacuum is called the boundary state, that first appeared in the literature [4] in the early days of string theory for factorizing the planar and non-planar loops of open string in the closed string channel. In the middle of the eighties after that the BRST invariant formulation of string theory became available the boundary states was considered again in a series of beautiful papers by Callan et al. [5], where, among other things, the ghost contribution was added and the boundary state with an external abelian gauge field was constructed. It was also used for deriving the gauge group of open string theories by requiring the tadpole cancellation [6]. Its extension to the case of Dirichlet boundary conditions was given in a series of beautiful papers written by M.Green et al. [7] for studying Dp-branes before it became clear that they were new states of string theory corresponding to the classical solution of the low-energy string effective action. In the last few years the boundary state has been widely used for studying properties of D branes in string theory.

In these lectures after a review of the main properties of perturbative string theory we discuss in detail T-duality for both open and closed string theories and we show how the requirement of T-duality in presence of open strings implies the existence of Dp-branes that are then identified with the new non-perturbative states obtained as classical solutions of the low-energy string effective action. Then, by requiring that the interaction between two Dp-branes gives the same result if we compute it in the open or in the closed string channel, we construct the boundary state that provides a stringy description of the simplest Dp brane solutions. Finally we show how to connect the boundary state to the supergravity classical solutions.

These lecture notes are partially based on the Ph.D. thesis of Antonella Liccardo.

2. Perturbative String Theory

The action of the bosonic string theory is

$$S = -\frac{T}{2}\int_M d^2\xi\sqrt{-h}\, h^{\alpha\beta}\partial_\alpha X^\mu \partial_\beta X_\mu \ , \tag{2.1}$$

where T is the string tension, M is the world-sheet of the string described by the world sheet coordinate $\xi^\alpha \equiv (\tau, \sigma)$, $h^{\alpha\beta}$ is the world sheet metric tensor and $h = \det h_{\alpha\beta}$. The string tension is related to the Regge slope by $T = (2\pi\alpha')^{-1}$.

The action in eq.(2.1) is invariant under local reparametrizations of the world sheet coordinates corresponding to $\xi^\alpha \to f^\alpha(\xi)$ and under local Weyl transformations corresponding to a local rescaling of the metric tensor $h_{\alpha\beta} \to \Lambda(\xi)h_{\alpha\beta}$. These symmetries allow one to bring the metric tensor in the form $h_{\alpha\beta} = e^\phi \eta_{\alpha\beta}$. This choice is referred to as the conformal gauge choice.

In this gauge the string action is still invariant under some residual local symmetries. It is in fact invariant under a combination of a Weyl rescaling and a local reparametrization $\xi^\alpha \to \xi^\alpha + \varepsilon^\alpha$ ($f^\alpha = 1 + \epsilon^\alpha$) satisfying the following condition

$$\partial^\alpha \varepsilon^\beta + \partial^\beta \varepsilon^\alpha = \Lambda(\sigma)\eta^{\alpha\beta}, \tag{2.2}$$

which corresponds to an infinitesimal conformal transformation. String theory in the conformal gauge is then conformal invariant.

The equation of motion for the string coordinate X^μ following from the action in eq.(2.1) in the conformal gauge is given by

$$(\partial_\sigma^2 - \partial_\tau^2)X^\mu = 0 \quad , \quad \mu = 0, ..., d-1, \tag{2.3}$$

while that for the metric implies the vanishing of the world sheet energy-momentum tensor

$$T_{\alpha\beta} \equiv \frac{2}{T\sqrt{-h}}\frac{\delta S}{\delta h^{\alpha\beta}} = \partial_\alpha X \cdot \partial_\beta X - \frac{1}{2}\eta_{\alpha\beta}\partial_\gamma X \cdot \partial^\gamma X = 0, \tag{2.4}$$

where d is the number of dimensions of the embedding space-time. By varying the action in eq.(2.1) in the conformal gauge, in addition to the previous eqs. of motion, we must also impose the following boundary conditions:

$$\int d\tau \left(\partial_\sigma X \cdot \delta X|_{\sigma=\pi} - \frac{1}{2}\partial_\sigma X \cdot \delta X|_{\sigma=0}\right) = 0, \tag{2.5}$$

where we have taken $\sigma \in [0, \pi]$

The previous boundary conditions can be satisfied in two different ways leading to two different theories. By imposing the periodicity condition

$$X^\mu(\tau, 0) = X^\mu(\tau, \pi), \tag{2.6}$$

and

$$\widetilde{X}^{\mu}(\bar{z}) = \hat{q}^{\mu} - i\sqrt{2\alpha'}\log\bar{z}\tilde{\alpha}_0^{\mu} + i\sqrt{2\alpha'}\sum_{n\neq 0}\frac{\tilde{\alpha}_n^{\mu}}{n}\bar{z}^{-n} \tag{3.66}$$

with $\alpha_0^{\mu} = \tilde{\alpha}_0^{\mu} = \sqrt{\alpha'/2}\,\hat{p}^{\mu}$. In the case of an open string theory one can introduce the variables:

$$z = e^{i(\tau-\sigma)} \quad ; \quad \bar{z} = e^{i(\tau+\sigma)} \tag{3.67}$$

and the string coordinate can be written as

$$X^{\mu}(z,\bar{z}) = \frac{1}{2}\left[X^{\mu}(z) + X^{\mu}(\bar{z})\right] \quad , \tag{3.68}$$

where X^{μ} is given in eq. (3.65) and $\sqrt{2\alpha'}\hat{p}^{\mu} = \alpha_0^{\mu}$. In superstring theory we must also introduce a conformal field with conformal dimension equal to 1/2 corresponding to the fermionic coordinate. In the closed string case we have two independent fields for the holomorphic and anti-holomorphic sectors which are obtained from eqs. (2.42) and (2.43) through the Wick rotation $\tau \to -i\tau$ and the conformal transformation $(\tau,\sigma) \to (z,\bar{z})$

$$\Psi^{\mu}(z) \sim \sum_t \psi_t z^{-t-1/2} \quad ; \quad \widetilde{\Psi}^{\mu}(\bar{z}) \sim \sum_t \tilde{\psi}_t \bar{z}^{-t-1/2} \tag{3.69}$$

In the open string case, starting from eq.(2.41) and applying the same operations we get again eqs.(3.69), but this time with the same oscillators.

In what follows we will explicitly consider only the holomorphic sector for the closed string. Analogous considerations hold for the antiholomorphic sector. In the case of an open string it is sufficient to consider the string coordinate at the string endpoint $\sigma = 0$. In both cases it is convenient to introduce a bosonic dimensionless variable:

$$x^{\mu}(z) \equiv X^{\mu}(z)/(\sqrt{2\alpha'}) = \tilde{q}^{\mu} - i\alpha_0^{\mu}\log z + i\sum_{n\neq 0}\frac{\alpha_n}{n}z^{-n} \quad , \tag{3.70}$$

where $\tilde{q} = \hat{q}/\sqrt{2\alpha'}$ and a fermionic one:

$$\psi^{\mu}(z) = -i\sum_t \psi_t z^{-t-1/2} \tag{3.71}$$

The theory can be quantized by imposing the following OPEs

$$x^{\mu}(z)x^{\nu}(w) = -\eta^{\mu\nu}\log(z-w) + \ldots; \quad \psi^{\mu}(z)\psi^{\nu}(w) = -\frac{\eta^{\mu\nu}}{z-w} + \ldots \quad , \tag{3.72}$$

where the dots denote finite terms for $z \to w$. Notice that these OPEs coincide with the 2-points Green's functions except for the Ramond case where the Green's function is equal to:

$$< \psi^{\mu}(z)\psi^{\nu}(w) >= -\frac{\eta^{\mu\nu}}{z-w}\frac{1}{2}\left[\sqrt{\frac{z}{w}}+\sqrt{\frac{w}{z}}\right] \quad . \tag{3.73}$$

Since, however, its singular behaviour when $z \to w$ is the same as in eq.(3.72), we use the contractions in eqs.(3.72) for both the NS and the R sector. In terms of the previous conformal fields we can define the generators of superconformal transformations:

$$G(z) = -\frac{1}{2}\psi \cdot \partial x \quad ; \quad T(z) = T^{x}(z) + T^{\psi}(z) = -\frac{1}{2}(\partial x)^2 - \frac{1}{2}\partial\psi \cdot \psi \quad . \tag{3.74}$$

Their mode expansion is given by:

$$L_n = \frac{1}{2\pi i}\oint dz z^{n+1} T(z) \qquad ; \qquad G_t = \frac{1}{2\pi i}\oint dz z^{t+1/2} G(z) \quad . \tag{3.75}$$

The conformal fields in eq.(3.74) satisfy the following OPEs:

$$T(z)T(w) = \frac{\frac{d}{dw}T(w)}{z-w} + 2\frac{T(w)}{(z-w)^2} + \frac{c/2}{(z-w)^4} + \ldots \quad , \tag{3.76}$$

$$T(z)G(w) = \frac{\partial/\partial w G(w)}{z-w} + \frac{3}{2}\frac{G(w)}{(z-w)^2} + \ldots \tag{3.77}$$

$$G(z)G(w) = \frac{2T(z)}{z-w} + \frac{d}{(z-w)^3} + \ldots \quad . \tag{3.78}$$

Using eqs.(3.75) it is easy to see that the previous OPEs imply the super Virasoro algebra in eq.(2.54) for both the NS and the R sector. But then the superconformal algebra that we get in the R sector differs from the one given in eq.(2.55). However, as we have noticed in the previous section, eq.(2.55) can be reduced to eq. (2.54) by translating L_0 in the R sector in eq.(2.55) by a constant:

$$L_0 \to L_0^{conf} \equiv L_0 + \frac{d}{16} = \sum_{n=1}^{\infty}(\alpha_{-n}\cdot\alpha_n + n\psi_{-n}\cdot\psi_n) + \alpha' p^2 + \frac{d}{16} \tag{3.79}$$

Therefore in the R sector we have two L_0 operators that are related by eq.(3.79). L_0 determines the spectrum of superstring through eq.(2.56), while L_0^{conf} that satisfies the algebra in eq.(2.54) encodes the correct conformal properties of the R sector.

In conformal field theory one introduces the concept of conformal or primary field $\Phi(z)$ with dimension h as the object that satisfies the following OPE with the energy-momentum tensor:

$$T(z)\Phi(w) = \frac{\partial_w \Phi(w)}{z-w} + h\frac{\Phi(w)}{(z-w)^2} + \ldots \quad . \tag{3.80}$$

From it one can compute the corresponding highest weight state $|\Phi\rangle$ by means of the following limiting procedure

$$|\Phi\rangle = \lim_{z\to 0} \Phi(z)|0\rangle \quad , \qquad \langle\Phi| = \lim_{z\to 0}\langle 0|\Phi^\dagger(z) \sim \lim_{z\to\infty}\langle 0|(z^2)^h \Phi(z) \tag{3.81}$$

The hermitian conjugate field $\Phi^\dagger$ in the previous expression has been defined as the field transformed under the conformal transformation $z \to 1/z$ apart from possibly a phase factor. Using the previous definition and the expression for L_0 given in eq.(3.75), it is easy to see that, if the conformal field has conformal dimension h, then the corresponding state is an eigenstate of L_0 with eigenvalue h and it is annihilated by all the Virasoro operators with $n > 0$, namely

$$L_0|\Phi\rangle = h|\Phi\rangle \quad ; \quad L_n|\Phi\rangle = 0 \quad . \tag{3.82}$$

In the bosonic case the physical conditions on the states given in eq.(2.27) imply that the vertex operators of the bosonic string theory $\mathcal{V}_\alpha(z)$ are conformal fields with conformal dimension equal to 1. This insures that the quantity $dz\mathcal{V}_\alpha(z)$ is invariant under conformal transformations. In the supersymmetric case the physical conditions in eq.(2.56) imply $L_0|\psi_{phys}\rangle = 1/2|\psi_{phys}\rangle$ in the NS-sector and $L_0|\psi_{phys}\rangle = 0$ in the R-sector. Therefore the corresponding vertex operators are not conformal fields with conformal dimension equal to 1 as in the bosonic string. On the other hand in the case of the R sector we have seen that, in order to determine the correct dimension of a vertex operator, we should use the operator $L_0^{\rm conf}$ that, for $d = 10$, acts on the spinorial ground state $|A\rangle$ as follows

$$L_0^{conf}|A\rangle = \frac{5}{8}|A\rangle \ , \tag{3.83}$$

This does not help, however, to get a vertex operator that has dimension equal to 1. In the case of superstring we will see that, in order to get the correct physical vertex operators in both NS and R sectors, one must add the contribution of the superghost degrees of freedom that will be discussed later on. The R vacuum state $|A\rangle$ in eq.(3.83) can be written in terms of the NS vacuum by introducing the spin field operator $S^A(z)$ satisfying the eq.:

$$\lim_{z\to 0} S^A(z)|0\rangle = |A\rangle \quad , \tag{3.84}$$

where $|0\rangle$ is the NS vacuum. Thus from eq. (3.83) we see that the spin field $S^A(z)$, which maps the NS vacuum into the R one, must have confromal dimension 5/8.

One can show that the spin field $S^A(z)$ satisfies the following OPEs [9]:

$$\psi^\mu(z)S_A(w) = (z-w)^{-1/2}(\Gamma^\mu)_{AB}S^B(w) + ...$$

$$S_A(w)S_B(w) = (z-w)^{-3/2}(\Gamma)_{AB}\psi^\mu + ... \quad . \tag{3.85}$$

Until now we have completely disregarded the analysis of the ghost and superghost degrees of freedom that, however, must be included in a correct Lorentz covariant quantization of string theory. They arise from the exponentiation of the Faddev-Popov determinant that is obtained when the string is quantized through the path-integral quantization. In particular in the bosonic case, choosing the conformal gauge, one gets the following ghost action [10]

$$S_{\text{ghosts}} \sim \int d^2z[b\bar{\partial}c + \text{c.c.}] \ , \tag{3.86}$$

where b and c are fermionic fields with conformal dimension equal respectively to 2 and -1. The ghost system of the bosonic string is a particular case of the fermionic bc system described in the Appendix corresponding to a screening charge $\mathcal{Q} = -3$ and a central charge of the Virasoro operator $c = -26$.

With the introduction of ghosts the string action in the conformal gauge becomes invariant under the BRST transformations and the physical states are characterized by the fact that they are annihilated by the BRST charge that in the bosonic case is given by

$$Q \equiv \oint \frac{dz}{2\pi i} J_{BRST}(z) \equiv \oint \frac{dz}{2\pi i} c(z)\left[T^x(z) + \frac{1}{2}T^{bc}(z)\right] \ , \tag{3.87}$$

where $T^x(z) = -1/2(\partial x)^2$, and T^{bc} is given in eq.(A.7) for $\lambda = 2$. It can be shown that Q is nilpotent if the space-time dimension $d = 26$. The physical states are annihilated by the BRST charge

$$Q|\psi_{\text{phys}}\rangle = 0 \tag{3.88}$$

This implies that the vertex operators corresponding to the physical states must satisfy the condition

$$[Q, \mathcal{W}(w)]_\eta = \oint_w \frac{dz}{2\pi i} J_{BRST}(z)\mathcal{W}(w) = 0 \ , \tag{3.89}$$

where $[,]_\eta$ means commutator ($\eta = -1$) [anticommutator ($\eta = 1$)] when the vertex operator is a bosonic [fermionic] quantity. By using the OPE it

Notice that the vertex in eq.(3.103) is in the picture -1, while the one in eq.(3.104) is in the picture $-1/2$. Vertex operators in different pictures are related through the picture changing procedure that we are now going to describe. Starting from a BRST invariant vertex $\mathcal{W}_t$ in the picture t (characterized by a value of P equal to t), where t is integer (half-integer) in the NS (R) sector, one can construct another BRST invariant vertex operator $\mathcal{W}_{t+1}$ in the picture $t+1$ through the following operation [9]

$$\mathcal{W}_{t+1}(w) = [Q, 2\xi(w)\mathcal{W}_t(w)]_\eta = \oint_w \frac{dz}{2\pi i} J_{BRST}(z)\; 2\xi(w)\mathcal{W}_t(w) \quad . \tag{3.106}$$

Using the Jacobi identity and the fact that $Q^2 = 0$ one can easily show that the vertex $\mathcal{W}_{t+1}(w)$ is BRST invariant:

$$[Q, \mathcal{W}_{t+1}]_\eta = 0 \tag{3.107}$$

On the other hand the vertex $\mathcal{W}_{t+1}(w)$ obtained through the construction in eq.(3.106) is not BRST trivial because the corresponding state contains the zero mode ξ_0 that is not contained in the Hilbert space of the $\beta\gamma$-system (see eq.(A.27)). In conclusion all the vertices constructed through the procedure given in eq.(3.106) are BRST invariant and non trivial in the sense that all give a non-vanishing result when inserted for instance in a tree-diagram correlator provided that the total picture number is equal to -2. Using the picture changing procedure from the vertex operator in eq.(3.103) we can construct the vertex operator in the 0 superghost picture which is given by [11]

$$\mathcal{W}_0(z) = c(z)\mathcal{V}_1(z) - \frac{1}{2}\gamma(z)\mathcal{V}_0(z) \quad . \tag{3.108}$$

Analogously starting from the massless vertex in the R sector in eq.(3.104) one can construct the corresponding vertex in an arbitrary superghost picture t.

In the closed string case the vertex operators are given by the product of two vertex operators of the open string. Thus for the massless NS-NS sector in the superghost picture $(-1,-1)$ we have

$$\mathcal{W}_{(-1,-1)} = \epsilon_{\mu\nu}\mathcal{V}^\mu_{-1}(k/2, z)\widetilde{\mathcal{V}}^\nu_{-1}(k/2, \bar{z}) \quad , \tag{3.109}$$

where $\mathcal{V}^\mu_{-1}(k/2, z) = c(z)\psi^\mu(z)e^{-\varphi(z)}e^{i\frac{k}{2}\cdot X(z)}$ and $\widetilde{\mathcal{V}}^\nu_{-1}$ is equal to an analogous expression in terms of the tilded modes. This vertex is BRST invariant if $k^2 = 0$ and $\epsilon_{\mu\nu}k^\nu = k^\mu\epsilon_{\mu\nu} = 0$.

In the R-R sector the vertex operator for massless states in the $(-\frac{1}{2},-\frac{1}{2})$ superghost picture is

$$\mathcal{W}_{(-1/2,-1/2)} = \frac{(C\Gamma^{\mu_1\ldots\mu_{m+1}})_{AB}\, F_{\mu_1\ldots\mu_{m+1}}}{2\sqrt{2}(m+1)!}\mathcal{V}^A_{-1/2}(k/2,z)\widetilde{\mathcal{V}}^B_{-1/2}(k/2,\bar{z}) \tag{3.110}$$

where $\mathcal{V}^A_{-1/2}(k/2,z) = c(z)S^A(z)e^{-\frac{1}{2}\varphi(z)}e^{i\frac{k}{2}\cdot X(z)}$ and

$$F_{\mu_1\ldots\mu_{m+1}} = \frac{(-1)^{m+1}}{2^5}u_D(k)(\Gamma_{\mu_1\ldots\mu_{m+1}}C^{-1})^{DE}\widetilde{u}_E(k) \ . \tag{3.111}$$

It is BRST invariant if $k^2 = 0$ and $F_{\mu_1\ldots\mu_m}$ is a field strength satisfying both the Maxwell equation ($dF = 0$) and the Bianchi identity ($d^*F = 0$). The two Weyl-Majorana spinors u_A and $\widetilde{u}_B$ may have the same or opposite chirality. In the first case one obtains type IIB theory while in the second case one obtains the type IIA theory. From eq.(3.111) one can see that the only field strengths which are allowed are those for $(m+1)$ odd in IIB theory and those for $(m+1)$ even in IIA theory. Moreover, from eq.(3.111) one can show that the field strenghts with values of m related by Hodge duality are not independent and one can restrict oneself to the values $(m+1) = 1,\ 3,\ 5$ in type IIB and $(m+1) = 2,4$ in type IIA string theory.

Since the physical state corresponding to the symmetric vertex given in eq.(3.110) cannot be used to compute its coupling with a D$-$ brane because the boundary state that we will construct in Sect. 7 is in an asymmetric picture, in the following we will explicitly write the vertex operator of a physical R-R state in the asymmetric picture $(-1/2,-3/2)$. It is given by [12]:

$$\mathcal{W}_{(-1/2,-3/2)} = \sum_{M=0}^{\infty}\frac{a_M}{2\sqrt{2}}\left(C\mathcal{A}^{(m)}\Pi_M\right)_{AB}\mathcal{V}^A_{-1/2+M}(k/2,z)\widetilde{\mathcal{V}}^B_{-3/2-M}(k/2,\bar{z}) \tag{3.112}$$

where

$$\left(C\mathcal{A}^{(m)}\right)_{AB} = \frac{(C\Gamma^{\mu_1\ldots\mu_m})}{m!}A_{\mu_1\ldots\mu_m} \ , \quad \Pi_q = \frac{1+(-1)^q\Gamma_{11}}{2} \tag{3.113}$$

and

$$\mathcal{V}^A_{-1/2+M}(k/2,z) = \partial^{M-1}\eta(z)\ldots\eta(z)c(z)S^A(z)e^{\left(-\frac{1}{2}+M\right)\varphi(z)}e^{i\frac{k}{2}\cdot X(z)} \tag{3.114}$$

$$\widetilde{\mathcal{V}}^B_{-3/2-M}(k/2,\bar{z}) = \bar{\partial}^M\widetilde{\xi}(\bar{z})\ldots\bar{\partial}\widetilde{\xi}(\bar{z})\widetilde{c}(\bar{z})\widetilde{S}^A(\bar{z})e^{\left(-\frac{3}{2}-M\right)\widetilde{\varphi}(\bar{z})}e^{i\frac{k}{2}\cdot\widetilde{X}(\bar{z})} \tag{3.115}$$

It can be shown that the vertex operator in eq.(3.112) is BRST invariant if $k^2 = 0$ and the following two conditions are satisfied

$$a_M = \frac{(-1)^{M(M+1)}}{[M!(M-1)!\ldots 1]^2} \ , \quad d^*A^{(m)} = 0 \ . \tag{3.116}$$

By acting with the picture changing operator on the vertex in eq.(3.112) it can be shown that one obtains the vertex in the symmetric picture in eq.(3.110). In particular one can show that only the first term in the sum in eq.(3.112) reproduces the symmetric vertex, while all the other terms give BRST trivial contributions.

4. T-Duality

The compactification of a dimension in string theory is characterised by the appearance of new interesting phenomena with respect to those already present in field theory. In fact, in the case of a closed string, together with the Kaluza-Klein (K-K) excitations, a new kind of states called winding states appear in the spectrum. It turns out that the bosonic closed string theory is invariant under the exchange of the winding modes with the K-K modes according to a transformation that is called T-duality. In the supersymmetric case, instead, this transformation is in general not a symmetry anymore but brings from a certain string theory to another string theory. For instance T-duality along a certain direction acts interchanging the IIA with the IIB theory. In the case of an open string, instead, this analysis naturally leads to the existence of other objects called Dp-branes.

Let us discuss in some detail the compactification and the T-duality invariance, starting with the bosonic closed string.

The most general solution of the eqs. of motion for the bosonic closed string in eq.(2.3) can be written as:

$$X^{\mu}(\tau,\sigma) = q^{\mu} + \sqrt{2\alpha'}\,(\alpha_0^{\mu} + \tilde{\alpha}_0^{\mu})\,\tau - \sqrt{2\alpha'}\,(\alpha_0^{\mu} - \tilde{\alpha}_0^{\mu})\,\sigma +$$

$$+ i\sqrt{\frac{\alpha'}{2}} \sum_{n\neq 0} \left(\frac{\alpha_n^{\mu}}{n} e^{-2in(\tau-\sigma)} + \frac{\tilde{\alpha}_n^{\mu}}{n} e^{-2in(\tau+\sigma)} \right) , \qquad (4.117)$$

where the momentum of the string is given by

$$p^{\mu} = \frac{1}{\sqrt{2\alpha'}}\,(\alpha_0^{\mu} + \tilde{\alpha}_0^{\mu}) \; . \qquad (4.118)$$

In the uncompactified case the two zero modes must be identified because the string coordinate must be invariant under $\sigma \to \sigma + \pi$ and the expression for the momentum in eq.(4.118) reduces to the one obtained just after eq.(2.13).

Let us compactify one of the space dimensions along a circle with radius R. This means that the string coordinate corresponding to this direction that we denote for simplicity with X without any index must be periodically identified as:

$$X \sim X + 2\pi R \; . \qquad (4.119)$$

As in the point particle case, the conjugate momentum corresponding to the compactified direction must be quantized as

$$p = \frac{n}{R} \quad \text{with} \quad n \in Z \, . \tag{4.120}$$

This is simply a consequence of the fact that the generator of the translations along the compact direction e^{ipa} must reduce to the identity for $a = 2\pi R$. Moreover in the compactified case the string coordinate X must be invariant under $\sigma \to \sigma + \pi$ apart from a factor $2\pi R w$ (w is an integer) as follows from eq.(4.119). This implies that

$$\pi\sqrt{2\alpha'}\,(\alpha_0 - \tilde{\alpha}_0) = 2\pi w R \qquad \text{with} \qquad w \in Z \, , \tag{4.121}$$

where w corresponds to the number of times that the closed string winds around the compact direction.

Eqs. (4.118) and (4.120) together with eq. (4.121) imply that the zero modes for the compact direction must have the following expression

$$\alpha_0 = \sqrt{\frac{\alpha'}{2}}\left(\frac{n}{R} + \frac{wR}{\alpha'}\right) \quad \text{and} \quad \tilde{\alpha}_0 = \sqrt{\frac{\alpha'}{2}}\left(\frac{n}{R} - \frac{wR}{\alpha'}\right) \, . \tag{4.122}$$

Inserting eq.(4.122) in eq.(2.23) and writing also the contribution of the uncompactified directions we get

$$L_0 = \frac{\alpha'}{4}\hat{p}^2 + \frac{1}{2}\alpha_0^2 + \sum_{n=1}^{\infty} \alpha_{-n}\cdot\alpha_n = \frac{\alpha'}{4}\hat{p}^2 + \frac{\alpha'}{4}\left(\frac{n}{R} + \frac{wR}{\alpha'}\right)^2 + \sum_{n=1}^{\infty} \alpha_{-n}\cdot\alpha_n \, , \tag{4.123}$$

and

$$\tilde{L}_0 = \frac{\alpha'}{4}\hat{p}^2 + \frac{\alpha'}{4}\left(\frac{n}{R} - \frac{wR}{\alpha'}\right)^2 + \sum_{n=1}^{\infty} \tilde{\alpha}_{-n}\cdot\tilde{\alpha}_n \, , \tag{4.124}$$

The mass operator becomes

$$M^2 = \frac{2}{\alpha'}\left[\sum_{n=1}^{\infty}(\alpha_{-n}\cdot\alpha_n + \tilde{\alpha}_{-n}\cdot\tilde{\alpha}_n) - 2\right] + \left(\frac{n}{R}\right)^2 + \left(\frac{wR}{\alpha'}\right)^2 \, . \tag{4.125}$$

From the previous expression we see that the spectrum of the closed string has been enriched by the appearance of two kinds of particles: the usual K-K modes which contribute to the energy with $\frac{n}{R}$ together with some new excitations that are called winding modes because they can be thought of as generated by the winding of the closed string around the compact direction which in fact contributes to the energy of the system as

$$T\, 2\pi R w = \frac{wR}{\alpha'} \, , \tag{4.126}$$

where $T = 1/(2\pi\alpha')$ is the string tension. All previous formulas can be trivially generalized to the case of a toroidal compactified theory in which more then one coordinate X^ℓ is compactified on circles with radii $R^{(\ell)}$. Of course in this case we will have K-K and winding modes corresponding to all the compactified directions.

From eq.(4.125) we see that the spectrum of the theory is invariant under the exchange of KK modes with winding modes together with an inversion of the radius of compactification:

$$w \leftrightarrow n \qquad ; \qquad R \leftrightarrow \hat{R} \equiv \frac{\alpha'}{R} . \tag{4.127}$$

This is called a T-duality transformation and $\hat{R}$ is the compactification radius of the T-dual theory. It can also be shown that both the partition function and the correlators are invariant under T-duality. This means that T-duality is a symmetry of the bosonic closed string theory. As a consequence of this invariance, whenever we have to consider compactified theories, we can limit ourselves to the case $R \geq \sqrt{\alpha'}$. That is the reason why $\sqrt{\alpha'}$ is often called the minimal length of the string theory.

Substituting eq.(4.127) into eq.(4.122) we obtain the action of T-duality on the zero modes

$$\alpha_0 \rightarrow \alpha_0 \qquad ; \qquad \widetilde{\alpha}_0 \rightarrow -\widetilde{\alpha}_0 . \tag{4.128}$$

This transformation, however, changes the operators $\widetilde{L}_n$ in eq.(2.17) and then does not leave invariant the physical subspace. In order to keep the physical subspace invariant we must extend the previous transformation property to all the non-zero oscillators:

$$\alpha_n \rightarrow \alpha_n \qquad ; \qquad \widetilde{\alpha}_n \rightarrow -\widetilde{\alpha}_n \qquad \forall n \in Z . \tag{4.129}$$

This transformation leaves obviously also invariant the mass of the states given in eq.(4.125).

Eqs.(4.128) and (4.129) allow us to define the action of T-duality directly on the string coordinate X. In fact writing

$$X = \frac{1}{2}\left(X_- + X_+\right), \tag{4.130}$$

where

$$X_- = q + 2\sqrt{2\alpha'}(\tau - \sigma)\alpha_0 + i\sqrt{2\alpha'}\sum_{n\neq 0}\frac{\alpha_n}{n}\, e^{-2in(\tau-\sigma)} , \tag{4.131}$$

and

$$X_+ = q + 2\sqrt{2\alpha'}(\tau + \sigma)\widetilde{\alpha}_0 + i\sqrt{2\alpha'}\sum_{n\neq 0}\frac{\widetilde{\alpha}_n}{n}\, e^{-2in(\tau+\sigma)} , \tag{4.132}$$

we see from eqs. (4.128) and (4.129) that the T-dual coordinate $\hat{X}$ must satisfy the conditions

$$\partial_\tau X \to \partial_\tau \hat{X} = -\partial_\sigma X \quad ; \quad \partial_\sigma X \to \partial_\sigma \hat{X} = -\partial_\tau X. \tag{4.133}$$

They are satisfied if the T-dual coordinate is equal to

$$\hat{X} = \frac{1}{2}(X_- - X_+) \ . \tag{4.134}$$

Therefore the T-duality transformation acts on the right sector as a parity operator changing sign of the right moving coordinate X_+ and leaving unchanged the left moving one X_-.

In an open string theory the string coordinate does not satisfy any periodicity requirement on σ. This implies that in its compactified version there are only K-K modes, while the winding modes are absent. This could suggest that T-duality is not a symmetry of the open string theory. Such a conclusion, however, leads to some problem when we remember that theories with open strings also contain closed strings. Let us consider a theory with open and closed strings with $d-p-1$ directions compactified on circles with radii R^ℓ and take the limit

$$R^\ell \to 0 \qquad \forall \text{ compact direction} \ , \tag{4.135}$$

In this limit the open string theory loses effectively $d - p - 1$ directions because all the K-K modes become infinitely massive decoupling from the spectrum and because there cannot be any open string oscillation along the directions with zero radii. Therefore in this limit the open string will appear to only be living in a $p + 1$-dimensional subspace of the entire d-dimensional target space. Let us analyze what happens in the same limit in the closed string sector. When the radii of the compact dimensions are vanishing the K-K modes again decouple, while the winding modes will appear as a continuum of states. This is a first indication that for closed strings the compact directions do not disappear from the theory as it happened for open strings! More precisely, in the closed string sector we can perform a T-duality transformation, that is allowed because it is a symmetry of this sector, and in so doing we can completely restore all the d space-time dimensions, as a consequence of the fact that in the limit in eq.(4.135) the T-dual radii go to infinity. But in this way we would end up with a theory in which open strings live in a $p + 1$-dimensional subspace of the entire space-time, while closed strings live in the entire d-dimensional target space. This mismatch can be solved by requiring that, in the T-dual picture, open string still can oscillate in d dimensions, while their endpoints are fixed on a $p + 1$-dimensional hyperplane that we call Dp-brane. Open strings with their endpoints fixed on these hyperplanes satisfy Dirichlet boundary conditions in the $d-p-1$ transverse directions. They are allowed boundary conditions as we have already seen in eq.(2.8) although they destroy the Poincarè invariance of the theory.

In conclusion, in order to avoid a different behaviour between the closed and the open sector of a string theory, we must require that the action of T-duality on an open string theory consists in transforming Neumann boundary conditions into Dirichlet ones. This can, in fact, be very naturally obtained if we extend the definition of the T-dual coordinate given in eq.(4.134) to the open string case. In this way we obtain the following T-dual open string coordinate:

$$\hat{X}^{\ell} = \frac{1}{2}\left[X_{-}^{\ell} - X_{+}^{\ell}\right] , \tag{4.136}$$

where now the left and right movers contain the same set of oscillators

$$X_{-}^{\ell} = q^{\ell} + c^{\ell} + \sqrt{2\alpha'}(\tau - \sigma)\alpha_0^{\ell} + i\sqrt{2\alpha'}\sum_{n\neq 0}\frac{\alpha_n^{\ell}}{n}\, e^{-in(\tau-\sigma)} , \tag{4.137}$$

and

$$X_{+}^{\ell} = q^{\ell} - c^{\ell} + \sqrt{2\alpha'}(\tau + \sigma)\alpha_0^{\ell} + i\sqrt{2\alpha'}\sum_{n\neq 0}\frac{\alpha_n^{\ell}}{n}\, e^{-in(\tau+\sigma)} , \tag{4.138}$$

From eqs.(4.136), (4.137) and (4.138) one can immediately see that T-duality has transformed a string coordinate satisfying Neumann boundary conditions and given by $1/2\left[X_{-}^{\ell} + X_{+}^{\ell}\right]$ into a T-dual one satisfying Dirichlet boundary conditions and given in eq.(4.136). Of course it is also true that, if we had started with a string coordinate satisfying Dirichlet boundary conditions we would have obtained a T-dual coordinate satisfying Neumann ones.

The fact that open strings satisfy Dirichlet boundary conditions implies the existence in the theory of objects, called the Dp-branes, that are characterized by the fact that open strings have their endpoints attached to them. From the three previous equations it also follows that

$$\hat{X}^{\ell}(\pi) - \hat{X}^{\ell}(0) = -2\pi\alpha' p^{\ell} = -\frac{2\pi\alpha' n^{(\ell)}}{R^{(\ell)}} = -2\pi n^{\ell}\hat{R}^{(\ell)} \Rightarrow \hat{X}^{\ell}(\pi) \sim \hat{X}^{\ell}(0) , \tag{4.139}$$

This means that in the T-dual theory the two endpoints of the open string are attached to the same D-brane.

From the previous construction it also follows that each Dp-brane can be transformed into a Dp'-brane through an appropriate sequence of compactifications and T-duality transformations. In fact let us start with a

Dp-brane embedded in a d-dimensional space-time and let us compactify one of the space directions X^α that lies in its world-volume on a circle with radius $R^{(\alpha)}$. The action of a T-duality transformation on this coordinate has the effect that an open string attached to the brane changes from Neumann to Dirichlet boundary conditions in that direction. Therefore, the brane 'loses' one longitudinal coordinate which becomes transverse and is transformed into a D$(p-1)$-brane embedded into a space-time which has the direction $\hat{X}^\alpha$ compactified on a circle of radius $\hat{R}^{(\alpha)} = \alpha'/R^{(\alpha)}$. To obtain a D$(p-1)$-brane living in the d-dimensional uncompactified space-time we need to make the decompactification limit in the T-dual theory, namely

$$\hat{R}^{(\alpha)} \to \infty, \quad \text{or equivalently} \quad R^{(\alpha)} \to 0\,. \tag{4.140}$$

In the same way, if instead of compactifying one of the space-time directions which are longitudinal to the Dp-brane, we compactify one of the directions transverse to the brane, and then we act with a T-duality transformation on this coordinate we will get a D$(p+1)$-brane embedded into a space-time with one compact direction. Then taking the limit in eq.(4.140) we get again the uncompactified theory.

To conclude we observe that open strings satisfying Neumann boundary conditions in all the directions can be thought as being attached to a space-filling brane that is a D25-brane in the bosonic string or a D9-brane in the superstring. Therefore, as a consequence of the previous discussion, starting with a space-filling brane through a T-duality transformation we can obtain an arbitrary Dp-brane. More precisely, a Dp-brane can be obtained from a space-filling brane by first compactifying $d-p-1$ directions, then performing a T-duality transformation and finally taking the decompactification limit.

Up to now we have treated a Dp-brane as a pure geometrical hyperplane to which open strings are attached and we have completely disregarded the excitations of the attached open strings. But we will see that, as soon as we let them come into play, they provide dynamical degrees of freedom to the Dp-brane.

Among all possible excitations of an open string the massless ones have the peculiarity of not changing the energy of the Dp-brane to which the open string is attached. Therefore from the brane point of view they can be interpreted as collective coordinates of the brane.

In absence of Chan-Paton factors, the massless excitations of an open string with Neumann boundary conditions in all directions are described by a d-dimensional abelian gauge potential A^μ. From the previous discussion it can be thought as a gauge field living on the space-filling brane. Then by compactifying $d_\perp = d-p-1$ dimensions and making a T-duality transformation in each of the compact directions, followed by a decompactification limit in the T-dual theory, we see that the vector potential $\hat{A}^\mu$ splits in a $(p+1)$-dimensional vector $\hat{A}^\alpha$ with $\alpha \in \{0, ..., p\}$ and $d_\perp$ scalars fields. The most natural interpretation of the T-dual version of the abelian gauge field is that the longitudinal coordinates $\hat{A}^\alpha$ still describe a gauge field living on the Dp-brane while the $d_\perp$ scalars coming from the transverse components $\hat{A}^\ell$, with $\ell \in \{p+1, ..., d-1\}$, appear as the transverse coordinates of the Dp-brane.

This interpretation becomes more clear as soon as we introduce a non-abelian $U(N)$ gauge group in the open string theory through the Chan-Paton factors and in addition we turn on Wilson lines. The Chan-Paton procedure for introducing non-abelian gauge degrees of freedom on an open string consists in adding non-dynamical degrees of freedom at each of its two endpoints. A generic string state will therefore be denoted with a ket $|\alpha, i, \bar{\jmath}\rangle$ where α describes the usual degrees of freedom of a string, while the indices i and $\bar{\jmath}$ refer to the gauge degrees of freedom. In the case of a $U(N)$ gauge group i transforms according to the fundamental representation N, while $\bar{\jmath}$ according to the complex conjugate representation $\bar{N}$:

$$|i'\rangle = U_{i'i}|i\rangle \quad ; \quad |\bar{\jmath}'\rangle = |\bar{\jmath}\rangle U^+_{jj'} \ . \tag{4.141}$$

If we now introduce a basis of $N \times N$ matrices λ^a_{ij}, expand an open string state as

$$|\alpha, a\rangle = \sum_{i,j=1}^{N} |\alpha, i, \bar{\jmath}\rangle \lambda^a_{ij} \ , \tag{4.142}$$

and use eq. (4.141) we see that the transformations in eq.(4.141) act on the matrices λ^a_{ij} as follows

$$\lambda^a_{ij} \rightarrow U_{i'i}\lambda^a_{ij}U^+_{jj'} = \left(U\lambda^a U^+\right)_{i'j'} \ . \tag{4.143}$$

This means that the matrix λ^a_{ij} transforms according to the adjoint representation $(N \times \bar{N})$ of $U(N)$ which is in fact the appropriate representation for a gauge field.

Let us now consider the effect of compactification in the presence of Chan-Paton factors. For the sake of simplicity we compactify just one coordinate that we denote with X without any index and turn on a pure gauge field of the form

$$(A)_{ij} = \frac{1}{2\pi R}\mathrm{diag}(\theta_1, ..., \theta_N) \ . \tag{4.144}$$

This corresponds to a pure gauge configuration generated by the matrix:

$$U = \mathrm{diag}\left(e^{iX\frac{\theta_1}{2\pi R}}, ..., e^{iX\frac{\theta_N}{2\pi R}}\right) \ , \tag{4.145}$$

because the gauge field configuration in eq.(4.144) can be written as

$$A = -iU^{-1}\partial U \ . \tag{4.146}$$

But in the case of a compact coordinate the presence of a pure gauge field affects the parallel transport along the compact dimension and we get non-zero Wilson lines:

$$e^{i\int_0^{2\pi R} Adx} = \mathrm{diag}(e^{i\theta_1}, ..., e^{i\theta_N}) \ . \tag{4.147}$$

In particular the parallel transport around the compact coordinate transforms $|i\rangle$ and $|\bar{j}\rangle$ as follows

$$|i\rangle \to e^{i\theta_i}|i\rangle \quad ; \quad |\bar{j}\rangle \to e^{-i\theta_j}|\bar{j}\rangle \ , \tag{4.148}$$

and therefore the open string state transforms as

$$|\alpha, a\rangle = \sum_{i,j=1}^{N} e^{i(\theta_i - \theta_j)}|\alpha, i, \bar{j}\rangle\lambda_{ij}^a \ . \tag{4.149}$$

The presence of Wilson lines changes the possible values of the momentum of the state $|\alpha, i, \bar{j}\rangle$. In fact in this case a translation of $2\pi R$ acts both on the string and the gauge degrees of freedom that are located at its endpoints. Requiring that this combined action leaves the state invariant:

$$e^{2\pi i R\hat{p}} e^{i(\theta_i - \theta_j)}|\alpha, i, \bar{j}\rangle = |\alpha, i, \bar{j}\rangle \ , \tag{4.150}$$

we get that the momentum of the state, that we call p, is equal to

$$p = \frac{n}{R} + \frac{(\theta_j - \theta_i)}{2\pi R} \ . \tag{4.151}$$

Let us now see what are the consequences of the presence of Wilson lines in the T-dual theory. Inserting eq.(4.151) in eq.(4.139) we get

$$\hat{X}(\pi) - \hat{X}(0) = -(2\pi n + \theta_j - \theta_i)\hat{R} \sim -(\theta_j - \theta_i)\hat{R} \ . \tag{4.152}$$

This means that the open string is stretching between two Dp-branes whose coordinates are $\theta_i\hat{R}$ and $\theta_j\hat{R}$. Moreover remembering eq.(4.144) we immediately see that

$$\theta_i\hat{R} = 2\pi\alpha'(A)_{ii} \quad ; \quad \theta_j\hat{R} = 2\pi\alpha'(A)_{jj} \ . \tag{4.153}$$

and we can conclude that turning on $U(N)$ Wilson lines in a theory of open strings along a compactified direction corresponds, in the T-dual theory, to introduce N Dp-branes located respectively at

$$X_1 = -2\pi\alpha'(A)_{11}, ..., X_N = -2\pi\alpha'(A)_{NN} \tag{4.154}$$

In this way the transverse components of a $U(N)$ gauge field carried by an open string are correctly interpreted as the coordinates of N Dp-branes.

In superstring theory the effect of T-duality on the bosonic coordinates is exactly the same as discussed for the bosonic string, namely T-duality acts as a parity transformation over the tilded sector

$$X = \frac{1}{2}(X_- + X_+) \to \hat{X} = \frac{1}{2}(X_- - X_+) \ . \tag{4.155}$$

For the fermionic coordinates the transformations under T-duality can be fixed by requiring the superconformal invariance of the theory which imposes

$$\psi_+ \to -\psi_+ \quad ; \quad \psi_- \to \psi_- \ , \tag{4.156}$$

or in terms of the oscillators

$$\widetilde{\psi}_t \to -\widetilde{\psi}_t \quad ; \quad \psi_t \to \psi_t \ , \tag{4.157}$$

This transformation propriety of the fermionic coordinates can also be understood as due to the requirement that the subspace of the physical states of the superstring, defined by eqs.(2.56) and by the analogous ones for the right sector, is left invariant by T-duality. Therefore, looking at the structure of the operator $\widetilde{G}_t$ given in eq.(2.48) and taking into account eq.(4.129), we obtain again eq.(4.157).

5. Classical Solutions Of The Low-Energy String Effective Action

In the previous sections we have seen that the requirement of invariance under T-duality transformations in presence of open strings implies the existence of p-dimensional objects called Dp-branes to which open strings can be attached determining their dynamics. Although these objects are required by T-duality their meaning is still rather obscure in the present framework. On the other hand, following a completely different line of research with the aim to get some non-perturbative information about string theories some people were investigating classical solutions of the low-energy string effective action. The underlying idea was in fact that, as the construction of 't Hooft-Polyakov monopoles in non abelian gauge theories teaches us many things about the non-perturbative structure of non-abelian gauge theories, so from the study of classical solutions of the low-energy string effective action one could learn a great deal on non-perturbative aspects of string theories. It turns out that starting from the low-energy string effective action one finds solutions of the classical equations of motion corresponding to p-dimensional objects. In the following we just want to remind their main properties.

The starting point is the low-energy string effective action containing the graviton, the dilaton and only one n-form potential, that written in the Einstein frame is given by:

$$S = \frac{1}{2\kappa^2}\int d^dx\sqrt{-g}\left[R - \frac{1}{2}(\nabla\varphi)^2 - \frac{1}{2(n+1)!}e^{-a\varphi}\left(F_{n+1}\right)^2\right] \quad , \qquad (5.158)$$

where $n = p+1$ and $F_{n+1} = dC$. For simplicity we have neglected all fermionic fields and the other NS-NS and R-R fields. An electric Dp-brane corresponds to the following ansatz:

$$ds^2 = [H(r)]^{2A}\left(\eta_{\alpha\beta}dx^\alpha dx^\beta\right) + [H(r)]^{2B}\left(\delta_{ij}dx^i dx^j\right) \quad , \qquad (5.159)$$

for the metric g, and

$$e^{-\varphi(x)} = [H(r)]^\tau \quad , \quad C_{01\ldots p}(x) = \pm\sqrt{2\sigma}[H(r)]^{-1} \quad , \qquad (5.160)$$

for the dilaton φ and for the R-R $(p+1)$-form potential C respectively. The two signs in C correspond to the brane and the anti-brane case and $H(r)$ is assumed to be only a function of the square of the transverse coordinates $r = x_\perp^2 = x_i x^i$. If the parameters are chosen as

$$A = -\frac{d-p-3}{2(d-2)} \quad , \quad B = \frac{p+1}{2(d-2)} \quad , \quad \tau = \frac{a}{2} \quad , \quad \sigma = \frac{1}{2} \quad , \qquad (5.161)$$

with a obeying the equation

$$\frac{2(p+1)(d-p-3)}{d-2} + a^2 = 4 \quad , \qquad (5.162)$$

then the function $H(r)$ satisfies the flat space Laplace equation. An extremal p-brane solution is constructed by introducing in the right hand side of the eqs. of motion following from the action in eq.(5.158) a δ-function source term in the transverse directions. If we restrict ourselves to the simplest case of just one p-brane, we obtain the following expression for $H(r)$

$$H(r) = 1 + 2\kappa T_p G(r) \quad , \qquad (5.163)$$

where

$$G(r) = \begin{cases} \left[(d-p-3)r^{(d-p-3)}\Omega_{d-p-2}\right]^{-1} & p < d-3 \ , \\ -\frac{1}{2\pi}\log r & p = d-3 \ , \end{cases} \qquad (5.164)$$

with

$$\Omega_q = 2\pi^{(q+1)/2}/\Gamma\big((q+1)/2\big) \qquad (5.165)$$

being the area of a unit q-dimensional sphere S_q. For future use it is convenient to introduce the quantity:

$$Q_p = \mu_p \frac{\sqrt{2}\,\kappa}{(d-p-3)\,\Omega_{d-p-2}} \quad ; \quad \mu_p \equiv \sqrt{2}T_p \ . \tag{5.166}$$

and (if $p < d-3$) to rewrite $H(r)$ in eq.(5.163) as follows:

$$H(x) = 1 + \frac{Q_p}{r^{d-3-p}} \ , \tag{5.167}$$

The classical solution has a mass per unit p-volume, M_p and an electric charge with respect to the R-R field, μ_p, given respectively by

$$M_p = \frac{T_p}{\kappa} \ , \qquad \mu_p = \pm\sqrt{2}T_p \ . \tag{5.168}$$

The fundamental observation made by Polchinski [3] has been to identify the Dp-branes required by T-duality with the p-branes obtained as classical solutions of the low-energy string effective action. Therefore, on the one hand the p-branes are new non-perturbative states of string theory and on the other hand have the important property that open string have their endpoints attached on them. The latter property will allow one to compute their interactions and more in general to study their properties by computing open string one-loop diagrams. On the other hand we should not be worried that the Dirichlet boundary conditions break Poincarè invariance because this happens in presence of any kind of solitonic state. In the next chapter we will introduce the boundary state and we will show that it provides a stringy description of these new states.

6. Bosonic Boundary State

As discussed in the previous section Dp-branes are extended p dimensional objects characterized by the fact that open strings can have their endpoints attached to them. In general they are dynamical and non rigid objects, that can fluctuate in shape and on which external fields can live. In these lectures we limit ourselves to treat them as static and rigid objects.

The open string with the endpoint at $\sigma = 0$ attached to a Dp-brane satisfies the usual Neumann boundary conditions along the directions longitudinal to the world volume of the brane

$$\partial_\sigma X^\alpha|_{\sigma=0} = 0 \qquad \alpha = 0, 1, ..., p \ , \tag{6.169}$$

and Dirichlet boundary conditions along the directions transverse to the brane

$$X^i|_{\sigma=0} = y^i \qquad i = p+1, ...d-1 \ , \tag{6.170}$$

where y^i are the coordinates of the brane and d is the dimension of the Minkowski space-time, that in the case of the bosonic string is equal to $d = 26$.

As the interaction between two superconducting plates is obtained by computing the vacuum fluctuation of the electromagnetic field that gives rise to the Casimir effect, so the interaction between two Dp-branes is given by the vacuum fluctuation of an open string that is stretching between them. This means that their interaction is simply given by the one-loop open string "free-energy" which is usually represented by the annulus or equivalently by the cylinder diagram. From either of those two diagrams it is easy to see that by exchanging the variables σ and τ the one-loop open string amplitude can also be viewed as a tree diagram of a closed string created from the vacuum, propagating for a while and then annihilating again into the vacuum. These two equivalent descriptions of the same diagram are called respectively the 'open-channel' and the 'closed-channel'. We want to stress that the physical content of the two descriptions is a priori completely different. In the first case we describe the interaction between two Dp-branes as a one-loop amplitude of open strings, which is the amplitude of a quantum theory of open strings, while in the second case we describe the same interaction as a tree-level amplitude of closed strings, which is instead a classical amplitude in a theory of closed strings. The fact that these two descriptions are equivalent is a consequence of the conformal symmetry of string theory that allows one to connect the two apriori different descriptions.

To show that, let us consider a one-loop diagram with an open string circulating in it and stretching between two parallel Dp-branes with coordinates respectively given by $(y^{p+1}, ..., y^{d-1})$ and $(w^{p+1}, ..., w^{d-1})$. The open string satisfies the boundary conditions in eq.(6.169) both at $\sigma = 0$ and $\sigma = \pi$ along the world-volume directions of the brane, while along the transverse directions satisfies the following equations:

$$X^i|_{\sigma=0} = y^i \qquad X^i|_{\sigma=\pi} = w^i \qquad i = p+1, ..., d-1 \ , \tag{6.171}$$

where we take σ and τ in the two intervals $\sigma \in [0, \pi]$ and $\tau \in [0, T]$.

We now want to find a conformal transformation acting on the previous open string boundary conditions in order to transform them into the boundary conditions for a closed string propagating between the two Dp-branes. In terms of the complex coordinate $\zeta \equiv \sigma + i\tau$, a conformal transformation simply transforms $\zeta \to f(\zeta)$, where $f(\zeta)$ is an arbitrary holomorphic function of ζ. Let us consider the following conformal transformation

$$\zeta = \sigma + i\tau \to -i\zeta = \tau - i\sigma \ . \tag{6.172}$$

After the inversion $\sigma \rightarrow -\sigma$ the previous conformal transformation simply amounts to exchange σ with τ and viceversa

$$(\sigma, \tau) \rightarrow (\tau, \sigma) \ . \tag{6.173}$$

Finally in order to have the closed string variables σ and τ to vary in the intervals $\sigma \in [0, \pi]$ and $\tau \in [0, \hat{T}]$ corresponding to a closed string propagating between the two D branes one must perform the following conformal rescaling

$$\sigma \rightarrow \frac{\pi}{T}\sigma \qquad \qquad \tau \rightarrow \frac{\pi}{T}\tau \ , \tag{6.174}$$

with $\hat{T} = \pi^2/T$. We have therefore constructed a conformal transformation that brings us from the open string to the closed string channel. In the closed string channel we need to construct the two boundary states $|B_X\rangle$ that describe the two Dp-branes respectively at $\tau = 0$ and $\tau = \hat{T}$. The equations that characterize these states are obtained by applying the conformal transformation previously constructed to the boundary conditions for the open string given in eqs.(6.169) and (6.171). At $\tau = 0$ we get the following conditions:

$$\partial_\tau X^\alpha|_{\tau=0}|B_X\rangle = 0 \qquad \qquad \alpha = 0, ..., p \ , \tag{6.175}$$

$$X^i|_{\tau=0}|B_X\rangle = y^i \qquad \qquad i = p+1, ..., d-1 \ . \tag{6.176}$$

Analogous conditions can be obtained for the Dp-brane at $\tau = \hat{T}$.

The previous equations can be easily written in terms of the closed string oscillators by making use of the expansion in eq.(2.13), obtaining

$$\begin{gathered}(\alpha_n^\alpha + \tilde{\alpha}_{-n}^\alpha)|B_X\rangle = 0 \ ; \ (\alpha_n^i - \tilde{\alpha}_{-n}^i)|B_X\rangle = 0 \ \ \forall n \neq 0 \\ \hat{p}^\alpha|B_X\rangle = 0 \qquad \qquad (\hat{q}^i - y^i)|B_X\rangle = 0 \ .\end{gathered} \tag{6.177}$$

Introducing the matrix

$$S^{\mu\nu} = (\eta^{\alpha\beta}, -\delta^{ij}) \ , \tag{6.178}$$

the equations for the non-zero modes can be rewritten as

$$(\alpha_n^\mu + S^\mu{}_\nu \tilde{\alpha}_{-n}^\nu)|B_X\rangle = 0 \qquad \forall \, n \neq 0 \ . \tag{6.179}$$

The state satisfying the previous equations can easily be determined to be

$$|B_X\rangle = N_p \delta^{d-p-1}(\hat{q}^i - y^i) \left(\prod_{n=1}^{\infty} e^{-\frac{1}{n}\alpha_{-n} S \cdot \tilde{\alpha}_{-n}} \right) |0\rangle_\alpha |0\rangle_{\tilde{\alpha}} |p = 0\rangle \ , \tag{6.180}$$

where N_p is a normalization constant to be fixed.

The previous boundary state describes only the degrees of freedom corresponding to the string coordinate X. In order to have a BRST invariant boundary state we have to supplement it with the boundary state for the ghost degrees of freedom obtaining the full boundary state

$$|B\rangle = |B_X\rangle|B_{gh}\rangle \ . \tag{6.181}$$

We will later on write the ghost part of the boundary state.

The overlap conditions for the conjugate boundary state can be easily obtained by taking the adjoint of eqs. (6.177) and (6.179) and are given by

$$\langle B_X|\left(\alpha^{\mu}_{-n} + S^{\mu}{}_{\nu}\tilde{\alpha}^{\nu}_{n}\right) \ ; \ \langle B_X|\hat{p}^{\alpha} = 0 \ ; \ \langle B_X|(\hat{q}^j - y^j) = 0 \ , \tag{6.182}$$

that imply

$$\langle B_X| = \langle p=0| \ {}_{\alpha}\langle 0| \ {}_{\tilde{\alpha}}\langle 0| \ N_p \delta^{d-p-1}(\hat{q}^i - y^i)\left(\prod_{n=1}^{\infty} e^{-\frac{1}{n}\alpha_n S\cdot\tilde{\alpha}_n}\right) \ , \tag{6.183}$$

In the following we will compute the interaction between two parallel Dp-branes both in the open and in the closed string channel. By comparing the two results we can determine the normalization factor N_p appearing in the boundary state in eq.(6.180). Let us start by computing the interaction in the closed string channel. For the sake of simplicity we perform this calculation considering only the part of the boundary state containing the string coordinate X and then adding by hand the contribution of the ghosts. With this simplification the free energy reads as

$$F = \langle B_X|D|B_X\rangle \ , \tag{6.184}$$

where D is the bosonic closed string propagator

$$D = \frac{\alpha'}{4\pi}\int_{|z|\le 1}\frac{d^2z}{|z|^2} z^{L_0-1}\bar{z}^{\tilde{L}_0-1} \ , \tag{6.185}$$

L_0 and $\tilde{L}_0$ are the usual Virasoro operators of the closed bosonic string given in eq. (2.23).

Inserting eqs.(6.180), (6.183) and (6.185) into eq. (6.184) we get

$$\langle B_X|D|B_X\rangle = (N_p)^2\frac{\alpha'}{4\pi}\int_{|z|\le 1}\frac{d^2z}{|z|^4}\langle 0|_{\alpha}\langle 0|_{\tilde{\alpha}}\langle p=0|\prod_{n=1}^{\infty}\left(e^{-\frac{1}{n}\alpha_n\cdot S\cdot\tilde{\alpha}_n}\right)$$

$$\delta^{d_\perp}(\hat{q}_i)\ z^{L_0}\bar{z}^{\tilde{L}_0}\ \delta^{d_\perp}(\hat{q}_i - y_i)\ \prod_{n=1}^{\infty}\left(e^{-\frac{1}{n}\alpha_{-n}\cdot S\cdot\tilde{\alpha}_{-n}}\right)|0\rangle_{\alpha}|0\rangle_{\tilde{\alpha}}|p=0\rangle \tag{6.186}$$

where $d_\perp \equiv d-p-1$ and $|y|$ is the distance between the two Dp-branes. The matrix element in the previous expression can be factorized in two parts containing respectively the contribution of the zero and non-zero modes. The contribution of the zero modes is given by:

$$\langle p=0|\delta^{d_\perp}(\hat{q}_i)|z|^{\frac{\alpha'}{2}\hat{p}^2}\delta^{d_\perp}(\hat{q}_i - y_i)|p=0\rangle =$$

$$= \int \frac{d^{d_\perp}Q}{(2\pi)^{d_\perp}} \int \frac{d^{d_\perp}Q'}{(2\pi)^{d_\perp}} \langle p=0|e^{iQ\cdot\hat{q}}|z|^{\frac{\alpha'}{2}\hat{p}^2}e^{iQ'\cdot(\hat{q}-y)}|p=0\rangle =$$

$$= \int \frac{d^{d_\perp}Q}{(2\pi)^{d_\perp}} \int \frac{d^{d_\perp}Q'}{(2\pi)^{d_\perp}} |z|^{\frac{\alpha'}{2}Q'^2} e^{-iQ'\cdot y}\langle p_\perp = -Q|p_\perp = Q'\rangle\langle p_\| = 0|p_\| = 0\rangle =$$

$$= V_{p+1}\int \frac{d^{d_\perp}Q}{(2\pi)^{d_\perp}}|z|^{\frac{\alpha'}{2}Q^2}e^{iQ\cdot y} \;, \tag{6.187}$$

where we have used the following normalization for each component of the momentum

$$\langle k|k'\rangle = 2\pi\delta(k-k') \;, \tag{6.188}$$

with

$$(2\pi)^d\delta^d(0) \equiv V_d \;. \tag{6.189}$$

Performing the gaussian integral, eq. (6.187) becomes

$$V_{p+1}e^{-y^2/(2\pi\alpha' t)}\left(2\pi^2 t\alpha'\right)^{-d_\perp/2} \quad , \quad |z| = e^{-\pi t} \;. \tag{6.190}$$

The contribution of the non-zero modes is instead given by

$$_\alpha\langle 0|_{\widetilde{\alpha}}\langle 0| \prod_{m=1}^{\infty}\left(e^{-\frac{1}{m}\alpha_m\cdot S\cdot\widetilde{\alpha}_m}\right) z^N \bar{z}^{\widetilde{N}} \prod_{n=1}^{\infty}\left(e^{-\frac{1}{n}\alpha_{-n}\cdot S\cdot\widetilde{\alpha}_{-n}}\right)|0\rangle_\alpha|0\rangle_{\widetilde{\alpha}} =$$

$$= {}_\alpha\langle 0|_{\widetilde{\alpha}}\langle 0| \prod_{m=1}^{\infty} e^{-\frac{1}{m}\alpha_m\cdot S\cdot\widetilde{\alpha}_m} \prod_{n=1}^{\infty} e^{-\frac{1}{n}\alpha_{-n}\cdot S\cdot\widetilde{\alpha}_{-n}|z|^{2n}}|0\rangle_\alpha|0\rangle_{\widetilde{\alpha}} \;, \tag{6.191}$$

where we have defined

$$N \equiv \sum_{n=1}^{\infty}\alpha_{-n}\cdot\alpha_n \quad ; \quad \widetilde{N} \equiv \sum_{n=1}^{\infty}\widetilde{\alpha}_{-n}\cdot\widetilde{\alpha}_n \;, \tag{6.192}$$

and we have used the following relations:

$$z^N e^{\alpha_{-n}} z^{-N} = e^{\alpha_{-n}z^n} \quad \text{and} \quad \bar{z}^N e^{\alpha_{-n}} \bar{z}^{-N} = e^{\alpha_{-n}\bar{z}^n} \quad \forall n \neq 0 \,. \tag{6.193}$$

By explicitly evaluating the contractions among the oscillators in eq.(6.191) we get

$$\prod_{n=1}^{\infty} \left(\frac{1}{1 - |z|^{2n}} \right)^{d-2} . \tag{6.194}$$

To be more precise the previous calculation leads to a power d instead of $d-2$ as we have written in eq.(6.194). The extra power (-2) comes from the ghost contribution that, for the sake of simplicity, we are not presenting here.

Inserting eqs.(6.187) and (6.194) in eq.(6.186) after having changed variables according to

$$|z| = e^{-\pi t} \qquad d^2 z = -\pi e^{-2\pi t} dt d\varphi , \tag{6.195}$$

we get

$$\langle B_X | D | B_X \rangle =$$

$$= (N_p)^2 V_{p+1} \frac{\alpha' \pi}{2} \int_0^{\infty} dt \, \left(2\pi^2 \alpha' t\right)^{-\frac{d_\perp}{2}} e^{-\frac{y^2}{2\pi\alpha' t}} e^{2\pi t} \prod_{n=1}^{\infty} \left(\frac{1}{1 - e^{-2\pi t n}} \right)^{d-2} =$$

$$= (N_p)^2 V_{p+1} \frac{\alpha' \pi}{2} \left(2\pi^2 \alpha'\right)^{-\frac{d_\perp}{2}} \int_0^{\infty} dt \, t^{-\frac{d_\perp}{2}} \, e^{-\frac{y^2}{2\pi\alpha' t}} [f_1(e^{-\pi t})]^{-24} \tag{6.196}$$

where we have introduced the function

$$f_1(q) \equiv q^{\frac{1}{12}} \prod_{n=1}^{\infty} (1 - q^{2n}) \; ; \tag{6.197}$$

The factor $\alpha' \pi / 2 = \alpha' / (4\pi) \pi (2\pi)$ in eq.(6.196) comes from the product of the factor $\alpha'/(4\pi)$ present in the propagator in eq.(6.185), the factor π obtained in eq.(6.195) and the factor 2π obtained by performing the trivial integration over the angular variable φ.

Let us now proceed to the calculation of the interaction between two Dp-branes in the open string channel. The one-loop planar free-energy for an open string with $d-p-1$ Dirichlet boundary conditions is equal to [4]

$$F = -\frac{1}{2} Tr \log \left[L_0 - 1 \right] = \int_0^{\infty} \frac{d\tau}{2\tau} Tr \left[e^{-2\pi (L_0 - 1)\tau} \right] , \tag{6.198}$$

where

$$L_0 = \alpha' k^2 + \frac{y^2}{(2\pi)^2 \alpha'} + \sum_{n=1}^{\infty} \alpha_{-n} \cdot \alpha_n , \tag{6.199}$$

[4]Note that here we use the regularized expression $\log(x) = -\lim_{\epsilon \to 0^+} \int_{\epsilon}^{\infty} \frac{d\tau}{\tau} e^{-\tau x}$

y represents the distance between the two parallel Dp-branes and k the momentum lying along the world volume of the two branes. Here the L_0 operator differs from the one in eq. (2.24) because in this case the open string satisfies Dirichlet boundary conditions in the $d-p-1$ transverse directions and Neumann boundary conditions in the longitudinal ones.

The trace in eq.(6.198) must be understood as an integration over the longitudinal loop momentum and a trace over the oscillators, namely

$$F = 2\frac{V_{p+1}}{2}\int\frac{d^{p+1}k}{(2\pi)^{p+1}}\times$$

$$\times\int_0^\infty\frac{d\tau}{\tau}\;e^{2\pi\tau}\;e^{-2\pi\tau\alpha' k^2}\;e^{-\frac{y^2\tau}{2\pi\alpha'}}\;e^{2\pi\tau}\;Tr\left(\prod_{n=1}^{\infty}e^{-2\pi\tau\alpha_{-n}\cdot\alpha_n}\right)=$$

$$= V_{p+1}\int_0^\infty\frac{d\tau}{\tau}\;(8\pi^2\alpha'\tau)^{-\frac{p+1}{2}}\;e^{-\frac{y^2\tau}{2\pi\alpha'}}\;Tr\left(\prod_{n=1}^{\infty}e^{-2\pi\tau\alpha_{-n}\cdot\alpha_n}\right)\;, \tag{6.200}$$

where we have performed the Gaussian integral over the longitudinal momentum circulating in the loop and we have inserted a factor 2 coming from the freedom of exchanging the two endpoints of the string. Evaluating explicitly the trace over the oscillators we get

$$Tr\left(\prod_{n=1}^{\infty}e^{-2\pi\tau\alpha_{-n}\cdot\alpha_n}\right)=\prod_{n=1}^{\infty}\left(\frac{1}{1-e^{-2\pi\tau n}}\right)^{d-2}. \tag{6.201}$$

Notice that also in this case we have introduced by hand the information that the ghosts contribution amounts to change the exponent from d to $d-2$. Finally, inserting eq. (6.201) into eq. (6.200) and writing it in terms of the function f_1 defined in eq.(6.197), the one-loop free-energy becomes

$$F = V_{p+1}\,(8\pi^2\alpha')^{-\frac{p+1}{2}}\int_0^\infty\frac{d\tau}{\tau}\tau^{-\frac{p+1}{2}}\;e^{-\frac{y^2\tau}{2\pi\alpha'}}\;\left(f_1(e^{-\pi\tau})\right)^{-24}=$$

$$= V_{p+1}\,(8\pi^2\alpha')^{-\frac{p+1}{2}}\int_0^\infty\frac{d\tau}{\tau}\;\tau^{12-\frac{p+1}{2}}e^{-\frac{y^2\tau}{2\pi\alpha'}}\;\left(f_1(e^{-\frac{\pi}{\tau}})\right)^{-24}\;, \tag{6.202}$$

where we have taken $d=26$ and we have used the modular transformation property of the function f_1

$$f_1(e^{-\frac{\pi}{t}})=\sqrt{t}f_1(e^{-\pi t}) \tag{6.203}$$

In order to compare eq.(6.202) with eq.(6.196) we must perform in the second one the change of variable $t=\frac{1}{\tau}$. In this way eq.(6.196) becomes

$$\langle B_X|D|B_X\rangle =$$

$$= (N_p)^2 V_{p+1} \frac{\alpha' \pi}{2} \left(2\pi^2 \alpha'\right)^{-\frac{d_\perp}{2}} \int_0^\infty \frac{d\tau}{\tau} \tau^{12-\frac{p+1}{2}} e^{-\frac{y^2 \tau}{2\pi\alpha'}} [f_1(e^{-\frac{\pi}{\tau}})]^{-24} . \quad (6.204)$$

By comparing eqs.(6.202) and (6.204) we can determine the normalization factor of the boundary state:

$$N_p = \frac{T_p}{2} \quad , \quad T_p = \frac{\sqrt{\pi}}{2^{\frac{d-10}{4}}} (2\pi\sqrt{\alpha'})^{\frac{d}{2}-2-p} . \quad (6.205)$$

In conclusion, by performing the calculation of F in the closed string channel with the normalization factor given in eq.(6.205) we get:

$$F = V_{p+1} \left(8\pi^2 \alpha'\right)^{-\frac{p+1}{2}} \int_0^\infty \frac{dt}{t} t^{\frac{p+1}{2}-12} e^{-\frac{y^2}{2\pi\alpha' t}} \left(f_1(e^{-\pi t})\right)^{-24} , \quad (6.206)$$

while performing the calculation in the open string channel we get:

$$F = V_{p+1} (8\pi^2 \alpha')^{-\frac{p+1}{2}} \int_0^\infty \frac{d\tau}{\tau} \tau^{-\frac{p+1}{2}} e^{-\frac{y^2 \tau}{2\pi\alpha'}} \left(f_1(e^{-\pi\tau})\right)^{-24} , \quad (6.207)$$

The two expressions are manifestly identical as one can see if one changes variable $\tau = 1/t$ in eq.(6.207) and uses the modular transformation in eq.(6.203).

Notice that the function $[f_1]^{-24}$ has the following expansion for large value of the argument ($x \to \infty$):

$$[f_1(e^{-\pi x})]^{-24} = \sum_{n=0}^{\infty} c_n e^{-2\pi x(n-1)} = e^{2\pi x} + 24 + 0(e^{-2\pi x}) . \quad (6.208)$$

In the open string channel the nth term of the previous expansion corresponds to the contribution in the loop of open string states with mass $\alpha' M^2 = n-1$, while in the closed string channel corresponds to the exchange between the two branes of closed string states with mass $\frac{\alpha'}{2} M^2 = 2(n-1)$. In particular from eq.(6.206) we can see that the dominant contribution to F at large distance ($y \to \infty$) comes from light closed string states:

$$F = V_{p+1} \left(8\pi^2 \alpha'\right)^{-\frac{p+1}{2}} \int_0^\infty \frac{dt}{t} t^{\frac{p+1}{2}-12} e^{-\frac{y^2}{2\pi\alpha' t}} \left(e^{2\pi t} + 24 + \ldots\right) , \quad (6.209)$$

where the first term corresponds to the exchange between the two branes of the closed string tachyon, the second term to the exchange of the massless closed string states and the additional terms to the exchange of closed string states with higher mass. The first term is obviously divergent, but it is due to the presence of the tachyon that will disappear in superstring. The second term, that is called massless tadpole, can be cancelled if we add the contribution of the non-orientable Moebius diagram and we choose a particular gauge group ($SO(2^{13})$ for the bosonic string or $SO(32)$ for the type I superstring). The requirement of tadpole cancellation is a convenient way of fixing the gauge group besides the one of anomaly cancellation. Both ways fix the gauge group in the type I theory to be $SO(32)$.

In the field theory limit ($\alpha' \to 0$) it is more convenient to use the expression for F written in the open string channel because in this case the dominant contribution comes from the lowest open string states circulating in the loop. This limit can be conveniently done by introducing in eq.(6.207) the dimensional Schwinger proper time s related to the modular parameter τ by the relation $s = \alpha'\tau$. Then using the expansion in eq.(6.208) we can rewrite eq.(6.207) as follows:

$$F = V_{p+1}(8\pi^2)^{-\frac{p+1}{2}} \int_0^\infty \frac{ds}{s} s^{-\frac{p+1}{2}} e^{-\frac{y^2 s}{2\pi(\alpha')^2}} \left(e^{2\pi s/\alpha'} + 24 + 0(e^{-2\pi s/(\alpha')^2} \right) . \tag{6.210}$$

The first term corresponds to the open string tachyon that will not be present in superstring and the second term corresponds to the open-string massless states. Finally the additional terms correspond to states with higher mass in open string theory that are negligible for $\alpha' \to 0$. Notice that, if we neglect the tachyon contribution that is absent in superstring, the massless states give a non vanishing contribution to F only if the distance between the two branes $y \to 0$..

As we have seen in eq.(6.181) the BRST invariant boundary state is the product of the boundary state $|B_X\rangle$ for the bosonic coordinate, that we have constructed in this section, and of $|B_{gh}\rangle$ that we are now going to construct. BRST invariance requires that the total boundary state satisfies the equation

$$(Q + \tilde{Q})|B\rangle = 0 , \tag{6.211}$$

where the BRST charge, given in eq.(3.87), is equal to

$$Q = \sum_n c_n L^X_{-n} + \sum_{n=-1}^{\infty} c_{-n} L^{gh}_n + \sum_{n=2}^{\infty} L^{gh}_{-n} c_n \tag{6.212}$$

$\tilde{Q}$ is given by an analogous expression in terms of the tilded variables. The overlap conditions in eq.(6.177) imply that the boundary state for the bosonic coordinate satisfies the following eqs.:

$$(L^X_m - \tilde{L}^X_{-m})|B_X\rangle = 0 . \tag{6.213}$$

Inserting the expression for Q and the corresponding expression for $\tilde{Q}$ in eq.(6.211) and using eq.(6.213) we can see that eq.(6.211) implies the following overlap conditions for the ghost boundary state

$$(c_n + \tilde{c}_{-n})|B_{gh}\rangle = 0 \quad ; \quad (b_n - \tilde{b}_{-n})|B_{gh}\rangle = 0 . \tag{6.214}$$

The second overlap condition in the previous equation follows from the first one and from the analogous of eq.(6.213) for the ghost boundary state:

$$(L_m^{gh} - \widetilde{L}_{-m}^{gh})|B_g h\rangle = 0 \; . \tag{6.215}$$

Eqs.(6.214) are satisfied by the state

$$|B_{gh}\rangle_{gh} = e^{\sum_{n=1}^{\infty}(c_{-n}\widetilde{b}_{-n} - b_{-n}\widetilde{c}_{-n})} \left(\frac{c_0 + \widetilde{c}_0}{2}\right) |q = 1\rangle|\widetilde{q = 1}\rangle \tag{6.216}$$

where $|q = 1\rangle$ is the state that is annihilated by the following oscillators

$$c_n|q = 1\rangle = 0 \quad \forall n \geq 1; \quad ; \quad b_m|q = 1\rangle = 0 \quad \forall m \geq 0 \; . \tag{6.217}$$

7. Fermionic Boundary State

In this section we want to generalize the previous construction to the superstring case where, together with the boundary state $|B_X\rangle$ corresponding to the bosonic coordinate X that we have already constructed we also need to determine the boundary state $|B_\psi\rangle$ corresponding to the fermionic coordinate ψ. The procedure that we follow for determining $|B_\psi\rangle$ is precisely the same used in the previous section. We perform on the boundary conditions for an open string stretching between two Dp-branes the conformal transformation that brings from the open string to the closed string channel. In this way we obtain the equations that the fermionic boundary state must satisfy. We then solve them finding the explicit expression for $|B_\psi\rangle$.

Let us consider the boundary conditions of an open superstring stretching between two Dp-branes and circulating in the planar loop describing the interaction between two parallel branes. If the bosonic degrees of freedom satisfy Neumann boundary conditions in all the directions we have the following boundary conditions for the fermionic coordinate:

$$\begin{cases} \psi_-(0,\tau) = \eta_1\psi_+(0,\tau) \\ \psi_-(\pi,\tau) = \eta_2\psi_+(\pi,\tau) \end{cases} \tag{7.218}$$

where η_1 and η_2 can take the values ± 1. If $\eta_1 = \eta_2$ we get the Ramond (R) sector, while if $\eta_1 = -\eta_2$ we get instead the Neveu-Schwarz (NS) sector.

In order to understand how they change when the bosonic coordinate satisfies Dirichlet boundary conditions in some of the directions, we compactify them and apply T-duality. A T-duality transformation along a direction i of an open string theory transforms Neumann into Dirichlet boundary conditions for the bosonic coordinate and, as discussed in sect. 4, changes the sign of the fermionic coordinate in the right sector leaving that of the left sector unchanged, i.e.

$$\psi^i_- \to \psi^i_- \qquad\qquad \psi^i_+ \to -\psi^i_+ \tag{7.219}$$

Therefore the boundary conditions in eq.(7.218) are generalized to the case of an open superstring satisfying Neumann boundary conditions in the directions longitudinal to the world-volume of the Dp-brane and Dirichelet boundary conditions in the transverse directions, as follows

$$\begin{cases} \psi^\mu_-(0,\tau) = \eta_1 S^\mu{}_\nu \psi_+{}^\nu(0,\tau) \\ \psi^\mu_-(\pi,\tau) = \eta_2 S^\mu{}_\nu \psi_+{}^\nu(\pi,\tau) \end{cases} \tag{7.220}$$

where the matrix S has been defined in eq.(6.178).

But, together with the assignment of the usual boundary conditions that connect the left and right modes at the endpoints of the open superstring, we must also give the periodicity or anti periodicity conditions for the fermionic degrees of freedom in going around the loop. These are chosen to be

$$\begin{cases} \psi_-(\sigma,0) = \eta_3 \psi_-(\sigma,T) \\ \psi_+(\sigma,0) = \eta_4 \psi_+(\sigma,T) \end{cases} \tag{7.221}$$

where η_3 and η_4 can take the values ± 1. From the boundary conditions in eqs.(7.220) and (7.221) we get

$$\psi^\mu_-(0,0) = \eta_1 S^\mu{}_\nu \psi^\nu_+(0,0) = \eta_1\eta_4 S^\mu{}_\nu \psi_+{}^\nu(0,T) \tag{7.222}$$

and

$$\psi^\mu_-(0,0) = \eta_3 \psi^\mu_-(0,T) = \eta_3\eta_1 S^\mu{}_\nu \psi_+{}^\nu(0,T) \tag{7.223}$$

But the two set of boundary conditions in eqs.(7.218) and (7.221) must be consistent with each other. This implies $\eta_3 = \eta_4$.

Let us now perform the conformal transformation given in eq.(6.172) on the previous open string boundary conditions in order to pass to the closed string channel. Since the right and left fermionic coordinates ψ_- and ψ_+ are two-dimensional conformal fields with conformal weight $h = \frac{1}{2}$ with respect to the their variables ζ and $\bar{\zeta}$ respectively, then under the conformal transformation

$$\zeta \to f(\zeta) = -i\zeta \qquad \text{and} \qquad \bar{\zeta} \to \bar{f}(\bar{\zeta}) = i\bar{\zeta} \tag{7.224}$$

they transform as

$$\psi_-(\zeta) \to \psi'_-(\zeta) = \left(\frac{\partial f(\zeta)}{\partial \zeta}\right)^{1/2} \psi_-(\zeta') = (-i)^{\frac{1}{2}} \psi_-(f(\zeta)) \tag{7.225}$$

and

$$\psi_+(\bar{\zeta}) \to \psi'_+(\bar{\zeta}) = \left(\frac{\partial f(\bar{\zeta})}{\partial \bar{\zeta}}\right)^{1/2} \psi_+(\bar{\zeta}) = (i)^{\frac{1}{2}} \psi_+(\bar{f}(\bar{\zeta})) \tag{7.226}$$

This implies that, performing the previous transformation on eq.(7.220), we get a relative factor i between the right and left modes. More specifically from the boundary conditions given in eqs.(7.220) and (7.221) after the conformal rescaling in eq.(6.174) we get

$$\begin{cases} \psi^\mu_-(0,\sigma) = i\eta_1 S^\mu{}_\nu \psi^\nu_+(0,\sigma) \\ \psi^\mu_-(\hat{T},\sigma) = i\eta_2 S^\mu{}_\nu \psi^\nu_+(\hat{T},\sigma) \end{cases} \tag{7.227}$$

and

$$\begin{cases} \psi^\mu_-(0,\tau) = \eta_3 \psi^\mu_-(\pi,\tau) \\ \psi^\mu_+(0,\tau) = \eta_3 \psi^\mu_+(\pi,\tau) \end{cases} \tag{7.228}$$

where we have explicitly put $\eta_4 = \eta_3$.

If we now compare the usual boundary conditions for the closed superstring theory given in eqs. (2.39) and (2.40) with eq. (7.228) we see that, as a consequence of the identity between η_3 and η_4, the fermionic boundary state has only the R-R and the NS-NS sectors.

As for the bosonic coordinate the boundary state for the fermionic coordinate at $\tau = 0$ is defined, from the first equation in (7.227), as the state that satisfies the equation:

$$(\psi^\mu_-(0,\sigma) - i\eta S^\mu{}_\nu \psi^\nu_+(0,\sigma))|B_\psi, \eta\rangle = 0 \tag{7.229}$$

where $\eta = \pm 1$. Using the mode expansion given in eqs. (2.42) and (2.43) we get the overlap conditions for the fermionic boundary state

$$\left(\psi^\mu_t - i\eta S^\mu{}_\nu \widetilde{\psi}^\nu_{-t}\right)|B_\psi, \eta\rangle = 0 \tag{7.230}$$

where the index t is integer [half-integer] in the R-R [NS-NS] sector.

In the case of the NS-NS-sector the determination of the fermionic boundary state satisfying eq.(7.230) is straightforward and leads to the following expression:

$$|B_\psi, \eta\rangle = -i \prod_{t=1/2}^{\infty} \left(e^{i\eta\psi_{-t}\cdot S\cdot\widetilde{\psi}_{-t}}\right)|0\rangle \tag{7.231}$$

In the R-R sector the boundary state has the same form as in the NS-NS sector for what the non-zero modes is concerned, but with integer instead of half-integer modes. We get therefore [5]

$$|B_\psi,\eta\rangle = -\prod_{t=1}^{\infty} e^{i\eta\psi_{-t}\cdot S\cdot\widetilde{\psi}_{-t}}|B_\psi,\eta\rangle^{(0)} \tag{7.232}$$

where the zero mode contribution $|B_\psi,\eta\rangle^{(0)}$ must satisfy the condition

$$\left(\psi_0^\mu - i\eta S^\mu{}_\nu\widetilde{\psi}_0^\nu\right)|B_\psi,\eta\rangle^{(0)} = 0 \tag{7.233}$$

The previous equation is satisfied by the state

$$|B_\psi,\eta\rangle^{(0)} = \mathcal{M}_{AB}|A\rangle|\widetilde{B}\rangle \tag{7.234}$$

where

$$\mathcal{M}_{AB} = \left(C\Gamma^0...\Gamma^p\frac{1+i\eta\Gamma^{11}}{1+i\eta}\right)_{AB} \tag{7.235}$$

where C is the charge conjugation matrix and Γ^μ are the Dirac Γ matrices in the 10-dimensional space (see Ref. [13] for some detail about the derivation of eqs.(7.234) and (7.235)).

The overlap conditions for the conjugate boundary state can be obtained from eq.(7.230) by taking the adjoint of this equation namely

$$\langle B_\psi,\eta|\left(\psi_{-t}^\mu + i\eta S^\mu{}_\nu\widetilde{\psi}_t^\nu\right) = 0 \tag{7.236}$$

and are solved by

$$\langle B_\psi,\eta|_{\mathrm{NS}} = i\langle 0|\prod_{t=1/2}^{\infty}\left(e^{i\eta\psi_t\cdot S\cdot\widetilde{\psi}_t}\right) \tag{7.237}$$

in the NS-NS sector and by

$$\langle B_\psi,\eta|_{\mathrm{R}} = -\langle B_\psi,\eta|_R^{(0)}\prod_{t=1}^{\infty} e^{i\eta\psi_t\cdot S\cdot\widetilde{\psi}_t} \tag{7.238}$$

in the R-R sector, where the zero mode is given by

$$\langle B_\psi,\eta|_{\mathrm{R}}^{(0)} = \langle A|\langle\widetilde{B}|\ \mathcal{N}_{AB} \tag{7.239}$$

[5]The unusual phases introduced in Eqs. (7.231) and (7.232) will turn out to be convenient to study the couplings of the massless closed string states with a D-brane and to find the correspondence with the classical D-brane solutions obtained from supergravity. Note that these phases are instead irrelevant when one computes the interactions between two D-branes.

with

$$\mathcal{N}_{AB} = (-1)^p \left(C\Gamma^0 ... \Gamma^p \frac{1 + i\eta\Gamma^{11}}{1 - i\eta} \right)_{AB} \tag{7.240}$$

Notice that the previous overlap conditions for the conjugate boundary state differ from those given in Ref. [12] by the exchange $\eta \to -\eta$. Our choice of η corresponds to keep also in the closed channel the same η_1, η_2 appearing in the open string boundary conditions (see eq.(7.220)).

As in the bosonic string we must also in this case introduce a boundary state for the reparametrization ghosts b, c. Moreover we must also add the boundary state for the superghosts β, γ. The complete boundary state for both the NS-NS and R-R sectors is given by:

$$|B, \eta\rangle_{R,NS} = \frac{T_p}{2} |B_{mat}, \eta\rangle |B_g, \eta\rangle \tag{7.241}$$

where

$$|B_{mat}\rangle = |B_X\rangle |B_\psi, \eta\rangle \qquad ; \qquad |B_g\rangle = |B_{gh}\rangle |B_{sgh}, \eta\rangle \tag{7.242}$$

The matter part of the boundary state consists of the boundary state for the bosonic coordinate X given in eq.(6.180) without the normalization factor N_p and of the one for the fermionic coordinate ψ given in eq.(7.231) for the NS-NS sector and in eq.(7.232) for the R-R sector. The ghost part $|B_g\rangle$ contains the boundary state corresponding to the ghosts (b, c) given in eq.(6.216) and the one corresponding to the superghosts (β, γ) that we now want to determine.

It is not difficult to check that the identifications (6.177) and (7.230) imply that $|B_{\rm mat}, \eta\rangle$ is annihilated by the following linear combinations of left and right generators of the super Virasoro algebra

$$\left(L_n^{\rm mat} - \tilde{L}_{-n}^{\rm mat} \right) |B_{\rm mat}, \eta\rangle = 0 \quad , \quad \left(G_m^{\rm mat} + i\eta \tilde{G}_{-m}^{\rm mat} \right) |B_{\rm mat}, \eta\rangle = 0 \ . \tag{7.243}$$

The boundary state $|B, \eta\rangle$ must be BRST invariant, that is

$$\left(Q + \tilde{Q} \right) |B, \eta\rangle = 0 \ , \tag{7.244}$$

where the BRST charge introduced in eq.(3.93) is equal to

$$Q = \oint \frac{dz}{2\pi i} \left[c(z) \left(T^{mat}(z) + \frac{1}{2} T^g(z) \right) - \gamma(z) \left(G^{mat}(z) + \frac{1}{2} G^g(z) \right) \right] . \tag{7.245}$$

If we write the previous expression in the following form

$$Q = Q^{(1)} + Q^{(2)} \ , \tag{7.246}$$

where

$$Q^{(1)} = \sum_n c_n L_{-n} - \sum_t \gamma_t G_{-t} \qquad (7.247)$$

and

$$2Q^{(2)} = \sum_{n=2}^{\infty} L^g_{-n} c_n + \sum_{n=-1}^{\infty} c_{-n} L^g_n - \sum_{t>3/2} G^g_{-t}\gamma_t - \sum_{t>-1/2} \gamma_{-t} G^g_t \, , \qquad (7.248)$$

it is easy to show that eqs.(7.243) and (7.244) imply

$$(c_n + \tilde{c}_{-n})\, |B_{\rm gh}\rangle = 0 \quad , \quad \left(b_n - \tilde{b}_{-n}\right) |B_{\rm gh}\rangle = 0 \; ,$$

$$(\gamma_t + {\rm i}\eta\tilde{\gamma}_{-t})\, |B_{\rm sgh}, \eta\rangle = 0 \quad , \quad \left(\beta_t + {\rm i}\eta\tilde{\beta}_{-t}\right) |B_{\rm sgh}, \eta\rangle = 0 \; . \qquad (7.249)$$

Those equations imply that the relations in eqs.(7.243) must be supplemented by the analogous ones in the ghost sector, namely

$$\left(L^{\rm g}_n - \tilde{L}^{\rm g}_{-n}\right) |B_{\rm g}, \eta\rangle = 0 \quad , \quad \left(G^{\rm g}_m + {\rm i}\eta\tilde{G}^{\rm g}_{-m}\right) |B_{\rm g}, \eta\rangle = 0 \; . \qquad (7.250)$$

The overlap equations involving the ghost fields b and c can be solved and one obtains the boundary state for the bc system given in eq.(6.216). On the other hand the overlap equations for the superghosts determine the superghost boundary state to be:

$$|B_{\rm sgh}, \eta\rangle_{\rm NS} = \exp\left[{\rm i}\eta \sum_{t=1/2}^{\infty} (\gamma_{-t}\tilde{\beta}_{-t} - \beta_{-t}\tilde{\gamma}_{-t})\right] |P = -1\rangle\, |\tilde{P} = -1\rangle \; , \qquad (7.251)$$

in the NS sector in the picture $(-1, -1)$ and

$$|B_{\rm sgh}, \eta\rangle_{\rm R} = \exp\left[{\rm i}\eta \sum_{t=1}^{\infty} (\gamma_{-t}\tilde{\beta}_{-t} - \beta_{-t}\tilde{\gamma}_{-t})\right] |B_{\rm sgh}, \eta\rangle^{(0)}_{\rm R} \; , \qquad (7.252)$$

in the R sector in the $(-1/2, -3/2)$ picture. The superscript $^{(0)}$ denotes the zero-mode contribution that, if $|P = -1/2\rangle\, |\tilde{P} = -3/2\rangle$ denotes the superghost vacuum that is annihilated by β_0 and $\tilde{\gamma}_0$, is given by [14]

$$|B_{\rm sgh}, \eta\rangle^{(0)}_{\rm R} = \exp\left[{\rm i}\eta\gamma_0\tilde{\beta}_0\right] |P = -1/2\rangle\, |\tilde{P} = -3/2\rangle \; . \qquad (7.253)$$

The conjugate boundary state for the superghost is equal to

$$_{\rm NS}\langle B_{\rm sgh}, \eta| = \langle P = -1|\, \langle\tilde{P} = -1|\, \exp\left[-{\rm i}\eta \sum_{t=1/2}^{\infty} (\beta_t\tilde{\gamma}_t - \gamma_t\tilde{\beta}_t)\right] \qquad (7.254)$$

in the NS sector and

$$
{}_{\rm R}\langle B_{\rm sgh},\eta| = \langle P=-3/2|\,\langle\tilde{P}=-1/2|\,\exp\left[-i\eta\beta_0\tilde{\gamma}_0\right]\times
$$

$$
\times\exp\left[-i\eta\sum_{t=1}^{\infty}(\beta_t\tilde{\gamma}_t-\gamma_t\tilde{\beta}_t)\right] \tag{7.255}
$$

in the R-R sector.

We would like to stress that the boundary states $|B\rangle_{\rm NS,R}$ are written in a definite picture $(P,\tilde{P})$ of the superghost system, where P is given in eq.(3.105) and $\tilde{P}=-2-P$ in order to soak up the anomaly in the superghost number. In particular we have chosen $P=-1$ in the NS sector and $P=-1/2$ in the R sector, even if other choices would have been in principle possible [14]. Since P is half-integer in the R sector, the boundary state $|B\rangle_{\rm R}$ has always $P\neq\tilde{P}$, and thus it can couple only to R-R states in the asymmetric picture $(P,\tilde{P})$. However, as we have seen in section 3 the massless R-R states in the $(-1/2,-3/2)$ picture contain a part that is proportional to the R-R *potentials* [15], as opposed to the standard massless R-R states in the symmetric picture $(-1/2,-1/2)$ that are always proportional to the R-R field strengths.

The boundary state in eq.(7.241) depends on the two values of $\eta=\pm1$. Actually, as we will now show, we have to take a combination of the two values of η corresponding to the GSO projection. Let us start with the NS sector. In the NS-NS sector the GSO projected boundary state is

$$
|B\rangle_{\rm NS}\equiv\frac{1+(-1)^{F+G}}{2}\,\frac{1+(-1)^{\widetilde{F}+\widetilde{G}}}{2}\,|B,+\rangle_{\rm NS}\ , \tag{7.256}
$$

where F and G are the fermion and superghost number operators

$$
F=\sum_{m=1/2}^{\infty}\psi_{-m}\cdot\psi_m-1\ ,\quad G=-\sum_{m=1/2}^{\infty}(\gamma_{-m}\beta_m+\beta_{-m}\gamma_m)\ . \tag{7.257}
$$

Their action on the boundary state corresponding to the fermionic coordinate ψ and to the superghosts can easily be computed and one gets:

$$
(-1)^F|B_\psi,\eta\rangle=-|B_\psi,-\eta\rangle\quad;\quad(-1)^{\widetilde{F}}|B_\psi,\eta\rangle=-|B_\psi,-\eta\rangle \tag{7.258}
$$

$$
(-1)^G|B_{sgh},\eta\rangle=|B_{sgh},-\eta\rangle\quad;\quad(-1)^{\widetilde{G}}|B_{sgh},\eta\rangle=|B_{sgh},-\eta\rangle \tag{7.259}
$$

Using the previous expressions after some simple algebra we get

$$
|B\rangle_{\rm NS}=\frac{1}{2}\Big(|B,+\rangle_{\rm NS}-|B,-\rangle_{\rm NS}\Big) \tag{7.260}
$$

Passing to the R-R sector the GSO projected boundary state is

$$|B\rangle_{\rm R} \equiv \frac{1+(-1)^p(-1)^{F+G}}{2}\,\frac{1-(-1)^{\widetilde{F}+\widetilde{G}}}{2}\,|B,+\rangle_{\rm R}\ . \tag{7.261}$$

where p is even for Type IIA and odd for Type IIB, and

$$(-1)^F = \psi_{11}(-1)^{\sum_{m=1}^{\infty}\psi_{-m}\cdot\psi_m}\ ,\quad G = -\gamma_0\beta_0 - \sum_{m=1}^{\infty}\left[\gamma_{-m}\beta_m + \beta_{-m}\gamma_m\right]\ . \tag{7.262}$$

From the previous expressions it is easy to see after some calculation that the action of the fermion number operators is given by:

$$(-1)^F|B_\psi,\eta\rangle = (-1)^p|B_\psi,-\eta\rangle \quad ; \quad (-1)^{\widetilde{F}}|B_\psi,\eta\rangle = |B_\psi,-\eta\rangle \tag{7.263}$$

and

$$(-1)^G|B_{sgh},\eta\rangle = |B_{sgh},-\eta\rangle \quad ; \quad (-1)^{\widetilde{G}}|B_{sgh},\eta\rangle = -|B_{sgh},-\eta\rangle \tag{7.264}$$

Using the previous expressions after some straightforward manipulations, one gets

$$|B\rangle_{\rm R} = \frac{1}{2}\Big(|B,+\rangle_{\rm R} + |B,-\rangle_{\rm R}\Big)\ . \tag{7.265}$$

8. Classical Solutions From Boundary State

In this section we want to connect the boundary state introduced in the previous sections to the Dirichlet branes intended as electric R-R charged p-brane solutions of the low-energy string effective action. In particular we will show that the large distance behaviour of the graviton, dilaton and R-R $p+1$-form fields that one obtains from the boundary state exactly agrees with that obtained from the classical solution in sect. 5.

The long distance behaviour of the classical massless fields generated by a Dp-brane can be determined by computing the projection of the boundary state along the various fields after having inserted a closed string propagator. This amounts to compute the following matrix element

$$\langle P_x|D|B\rangle \tag{8.266}$$

where P_x runs over all the projectors of the closed superstring massless sector listed in Ref. [16], D is the propagator in eq.(6.185) if we perform the calculation in the bosonic string or is given by

$$D = \frac{\alpha'}{4\pi}\int\frac{d^2z}{|z|^2}z^{L_0-a}\bar{z}^{\tilde{L}_0-a} \tag{8.267}$$

if we more correctly perform the calculation in superstring, where the constant $a = 1/2$ in the NS-NS sector and $a = 0$ in the R-R sector.

Let us start by computing the expression for the generic NS-NS massless field which is given by

$$J^{\mu\nu} \equiv {}_{-1}\langle\widetilde{0}|_{-1}\langle 0|\psi^{\nu}_{1/2}\,\widetilde{\psi}^{\mu}_{1/2}|D|B\rangle_{NS} = -\frac{T_p}{2k_{\perp}^2} V_{p+1} S^{\nu\mu} \tag{8.268}$$

This equation is exactly the same of the one that one gets in the bosonic string (except that in this case $d = 10$ and not $d = 26$) if we use the propagator in eq.(6.185), the boundary state in eq.(6.180) and the bosonic massless closed string state $\langle\widetilde{0}|\langle 0|\alpha_1^{\nu}\,\widetilde{\alpha}_1^{\mu}$. Because of this we keep the value of the space-time dimension d arbitrary in such a way that our calculation is valid in both cases. Specifing the different polarizations corresponding to the various fields (see Refs. [13, 16] for details) we get

$$\delta\phi = \frac{1}{\sqrt{d-2}}\left(\eta^{\mu\nu} - k^{\mu}\ell^{\nu} - k^{\nu}\ell^{\mu}\right) J_{\mu\nu} = \frac{d-2p-4}{2\sqrt{2(d-2)}}\mu_p \frac{V_{p+1}}{k_{\perp}^2} \tag{8.269}$$

for the dilaton,

$$\delta h_{\mu\nu}(k) = \frac{1}{2}\left(J_{\mu\nu} + J_{\nu\mu}\right) - \frac{\delta\phi}{\sqrt{d-2}}\,\eta_{\mu\nu} =$$

$$= \sqrt{2}\mu_p \frac{V_{p+1}}{k_{\perp}^2}\,\mathrm{diag}\left(-A, A \ldots A, B \ldots B\right) \;, \tag{8.270}$$

for the graviton, where A and B are given in eq. (5.161); and

$$\delta B_{\mu\nu}(k) = \frac{1}{\sqrt{2}}\left(J_{\mu\nu} - J_{\nu\mu}\right) = 0 \tag{8.271}$$

for the antisymmetric tensor. In the R-R sector we get instead

$$\delta C_{01\ldots p}(k) \equiv \langle P^{(C)}_{01\cdots p}|\, D\,|B\rangle_{\mathrm{R}} = \mp\,\mu_p \frac{V_{p+1}}{k_{\perp}^2} \;. \tag{8.272}$$

Expressing the previous fields in configuration space using the following Fourier transform valid for $p < d-3$

$$\int d^{(p+1)}x\, d^{(d-p-1)}x \frac{e^{ik_{\perp}\cdot x_{\perp}}}{(d-p-3)\, r^{d-p-3}\,\Omega_{d-p-2}} = \frac{V_{p+1}}{k_{\perp}^2} \;, \tag{8.273}$$

remembering the expression Q_p defined in eq.(5.166) and rescaling the fields according to

$$\varphi = \sqrt{2}\kappa\phi \quad , \quad \tilde{h}_{\mu\nu} = 2\kappa h_{\mu\nu} \quad , \quad C_{01\ldots p} = \sqrt{2}\kappa C_{01\ldots p} \;, \tag{8.274}$$

we get the following large distance behaviour

$$\delta\varphi(r) = \frac{d-2p-4}{2\sqrt{2(d-2)}} \frac{Q_p}{r^{d-p-3}} \tag{8.275}$$

for the dilaton,

$$\delta\tilde{h}_{\mu\nu}(r) = 2\frac{Q_p}{r^{d-p-3}} \, \mathrm{diag}\,(-A, \ldots A, B \ldots B) \quad , \tag{8.276}$$

for the graviton and

$$\delta C_{01\ldots p} = \mp \frac{Q_p}{r^{d-p-3}} \tag{8.277}$$

for the R-R form potential.

The previous equations reproduce exactly the behavior for $r \to \infty$ of the metric in eq.(5.159) and of the R-R potential given in eq.(5.160). In fact at large distance their fluctuations around the background values are exactly equal to $\delta\tilde{h}_{\mu\nu}$ and $\delta C_{01\ldots p}$. In the case of the dilaton, in order to find agreement between the boundary state and the classical solution, we have to take $d = 10$. This strongly suggests that, as expected, the calculation has to be performed in superstring. As a matter of fact, a comparison between the p-brane solution of the classical eqs. of motion that follow from the action in (5.158) and a string calculation, does make sense only in the superstring case where the graviton, dilaton and Kalb-Ramond field come from the NS-NS sector and the antisymmetric gauge potentials like $C_{\mu_1\ldots\mu_n}$ from the R-R sector. Nonetheless, the bosonic case we have also considered in this section already tells us what are the distinctive features of the boundary state and how the long-distance behavior of the massless fields is encoded in it.

9. Interaction Between a p and a p' Brane

In this section we study the static interaction between a Dp-brane located at y_1, and a Dp'-brane located at y_2, with $NN \equiv \min\{p, p'\} + 1$ directions common to the brane world-volumes, $DD \equiv \min\{d-p-1, d-p'-1\}$ directions transverse to both, and $\nu = (d - NN - DD)$ directions of mixed type. We will not consider instantonic D-branes, hence also $NN \geq 1$. The two D-branes simply interact via tree-level exchange of closed strings whose propagator is

$$D = \frac{\alpha'}{4\pi} \int \frac{d^2z}{|z|^2} z^{L_0} \, \bar{z}^{\tilde{L}_0} \quad , \tag{9.278}$$

so that indicating with $|B_1\rangle$ and $|B_2\rangle$ the boundary states describing the two D-branes the static amplitude is given by

$$A = \langle B_1 | \, D \, | B_2 \rangle = \frac{T_p T_{p'}}{4} \frac{\alpha'}{4\pi} \int_{|z|<1} \frac{d^2z}{|z|^2} \mathcal{A} \, \mathcal{A}^{(0)} \quad , \tag{9.279}$$

where we have indicated with $\mathcal{A}$ and $\mathcal{A}^{(0)}$ respectively the non zero mode and the zero mode contribution in which the previous amplitude can be factorized. We do not have any intercept as we had in eq.(6.185) for the bosonic string because we assume that both L_0 and $\tilde{L}_0$ contain the ghost degrees of freedom. The details of the computation of the quantity in eq.(9.279) can be found in Ref. [12]. Here we just give the results of the various terms starting from the non-zero modes. In the NS-NS sector after the GSO projection we get

$$\mathcal{A}_{\text{NS-NS}} = \frac{1}{2}\left[\left(\frac{f_3}{f_1}\right)^{8-\nu}\left(\frac{f_4}{f_2}\right)^{\nu} - \left(\frac{f_4}{f_1}\right)^{8-\nu}\left(\frac{f_3}{f_2}\right)^{\nu}\right] , \tag{9.280}$$

In the R-R sector instead before the GSO projection we get

$$\mathcal{A}_{\text{R-R}}(\eta_1,\eta_2) = \left[2^{\nu-4}\left(\frac{f_2}{f_1}\right)^{8-2\nu}\delta_{\eta_1\eta_2,1} + \delta_{\eta_1\eta_2,-1}\right] , \tag{9.281}$$

where the functions f_i are equal to

$$f_1 \equiv q^{\frac{1}{12}}\prod_{n=1}^{\infty}(1-q^{2n}) \quad ; \quad f_2 \equiv \sqrt{2}q^{\frac{1}{12}}\prod_{n=1}^{\infty}(1+q^{2n}) \; ; \tag{9.282}$$

$$f_3 \equiv q^{-\frac{1}{24}}\prod_{n=1}^{\infty}(1+q^{2n-1}) \quad ; \quad f_4 \equiv q^{-\frac{1}{24}}\prod_{n=1}^{\infty}(1-q^{2n-1}) \; . \tag{9.283}$$

They transform as follows under the modular transformation $t \to 1/t$ ($q = e^{-\pi t}$):

$$f_1(e^{-\frac{\pi}{t}}) = \sqrt{t}f_1(e^{-\pi t}) \; ; \; f_2(e^{-\frac{\pi}{t}}) = f_4(e^{-\pi t}) \; ; f_3(e^{-\pi t}) = f_3(e^{-\frac{\pi}{t}}) \; . \tag{9.284}$$

The zero modes contribution in the NS-NS sector comes only from the bosonic coordinate and can be obtained by following the same procedure outlined in eqs.(6.187)-(6.190) for the bosonic string. Also in this way one gets eq.(6.190). If we insert the contributions in eqs.(9.280) and (6.190) in eq.(9.279) we get the total NS-NS contribution to the interaction between a Dp and a Dp' brane

$$\begin{aligned}\mathcal{A}_{\text{NS-NS}} &= V_{NN}\,(8\pi^2\alpha')^{-\frac{NN}{2}}\int_0^{\infty} dt\left(\frac{1}{t}\right)^{\frac{DD}{2}} e^{-y^2/(2\alpha'\pi t)} \\ &\times\frac{1}{2}\left[\left(\frac{f_3}{f_1}\right)^{8-\nu}\left(\frac{f_4}{f_2}\right)^{\nu} - \left(\frac{f_4}{f_1}\right)^{8-\nu}\left(\frac{f_3}{f_2}\right)^{\nu}\right] ,\end{aligned} \tag{9.285}$$

where V_{NN} is the common world-volume of the two D-branes, $|y|$ is the transverse distance between them.

It is interesting to notice that the two terms in the square brackets of Eq. (9.285) come respectively from the NS-NS and the NS-NS$(-1)^{(F+G)}$ sectors of the exchanged closed string, which, under the transformation $t = 1/\tau$, are mapped into the NS and R sectors of the open string suspended between the branes. Notice that $A_{\text{NS-NS}} = 0$ if $\nu = 4$.

The evaluation of the zero mode contribution in the R-R sector requires more care due to the presence of zero modes both in the fermionic matter fields and the bosonic superghosts. Inserting eq. (9.281) into eq. (9.279) we can write the total R-R contribution as

$$A_{\text{R-R}}(\eta_1, \eta_2) = V_{NN}\,(8\pi^2\alpha')^{-\frac{NN}{2}} 2^{-\frac{\nu}{2}} \int_0^\infty dt \left(\frac{1}{t}\right)^{\frac{DD}{2}} \mathrm{e}^{-y^2/(2\pi\alpha' t)}$$

$$\times \left[2^{\nu-4} \left(\frac{f_2}{f_1}\right)^{8-2\nu} \delta_{\eta_1\eta_2,+1} + \delta_{\eta_1\eta_2,-1} \right] {}^{(0)}_{\text{R}}\langle B^1, \eta_1 | B^2, \eta_2 \rangle^{(0)}_{\text{R}} \;, \qquad (9.286)$$

where

$$|B, \eta\rangle^{(0)}_{\text{R}} = |B_\psi, \eta\rangle^{(0)}_{\text{R}}\; |B_{\text{sgh}}, \eta\rangle^{(0)}_{\text{R}} \;. \qquad (9.287)$$

Note that in Eq. (9.286) it is essential *not* to separate the matter and the superghost zero-modes. In fact, a naïve evaluation of ${}^{(0)}_{\text{R}}\langle B^1, \eta_1 | B^2, \eta_2 \rangle^{(0)}_{\text{R}}$ would lead to a divergent or ill defined result: after expanding the exponentials in ${}^{(0)}_{\text{R}}\langle B^1_{\text{sgh}}, \eta_1 | B^2_{\text{sgh}}, \eta_2 \rangle^{(0)}_{\text{R}}$, all the infinite terms with any superghost number contribute, and yield the divergent sum $1+1+1+\ldots$ if $\eta_1\eta_2 = -1$, or the alternating sum $1-1+1-\ldots$ if $\eta_1\eta_2 = 1$. This problem has already been addressed in Ref. [14] and solved by introducing a regularization scheme for the pure Neumann case ($NN = 10$). This method has been extended to the most general case with D-branes in Ref. [12]. Here, we give the final result for the fermionic zero mode part of the R-R sector:

$${}^{(0)}_{\text{R}}\langle B^1, \eta_1 | B^2, \eta_2 \rangle^{(0)}_{\text{R}} = -16\,\delta_{\nu,0}\,\delta_{\eta_1\eta_2,1} + 16\,\delta_{\nu,8}\,\delta_{\eta_1\eta_2,-1} \;. \qquad (9.288)$$

We can now write the final expression for the R-R amplitude. Inserting Eq. (9.288) into Eq. (9.286), after performing the GSO projection we get

$$A_{\text{R-R}} = V_{NN}(8\pi^2\alpha')^{-\frac{NN}{2}} \cdot$$

$$\int_0^\infty dt \left(\frac{1}{t}\right)^{\frac{DD}{2}} \mathrm{e}^{-y^2/(2\pi\alpha' t)}\, \frac{1}{2} \left[-\left(\frac{f_2}{f_1}\right)^8 \delta_{\nu,0} + \delta_{\nu,8} \right] \;. \qquad (9.289)$$

The $\nu = 0$ and $\nu = 8$ terms in Eq. (9.289) come respectively from the R-R and the R-R$(-1)^{(F+G)}$ sectors of the exchanged closed string, which, under the transformation $t \to 1/t$, are mapped into the NS$(-1)^{(F+G)}$ and

R$(-1)^{(F+G)}$ sectors of the open string suspended between the branes. Due to the "abstruse identity", the total D-brane amplitude

$$A = A_{\rm NS-NS} + A_{\rm R-R} \tag{9.290}$$

vanishes if $\nu = 0, 4, 8$; these are precisely the configurations of two D-branes which break half of the supersymmetries of the Type II theory and satisfy the BPS no-force condition.

Acknowledgements

We would like to thank M. Frau, T. Harmark, A. Lerda, R. Marotta, I. Pesando and R. Russo for many discussions on the subject of these lectures.

Appendix A

In this appendix we will describe the properties of bosonic and fermionic bc systems that enter in the covariant quantization of string theories. Their dynamics is described by the action

$$S[b,c] \sim \int d^2z \; [b\bar{\partial}c + \bar{b}\partial\bar{c}] \ , \tag{A.1}$$

which implies the equations of motion

$$\bar{\partial}b = 0 \qquad ; \qquad \bar{\partial}c = 0 \ , \tag{A.2}$$

and their conjugate ones. Thus, the fields b and c are functions only of z and they admit the following holomorphic expansions [6]

$$b(z) = \sum b_n z^{-n-\lambda} \qquad ; \qquad c(z) = \sum c_n z^{-n+\lambda-1} \ . \tag{A.3}$$

where the variable n is integer for integer values of the conformal dimension λ, while for half-integer values of λ different spin structures are possible. In particular for periodic boundary conditions (R-sector) n is integer while for anti-periodic boundary conditions (NS-sector) n is half-integer. The oscillators in eq. (A.3) satisfy the following hermiticity properties

$$c_n^\dagger = c_{-n} \quad ; \quad b_n^\dagger = \epsilon b_{-n} \ , \tag{A.4}$$

where $\epsilon = 1$ for fermions and $\epsilon = -1$ for bosons.

[6] To avoid repetition, we write all definitions for the holomorphic sector of the theory only; similar expressions hold for the antiholomorphic sector.

The theory can be quantized by either requiring canonical commutation relations that on the mode expansion in eq.(A.3) read as

$$[c_n, b_m]_\epsilon = \delta_{n+m,0} \qquad ; \qquad [b_n, b_m]_\epsilon = [c_n, c_m]_\epsilon = 0 \ , \tag{A.5}$$

where $[\ ,\ ]_\epsilon$ means commutator [anticommutator] for bosonic [fermionic] fields or by imposing the OPE

$$c(z)b(w) = \frac{1}{z-w} \quad , \quad b(z)c(w) = \frac{\epsilon}{z-w} \ . \tag{A.6}$$

The energy-momentum tensor $T(z)$ and the ghost number current $j(z)$ of the theory are given by

$$T(z) =: [-\lambda b\partial c + (1-\lambda)\partial b c] : = \sum_n L_n z^{-n-2} \ , \tag{A.7}$$

$$j(z) = - : b(z)c(z) : \ = \epsilon \ : c(z)b(z) := \sum_n j_n z^{-n-1} \ , \tag{A.8}$$

where the normal ordering is explicitly given by

$$: b_n c_{-n} := \begin{cases} b_n c_{-n} & \text{if } n < 1-\lambda \\ -\epsilon c_{-n} b_n & \text{if } n \geq 1-\lambda \end{cases} \ . \tag{A.9}$$

The Fourier coefficients L_n and j_n take the form

$$\begin{aligned} L_n &= \oint \frac{dz}{2\pi i} T(z) z^{n+1} = \sum_m (\lambda n - m) : b_m c_{n-m} : \\ j_n &= \oint \frac{dz}{2\pi i} j(z) z^n = -\sum_m : b_m c_{n-m} : \ , \end{aligned} \tag{A.10}$$

and as a consequence of eq. (A.4) they satisfy the following hermiticity properties

$$L_n^\dagger = L_{-n} \quad ; \quad j_n^\dagger = -j_{-n} \ . \tag{A.11}$$

From the ghost number current we can obtain the ghost number j_0 as

$$j_0 = \oint dz j(z) = -\sum_{n=-\infty}^{\infty} : b_n c_{-n} : \ , \tag{A.12}$$

By using the OPE in eq.(A.6) one gets

$$T(z)b(w) = \frac{\lambda b(w)}{(z-w)^2} + \frac{\partial_w b(w)}{(z-w)} + .. ; \ T(z)c(w) = \frac{(1-\lambda)c(w)}{(z-w)^2} + \frac{\partial_w c(w)}{(z-w)} + .. \ , \tag{A.13}$$

that are consistent with the fact that b and c are conformal fields with conformal weights λ and $1-\lambda$ respectively and

$$j(z)b(w) = -\frac{b(w)}{(z-w)} + \ldots \quad ; \quad j(z)c(w) = \frac{c(w)}{(z-w)} + \ldots \ , \tag{A.14}$$

that imply that b and c have ghost charge -1 and 1 respectively. Moreover one can see that $T(z)$ and $j(z)$ satisfy the OPE

$$\begin{aligned} T(z)T(w) &= \frac{c/2}{(z-w)^4} + \frac{2T(w)}{(z-w)^2} + \frac{\partial_w T(w)}{z-w} + \cdots \\ T(z)j(w) &= \frac{Q}{(z-w)^3} + \frac{j(w)}{(z-w)^2} + \frac{\partial_w j(w)}{z-w} + \cdots \\ j(z)j(w) &= \frac{\epsilon}{(z-w)^2} + \cdots \ , \end{aligned} \tag{A.15}$$

where the "screening charge" Q and the c-number of the Virasoro algebra are respectively given by

$$Q = \epsilon(1-2\lambda) \quad , \quad c = \epsilon(1-3Q^2) \ . \tag{A.16}$$

Using eqs. (A.10) the OPEs given in eq. (A.15) imply the commutation relations

$$[L_n, L_m] = (n-m)L_{n+m} + \frac{1-3Q^2}{12} n(n^2-1)\delta_{n+m,0} \ , \tag{A.17}$$

$$[L_n, j_m] = -m j_{n+m} + \frac{Q}{2} n(n+1)\delta_{n+m,0} \ , \tag{A.18}$$

$$[j_n, j_m] = n\delta_{n+m,0} \ . \tag{A.19}$$

We observe that the Q-dependent term in eq.(A.18) (or equivalently in the second OPE in eq.(A.15)), which makes the current $j(w)$ not quite a good conformal primary field, is a consequence of the anomaly appearing in the conservation law of the ghost number current [9]

$$\bar{\partial} j(z) = \frac{1}{8} Q\sqrt{g} R^{(2)} \ , \tag{A.20}$$

where g and $R^{(2)}$ are respectively, the determinant of the metric and the scalar curvature of the two dimensional world-sheet Σ, on which the theory is defined.

Comparing the two equations obtained from (A.18) for $n=-m=1$ and $n=-m=-1$ and using the hermiticity properties in eq.(A.11) we get that j_0 is neither hermitian nor antihermitian, but satisfies the following property

$$j_0 + j_0^\dagger + Q = 0 \ . \tag{A.21}$$

Let us introduce the q-vacuum, $|q>$, defined by the relations

$$\begin{aligned} b_n|q> &= 0 \qquad \text{if} \qquad n > \epsilon q - \lambda \ , \\ c_n|q> &= 0 \qquad \text{if} \qquad n \geq -\epsilon q + \lambda \ . \end{aligned} \tag{A.22}$$

Then, from eq.(A.10) one can easily show that $|q>$ is an eigenstate of both j_0 and L_0 with eigenvalues given respectively by the following equations

$$j_0|q>= q|q> \quad , \quad L_0|q>= \frac{1}{2}\epsilon q(q+Q)|q> \ , \tag{A.23}$$

and that, as a consequence of eq. (A.21), it is normalized as follows

$$<q'|q>= \delta(q' + q + Q) \tag{A.24}$$

We observe that the state $|q = 0>$ is the only $SL(2,R)$ invariant vacuum since it is the only one which is annihilated simultaneously by L_0, L_1 and L_{-1}.

Because of eq.(A.24), in order not to get a vanishing result when we compute correlation functions involving b and c, we must make sure that the total ghost number of the correlator be equal $-Q$. For instance the following correlation function

$$<-q-Q|c(z)b(w)|q>= \left(\frac{z}{w}\right)^{\epsilon q} \frac{1}{z-w} \ , \tag{A.25}$$

is different from zero. The contraction given in eq.(A.6) can be obtained from the previous equation by choosing the $SL(2,R)$ invariant vacuum $|q = 0\rangle$.

By using the mode expansion in eq.(A.3) and the anticommutation relations in eq. (A.5), or equivalently the contraction in eq.(A.6) together with the Wick theorem, one can very easily compute any correlation function of b and c fields on the sphere.

A fermionic bc system can be bosonized in terms of a scalar field with a background charge Q through the following relations

$$b(z) =: e^{-\varphi(z)} : \qquad\qquad c(z) =: e^{\varphi(z)} : \ , \tag{A.26}$$

while a bosonic bc system can be "bosonized" in terms of a scalar field φ with background charge Q and a fermionic bc system with $\lambda = 1$, that we call $\xi\eta$ system, through the following relations

$$\beta(z) = \partial\xi(z)e^{-\phi(z)} \qquad\qquad \gamma(z) = e^{\phi(z)}\eta(z) \ . \tag{A.27}$$

where we have called β, γ the bosonic b, c fields.

Let us give some detail about this bosonization procedure. The action of a scalar field with background charge $\mathcal{Q}$ is given by

$$S[\varphi] \sim \int_\Sigma d^2z[-\epsilon\bar{\partial}\varphi\partial\varphi - \frac{1}{4}\mathcal{Q}\sqrt{g}R^{(2)}\varphi] \tag{A.28}$$

where g and $R^{(2)}$ are respectively, the determinant of the metric and the scalar curvature of the two dimensional world-sheet Σ, on which the theory is defined. The equation of motion for this field is

$$\partial\bar{\partial}\varphi(z) = \frac{1}{8}\epsilon\mathcal{Q}\sqrt{g}R^{(2)} \quad . \tag{A.29}$$

Notice that $\epsilon\partial\varphi(z)$ satisfies exactly the same equation as the anomalous current $j(z)$ of the previous system (see eq. (A.20)). This system is invariant under the conformal transformations generated by the energy-momentum tensor

$$T(z) =: \frac{1}{2}[\epsilon(\partial\varphi)^2 - \mathcal{Q}\partial^2\varphi] : (z) \quad , \tag{A.30}$$

and under a $U(1)$ Kac-Moody algebra generated by the current

$$j(z) = \epsilon\partial\varphi(z) \quad . \tag{A.31}$$

The theory can be quantized by requiring the standard OPE for a free scalar field, namely

$$\varphi(z)\varphi(w) = \epsilon\log(z-w) \quad . \tag{A.32}$$

By using it one can easily check that $T(z)$ and $j(z)$ satisfy the following OPEs

$$T(z)T(w) = \frac{c/2}{(z-w)^4} + 2\frac{T(w)}{(z-w)^2} + \frac{\partial_w T(w)}{z-w} + \cdots \quad , \tag{A.33}$$

$$T(z)j(w) = \frac{\mathcal{Q}}{(z-w)^3} + \frac{j(w)}{(z-w)^2} + \frac{\partial_w j(w)}{z-w} + \cdots \quad , \tag{A.34}$$

$$j(z)j(w) = \frac{\epsilon}{(z-w)^2} + \cdots \quad , \tag{A.35}$$

where the central charge c of the Virasoro algebra is equal to

$$c = 1 - 3\epsilon\mathcal{Q}^2 \quad . \tag{A.36}$$

The presence of a third-order pole in (A.34) is a signal of the fact that $j(z)$ is not really a good conformal field of weight 1, when there is a non-vanishing background charge $\mathcal{Q}$. Notice that eqs. (A.33) - (A.35) reproduce the OPE given in eq. (A.15) except for the value of the central charge c.

The field φ admits the following expansion

$$\varphi(z) = x + N \log z + \sum_{n \neq 0} \frac{\alpha_n}{n} z^{-n} \quad , \tag{A.37}$$

where the harmonic oscillators satisfy the usual commutation relations

$$[\alpha_n, \alpha_m] = n\epsilon\delta_{n+m,0} \qquad [x, N] = -\epsilon \ . \tag{A.38}$$

Using eq.(A.37) in eqs.(A.30) and (A.31), one can easily obtain the oscillator expressions for the Virasoro generators L_n and for the Fourier components j_n of the current $j(z)$, namely

$$L_n = \frac{1}{2} \sum_m : \alpha_m \alpha_{n-m} : -\frac{1}{2} Q(n+1)\alpha_n \quad ,$$

$$j_n = -\epsilon\alpha_n \quad , \tag{A.39}$$

with $\alpha_0 = -N$ and where the symbol $: \quad :$ is the usual normal ordering of harmonic oscillators. The OPEs in eqs.(A.33), (A.34) and (A.35) are then equivalent to the following commutation relations

$$[L_n, L_m] = (n-m)L_{n+m} + \frac{1-3Q^2}{12} n(n^2-1)\delta_{n+m,0} \quad ,$$

$$[L_n, j_m] = -m j_{n+m} - \frac{Q}{2} n(n+1)\delta_{n+m,0} \ ; \quad [j_n, j_m] = n\delta_{n+m,0} \ . \tag{A.40}$$

As in the case of a bc system the zero mode of the fermionic number current is not hermitian, but satisfies the following hermiticity properties

$$j_0 + j_0^\dagger + Q = 0 \quad , \tag{A.41}$$

which implies

$$< q \,|\, q' >= \delta(q + q' + Q) \ . \tag{A.42}$$

where $|q>$ and $|q'>$ are eigenstates of N with eigenvalues q and q' respectively

Due to the presence of the zero mode logarithmic term in eq.(A.37), the field $\varphi(z)$ does not transform properly under a conformal transformation, whereas $: e^{q\varphi(z)} :$ behaves as a primary conformal field of weight $\frac{1}{2}\epsilon q(q+Q)$. In addition it transforms as a field with charge q under the ghost number current generated by $j(z)$. This can be checked by computing the following OPE

$$T(z) : e^{q\varphi(w)} := \frac{1}{2}\epsilon q(q+Q) \frac{: e^{q\varphi(w)} :}{(z-w)^2} + \frac{\partial_w : e^{q\varphi(w)} :}{(z-w)} + \cdots \quad , \tag{A.43}$$

$$j(z) : e^{q\varphi(w)} := q\frac{: e^{q\varphi(w)} :}{z - w} \quad . \tag{A.44}$$

Introducing the corresponding highest weight state according to

$$|q\!> = \lim_{z\to 0} : e^{q\varphi(z)} : |0\!> \quad , \tag{A.45}$$

it is easy to see that $|q\!>$ is an eigenstate of the ghost number j_0 and of L_0 with eigenvalues given respectively by

$$L_0|q\!> = \frac{1}{2}\epsilon q(q+\mathcal{Q})|q\!> \quad , \quad j_0|q\!> = q|q\!> \quad . \tag{A.46}$$

If we consider the case $\epsilon = 1$ and we takes $\mathcal{Q} = 1 - 2\lambda$ we immediately see that the central charge in eq. (A.36) reproduces exactly the one given in eq. (A.16) and that the OPEs in eqs.(A.15) and in eqs.(A.33), (A.34) and (A.35) are coincident. Moreover if we consider eqs. (A.43) and (A.44) for $\epsilon = 1$ and $\mathcal{Q} = 1-2\lambda$ and put $q = \pm 1$ they reproduce eqs. (A.13) and (A.14) respectively for b and c. This is consistent with the fact that a fermionic bc system is completely equivalent to a scalar field with a background charge $\mathcal{Q} = 1 - 2\lambda$ and with $\epsilon = 1$. The fields b and c can be expressed in terms of the scalar field through the bosonization eqs.(A.26) and the current $j(z)$ in eq.(A.31) turns out to be the bosonized version of the fermionic number current in eq.(A.8). Consequently the zero mode N in eq.(A.37) is just the bosonized version of the fermionic number, as one can see from

$$N = \oint dz j(z) = j_0 \quad . \tag{A.47}$$

In the case of a bosonic bc system the central charge of the Virasoro algebra in eq.(A.16) can be written as:

$$c = -1 + 3\mathcal{Q}^2 = (1 + 3\mathcal{Q}^2) - 2 \quad , \tag{A.48}$$

that corresponds to the sum of the central charges of a scalar field with $\epsilon = -1$ given by $c = 1 + 3\mathcal{Q}^2$ and of a fermionic bc system with $\lambda = 1$ given by $c = -2$. In this case the "bosonization" rules are given in eqs.(A.27). Introducing the new energy momentum tensor as

$$T(z) = T_\varphi(z) + T_{\eta\xi}(z) \quad ; \tag{A.49}$$

where T_φ is given in eq. (A.30) for $\epsilon = -1$, $\mathcal{Q} = (-1 + 2\lambda)$ and $T_{\eta\xi}$ is given in eq. (A.7) for $\epsilon = 1$ and $\lambda = 1$, it is easy to verify that the fields in the r.h.s. of eqs.(A.27) have exactly the same conformal weights of a bosonic

(b, c) system. Moreover, if we introduce the sum of $U(1)$ number currents of the scalar field φ and of the fermionic ξ, η system:

$$j(z) = j_{\varphi}(z) + j_{\eta\xi}(z) = -\partial\varphi(z) + \xi(z)\eta(z) \quad , \tag{A.50}$$

it is easy to verify that the OPE of $j(z)$ with $\beta(z)$ and $\gamma(z)$ has no simple pole term implying that both $\beta(z)$ and $\gamma(z)$ have charge zero with respect to the total $U(1)$ number given by

$$P = (j_0)_{\varphi} + (j_0)_{\eta\xi} = \oint \frac{dz}{2\pi i}(-\partial\phi + \xi\,\eta) \tag{A.51}$$

On the other hand the $U(1)$ current for the bosonic b, c system given in eq.(A.12) is instead reproduced in the "bosonized" system by only the term $(j_0)_{\varphi}$ in eq.(A.51).

Acknowledgements

This work has been partially supported by the European Commission TMR programme ERBFMRX-CT96-0045 and by ‘Programma di breve mobilita per scambi internazionali dell’Universita di Napoli “Federico II” ’.

References

1. J. Polchinski, S. Chaudhuri and C.V. Johnson, “Notes on D-branes,” hep-th/9602052; J. Polchinski, “TASI lectures on D-branes,” hep-th/9611050.
2. M.J. Duff, R.R. Khuri and J.X. Lu, “String solitons, ” *Phys. Rep.* **259** (1995) 213; K.S. Stelle, “BPS Branes in Supergravity,” hep-th/9803116; R. Argurio, “Brane Physics in M-Theory,” hep-th/9807171.
3. J. Polchinski, “Dirichlet-branes and Ramond-Ramond Charges,” *Phys. Rev. Lett.* **75** (1995) 4724, hep-th/9510017.
4. E. Cremmer and J. Scherk, “Factorization of the Pomeron sector and currents in the dual resonance model,” *Nucl. Phys.* **B50** (1972) 222; L. Clavelli and J. Shapiro, “Pomeron factorization in general dual models,” *Nucl. Phys.* **B57** (1973) 490; M. Ademollo et al., “Soft dilatons and scale renormalization in dual theories,” *Nucl. Phys.* **B94** (1975) 221.
5. C.G. Callan, C. Lovelace, C.R. Nappi and S.A. Yost, “String loops corrections to beta functions,” *Nucl. Phys.* **B288** (1987) 525; “Adding holes and crosscaps to the superstring,” *Nucl. Phys.* **B293** (1987) 83; “Loop corrections to superstring equations of motion ,” *Nucl. Phys.* **B308** (1988) 221.
6. J. Polchinski and Y. Cai, “Consistency of open superstring theories,” *Nucl. Phys.* **B296** (1988) 91.
7. M.B. Green and P. Wai, “The insertion of boundaries in world-sheets,” *Nucl. Phys.* **B431** (1994) 131; M.B. Green, “Wilson-Polyakov loops for critical strings and superstrings at finite temperature,” *Nucl. Phys.* **B381** (1992) 201; M.B. Green and M. Gutperle, “Light-cone supersymmetry and D-branes,” *Nucl. Phys.* **B476** (1996) 484, hep-th/9604091.
8. E.F. Corrigan and D.B. Fairlie, “Off-Shell States in Dual Resonance Theory, ” *Nucl. Phys.* **B91** (1975) 527-545.
9. D. Friedan, E. Martinec and S. Shenker, “Conformal invariance, supersymmetry and string theory,” *Nucl. Phys.* **B271** (1986) 93.

10. M.B. Green, J.H. Schwarz and E. Witten, "Superstring Theory," Cambridge University Press, 1987.
11. R. Pettorino and F. Pezzella,"More about picture changed vertices in superstring theory," *Phys. Lett.* **B269** (1991) 77.
12. M. Billó, P. Di Vecchia, M. Frau, A. Lerda, I. Pesando, R. Russo and S. Sciuto, "Microscopic string analysis of the D0-D8 brane system and dual R-R states," *Nucl. Phys.* **B526** (1998) 199, hep-th/9802088.
13. P. Di Vecchia, M. Frau, I. Pesando, S. Sciuto, A. Lerda, R. Russo, "Classical p-branes from boundary state," *Nucl. Phys.* **B507** (1997) 259, hep-th/9707068.
14. S.A. Yost, "Bosonized superstring boundary state and partition functions," *Nucl. Phys.* **B321** (1989) 629.
15. M. Bianchi, G. Pradisi and A. Sagnotti,"Toroidal compactifications and symmetry breaking in open-string theories," *Nucl. Phys.* **B376** (1992) 365.
16. P. Di Vecchia, M. Frau, A. Lerda and A. Liccardo, "(F, Dp) bound states from the boundary state," hep-th/9906214. To be published in Nuclear Physics.

8'99

8'99
PEPSI

MODULI SPACES OF CALABI-YAU COMPACTIFICATIONS

JAN LOUIS
Martin-Luther-Universität Halle-Wittenberg,
FB Physik, D-06099 Halle, Germany

Abstract. We review properties of Calabi-Yau compactifications of string theory, M-theory and F-theory.

1. Introduction

Calabi-Yau compactifications have played an important role in studying supersymmetric vacua of string theory. More recently they have also featured in compactifications of M-theory and F-theory. The moduli of the Calabi-Yau metric appear in the four-dimensional effective Lagrangian as scalar fields which are flat directions of the effective potential. In these lectures we focus on their moduli spaces and the corresponding couplings in in the low energy effective action and neglect other parts of the massless spectrum in our considerations.

2. A short story about string theory, F-theory and M-theory

2.1. STRING THEORY

In string theory the fundamental objects are one-dimensional strings which, as they move in time, sweep out a 2-dimensional worldsheet Σ [1]. Strings can be open or closed and their worldsheet is embedded in a D-dimensional target space of Minkowskian signature which is identified with spacetime. States in the target space appear as eigenmodes of the string and their scattering amplitudes are described by appropriate scattering amplitudes of strings. These scattering amplitudes are built from a fundamental vertex, which for closed strings is depicted in Fig. 1. It represents the splitting of a string or the joining of two strings and the strength of this interaction is governed by a dimensionless string coupling constant g_s. Out of the funda-

L. Thorlacius and T. Jonsson (eds.), M-Theory and Quantum Geometry, 61-89.

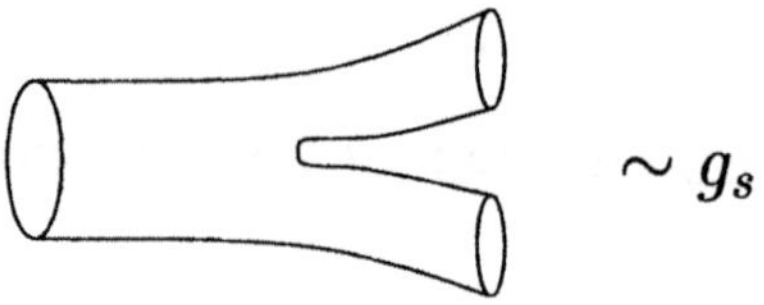

Figure 1. The fundamental closed string vertex.

mental vertex one composes all possible closed string scattering amplitudes $\mathcal{A}$, for example the four-point amplitude shown in Fig. 2. The expansion

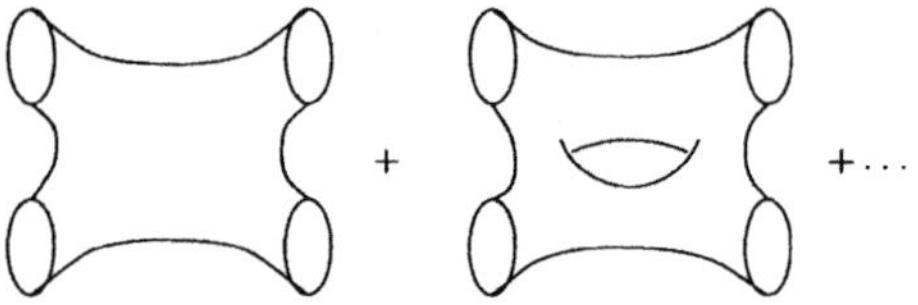

Figure 2. The perturbative expansion of string scattering amplitudes. The order of g_s is governed by the number of holes in the world sheet.

in the topology of the Riemann surface (i.e. the number of holes in the surface) coincides with a power series expansion in the string coupling constant formally written as

$$\mathcal{A} = \sum_{n=0}^{\infty} g_{\mathrm{s}}^{-\chi} \mathcal{A}^{(n)} , \tag{1}$$

where $\mathcal{A}^{(n)}$ is the scattering amplitude on a Riemann surface Σ of genus n and $\chi(\Sigma)$ is the Euler characteristic of the Riemann surface

$$\chi(\Sigma) = \frac{1}{4\pi} \int_{\Sigma} R^{(2)} = 2 - 2n - b . \tag{2}$$

$R^{(2)}$ is the curvature on Σ and b the number of boundaries of the Riemann surface (for the four-point amplitude of Fig. 2 one has $b = 4$).[1]

In all string theories there is a massless scalar field ϕ called the dilaton which couples to $R^{(2)}$ and therefore its vacuum-expectation value determines the size of the string coupling; one finds [2, 1]

$$g_{\mathrm{s}} = e^{\langle\phi\rangle} . \tag{3}$$

[1]For open strings different diagrams contribute at the same order of the string loop expansion. See [1] for further details.

g_s is a free parameter since ϕ is a flat direction (a modulus) of the effective potential. String perturbation theory is defined in that region of the parameter space (which is also called the moduli space) where $g_s < 1$ and the tree-level amplitude (genus-0) is the dominant contribution with higher-loop amplitudes suppressed by higher powers of g_s. Until 1995 this was the only regime accessible in string theory.

Unitarity of the amplitudes imposes a restriction on the maximal number of spacetime dimensions and the spacetime spectrum. In these lectures we exclusively focus on string theories defined in spacetime supersymmetric backgrounds and all such string theories necessarily have $D \leq 10$. They are particularly simple in their maximal possible dimension $D = 10$ where one has only five consistent string theories: type-IIA, type-IIB, heterotic $E_8 \times E_8$ (HE8), heterotic SO(32) (HSO) and the type-I SO(32) string.[2] The first two have 32 supercharges ($q = 32$) while the other three string theories all have 16 supercharges ($q = 16$). The massless spectrum of all 5 theories is summarized as follows:

type	q	bosonic spectrum	
IIA	32	NS-NS	$G_{\mu\nu}$, $B_{\mu\nu}$, ϕ
		R-R	V_μ, $C_{\mu\nu\rho}$
IIB	32	NS-NS	$G_{\mu\nu}$, $B_{\mu\nu}$, ϕ
		R-R	$c^*_{\mu\nu\rho\sigma}$, $B'_{\mu\nu}$, ϕ'
HE8	16		$G_{\mu\nu}$, $B_{\mu\nu}$, ϕ
			A_μ in adjoint of $E_8 \times E_8$
HSO	16		$G_{\mu\nu}$, $B_{\mu\nu}$, ϕ
			A_μ in adjoint of SO(32)
I	16	NS-NS	$G_{\mu\nu}$, ϕ
		R-R	$B_{\mu\nu}$
		open string	A_μ in adjoint of SO(32)

[2]For closed strings an additional constraint arises from the requirement of modular invariance of one-loop amplitudes which results in an anomaly-free spectrum of the corresponding low-energy effective theory [3]. For open strings anomaly cancellation is a consequence of the absence of tadpole diagrams [1].

2.2. CALABI-YAU COMPACTIFICATIONS

String theories in backgrounds with $D < 10$ can be constructed either as geometrical compactifications or by specifying an appropriate conformal field theory on the string worldsheet. In these lectures we only discuss the geometrical constructions, that is we choose to compactify the 10-dimensional Minkowski space $\mathcal{M}^{(10)}$ on a compact manifold K

$$\mathcal{M}^{(10)} = \mathcal{M}^{(D)} \times K^{(10-D)} . \tag{4}$$

Consistency requires K to be Ricci-flat while preserving some supercharges implies a constraint on the holonomy group of K. One finds that part of the supersymmetry is preserved if one compactifies on Calabi-Yau manifolds Y_n . These are complex n-dimensional Ricci-flat compact Kähler manifolds with holonomy group $SU(n)$. One has

n	manifold	χ	SUSY preserved
1	Torus T^2	0	all
2	K3-surface	24	$q/2$
3	Calabi-Yau threefold Y_3	not fixed	$q/4$
4	Calabi-Yau fourfold Y_4	not fixed	$q/8$

χ is the Euler number of the Calabi-Yau manifold and we see that for $n = 1, 2$ they are topologically unique. More properties of Calabi–Yau manifolds are assembled in the Appendix.

2.3. STRING DUALITIES

The past few years have shown [4] that various string theories are interrelated by a complicated 'web' of duality relations . One distinguishes perturbative and non-perturbative dualities. Perturbative dualities already hold at weak string coupling and the map which identifies the perturbative theories does not involve the dilaton. An example is T-duality [5] which identifies different (perturbative) regions of toroidal compactifications. On the other hand non-perturbative dualities identify regions of the parameter space which are not simultaneously at weak coupling and the duality map involves the dilaton in a nontrivial way. Such non-perturbative dualities are of utmost importance since they map the strong-coupling region of a given (string) theory to the weak-coupling region of a dual theory where perturbative methods are applicable and hence the strong-coupling limit gets (at least partially) under quantitative control. The non-perturbative dualities cannot be proven at present. Rather their validity has only been checked for quantities or couplings which do not receive quantum corrections. Such couplings do exist in supersymmetric (string) theories and it is precisely for this reason that supersymmetry has played such an important (technical) role in establishing non-perturbative dualities.

Let A and B be two perturbatively distinct string theories each with its own string coupling g_A and g_B, respectively. However, it is possible that once all quantum corrections (including the non-perturbative corrections) are taken into account A and B are equivalent as quantum theories and one has $A \equiv B$. This situation can occur in two different ways: The strong-coupling limit of A is mapped to the weak coupling limit of B or in other words $g_A \sim g_B^{-1}$. Along with this strong-weak coupling relation goes a map of the elementary excitations of theory A to the non-perturbative, solitonic excitations of theory B and vice versa. Some of these solitonic excitations have a description in string theory as open strings with Dirichlet boundary conditions ending on a fixed spatial p-dimensional hyper-plane – a Dp-brane [6]. Such Dp-branes must be regarded as dynamical objects with degrees of freedom induced by the attached open strings. A careful analysis shows that the corresponding states in spacetime are not part of the perturbative spectrum but rather correspond to non-perturbative solitonic type excitations[3]. It is precisely these states which dramatically affect the properties of string theory in its non-perturbative regime. The theories A and B are called S–dual and one also refers to this situation as a 'string–string duality'.

There is a variant of the above situation where the dilaton of theory A is not mapped to the dilaton of theory B but rather to any of the other perturbative moduli R_B of theory B. In this case one has the identifications $\phi_A \sim R_B$, $\phi_B \sim R_A$, or in other words the strong-coupling limit of A is independent of g_B. Thus the strong-coupling limit of A is again controlled by the perturbative regime of theory B and hence accessible in perturbation theory (at least in principle). The known S–dualities are summarized in the following table

D	q	duality
10	16	HSO $\sim$ I
6	16	IIA/$K3 \sim$ H/T^4
4	8	IIA/$Y_3 \sim$ H/$K3 \times T^2$
2	4	IIA/$Y_4 \sim$ H/$Y_3 \times T^2$

Another situation is encountered when the strong-coupling limit of a theory A is controlled not by a distinct theory B, but rather by a different

[3]They are non-perturbative in that their mass (or rather their tension for higher-dimensional D-branes) goes to infinity in the weak coupling limit $g_s \to 0$.

perturbative region of the same theory A. That is, the strong-coupling regime of A has an alternative weakly-coupled description within the same theory A but in terms of a different set of elementary degrees of freedom. For such a self-duality to hold the theory A has to have a nontrivial (discrete) symmetry Γ_S which maps the strong-coupling region to a region of weak coupling and simultaneously the different elementary excitations onto each other. An example of this situation is believed to be the type-IIB string in $D = 10$ which is conjectured to have $\Gamma_S = \mathrm{SL}(2, \mathbf{Z})$ [7, 8]. This exact symmetry predicts an infinite number of equivalent weakly coupled type-IIB strings which carry R-R charge; such strings have indeed been identified as appropriate D-strings [9, 10]. For later reference we need to record that the $\mathrm{SL}(2, \mathbf{Z})$ acts on the complex scalar

$$\tau \equiv \phi' + ie^{-\phi} \tag{5}$$

which is composed out of the two scalars ϕ, ϕ' of IIB. The transformation is

$$\tau \mapsto \tau' = \frac{a\tau + b}{c\tau + d}\,, \tag{6}$$

where $ad - bc = 1,\ a, b, c, d \in \mathbf{Z}$.

2.4. F-THEORY

The exact $\mathrm{SL}(2, \mathbf{Z})$ symmetry of IIB string theory inspired Vafa to construct non-perturbative string backgrounds where the dilaton is not constant [11]. More precisely he proposed to compactify IIB on the base B_n of an elliptically fibred Calabi-Yau manifold Y_{n+1}. Elliptically fibred Calabi-Yau manifolds are locally a fibre bundle with a two-torus T^2 fibred over the base B_n but on over codimension one loci the torus can degenerate. As a consequence nontrivial closed loops on B_n can induce a $\mathrm{SL}(2, \mathbf{Z})$ transformation of the complex structure of the fibre. The complex dilaton τ_{IIB} of IIB is identified with the complex structure modulus of the torus

$$\tau_{IIB} \equiv \tau_{T^2} \tag{7}$$

and thus is not constant over the compactification manifold B_n but can have $\mathrm{SL}(2, \mathbf{Z})$ monodromy [12]. It is precisely this fact which results in nontrivial (non-perturbative) string vacua inaccessible in string perturbation theory. Such vacua are termed F-theory compactifications on elliptic Calabi–Yau manifolds and each such compactification is conjectured to capture part of the non-perturbative physics of an appropriate string vacuum.[4] One finds

[4]One can alternatively define F-theory as a type IIB in a background of D7-branes or as a particular decompactification limit of type IIA [13].

[11, 14]

D	q	duality
8	16	$F/K3 \sim H/T^2$
6	8	$F/Y_3 \sim H/K3$
4	4	$F/Y_4 \sim H/Y_3$

2.5. M-THEORY

The various dualities discussed so far relate different perturbative string theories. In these cases the strong-coupling limit of a given string theory is controlled by another (or the same) perturbative string theory. However, not all strong-coupling limits are of this type. Instead it is possible that the strong-coupling limit of a given theory is something entirely new, not any of the other string theories [8]. This was first proposed for the strong-coupling limit of the type-IIA theory in $D = 10$. The Kaluza-Klein BPS-spectrum of this theory obeys (in the string frame)

$$M^{\rm KK} \sim \frac{|n|}{g_{\rm s}} , \tag{8}$$

where n is an arbitrary integer. These KK-states are not part of the perturbative type-IIA spectrum since they become heavy in the weak-coupling limit $g_{\rm s} \to 0$. However, in the strong-coupling limit $g_{\rm s} \to \infty$ they become light and can no longer be neglected in the effective theory. This infinite number of light states (which can be identified with D-particles of type-IIA string theory, or extremal black holes of IIA supergravity) signals that the theory effectively decompactifies with the radius R_{11} of the extra dimension being the string coupling constant

$$R_{11} \sim g_{\rm s}^{\frac{2}{3}} . \tag{9}$$

Supersymmetry is unbroken in this limit and hence the KK-states assemble in supermultiplets of the 11-dimensional supergravity. In particular the massless multiplet contains as bosonic components the 11-dimensional metric G_{MN} and a 3-form A_{MNP}. Since there is no string theory which has 11-dimensional supergravity as the low-energy limit, the strong-coupling limit of type-IIA string theory has to be a new theory, called M-theory, which cannot be a theory of (only) strings. Only limited amount of information is so far known about M-theory but it is supposed to capture all degrees of freedom of all known string theories, both at the perturbative and the non-perturbative level [15, 8, 16].[5]

[5]There exists a conjecture according to which the degrees of freedom of M-theory are captured in $U(N)$ supersymmetric matrix models in the $N \to \infty$ limit [17]. These

Calabi-Yau compactifications of M-theory also correspond to particular non-perturbative limits of string theories. One has

D	q	duality
10	32	$\mathrm{M}/S^1 \sim \mathrm{IIA}$
7	16	$\mathrm{M}/K3 \sim \mathrm{H}/T^3$
5	8	$\mathrm{M}/Y_3 \sim \mathrm{H}/K3 \times S^1$
3	4	$\mathrm{M}/Y_4 \sim \mathrm{H}/Y_3 \times S^1$

2.6. THREE TRIPLETS OF DUALITIES

In the next section we will discuss some of these dualities in more detail and with particular emphasis on the map between the moduli spaces. We will organize our discussion by the number of supercharges and discuss the first two of the following three triplets of dualities:[6]

D	q	duality
8	16	$\mathrm{F}/K3 \sim \mathrm{H}/T^2$
7	16	$\mathrm{M}/K3 \sim \mathrm{H}/T^3$
6	16	$\mathrm{IIA}/K3 \sim \mathrm{H}/T^4$
6	8	$\mathrm{F}/Y_3 \sim \mathrm{H}/K3$
5	8	$\mathrm{M}/Y_3 \sim \mathrm{H}/K3 \times S^1$
4	8	$\mathrm{IIA}/Y_3 \sim \mathrm{H}/K3 \times T^2$
4	4	$\mathrm{F}/Y_4 \sim \mathrm{H}/Y_3$
3	4	$\mathrm{M}/Y_4 \sim \mathrm{H}/Y_3 \times S^1$
2	4	$\mathrm{IIA}/Y_4 \sim \mathrm{H}/Y_3 \times T^2$

3. The $q = 16$ triplet

Let us first discuss toroidal compactifications of the heterotic string H/T^n where $n = 10 - D$. The massless multiplets are the gravitational multiplet GR containing the spacetime metric $G_{\mu\nu}$ an antisymmetric tensor $B_{\mu\nu}$, n Abelian graviphotons γ_μ and a real scalar ϕ. The second massless multiplet is the vector multiplet V which contains a (non-Abelian) vector A_μ and n

matrix models have been known for some time [18] and were also known to describe supermembranes [19] in the light-cone gauge [20]. The same quantum-mechanical models describe the short-distance dynamics of N D-particles, caused by the exchange of open strings [10]. For a review see, for example, [21].

[6]The case $q = 4$ is still under active investigation and therefore its discussion is postponed to some later occasion.

real scalars Z in the adjoint representation of the gauge group G

$$\begin{aligned} GR: &\quad (G_{\mu\nu}, B_{\mu\nu}, n\times\gamma_\mu, \phi) \ , \\ V: &\quad (A_\mu, n\times Z) \ , \qquad\qquad \mu = 0,\dots,D-1 \ . \end{aligned}$$

The scalars in the Cartan-subalgebra of G are flat directions of the effective potential and parameterize the Coulomb-branch of the theory where G is broken to its maximal Abelian subgroup $G \to U(1)^r$ ($r = \mathrm{rank}(G)$). In the heterotic string one has $r = 16$ and therefore a massless spectrum of

$$1\, GR \ + \ (16+n)\, V \ . \tag{10}$$

The moduli space spanned by the scalars in the Cartan-subalgebra is given by [22]

$$\mathcal{M}_{H/T^n} = R^+ \times \frac{SO(16+n,n)}{SO(16+n)\times SO(n)} \Big/ \Gamma_T \ , \tag{11}$$

where Γ_T is the T-duality group

$$\Gamma_T = SO(16+n, n, \mathbf{Z}) \ . \tag{12}$$

The factor R^+ is spanned by the dilaton ϕ (the scalar in the gravitational multiplet) and supersymmetry does not allow any mixing with the other moduli. At special points in this moduli space the gauge group is enhanced to non-Abelian subgroups of $E_8\times E_8$ or $SO(32)$.

Let us now turn to type IIA compactified on K3 or IIA/K3 for short. These theories live in $D = 6$ and the massless spectrum is determined by the zero modes of the Laplacian on K3. Some details are collected in the Appendix or better in ref. [23]. One finds in the NS-NS sector the graviton $G_{\mu\nu}$, the antisymmetric tensor $B_{\mu\nu}$, 20 scalars from the $(1,1)$-deformations of the Calabi-Yau metric $\delta G_{i\bar{j}}$, 38 scalars from the deformations of the complex structure δG_{ij}, 20 scalars from the $(1,1)$-forms $B_{i\bar{j}}$, 2 scalars from the $(2,0)$ and $(0,2)$-forms B_{ij} and the dilaton ϕ. These are altogether 81 scalars in the NS-NS sector. In the R-R sector one has a vector A_μ, a three-form $A_{\mu\nu\rho}$, 20 vectors $A_{\mu i\bar{j}}$, and 2 vectors $A_{\mu ij}$. In $D = 6$ a three-form is Poincare dual to a vector

$$\epsilon^{\mu_1\dots\mu_6}\partial_{\mu_1} A_{\mu_2\mu_3\mu_4} \sim \partial^{\mu_5} A^{\mu_6} \ , \qquad \text{or} \qquad dA_3 \sim *dA_1 \ , \tag{13}$$

so that there are altogether 24 gauge fields in the R-R sector. These fields nicely assemble into $1\,GR + 20\,V$ with a moduli space [24, 23]

$$\mathcal{M}_{IIA} = R^+ \times \frac{SO(20,4)}{SO(20)\times SO(4)} / \Gamma_T \ , \tag{14}$$

where

$$\Gamma_T = SO(20, 4, \mathbf{Z}) \ , \tag{15}$$

and the R^+-factor is again spanned by the field in the GR multiplet – the type IIA dilaton ϕ_{IIA}.

The string theories IIA/K3 and H/T^4 are conjectured to be S–dual [25, 7, 8, 26, 27]. Both theories have the same representation of supersymmetry with exactly the same massless spectrum. Furthermore, from (11), (14) one learns that also the moduli spaces (including the discrete identifications Γ_T) of the two string compactifications coincide

$$\mathcal{M}_{H/T^4} = \mathcal{M}_{IIA} \ . \tag{16}$$

The effective actions of the two perturbative theories agree if one identifies [8]

$$\begin{aligned} \phi_{\rm H} &= -\phi_{\rm IIA} \ , \\ H_{\rm H} &= e^{-2\phi_{\rm IIA}} * H_{\rm IIA} \ , \\ (g_{\mu\nu})_{\rm H} &= e^{-2\phi_{\rm IIA}} (g_{\mu\nu})_{\rm IIA} \ , \end{aligned} \tag{17}$$

where H is the field strength of the antisymmetric tensor. The first equation in (17) implies a strong-weak coupling relation while the second is the equivalent of an electric-magnetic duality. Further evidence for this S-duality arises from the observation that the zero modes in a solitonic string background of the type-IIA theory compactified on $K3$ have the same structure as the Kaluza–Klein modes of the heterotic string compactified on T^4 [26, 27].

The non-Abelian gauge symmetry enhancement is a simple Higgs mechanism in the heterotic vacuum. In the type IIA vacuum it is more intriguing and related to the singularities of the K3. Whenever an effective theory becomes singular at special points (or submanifolds) of the moduli space it signals the breakdown of the effective description. Heavy modes can become light and should no longer be excluded from the low energy effective theory. This is precisely what happens at the orbifold singularities of K3 where 2-cycles collapse. A D2-brane can wrap around such a 2-cycle generating a non-Abelian gauge boson. These singularities follow an A-D-E classification and thus the corresponding gauge bosons can be mapped to the gauge bosons of the heterotic string.

Let us now turn to the next duality in one dimension higher $D = 7$. On the heterotic side of the previous duality it is simple to decompactify one dimension. On the type IIA side this is impossible for K3 but recall that the strong coupling limit of the ten-dimensional type IIA string is governed by a theory in one dimension higher, M-theory. Thus one is led to consider $M/K3$ as the possible dual of H/T^3.

The massless spectrum of $M/K3$ containsthe 7-dimensional spacetime metric $G_{\mu\nu}$, 58 deformations of the Calabi-Yau metric $\delta G_{i\bar{j}}, \delta G_{ij}$, a 3-form $A_{\mu\nu\rho}$ and 22 vectors $A_{\mu i\bar{j}}, A_{\mu ij}$. Note that there is no dilaton and no antisymmetric tensor $B_{\mu\nu}$ in this compactification. However, in $D=7$ a 3-form is dual to an antisymmetric tensor

$$dA_3 = *dB_2 \ . \tag{18}$$

The massless fields of M/K3 assemble into 1 GR including $(G_{\mu\nu}, B_{\mu\nu}, 3A_\mu, \phi)$ and 19 V including $(A_\mu, 3Z)$. The moduli space is determined by the moduli space of K3-surfaces [23]

$$\mathcal{M}_{M/K3} = R^+ \times \frac{SO(19,3)}{SO(19)\times SO(3)} \Big/ SO(19,3,\mathbf{Z}) \ , \tag{19}$$

where the R^+-factor is spanned by the ϕ in the gravitational multiplet which is related to the volume of K3. From eqs. (11), (19) we see

$$\mathcal{M}_{M/K3} = \mathcal{M}_{H/T^3} \ , \tag{20}$$

including the discrete identifications. A more detailed comparison of the respective effective actions [8] reveals that the 7-dimensional heterotic string coupling g_H^7 is related to the volume of K3 measured in the 11-dimensional M-theory metric by

$$\left(g_H^7\right)^{4/3} = Vol_M\,(K3) \ . \tag{21}$$

g_H^7 in turn can be related to the heterotic string couplings in $D=6,10$ via

$$\frac{1}{\left(g_H^6\right)^2} = \frac{R}{\left(g_H^7\right)^2} = \frac{Vol\,(T^4)}{\left(g_H^{10}\right)^2} \ , \tag{22}$$

where R is the radius of 7th dimension measured in the 7-dimensional string metric. The low energy description of M-theory in terms of 11-dimensional supergravity is valid for large $Vol_M\,(K3)$. From eq. (21) one infers that in this limit the heterotic string becomes strongly coupled.

The non-Abelian gauge symmetry enhancement in M/K3 has the same explanation as before: a shrinking 2-cycle of K3 corresponds to a massless gauge boson on the heterotic side.

We already discussed F-theory compactifications in section 2.4 where we defined them IIB compactifications on the base of an elliptic Calabi-Yau manifold. Let us also relate them to M-theory compactifications. Consider M/T^2 which is the strong coupling limit of IIA/S^1. The latter theory is

T-dual to IIB/S^1 with the following relation of parameters (measured in the M-theory metric)

$$g_{IIA} = R_{11}^{3/2}, \qquad g_{IIB} = \frac{R_{11}}{R_{10}}, \qquad R_{IIB} = \frac{1}{\sqrt{R_{11}R_{10}}} . \tag{23}$$

Thus we can view 10-dimensional IIB theory as the following limit

$$IIB \sim \lim_{R_{10},R_{11}\to 0} M/T^2 \qquad \text{with } g_{IIB} \text{ fixed} . \tag{24}$$

Thus the size of T^2 shrinks but the complex structure $\tau_{T^2} = \tau_{IIB}$ is kept finite (c.f. (7)).

With this relation in mind one can employ what is called the adiabatic argument [28]. Consider the compactification $IIB/B_n \times S^1$. By the previous argument this theory is related to $M/B_n \times T^2$. For large B_n the manifold $B_n \times T^2$ is locally the same as an elliptic Calabi-Yau Y_{n+1} and thus adiabatically one has

$$F/Y_{n+1} := IIB/B_n = \lim_{T^2\to 0} M/Y_{n+1} . \tag{25}$$

This can be immediately related to T^2-compactifications of the heterotic string. We already established $M/K3 \sim H/T^3$ with $Vol_M\,(K3) = (g_H^7)^{4/3}$. For an elliptic K3 and using the adiabatic argument this implies

$$\left(g_H^7\right)^{4/3} = \left(\frac{g_H^8}{\sqrt{R_8}}\right)^{4/3} = Vol_M(B)\cdot Vol(T^2) . \tag{26}$$

Thus a shrinking T^2 corresponds to the decompactification $R_8 \to \infty$ and therefore

$$F/K3 \sim \lim_{T^2\to 0} M/K3 \sim H/T^2 . \tag{27}$$

Or in words: F-theory compactified on an elliptic $K3$ yields an 8-dimensional vacuum with 16 supercharges which is dual to the heterotic string compactified on T^2 [11, 29].

The previous 'back-of-an-envelope' argument can be made more precise [11, 30]. The moduli space of elliptic K3's with a zero size fibre is [30]

$$\mathcal{M}_{F/K3} \;=\; R^+ \times \frac{SO\,(18,2)}{SO\,(18) \times SO\,(2)}\Big/\, SO\,(18,2,\mathbf{Z}) \;, \tag{28}$$

which coincides with the moduli space of H/T^2 (c.f. (11)). The R^+-factor is spanned by the volume of the base B which for elliptic K3 necessarily is a $\mathbf{P}^1$.

4. The $q = 8$ triplets

The important new feature for string vacua with $q = 8$ is the fact that the massless spectrum and the gauge group is no longer uniquely fixed. Let us again first discuss the three heterotic theories $H/K3 \times T^{0,1,2}$ and then the corresponding dualities.

The massless multiplets in $D = 6, q = 8$ are the gravitational multiplet which contains the metric and a selfdual antisymmetric tensor $B^+_{\mu\nu}$, the vector multiplet V which only contains a vector and no scalars,[7] the tensor multiplet T containing an anti-selfdual antisymmetric tensor $B^-_{\mu\nu}$ and a real scalar ϕ and the hypermultiplets H featuring 4 real scalars q

$$\begin{aligned} GR: &\quad (G_{\mu\nu}, B^+_{\mu\nu}) \\ V: &\quad (A_\mu) \\ T: &\quad \left(B^-_{\mu\nu}, \phi\right) \\ H: &\quad (4q)\ . \end{aligned}$$

In order to preserve 8 supercharges the compactification manifold must be a K3 but in addition the vector bundle on K3 has to be holomorphic and stable, i.e. [1]

$$F_{ij} = F_{\bar{i}\bar{j}} = 0 = g_{i\bar{j}} F^{i\bar{j}}\ . \tag{29}$$

On K3 these conditions coincide with the instanton condition $F = \tilde{F}$.

In addition the chiral fermions in $D = 6$ lead to gauge and gravitational anomalies. The anomalies cancel if the following conditions are satisfied [31, 32]

- the number of hypermultiplets n_H, vector multiplets n_V and tensor multiplets n_T satisfy

$$n_H - n_V + 29 n_T - 273 = 0\ . \tag{30}$$

- A Green-Schwarz anomaly cancellation mechanism can be employed with a modified 2-form field strength

$$H = dB + \omega_L - \sum_a v_a\, \omega^a_{YM}\ , \tag{31}$$

  where $\omega_L(\omega^a_{YM})$ are gravitational (Yang-Mills) Chern-Simons terms and v_a some numerical coefficients. The modified definition of H implies the consistency condition

$$0 = \int_{K3} dH = \int_{K3} tr R \wedge R - \sum_a \int_{K3} tr\,(F \wedge F)_a = 24 - \sum_a n_a\ , \tag{32}$$

[7] Thus there is no Coulomb branch in $D = 6$.

where n_a is the instanton number. Thus there necessarily has to be a a non-trivial instanton background on K3.

In perturbative heterotic vacua there is only one dilaton and one antisymmetric tensor and thus one always has $n_T = 1$. The moduli space for this class of vacua reads

$$\mathcal{M} = R^+ \times \mathcal{M}_H \ , \tag{33}$$

where R^+ is the factor spanned by the scalar in the tensor multiplet (the heterotic dilaton) while $\mathcal{M}_H$ is the moduli space spanned by the scalars in the hypermultiplets. It includes the 80 moduli of K3 and the moduli of Yang-Mills instantons on K3 (which are parameterized by their size, position on K3 and orientation in the gauge group G). Supersymmetry requires $\mathcal{M}_H$ to be a quaternionic manifold. For $n_T \neq 1$ (which can occur in non-perturbative vacua of the heterotic string) one finds [33]

$$\mathcal{M} = \frac{O\,(1, n_T)}{O\,(n_T)} \times \mathcal{M}_H \ . \tag{34}$$

$\mathcal{M}_H$ has singularities when the size ρ of an instanton shrinks $\rho \to 0$ [34]. For the $SO(32)$ heterotic string these singularities are caused by non-perturbative gauge bosons becoming massless [35]. For k instantons shrinking at the same point on K3 the perturbative gauge group $SO(32)$ is enhanced beyond the perturbatively allowed rank to

$$G_{NP} = SO\,(32) \times Sp\,(k) \tag{35}$$

with an additional hypermultiplet in the $(\mathbf{32}, \mathbf{2k})$ representation of G_{NP}.

For the $E_8 \times E_8$ heterotic string a different explanation is employed. A small instanton is associated with a five-brane of M-theory (more precisely of $M/K3 \times S^1/\mathbf{Z}_2$) with additional tensor multiplets living on the worldvolume of the five-brane. In this case one has $n_T \geq 1$ and a non-perturbatively different situation compared to the heterotic $SO(32)$ case [36, 14, 32, 30].

Next we turn to $H/K3 \times S^1$ in $D = 5$. The massless multiplets in this case are:

$$\begin{aligned} GR: &\quad (G_{\mu\nu}, A_\mu) \\ V: &\quad (A_\mu, Z) \\ T: &\quad (B_{\mu\nu}, \phi) \\ H: &\quad (4q)\ . \end{aligned}$$

In $D = 5$ an antisymmetric tensor is dual to a vector $dB_2 \sim *dA_1$ and thus the tensor multiplet is dual to a vector multiplet $T \sim V$. Furthermore, due to the presence of the scalar in the vector multiplet a Coulomb branch exists and the moduli space has the form

$$\mathcal{M} = \mathcal{M}_H \times \mathcal{M}_V \ . \tag{36}$$

Supersymmetry dictates that locally the moduli space is a direct product. Since the hypermultiplets are the same as in $D = 6$ also $\mathcal{M}_H$ is unchanged. $\mathcal{M}_V$ is known at the tree level and one has [37]

$$\mathcal{M}_V^{(0)} = R^+ \times \frac{SO(1, r+1)}{SO(r+1)} \ , \tag{37}$$

where $r = \text{rank}(G)$ and the extra vector multiplet corresponds to the radius of S^1. At the quantum level only $\mathcal{M}_V$ is corrected in string theory since the dilaton is part of a tensor multiplet (or the dual vector multiplet). The corrections are such that there is only a perturbative correction at 1-loop and non-perturbative corrections [38, 37]. This non-renormalization theorem is dictated by supersymmetry.

Finally, we consider $H/K3 \times T^2$ which has $D = 4, q = 8\,(N = 2)$. The massless multiplets in this case are

$$\begin{aligned}
GR: &\quad (G_{\mu\nu}, A_\mu) \\
V: &\quad (A_\mu, Z) \\
H: &\quad (4q) \\
VT: &\quad (B_{\mu\nu}, A_\mu, \phi) \ .
\end{aligned}$$

where Z now is a complex scalar in the adjoint representation of G. VT is the vector-tensor multiplet which contains the heterotic dilaton ϕ [39]. In $D = 4$ the duality relates $dB_2 = *da_0$ and thus the vector-tensor multiplet is dual to an (Abelian) vector multiplet $VT \sim V$. The moduli space reads

$$\mathcal{M} = \mathcal{M}_H \times \mathcal{M}_V \ , \tag{38}$$

where $\mathcal{M}_H$ is the same (quaternionic) space as in $D = 6, 5$ while $\mathcal{M}_V$ is a special Kähler manifold [40]. Special Kähler manifolds have a Kähler metric

$$G_{i\bar{j}} = \frac{\partial}{\partial Z^i} \frac{\partial}{\partial \bar{Z}^{\bar{j}}} K(Z, \bar{Z}) \ , \qquad i, j = 1, \ldots, n_V \ , \tag{39}$$

with a Kähler potential K determined by a holomorphic prepotential F

$$K = -\ln\left[X^I \bar{F}_I(\bar{X}) + \bar{X}^I F_I(X)\right] , \qquad I = 0, 1, \ldots, n_V \ , \tag{40}$$

where

$$F_I \equiv \frac{\partial F}{\partial X^I} \ , \qquad Z^i = \frac{X^i}{X^0} \ , \qquad F(\lambda X) = \lambda^2 F(X) \ . \tag{41}$$

The gauge kinetic terms have the following structure

$$\mathcal{L} = -\frac{1}{4} g_{IJ}^{-2} F_{\mu\nu}^{I} F_{\mu\nu}^{J} + \frac{\theta_{IJ}}{2\pi} F_{\mu\nu}^{I} \tilde{F}_{\mu\nu}^{J} + \ldots \tag{42}$$

where

$$\begin{aligned} g_{IJ}^{-2} &\sim \mathcal{N}_{IJ} - \bar{\mathcal{N}}_{IJ} \,, \qquad \theta_{IJ} \sim \mathcal{N}_{IJ} + \bar{\mathcal{N}}_{IJ} \,, \\ \mathcal{N}_{IJ} &= \frac{1}{4}\bar{F}_{IJ} - \frac{N_{IK} Z^K N_{JL} Z^L}{Z^K N_{KZ} Z^L} \,, \qquad N_{IJ} = \frac{1}{4}(F_{IJ} + \bar{F}_{IJ}) \,. \end{aligned} \tag{43}$$

The gauge group in $H/K3 \times T^2$ string vacua contains three $U(1)$ vector multiplets, two from T^2 (denoted by T, U) and the dual of the vector-tensor multiplet denoted by S. At the tree level the holomorphic prepotential is uniquely determined by [41, 39]

$$F^{(0)} = (X^0)^2 \cdot S\,(TU - \Phi^a \Phi^a) \ , \tag{44}$$

where Φ^a are the $U(1)$ multiplets which span the Cartan subalgebra of the heterotic gauge group. This prepotential corresponds to the moduli space

$$\mathcal{M}_V^{(0)} = \frac{SU\,(1,1)}{U\,(1)} \times \frac{SO\,(2, r+2)}{SO\,(r+2) \times SO\,(2)} \,, \tag{45}$$

where the first factor is spanned by the dilaton multiplet. Quantum corrections induce only one-loop and non-perturbative corrections to F [39]

$$F = F^{(0)} + F^{(1)} + F^{(NP)} \ . \tag{46}$$

The quaternionic moduli space of the hypermultiplets is the same as in D=6,5 and receives no quantum corrections since the dilaton cannot couple to $\mathcal{M}_H$.

Let is summarize the moduli spaces which appear in K3 compactifications of the heterotic string

$$\begin{aligned} D = 6: &\quad \mathcal{M} = \mathcal{M}_H \times R^+ \\ D = 5: &\quad \mathcal{M} = \mathcal{M}_H \times \left(R^+ \times \tfrac{SO(1,r+1)}{SO(r+1)} + q.c.\right) \\ D = 4: &\quad \mathcal{M} = \mathcal{M}_H \times \left(\tfrac{SU(1,1)}{U(1)} \times \tfrac{SO(2,r+2)}{SO(r+2)\times SO(2)} + q.c.\right) \,, \end{aligned}$$

where $q.c$ indicates that there are quantum corrections.[8]

Let us now turn to the discussion of the dual vacua. In $D = 4$ the dual of the heterotic vacua are vacua constructed as IIA string theory compactified

[8]These quantum corrections are not additive; rather they generically destroy the tree level factorization.

on Calabi-Yau threefolds Y_3, IIA/Y_3 for short [42]. In the NS-NS sector the massless spectrum contains $G_{\mu\nu}, B_{\mu\nu}, \phi$, the deformations of the Calabi-Yau metric and the antisymmetric tensor on the Calabi-Yau manifold. The deformations of the metric are given by the deformations of the Kähler form and the deformations of the complex structure. The former can be expanded in terms of harmonic $(1,1)$-forms $e^A_{i\bar{j}}$ on Y_3 [43]

$$\delta G_{i\bar{j}} = \sum_{A=1}^{h_{1,1}} M^A(x)\, e^A_{i\bar{j}} \,. \tag{47}$$

The deformations of complex structure are expanded as

$$\delta G_{ij} = \sum_{\alpha=1}^{h_{1,2}} q^\alpha(x) b^\alpha_{ij} \,, \tag{48}$$

where $b^\alpha_{ij} = \Omega_j^{\bar{j}\bar{k}} \chi_{i\bar{j}\bar{k}}$ and $\chi_{i\bar{j}\bar{k}}$ are the $(1,2)$-forms, Ω_{ijk} is the $(3,0)$-form. For the antisymmetric tensor one has

$$B_{i\bar{j}} = \sum_{A=1}^{h_{1,1}} B^A(x)\, e^A_{i\bar{j}} \,. \tag{49}$$

In the R-R sector one finds

$$A_{ijk} = C^0(x)\Omega_{ijk} \,, \quad A_{i\bar{j}\bar{k}} = \sum_{\alpha=1}^{h_{1,2}} C^\alpha(x)\chi^\alpha_{i\bar{j}\bar{k}} \,, \quad A_{\mu i\bar{j}} = \sum_{A=1}^{h_{1,1}} A^A_\mu(x) e^A_{i\bar{j}} \,, \tag{50}$$

where C^0, C^α are complex.

The massless fields assemble in the following $N=2$ supermultiplets

$$\begin{array}{ll} GR: & (G_{\mu\nu}, A_\mu) \\ T: & (B_{\mu\nu}, \phi, C^0) \\ V: & (A^A_\mu, Z^A) \\ H: & (q^\alpha, C^\alpha) \end{array}$$

where $Z^A = M^A + iB^A$. Thus altogether one has a spectrum of

$$1GR + 1T + h_{1,1}V + h_{1,2}H \,. \tag{51}$$

Using the (Poincare) duality $T \sim H$ one can express the moduli space as

$$\mathcal{M} = \mathcal{M}_H \times \mathcal{M}_V \,. \tag{52}$$

$\mathcal{M}_V$ is not quantum corrected since the dilaton ϕ sits in T or H, respectively. This component of the moduli space is "known" in the sense that the prepotential obeys the following general structure [44, 45]

$$F = (X^0)^2 \Big[d_{ABC} Z^A Z^B Z^C + \chi\zeta(3) + \sum_{d_A} n_{d_A} Li_3(e^{-2\pi d_A Z^A}) \Big] \, , \tag{53}$$

where

$$d_{ABC} = \int_{Y_3} e^A \wedge e^B \wedge e^C \, , \qquad Li_3\,(x) = \sum_{j=1}^{\infty} \frac{x^j}{j^3} \; . \tag{54}$$

The d_{ABC} are the Calabi-Yau intersection numbers and the n_{d_A} are integers which count the number of rational curves on Y_3.

On the hypermultiplet side only $\mathcal{M}_H^{(0)}$ is known. One has [46]

$$\begin{aligned} \mathcal{L} &= e^{-2\phi}\left(-\frac{1}{2}R + 2\,(\partial\phi)^2 - G_{\alpha\beta}\partial q^\alpha \partial \bar{q}^\beta - \frac{1}{6}H^2_{\mu\nu\rho}\right) \\ &+ \epsilon^{\mu\tau\rho\sigma} H_{\mu\tau\rho}\left(CR^{-1}\partial_\sigma \bar{C} - \frac{1}{2}CR^{-1}\partial_\sigma N R^{-1}\,(C+\bar{C}) - C \leftrightarrow \bar{C}\right) \\ &- \frac{1}{2}\left(\partial_\mu C - \frac{1}{2}\,(C+\bar{C})\,R^{-1}\partial_\mu \mathcal{N}\right) R^{-1}\left(\partial_\mu \bar{C} - \frac{1}{2}\,(C+\bar{C})\,R^{-1}\partial_\mu \bar{\mathcal{N}}\right) . \end{aligned} \tag{55}$$

$G_{\alpha\beta}$ is a special Kähler metric and $R \equiv \mathrm{Re}\mathcal{N}$ so that both are determined by a holomorphic prepotential. The reason for this special feature is that in IIB compactifications one finds a massless spectrum

$$G + T + h_{1,2}\; V + h_{1,1}\; H \; . \tag{56}$$

The role of $h_{1,1}$ and $h_{1,2}$ is exactly reversed compared to the IIA case. It is believed that for any given Y_3 there exists a mirror partner $\widetilde{Y}_3$ with the property [47, 48]

$$h_{1,1}(\widetilde{Y}_3) = h_{1,2}(Y_3) \, , \qquad h_{1,2}(\widetilde{Y}_3) = h_{1,1}(Y_3) \tag{57}$$

such that the Euler number is reversed $\chi(Y_3) = -\chi(\widetilde{Y}_3)$. Thus, in string theory this mirror symmetry leads to a perturbative duality

$$IIB/\widetilde{Y}_3 \equiv IIA/Y_3 \; . \tag{58}$$

In IIB compactifications the q^α reside in vector multiplets and thus one infers that they have to be coordinates on a special Kähler manifold. More generally, the duality implies a map (the c-map) between the moduli spaces which acts as [49]

$$c : \mathcal{M}_V \to \mathcal{M}_H^{(0)} \; . \tag{59}$$

The quantum corrections to the hypermultiplet geometry are not fully known yet. Let us start our discussion with the case $h_{1,2} = 0$ so that only the (universal) tensor multiplet is present. A string 1-loop computation has been performed which determined the correction to the Einstein term [50]

$$\mathcal{L} = -\frac{1}{2}(e^{-2\phi} + \chi)R + \dots \,. \tag{60}$$

Then supersymmetry uniquely determines the corrections [51]

$$\begin{aligned} \mathcal{L} \;=\; & -\frac{1}{2}(e^{-2\phi} + \chi)(R - \frac{1}{6}H^2_{\mu\nu\rho}) + 2e^{-4\phi}(\chi + e^{-2\phi})^{-1}(\partial\phi)^2 \\ & -\partial_\mu C \partial_\mu \bar{C} + \epsilon^{\mu\tau\rho\sigma} H_{\mu\tau\rho} \left(C\partial_\sigma \bar{C} - \bar{C}\partial_\sigma C\right) \,. \end{aligned} \tag{61}$$

This Lagrangian has a perturbative Peccei-Quinn symmetry which originates from the fact that C appears in the R-R sector

$$C \to C + \text{const.} \,. \tag{62}$$

This symmetry is exact in string perturbation theory and forbids any higher loop correction. Thus the 1-loop Lagrangian is perturbatively exact and we have a "new" non-renormalization theorem. (Non-perturbative corrections do exist [52].)

The antisymmetric tensor in the Lagrangian (61) can be dualized to a scalar $\tilde{\phi}$ and in this dual basis the 1-loop corrected metric is found to be Kähler with a Kähler potential

$$K = -\ln\left(S + \bar{S} + 2\chi - C\bar{C}\right), \qquad S \equiv e^{-2\phi} + i\tilde{\phi} + C\bar{C} \,. \tag{63}$$

In this dual basis the metric appears to be corrected at all loops and does agree with the metric conjectured by Strominger [50].

For $h_{1,2} \neq 0$ the quantum corrections are not fully known. One does know [50, 51]

$$\begin{aligned} \mathcal{L} \;=\; & -\frac{1}{2}(e^{-2\phi} + \chi)R - \frac{1}{2}\left(e^{-2\phi} - \chi\right) G_{\alpha\beta}\partial q^\alpha \partial \bar{q}^\beta \\ & -2\chi\epsilon^{\mu\nu\rho\sigma} H_{\mu\nu\rho} V_\sigma + \dots \end{aligned} \tag{64}$$

where

$$V_\sigma \equiv \frac{\partial K}{\partial q^\alpha}\partial_\sigma q^\alpha - \frac{\partial K}{\partial \bar{q}^\alpha}\partial_\sigma \bar{q}^\alpha \,. \tag{65}$$

However, the $N = 2$ supersymmetric (i.e. quaternionic) completion of $\mathcal{L}$ is not completely known yet [53]. Nevertheless, the presence of $2(h_{1,2} + 1)$ continuous PQ-symmetries suggests that also for the case $h_{1,2} \neq 0$ the 1-loop correction is exact in perturbation theory and there is a perturbative non-renormalization theorem.

It is conjectured that IIA/Y_3 is dual to $H/K3 \times T^2$ [42]. In particular this implies that spectrum and the respective moduli spaces have to agree

$$\mathcal{M}_H^{IIA} = \mathcal{M}_H^{\text{Het}} , \qquad \mathcal{M}_V^{IIA} \equiv \mathcal{M}_V^{\text{Het}} . \tag{66}$$

There have been very little checks on $\mathcal{M}_H$ so far. The validity of this duality has been only been checked for $\mathcal{M}_V$ but for quite a number of dual string vacua [42, 54, 55]. One has to find

$$F^{IIA} = F^{\text{het}} , \tag{67}$$

which implies

$$d_{ABC} Z^A Z^B Z^C + \chi\zeta(3) + \sum_{d_A} n_{d_A} Li_3 = S(TU - \Phi^a\Phi^a) + F^{(1)} + F^{(NP)} . \tag{68}$$

This is a condition on the intersection numbers d_{ABC}. They have to obey

$$d_{SSS} = d_{SS\hat{A}} = 0 , \qquad \text{sign}(d_{S\hat{A}\hat{B}}) = (+, -, ..., -) , \tag{69}$$

where $\hat{A}$ denotes all moduli except the dual of the heterotic dilaton. These conditions are the statement that the Calabi–Yau manifold has to be a $K3$-fibration [54, 28, 56]. That is, the Y_3 manifold is fibred over a $\mathbf{P}^1$ base with fibres that are $K3$ manifolds. The size of the $\mathbf{P}^1$ is parameterized by the modulus which is the type II dual of the heterotic dilaton. Over a finite number of points on the base, the fibre can degenerate to something other than $K3$ and such fibres are called singular. The other Kähler moduli are either moduli of the $K3$ fibre or of the singular fibres. In general one finds

$$\text{sign}(d_{S\hat{A}\hat{B}}) = (+, -, \ldots, -, 0, \ldots, 0) , \tag{70}$$

where the non-vanishing entries correspond to moduli from generic $K3$ fibres while the zeros arise from singular fibres. Since a $K3$ has at most 20 Kähler moduli the non-vanishing entries have to be less than 20. From eq. (68) one concludes that type II Calabi-Yau compactifications in the large radius limit can be the dual of perturbative heterotic vacua if they are $K3$-fibrations with all moduli corresponding to generic fibres. This class of type II vacua is automatically consistent with the heterotic bound on the rank of the gauge group. The $(1,1)$ moduli of singular fibres have no counterpart in perturbative heterotic vacua. If there were heterotic moduli with such couplings they would not couple properly to the (heterotic) dilaton and furthermore violate the bound on the rank of the gauge group. However, we already discussed the possibility that in $D = 6$ the gauge group can be non-perturbatively enhanced at singular points in the moduli space [35]. It was further shown that these non-perturbative gauge fields do not share the canonical coupling to the dilaton. Upon compactification to $D = 4$ the scalars of these non-perturbative vector multiplets couple precisely like type II moduli corresponding to singular fibres [57, 58].

Let us now turn to vacua of M/Y_3 (which have $D = 5, q = 8$). We expect this theory to be dual to $H/K3 \times S^1$ by the following chain of arguments. M/S^1 is the dual of IIA in $D = 10$. Thus compactifying both theories on Y_3 one expects the dual pair $M/Y_3 \times S^1 \sim IIA/Y_3$. However, the previous discussion also suggests $IIA/Y_3 \sim H/K3 \times T^2$ and thus one is lead to conjecture $M/Y_3 \sim H/K3 \times S^1$ [38, 37].[9]

The massless spectrum of M/Y_3 contains the metric $G_{\mu\nu}$, $h_{1,1}$ deformations of the Kähler form $\delta G_{i\bar{j}}$, $h_{1,2}$ (complex) deformations of the complex structure δG_{ij}, a three-form $A_{\mu\nu\rho}$, $h_{1,1}$ vectors $A_{\mu i\bar{j}}$, one complex scalar A_{ijk} and $h_{1,2}$ complex scalars $A_{ij\bar{k}}$. The duality in $D = 5$ relates the 3-form to a scalar

$$dA_3 = *d\tilde{\phi} \, , \tag{71}$$

and so altogether one has the spectrum

$$1GR + (h_{1,1} - 1)\, V + (h_{1,2} + 1)\, H \, , \tag{72}$$

where

$$\begin{array}{ll} GR: & (G_{\mu\nu}, A_\mu) \\ V: & (A_\mu, Z) \\ H: & (4q) \quad . \end{array}$$

Z is real and the volume of Y_3 resides not in a vector multiplet but rather in a hypermultiplet. (This accounts for the ± 1 in the counting.) The moduli space is again

$$\mathcal{M} = \mathcal{M}_V \times \mathcal{M}_H \, . \tag{73}$$

A more detailed comparison of the effective actions of M/Y_3 and $H/K3 \times S^1$ reveals

$$\frac{1}{(g_h^5)^2} = \frac{[Vol_M(\mathbf{P}_1)]^{\frac{3}{2}}}{[Vol_M(Y_3)]^{\frac{1}{2}}} = \frac{Vol_M(\mathbf{P}_1)}{[Vol_M(K3)]^{\frac{1}{2}}} \, , \tag{74}$$

where $\mathbf{P}_1$ is again the base of the K3-fibrations and the second equation used the adiabatic argument. From (74) we immediately infer that large $\mathbf{P}_1$ corresponds to weak heterotic coupling while a large K3 corresponds to strong heterotic coupling. Eq. (74) can also be 'derived' by 'fibering' the duality $M/K3 \sim H/T^3$ over $\mathbf{P}_1$. This implies $M/K3 \times \mathbf{P_1} \sim H/T^3 \times \mathbf{P}_1$

[9]For perturbative heterotic vacua one expects again that Y_3 is K3-fibred.

and using the adiabatic argument also $M/Y_3 = H/K3 \times S^1$. In terms of the couplings one has

$$\frac{1}{\left(g_h^5\right)^2} = \frac{Vol_H\left(\mathbf{P}_1\right)}{\left(g_h^7\right)^2} = \frac{Vol_M\left(\mathbf{P}_1\right)}{\left[Vol_M\left(K3\right)\right]^{\frac{1}{2}}}\,, \tag{75}$$

where we used $\left(g_h^7\right)^2 = \left[Vol_M\left(K3\right)\right]^{\frac{3}{2}}$.

Finally we turn to F-theory compactified on an elliptic Calabi–Yau threefold F/Y_3. Such vacua have $D = 6, q = 8$ and are conjectured to be dual to the heterotic string compactified on $K3$ [14]. Recall that via the adiabatic argument one has for any elliptic Calabi-Yau manifold

$$F/Y_n = \lim_{T^2 \to 0} M/Y_n\,. \tag{76}$$

For threefolds Y_3 one also has the duality $M/Y_3 = H/K3 \times S^1$ and thus from (74)

$$\frac{1}{\left(g_h^5\right)^2} = \frac{R_6}{\left(g_h^6\right)^2} = \frac{\left[Vol_M\left(\mathbf{P}_1\right)\right]^{\frac{3}{2}}}{\left[Vol_M\left(Y_3\right)\right]^{\frac{1}{2}}} = \frac{\left[Vol_M\left(\mathbf{P}_1\right)\right]^{\frac{3}{2}}}{\left[Vol_M\left(B_2\right)\right]^{\frac{1}{2}}\left[Vol_M\left(T^2\right)\right]^{\frac{1}{2}}}\,. \tag{77}$$

Thus the limit $T^2 \to 0$ sends $R_6 \to \infty$ with

$$\frac{1}{\left(g_h^6\right)^2} = \frac{\left[Vol_M\left(P_1\right)\right]^{3/2}}{\left[Vol_M\left(B_2\right)\right]^{\frac{1}{2}}}\,. \tag{78}$$

As before there is an alternative 'derivation' by fibering the $D = 8$ duality $F/K3 \sim H/T^2$ over $\mathbf{P}_1$. This gives $F/K3 \times \mathbf{P}_1 \sim H/T^2 \times \mathbf{P}_1$ and via the adiabatic argument $F/Y_3 \sim H/K3$.

The spectrum of F/Y_3 features

$$1GR + \left(h_{1,2}(Y_3) + 1\right)H + \left(h_{1,1}(B_2) - 1\right)T + n_V V \tag{79}$$

with B_2 being the base of the elliptic fibration. Since the gauge group G can be non-Abelian n_V is generically not determined. However, for the rank of G one has [14][10]

$$\mathrm{r}(G) = h_{1,1}\left(Y_3\right) - h_{1,1}\left(B_2\right) - 1. \tag{80}$$

The F-theory duals of the perturbative heterotic string are constructed from threefolds Y_3 which are elliptically and K3-fibred at the same time

[10]This can be derived from compactification to $D = 5$ where one has a Coulomb branch and $n_V^5 = \mathrm{r}(G) + n_T + 1 = h_{1,1}\left(Y_3\right) - 1$ vector multiplets.

[14]. This determines the base B_2 to be the Hirzebruch surface $\mathbb{F}_k$. Such manifolds have $h_{1,1}(\mathbb{F}_k) = 2$ and thus $n_T = 1$ as required for the perturbative heterotic string. In fact there is a beautiful correspondence between the heterotic vacua labelled by the instanton numbers (n_1, n_2) and elliptically fibred Calabi-Yau manifolds with the base being the Hirzebruch surfaces $\mathbb{F}_{n_2-12}$ [14].

Blown up $\mathbb{F}_k$ have $h_{1,1}(\mathbb{F}_k) > 2$ and thus $n_T > 1$. These F-theory vacua thus capture non-perturbative physics of the heterotic vacua including the possibility of additional tensor multiplets, the transitions between the various branches of moduli space and subspaces of symmetry enhancement [14, 57, 59].

A. Calabi-Yau manifolds

In this appendix we briefly recall a few facts about Calabi–Yau manifolds which we frequently use in the main text. (For a more extensive review see for example [1, 23, 48].)

A Calabi–Yau manifold Y_n is a Ricci-flat Kähler manifold of vanishing first Chern-class. Its holonomy group is $SU(n)$ where n is the complex dimension of Y_n. The simplest Calabi–Yau manifolds are tori of complex dimension 1. For $n = 2$ all Calabi–Yau manifolds are topologically equivalent to the $K3$ surface, while for $n = 3, 4$ one finds many topologically distinct Calabi–Yau manifolds. Such manifolds are of interest in string theory since they break some of the supersymmetries when a ten-dimensional string theory is compactified on Y_n.

The massless modes of a string vacuum are directly related to the zero modes of the Laplace operator on Y_n. These zero modes are the non-trivial differential k-forms on Y_n and they are elements of the cohomology groups $H^k(Y)$. On a compact Kähler manifold one can decompose any k-form into (p,q)-forms with p holomorphic and q antiholomorphic differentials $(p + q = k)$. Analogously, the associated cohomology groups decompose according to

$$H^k(Y) = \oplus_{p+q=k} H^{p,q}(Y). \tag{81}$$

The dimension of $H^{p,q}(Y)$ is called the Hodge number $h_{p,q}$ ($h_{p,q} = \dim H^{p,q}$); it is symmetric under the exchange of p and q, i.e. $h_{p,q} = h_{q,p}$, and Poincaré duality identifies $h_{p,q} = h_{n-p,n-q}$. Finally, the Euler number is given by

$$\chi = \sum_{p,q} (-1)^{p+q} h_{p,q} \,. \tag{82}$$

For K3 the Hodge numbers are

$$\begin{array}{ccccc} & & h^{0,0} & & \\ & h^{1,0} & & h^{0,1} & \\ h^{2,0} & & h^{1,1} & & h^{0,2} \\ & h^{2,1} & & h^{1,2} & \\ & & h^{2,2} & & \end{array} = \begin{array}{ccccc} & & 1 & & \\ & 0 & & 0 & \\ 1 & & 20 & & 1 \\ & 0 & & 0 & \\ & & 1 & & \end{array} .$$

The Euler number is $\chi = 24$.

The moduli space of non-trivial metric deformations which preserve the Calabi-Yau condition is parameterized by 20 deformations of Kähler form $\delta G_{i\bar{j}}$ and 19 + 19 deformations of complex structure δG_{ij}. They are the coordinates of the homogeneous space

$$\mathcal{M} = R^+ \times \frac{SO\,(3,19)}{SO\,(3) \times SO\,(19)} / SO\,(3,19,\mathbf{Z}) \ , \tag{83}$$

where R^+ is spanned by the volume of K3.

For a Calabi-Yau threefold Y_3 one has the Hodge diamond

$$\begin{array}{ccccccc} & & & h^{0,0} & & & \\ & & h^{1,0} & & h^{0,1} & & \\ & h^{2,0} & & h^{1,1} & & h^{0,2} & \\ h^{3,0} & & h^{2,1} & & h^{1,2} & & h^{0,3} \\ & h^{3,1} & & h^{2,2} & & h^{1,3} & \\ & & h^{3,2} & & h^{2,3} & & \\ & & & h^{3,3} & & & \end{array} = \begin{array}{ccccccc} & & & 1 & & & \\ & & 0 & & 0 & & \\ & 0 & & h^{1,1} & & 0 & \\ 1 & & h^{1,2} & & h^{1,2} & & 1 \\ & 0 & & h^{1,1} & & 0 & \\ & & 0 & & 0 & & \\ & & & 1 & & & \end{array}$$

where $h_{1,1}$ and $h_{1,2}$ are arbitrary and the Euler number is $\chi(Y_3) = 2(h_{1,1} - h_{1,2})$. $h_{1,1}$ counts the number of Kähler deformations of the metric while $h_{1,2}$ counts the number of deformations of the complex structure. The moduli space is locally a direct product of the Kähler moduli space and the complex structure moduli space

$$\mathcal{M} = \mathcal{M}_{h_{1,1}} \times \mathcal{M}_{h_{1,2}} \ . \tag{84}$$

Each factor is a special Kähler manifold and the corresponding K obeys eq. (40). For $\mathcal{M}_{h_{1,2}}$ one finds

$$K_{1,2} = -\ln \int_{Y_3} \Omega \wedge \bar{\Omega} \tag{85}$$

where $\Omega(\bar{\Omega})$ is the unique $(3,0)$-form ($(0,3)$-form) on Y_3. In the large volume limit one has for $\mathcal{M}_{h_{1,1}}$

$$K_{1,1} = -\ln Vol\,(Y_3) = -\ln d_{ABC} M^A M^B M^C \ . \tag{86}$$

It is believed that most Calabi–Yau threefolds (if not all) have a mirror partner [47, 48]. That is, for a given Calabi–Yau threefold Y_3 with given $h_{1,1}(Y_3)$ and $h_{1,2}(Y_3)$ there exists a mirror manifold $\tilde{Y}_3$ with $h_{1,1}(\tilde{Y}_3) = h_{1,2}(Y_3)$ and $h_{1,2}(\tilde{Y}_3) = h_{1,1}(Y_3)$ which implies $\chi(Y_3) = -\chi(\tilde{Y}_3)$.

Acknowledgements

We would like to thank the organizers and in particular Larus Thorlacius for organizing a stimulating summer school in a spectacular surrounding.

We also thank Michael Haack and Holger Günther for discussions and a careful reading of the manuscript.

References

1. For a review of perturbative string theory see, for example,
M. Green, J. Schwarz and E. Witten, *"Superstring theory"*, Vol. 1&2, Cambridge University Press, (1987);
D. Lüst and S. Theisen, *"Lectures on string theory"*, Springer, (1989);
J. Polchinski, *"String Theory"*, Vol. 1&2, Cambridge University Press, (1998).
2. For a review see, for example, A.A. Tseytlin, in *Superstrings '89*, eds. M. Green, R. Iengo, S. Randjbar-Daemi, E. Sezgin and A. Strominger, World Scientific (1990).
C. Callan and L. Thorlacius, in *Particles, Strings and Supernovae*, eds. A. Jevicki and C.-I. Tan World Scientific, (1989).
3. A. Schellekens and N. Warner, *"Anomalies and modular invariance in string theory"*, Phys. Lett. **177B** (1986) 317.
4. For a review see, for example,
J. Schwarz, *"Lectures on superstring and M-theory dualities"*, hep-th/9607021;
A. Sen, *"Unification of string dualities"*, hep-th/9609176;
W. Lerche, *"Introduction to Seiberg-Witten Theory and its Stringy Origin"*, Nucl. Phys. Proc. Suppl. **55B** (1997) 83, hep-th/9611190;
P.K. Townsend, *"Four Lectures on M-theory"*, lectures given at the 1996 ICTP Summer School in High Energy Physics and Cosmology, Trieste, hep-th/9612121;
S. Förste and J. Louis, *" Duality in String Theory"*, in *Gauge Theories, Applied Supersymmetry and Quantum Gravity II*, ed. A. Sevrin, K.S. Stelle K. Thielemans and A. Van Proeyen, Imperial College Press, 1997, hep-th/9612192;
C. Vafa, *"Lectures on Strings and Dualities"*, hep-th/9702201;
A. Klemm, *"On the Geometry behind N=2 Supersymmetric Effective Actions in Four Dimensions"*, hep-th/9705131;
E. Kiritsis, *"Introduction to non-perturbative String Theory"*, hep-th/9708130;
W. Lerche, *"Recent Developments in String Theory"*, hep-th/9710246;
P.K. Townsend, *"M-theory from its Superalgebra"*, hep-th/9712004;
B. de Wit und J. Louis, *"Supersymmetry and Dualities in Various Dimensions"*, hep-th/9801132, lectures at the NATO Advanced Study Institute on "Strings, Branes and Duality", Cargèse, France, June 1997, published in the Proccedings.
5. For a review, see, for example, A. Giveon, M. Porrati and E. Rabinovici, *"Target space duality in string theory"*, Phys. Rept. **244** (1994) 77, hep-th/9401139.
6. For a review see, for example, J. Polchinski, *"TASI lectures on D-branes"*, hep-th/9611050;
C. Bachas, *"(Half) a Lecture on D-branes"*, in *Gauge Theories, Applied Supersymmetry and Quantum Gravity II*, ed. A. Sevrin, K.S. Stelle K. Thielemans and A. Van Proeyen, Imperial College Press, 1997, hep-th/9701019.

7. C.M. Hull and P.K. Townsend, *"Unity of superstring dualities"*, Nucl. Phys. **B438** (1995) 109, hep-th/9410167;
C.M. Hull, *"String-string duality in ten-dimensions"*, Phys. Lett. **B357** (1995) 545, hep-th/9506194.
8. E. Witten, *"String theory dynamics in various dimensions"*, Nucl. Phys. **B443** (1995) 85, hep-th/9503124.
9. J. Schwarz, *"An $SL(2, Z)$ multiplet of type-IIB superstrings"*, Phys. Lett. **B360** (1995) 13, hep-th/9508143.
10. E. Witten, *"Bound states of strings and p-branes"*, Nucl. Phys. **B460** (1996) 335, hep-th/9510135.
11. C. Vafa, *"Evidence for F theory"*, Nucl. Phys. **B469** (1996) 403, hep-th/9602022.
12. B.R. Greene, A. Shapere, C. Vafa and S.-T. Yau, *"Stringy cosmic strings and noncompact Calabi-Yau Manifolds"*, Nucl. Phys. **B337** (1990) 1.
13. P. Berglung and P. Mayr, *"Heterotic String/F-theory Duality from Mirror Symmetry"*, Adv. Theor. Math. Phys. **2** (1999) 1307, hep-th/9811217.
14. D.R. Morrison and C. Vafa, *"Compactifications of F-theory on Calabi-Yau threefolds - I"*, Nucl. Phys. **B473** (1996) 74, hep-th/9602111; *"Compactifications of F-theory on Calabi-Yau threefolds - II"*, Nucl. Phys. **B476** (1996) 437, hep-th/9603161.
15. P.K. Townsend, *"The eleven-th dimensional supermembrane revisited"*, Phys. Lett. **B350** (1995) 184, hep-th/9501068, *"D-branes from M-branes"*, Phys. Lett. **B373** (1996) 68, hep-th/9512062.
16. P. Hořava and E. Witten, *"Heterotic and type-I string dynamics from eleven-dimensions"*, Nucl. Phys. **B460** (1996), 506, hep-th/9510209; *"Eleven-dimensional supergravity and a manifold with boundary"*, Nucl. Phys. **B475** (1996)94, hep-th/9603142.
17. T. Banks, W. Fischler, S.H. Shenker and L. Susskind, *"M Theory As A Matrix Model: A Conjecture"*, Phys. Rev. **D55** (1997) 5112, hep-th/9610043.
18. M. Claudson and M.B. Halpern, *"Supersymmetric ground state wave functions"*, Nucl. Phys. **B250** (1985) 689;
R. Flume, *"On quantum mechanics with extended supersymmetry and nonabelian gauge constraints"*, Ann. Phys. **164** (1985) 189;
M. Baake, P. Reinicke, and V. Rittenberg, *"Fierz identities for real Clifford algebras and the number of supercharges"*, J. Math. Phys. **26** (1985) 1070.
19. E.A. Bergshoeff, E. Sezgin and P.K. Townsend, *"Supermembranes and eleven-dimensional supergravity"*, Phys. Lett. **B189** (1987) 75; *"Properties of the eleven-dimensional supermembrane theory"*, Ann. Phys. **185** (1988) 330.
20. B. de Wit, J. Hoppe and H. Nicolai, *"On the quantum mechanics of supermembranes"*, Nucl. Phys. **B305** (1988) 545.
21. For a review, see, for example, R. Dijkgraaf, E. Verlinde and H. Verlinde, *"Notes on Matrix and Micro Strings"*, hep-th/9709107;
N. Seiberg, *"Why is the Matrix Model correct?"*, hep-th/9710009;
A. Bilal, *"M(atrix) Theory: a Pedagogical Introduction"*, hep-th/9710136;
T. Banks, *"Matrix Theory"*, hep-th/9710231;
D. Bigatti and L. Susskind, *"Review of Matrix Theory"*, hep-th/9712072.
22. K.S. Narain, *"New heterotic string theories in uncompactified dimensions < 10"*, Phys. Lett. **169B** (1986) 41;
K.S. Narain, M.H. Samadi and E. Witten, *"A note on toroidal compactification of heterotic string theory"*, Nucl. Phys. **B279** (1987) 369.
23. For a review, see P. Aspinwall, *"$K3$ surfaces and string duality"*, hep-th/9611137.
24. N. Seiberg, *"Observations on the moduli space of superconformal field theories"*, Nucl. Phys. **B303** (1988) 286;
P.S. Aspinwall and D.R. Morrison, *"String theory on $K3$ surfaces"*, in *Mirror Symmetry II*, eds. B. Greene and S.-T. Yau, International Press, Cambridge, 1997, hep-th/9404151.

25. M.J. Duff and J.X. Lü, *"Black and super p-branes in diverse dimensions"*, Nucl. Phys. **B416** (1994) 301;
M.J. Duff and R. Minasian, *"Putting string/string duality to the test"*, Nucl. Phys. **B436** (1995) 261, hep-th/9406198.
26. A. Sen, *"String-string duality conjecture in six dimensions and charged solitonic strings"*, Nucl. Phys. **B450** (1995) 103, hep-th/9504027.
27. J.A. Harvey and A. Strominger, *"The heterotic string is a soliton"*, Nucl. Phys. **B449** (1995) 456, hep-th/9504047.
28. C. Vafa and E. Witten, *"Dual string pairs with $N = 1$ and $N = 2$ supersymmetry in four-dimensions"*, Nucl. Phys. Proc. Suppl. **46** (1996) 225, hep-th/9507050.
29. A. Sen, *"F-theory and orientifolds"*, hep-th/9605150.
30. P.S. Aspinwall and D.R. Morrison, *"Point-like Instantons on K3 Orbifolds"*, Nucl. Phys. **B503** (1997) 533, hep-th/9705104.
31. M.B. Green, J. Schwarz and P. C. West, *"Anomaly-free chiral theories in six dimensions"*, Nucl. Phys. **B254** (1985) 327;
J. Erler, *"Anomaly cancellation in six dimensions"*, J. Math. Phys. **35** (1994) 1819, hep-th/9304104;
J. Schwarz, *"Anomaly-free supersymmetric models in six dimensions"*, Phys. Lett. **B371** (1996) 223, hep-th/9512053.
32. M. Berkooz, R. Leigh, J. Polchinski, J. Schwarz, N. Seiberg and E. Witten, *"Anomalies, Dualities, and Topology of $D = 6, N = 1$ superstring vacua"*, Nucl. Phys. **B475** (1996) 115, hep-th/9605184.
33. S. Ferrara, R. Minasian and A. Sagnotti, *"Low-Energy Analysis of M and F Theories on Calabi-Yau Threefolds"*, Nucl. Phys. **B474** (1996) 323, hep-th/9604097.
34. C.G. Callan, J.A. Harvey and A. Strominger, *"World-sheet approach to heterotic instantons and solitons"*, Nucl. Phys. **B359** (1991) 611; *"World brane actions for string solitons"*, Nucl. Phys. **B367** (1991) 60; *"Supersymmetric string solitons"*, hep-th/9112030.
35. E. Witten, *"Small instantons in string theory."*, Nucl. Phys. **B460** (1996) 541, hep-th/9511030.
36. N. Seiberg and E. Witten, *"Comments on string dynamics in six dimensions"*, Nucl. Phys. **B471** (1996) 121, hep-th/9603003.
37. I. Antoniadis, S. Ferrara and T.R. Taylor, *"$N = 2$ Heterotic Superstring and its Dual Theory in Five Dimensions"*, Nucl. Phys. **B460** (1996) 489, hep-th/9511108.
38. E. Witten, *"Phase Transitions In M-Theory And F-Theory"*, Nucl. Phys. **B471** (1996) 195, hep-th/9603150.
39. B. de Wit, V. Kaplunovsky, J. Louis and D. Lüst, *"Perturbative Couplings of Vector Multiplets in $N = 2$ Heterotic String Vacua"*, Nucl. Phys. **B451** (1995) 53, hep-th/9504006.
40. For recent reviews on special geometry, see, for example, B. de Wit and A. van Proeyen, *"Special geometry and symplectic transformations"*, Nucl. Phys. B (Proc. Suppl.) **45B,C** (1996) 196, hep-th/9510186;
A. van Proeyen, *"Vector multiplets in $N = 2$ supersymmetry and its associated moduli spaces"*, hep-th/9512139.
41. S. Ferrara and A. Van Proeyen, *"A Theorem on $N = 2$ Special Kähler Product Manifolds"* Class. Quant. Grav. **6** (1989) 243.
42. S. Kachru and C. Vafa, *"Exact results for $N = 2$ compactifications of heterotic strings"*, Nucl. Phys. **B450** (1995) 69, hep-th/9505105;
S. Ferrara, J.A. Harvey, A. Strominger and C. Vafa, *"Second-quantized mirror symmetry"*, Phys. Lett. **B361** (1995) 59, hep-th/9505162.
43. P. Candelas and X. de la Ossa, *"Moduli space of Calabi-Yau manifolds"*, Nucl. Phys. **B355** (1991), 455; A. Strominger, *"Special Geometry"*, Commun. Math. Phys. **133** (1990) 163.
44. P. Candelas, X. de la Ossa, P. Green and L. Parkes , Nucl. Phys. **B359** (1991) 21;
P. Candelas, X. De la Ossa, A. Font, S. Katz and D. Morrison, Nucl. Phys. **B416**

(1994) 481, hep-th/9308083;
P. Candelas, A. Font, S. Katz and D. Morrison, Nucl. Phys. **B429** (1994) 626, hep-th/9403187.

45. J.A. Harvey and G. Moore, Nucl. Phys. **B463** (1996) 315, hep-th/9510182.
46. S. Ferrara and S. Sabharwal, *"Quaternionic Manifolds for Type II Superstrings Vacua of Calabi–Yau Spaces"* Class. Quant. Grav. **6** (1989) L77, Nucl. Phys. **B332** (1990) 317;
M. Bodner, A.C. Cadavid and S. Ferrara, Class. Quant. Grav. **8** (1991) 789.
47. L. Dixon, *"Some Worldsheet Properties of Superstring Compactifications, on Orbifolds and Otherwise"*, in Superstrings, Unified Theories, and Cosmology ed. G. Furlan et al., World Scientific (1988);
P. Candelas, M. Lynker and R. Schimmrigk, *"Calabi–Yau manifolds in weighted $P(4)$"*, Nucl. Phys. B341 (1990) 383;
B. Greene and R. Plesser, *"Duality in Calabi–Yau Moduli Space"*, Nucl. Phys. B338 (1990) 15.
48. For a review, see, for example, S. Hosono, A. Klemm and S. Theisen, *"Lectures on mirror symmetry"*, hep-th/9403096;
B. Greene, *"String Theory on Calabi-Yau Manifolds"*, hep-th/9702155.
49. S. Cecotti, S. Ferrara and L. Girardello, *"Geometry of type-II superstrings and the moduli of superconformal field theories"*, Int. J. Mod. Phys. **A4** (1989), 2475.
50. E. Kiritsis and C. Kounnas, *"Infrared regularization of superstring theory and the one loop calculation of coupling constants"*, Nucl. Phys. **B442** (1995) 472, hep-th/9501020;
A. Strominger *"Loop Corrections to the Universal Hypermultiplet"* Phys. Lett. **B421** (1998) 139, hep-th/9706195;
I. Antoniadis, S. Ferrara, R. Minasian and K.S. Narain, *"R^4 couplings in M and type II theories on Calabi-Yau spaces"*, Nucl. Phys. **B507** (1997) 571, hep-th/9707013.
51. H. Günther, C. Herrmann and J. Louis, *"Quantum Corrections in the Hypermultiplet Moduli Space"*, hep-th/9901137.
52. H. Ooguri and C. Vafa, Phys. Rev. Lett. **77** (1996) 3296, hep-th/9608079.
53. H. Günther, C. Herrmann and J. Louis, in preperation.
54. A. Klemm, W. Lerche and P. Mayr, *"K3 fibrations and heterotic type-II string duality"*, Phys. Lett. **B357** (1995) 313, hep-th/9506112.
55. V. Kaplunovsky, J. Louis and S. Theisen, *"Aspects of duality in $N = 2$ string vacua"*, Phys. Lett. **B357** (1995) 71, hep-th/9506110;
I. Antoniadis, E. Gava, K.S. Narain and T.R. Taylor, *"$N = 2$ type-II - heterotic duality and higher derivative F-terms"*, Nucl. Phys. **B455** (1995) 109, hep-th/9507115;
G. Aldazabal, A. Font, L .E. Ibañez and F. Quevedo, *"Chains of $N = 2, D = 4$ heterotic type-II duals"*, Nucl. Phys. **B461** (1996) 85, hep-th/9510093;
G. Lopes Cardoso, G. Curio, D. Lüst, T. Mohaupt and S.-J. Rey, *"BPS spectra and non-perturbative couplings in $N = 2,4$ supersymmetric string theories"*, Nucl. Phys. **B464** (1996) 18, hep-th/9512129;
P. Candelas and A. Font, *"Duality Between the Webs of Heterotic and Type II Vacua*, hep-th/9603170;
P. Berglund, S. Katz, A. Klemm and P. Mayr, *"New Higgs transitions between dual $N = 2$ string models"*, Nucl. Phys. **B483** (1997) 209, hep-th/9605154;
M. Bershadsky, K. Intriligator, S. Kachru, D.R. Morrison, V. Sadov and C. Vafa, *"Geometric singularities and enhanced gauge symmetries"*, Nucl. Phys. **B481** (1996) 215, hep-th/9605200;
J. Louis, J. Sonnenschein, S. Theisen and S. Yankielowicz, *"Nonperturbative properties of heterotic string vacua"*, Nucl. Phys. **B480** (1996) 185, hep-th/9606049;
B. de Wit, G. Lopes Cardoso, D. Lüst, T. Mohaupt and S.-J. Rey, *"Higher order gravitational couplings and modular forms in $N = 2$, $D = 4$ heterotic string com-*

pactifications", Nucl. Phys. **B481** (1996) 353hep-th/9607184;
for a review see, J. Louis and K. Förger, "*Holomorphic couplings in string theory*", Nucl. Phys. (Proc. Suppl.) **55B** (1997) 33, hep-th/9611184.

56. P.S. Aspinwall and J. Louis, "*On the ubiquity of $K3$ fibrations in string duality*", Phys. Lett. **B369** (1996) 233, hep-th/9510234.
57. P.S. Aspinwall and M. Gross, Phys. Lett. **B382** (1996) 81, hep-th/9602118.
58. J. Louis, J. Sonnenschein, S. Theisen and S. Yankielowicz, "*Non-perturbative Properties of Heterotic String Vacua compactified on $K3 \times T^2$*", Nucl. Phys. **B451** (1995) 53, hep-th/9606049.
59. M. Bershadsky, K. Intriligator, S. Kachru, D.R. Morrison, V. Sadov and C. Vafa, "*Geometric singularities and enhanced gauge symmetries*", hep-th/9605200.

17 8'99

13 8'99

THE M(ATRIX) MODEL OF M-THEORY

WASHINGTON TAYLOR
Center for Theoretical Physics
MIT, Bldg. 6-306
Cambridge, MA 02139, U.S.A.
email: wati@mit.edu

Abstract. These lecture notes give a pedagogical and (mostly) self-contained review of some basic aspects of the Matrix model of M-theory. The derivations of the model as a regularized supermembrane theory and as the discrete light-cone quantization of M-theory are presented. The construction of M-theory objects from matrices is described, and gravitational interactions between these objects are derived using Yang-Mills perturbation theory. Generalizations of the model to compact and curved space-times are discussed, and the current status of the theory is reviewed.

1. Introduction

This series of lectures describes the matrix model of M-theory, also known as M(atrix) Theory. Matrix theory is a supersymmetric quantum mechanics theory with matrix degrees of freedom. It has been known for over a decade [1, 2] that matrix theory arises as a regularization of the 11D supermembrane theory in light-front gauge. It was conjectured in 1996 [3] that when the size of the matrices is taken to infinity this theory gives a microscopic second-quantized description of M-theory in light-front coordinates.

These lectures focus on some basic aspects of matrix theory. We begin by describing in some detail the two alternative definitions of the theory in terms of a quantized and regularized supermembrane theory and as a compactification of M-theory on a lightlike circle. Given these definitions of the theory, we then focus on the question of whether the physics of M-theory and 11-dimensional supergravity can be described constructively using finite size matrices. We show that all the objects of M-theory, including the supergraviton, membrane and 5-brane can be constructed explicitly

L. Thorlacius and T. Jonsson (eds.), M-Theory and Quantum Geometry, 91-177.

from configurations of matrices, although these results are not yet complete in the case of the 5-brane. We then turn to the gravitational interactions between these objects, and review what is known about the connection between perturbative calculations in the matrix quantum mechanics theory and supergravity interactions. In the last part of the lectures, some discussion is given of how the matrix theory formalism may be generalized to describe compact or curved space-times.

Previous reviews of matrix theory and related work have appeared in [4, 5, 6, 7, 8, 9, 10].

In Section 2 we show how matrix theory can be derived from the light-front quantization of the supermembrane theory in 11 dimensions. We discuss in Section 3 the conjecture of Banks, Fischler, Shenker and Susskind that matrix theory describes light-front M-theory in flat space, and we review an argument of Seiberg and Sen showing that finite N matrix theory describes the discrete light-cone quantization (DLCQ) of M-theory. In Section 4 we show how the objects of M-theory (the supergraviton, supermembrane and M5-brane) can be described in terms of matrix theory degrees of freedom. Section 5 reviews what is known about the interactions between these objects. We discuss the problem of reproducing N-body interactions in 11D classical supergravity from matrix theory, beginning with two-body interactions in the linearized theory and then discussing many-body interactions and nonlinear terms as well as quantum corrections to the supergravity theory. Section 6 contains a discussion of the problems of formulating matrix theory on a compact or curved background geometry. Finally, we conclude in section 7 with a summary of the current state of affairs and the outlook for the future of this theory.

Even if in the long run matrix theory turns out not to be the most useful description of M-theory, there are many features of this theory which make it well worth studying. It is the simplest example of a quantum supersymmetric gauge theory which seems to correspond to a theory of gravity in a fixed background in some limit. It is the only known example of a well-defined quantum theory which has been shown explicitly to give rise to long-range interactions which agree with gravity at the linearized level and which also contain some nonlinearity. Finally, it provides simple examples of many of the remarkable connections between D-brane physics and gauge theory, giving intuition which may be applicable to a wide variety of situations in string theory and M-theory.

2. Matrix theory from the quantized supermembrane

In this section we show that supersymmetric matrix quantum mechanics arises naturally as a regularization of the supermembrane action in 11 dimensions. We begin our discussion with some motivational remarks.

In retrospect, the supermembrane is a natural place to begin when trying to construct a microscopic description of M-theory. There are several distinct 10-dimensional supersymmetric theories of gravity. These theories are well-defined classically but, as with all theories of gravity, are difficult to quantize directly. Each of these theories has a bosonic antisymmetric 2-form tensor field $B_{\mu\nu}$. This field is analogous to the 1-form field A_μ of electromagnetism, but carries an extra space-time index. Each of these 10D supergravity theories admits a classical stringlike black hole solution which is "electrically" charged under the 2-form field, in the sense that the two-dimensional world-volume Σ of the string couples to the B field through a term

$$\int_\Sigma B_{\mu\nu}\epsilon^{ab}(\partial_a X^\mu)(\partial_b X^\nu).$$

where X^μ are the embedding functions of the string world-volume in 10 dimensions. This is the higher-dimensional analog of the usual coupling of a charged particle to a gauge field through $\int A_\mu \dot{X}^\mu$.

The quantization of strings in 10-dimensional background geometries can be carried out consistently in only a limited number of ways. These constructions lead to the perturbative descriptions of the five superstring theories known as the type I, IIA, IIB and heterotic $E_8 \times E_8$ and $SO(32)$ theories. These quantum superstring theories are first-quantized from the point of view of the target space—that is, a state in the string Hilbert space corresponds to a single particle-like state in the target space consisting of a single string. Although the quantized string spectrum naturally contains states corresponding to quanta of the supergravity fields (including the NS-NS field $B_{\mu\nu}$), it is not possible to give a simple description in terms of the string Hilbert space for extended objects such as D-branes and the NS 5-brane. These objects are essentially nonperturbative phenomena in the superstring theories.

One of the most important developments in the last few years has been the discovery of a network of duality symmetries which relates the five superstring theories to each other and to 11-dimensional supergravity. Of these six theories, the quantized superstring gives a microscopic description of the five 10-dimensional theories. It has been hypothesized that there is a microscopic 11-dimensional theory, dubbed M-theory, underlying this structure which reduces in the low-energy limit to 11D supergravity [11]. To date, however, a precise description of this theory is lacking. Such a theory cannot be described by a quantized string since there is no antisymmetric 2-form field in the 11D supergravity multiplet and hence no stringlike solution of the gravity equations. The 11D supergravity theory contains, however, an antisymmetric 3-form field A_{IJK}, and the classical theory admits membrane-like solutions which couple electrically to this field. It is easy to imagine that a microscopic description of M-theory might be found by quantizing this supermembrane. This idea was explored extensively in the 80's, when it was first realized that a consistent classical theory of a supermembrane could be realized in 11 dimensions. At that time, while no satisfactory covariant quantization of the membrane theory was found, it was shown that the supermembrane could be quantized in light-front coordinates. In fact, an elegant regularization of this theory was suggested by Goldstone and Hoppe [1] in 1982. They showed that for the bosonic membrane the regularized quantum theory is a simple quantum-mechanical theory of $N \times N$ matrices which leads to the membrane theory in the large N limit. This approach was generalized to the supermembrane by de Wit, Hoppe and Nicolai [2], who showed that the regularized supermembrane theory is precisely the supersymmetric matrix quantum mechanics now known as Matrix Theory. A remarkable feature of the quantum supermembrane theory is that unlike the quantized string theories, the membrane theory automatically gives a second quantized theory from the point of view of the target space. This issue will be discussed in more detail in Section 2.

In this section we describe in some detail how matrix theory arises from the quantization of the supermembrane. In 2.1 we review how the bosonic string is quantized in the light-front formalism. This will be a useful reference point for our discussion of membrane quantization. In 2.2 we describe the theory of the relativistic bosonic membrane in flat space. The light-front description of this theory is discussed in 2.3, and the matrix regularization of the theory is described in 2.4. In 2.5 we discuss briefly the description of the bosonic membrane moving in a general background geometry. In 2.6 we extend the discussion to the supermembrane. We discuss the supermembrane in an arbitrary background geometry. We discuss the κ-symmetry of the supermembrane theory which leads, even at the classical level, to the condition that the background geometry satisfies the classical 11D supergravity equations of motion. The matrix theory Hamiltonian is derived from the regularized supermembrane theory. The problem of finding a covariant membrane quantization is discussed in 2.7.

The material in this section roughly follows the original papers [1, 2, 12]. Note, however, that the original derivation of the matrix quantum mechanics theory was done in the Nambu-Goto-type membrane formalism, while we use here the Polyakov-type approach. We only consider closed membranes in the discussion here; little is known about the open membrane which must end on the M-theory 5-brane, but it would be very interesting to generalize the discussion here to the open membrane.

2.1. REVIEW OF LIGHT-FRONT STRING

We begin with a brief review of the bosonic string. This will be a useful model to compare with in our discussion of the supermembrane.

The Nambu-Goto action for the relativistic bosonic string moving in a flat background space-time is

$$S = -T_s \int d^2\sigma \sqrt{-\det h_{ab}} \tag{2.1}$$

where $T_s = 1/(2\pi\alpha')$ and

$$h_{ab} = \partial_a X^\mu \partial_b X_\mu. \tag{2.2}$$

It is convenient to use the Polyakov formalism in which an auxiliary world-sheet metric γ is introduced

$$S = -\frac{1}{4\pi\alpha'} \int d^2\sigma \sqrt{-\gamma}\gamma^{ab}\partial_a X^\mu \partial_b X_\mu \tag{2.3}$$

Solving the equation of motion for γ_{ab} leads to

$$\gamma_{ab} = h_{ab} = \partial_a X^\mu \partial_b X_\mu \tag{2.4}$$

and replacing this in (2.3) gives (2.1).

The action (2.3) is simplified by going to the gauge

$$\gamma_{ab} = \eta_{ab}. \tag{2.5}$$

In this gauge we simply have the free field action

$$S = -\frac{1}{4\pi\alpha'} \int d^2\sigma \eta^{ab}\partial_a X^\mu \partial_b X_\mu. \tag{2.6}$$

The fields X^μ satisfy the equation of motion $\Box X^\mu = 0$ and are subject to the auxiliary Virasoro constraints

$$\begin{aligned} \dot{X}^\mu(\partial X_\mu) &= 0 \\ \dot{X}^\mu \dot{X}_\mu &= -(\partial X^\mu)(\partial X_\mu) \end{aligned} \tag{2.7}$$

(we denote τ derivatives by a dot and σ derivatives by ∂). Because this is a free theory it is fairly straightforward to quantize. The approaches to quantizing this theory include the BRST and light-front formalisms. The Virasoro constraints can be explicitly solved in light-front gauge

$$X^+(\tau, \sigma) = x^+ + p^+\tau. \tag{2.8}$$

In the classical theory we have

$$\begin{aligned} \dot{X}^- &= \frac{1}{2p^+}\left(\dot{X}^i\dot{X}^i + \partial X^i \partial X^i\right) \\ \partial X^- &= \frac{1}{p^+}\dot{X}^i\partial X^i \end{aligned} \tag{2.9}$$

The transverse degrees of freedom X^i have Fourier modes with the commutation relations of simple harmonic oscillators. These are straightforward to quantize. The string spectrum is then given by the usual mass-shell condition

$$M^2 = 2p^+p^- - p^ip^i = \frac{1}{\alpha'}(N-a) \tag{2.10}$$

2.2. THE BOSONIC MEMBRANE THEORY

We now discuss the relativistic bosonic membrane moving in an arbitrary number D of space-time dimensions. The story begins in a very similar fashion to the relativistic string. We want to use a Nambu-Goto-style action

$$S = -T\int d^3\sigma\sqrt{-\det h_{\alpha\beta}} \tag{2.11}$$

where T is the membrane tension

$$T = \frac{1}{(2\pi)^2 l_p^3} \tag{2.12}$$

and

$$h_{\alpha\beta} = \partial_\alpha X^\mu \partial_\beta X_\mu \tag{2.13}$$

is the pullback of the metric to the three-dimensional membrane world-volume, with coordinates $\sigma_\alpha, \alpha \in \{0,1,2\}$. We will use the notation $\tau = \sigma_0$ and use indices $a, b, \ldots$ to describe "spatial" indices $a \in \{1,2\}$ on the membrane world-volume.

We again wish to use a Polyakov-type formalism in which an auxiliary world-sheet metric $\gamma_{\alpha\beta}$ is introduced

$$S = -\frac{T}{2}\int d^3\sigma\sqrt{-\gamma}\left(\gamma^{\alpha\beta}\partial_\alpha X^\mu \partial_\beta X_\mu - 1\right). \tag{2.14}$$

The need for the extra "cosmological" term arises from the absence of scale invariance in the theory. Computing the equations of motion from varying $\gamma_{\alpha\beta}$, and using $\delta\sqrt{-\gamma} = \frac{1}{2}\sqrt{-\gamma}\gamma^{\alpha\beta}\delta\gamma_{\alpha\beta}, \delta\gamma^{\epsilon\phi} = -\gamma^{\alpha\epsilon}\gamma^{\phi\beta}\delta\gamma_{\alpha\beta}$, we get

$$-\gamma^{\alpha\gamma}\gamma^{\beta\delta}h_{\gamma\delta} + \frac{1}{2}\gamma^{\alpha\beta}t - \frac{1}{2}\gamma^{\alpha\beta} = 0 \tag{2.15}$$

where $t = \gamma^{\alpha\beta}h_{\alpha\beta}$. Lowering all indices gives

$$\frac{1}{2}\gamma_{\alpha\beta}(t-1) = h_{\alpha\beta} \tag{2.16}$$

or

$$\gamma_{\alpha\beta} = \frac{2h_{\alpha\beta}}{t-1}. \tag{2.17}$$

Contracting indices gives

$$3 = \frac{2t}{t-1} \tag{2.18}$$

so $t = 3$ and

$$\gamma_{\alpha\beta} = h_{\alpha\beta} = \partial_\alpha X^\mu \partial_\beta X_\mu. \tag{2.19}$$

Replacing this in (2.14) again gives (2.11). The equation of motion which arises from varying X^μ is

$$\partial_\alpha\left(\sqrt{-\gamma}\gamma^{\alpha\beta}\partial_\beta X^\mu\right) = 0. \tag{2.20}$$

To follow the procedure we used for the bosonic string theory, we would now like to use the symmetries of the theory to gauge-fix the metric $\gamma_{\alpha\beta}$. Unfortunately, whereas for the string we had three components of the metric and three continuous symmetries (two diffeomorphism symmetries and a scale symmetry), for the membrane we have six independent metric components and only three diffeomorphism symmetries. We can use these symmetries to fix the components $\gamma_{0\alpha}$ of the metric to be

$$\begin{aligned} \gamma_{0a} &= 0 \\ \gamma_{00} &= -\frac{4}{\nu^2}\bar{h} \equiv -\frac{4}{\nu^2}\det h_{ab} \end{aligned} \tag{2.21}$$

where ν is a constant whose normalization has been chosen to make the later matrix interpretation transparent. Once we have chosen this gauge, no further components of the metric γ_{ab} can be fixed. This gauge can only be chosen when the membrane world-volume is of the form $\Sigma \times \mathbb{R}$ where Σ is a Riemann surface of fixed topology. The membrane action becomes

$$S = \frac{T\nu}{4}\int d^3\sigma\left(\dot{X}^\mu\dot{X}_\mu - \frac{4}{\nu^2}\bar{h}\right). \tag{2.22}$$

It is natural to rewrite this theory in terms of a canonical Poisson bracket on the membrane at constant τ where $\{f,g\} \equiv \epsilon^{ab}\partial_a f\partial_b g$ with $\epsilon^{12} = 1$. We will assume that the coordinates σ are chosen so that with respect to the symplectic form associated to this canonical Poisson bracket the volume of the Riemann surface Σ is $\int d^2\sigma = 4\pi$. In terms of this metric we have the handy formulae

$$\begin{aligned}\bar{h} = \det h_{ab} &= \partial_1 X^\mu \partial_1 X^\mu \partial_2 X^\nu \partial_2 X^\nu - \partial_1 X^\mu \partial_2 X^\mu \partial_1 X^\nu \partial_2 X^\nu \\ &= \frac{1}{2}\{X^\mu, X^\nu\}\{X_\mu, X_\nu\} \quad (2.23)\\ \partial_a(\bar{h}h^{ab}\partial_b X^\mu) &= \{\{X^\mu, X^\nu\}, X_\nu\} \quad (2.24)\\ \bar{h}h^{ab}\partial_a X^\mu \partial_b X^\nu &= \{X^\mu, X^\lambda\}\{X_\lambda, X^\nu\} \quad (2.25)\end{aligned}$$

In terms of the Poisson bracket, the membrane action becomes

$$S = \frac{T\nu}{4}\int d^3\sigma \left(\dot{X}^\mu \dot{X}_\mu - \frac{2}{\nu^2}\{X^\mu, X^\nu\}\{X_\mu, X_\nu\}\right). \quad (2.26)$$

The equations of motion for the fields X^μ are

$$\begin{aligned}\ddot{X}^\mu &= \frac{4}{\nu^2}\partial_a\left(\bar{h}h^{ab}\partial_b X^\mu\right)\\ &= \frac{4}{\nu^2}\{\{X^\mu, X^\nu\}, X_\nu\} \quad (2.27)\end{aligned}$$

The auxiliary constraints on the system are

$$\begin{aligned}\dot{X}^\mu \dot{X}_\mu &= -\frac{4}{\nu^2}\bar{h}\\ &= -\frac{2}{\nu^2}\{X^\mu, X^\nu\}\{X_\mu, X_\nu\} \quad (2.28)\end{aligned}$$

and

$$\dot{X}^\mu \partial_a X_\mu = 0. \quad (2.29)$$

It follows directly from (2.29) that

$$\{\dot{X}^\mu, X_\mu\} = 0. \quad (2.30)$$

We have thus expressed the bosonic membrane theory as a constrained Hamiltonian system. The degrees of freedom are D functions X^μ on the 3-dimensional world-volume of a membrane which has topology $\Sigma\times\mathbb{R}$ where Σ is a Riemann surface. This theory is still completely covariant. It is difficult to quantize, however, because of the constraints and the nonlinearity of the equations of motion. The direct quantization of this covariant theory will be discussed further in Section 2.7.

2.3. THE LIGHT-FRONT BOSONIC MEMBRANE

As we did for the bosonic string, we now consider the theory in light-front coordinates

$$X^\pm = (X^0 \pm X^{D-1})/\sqrt{2}. \quad (2.31)$$

Just as in the case of the string, the constraints (2.28,2.29) can be explicitly solved in light-front gauge

$$X^+(\tau, \sigma_1, \sigma_2) = \tau. \tag{2.32}$$

We have

$$\begin{aligned}\dot{X}^- &= \frac{1}{2}\dot{X}^i\dot{X}^i + \frac{2\bar{h}}{\nu^2} \\ &= \frac{1}{2}\dot{X}^i\dot{X}^i + \frac{1}{\nu^2}\{X^i, X^j\}\{X^i, X^j\} \\ \partial_a X^- &= \dot{X}^i\partial_a X^i\end{aligned} \tag{2.33}$$

We can go to a Hamiltonian formalism by computing the canonically conjugate momentum densities.

$$\begin{aligned}P^+ &= -\frac{\delta\mathcal{L}}{\delta(\dot{X}^-)} = \frac{\nu T}{2} \\ P^i &= \frac{\delta\mathcal{L}}{\delta(\dot{X}^i)} = \frac{\nu T}{2}\dot{X}^i\end{aligned} \tag{2.34}$$

The total momentum in the direction P^+ is then

$$p^+ = \int d^2\sigma P^+ = 2\pi\nu T. \tag{2.35}$$

The Hamiltonian of the theory is given by

$$\begin{aligned}H &= \int d^2\sigma \left(P^i\dot{X}^i - P^+\dot{X}^- - \mathcal{L}\right) \\ &= \frac{\nu T}{4}\int d^2\sigma \left(\dot{X}^i\dot{X}^i + \frac{4\bar{h}}{\nu^2}\right) \\ &= \frac{\nu T}{4}\int d^2\sigma \left(\dot{X}^i\dot{X}^i + \frac{2}{\nu^2}\{X^i, X^j\}\{X^i, X^j\}\right).\end{aligned} \tag{2.36}$$

The only remaining constraint is that the transverse degrees of freedom must satisfy

$$\{\dot{X}^i, X^i\} = 0 \tag{2.37}$$

This theory has a residual invariance under time-independent area-preserving diffeomorphisms. Such diffeomorphisms do not change the symplectic form and thus manifestly leave the Hamiltonian (2.36) invariant.

We now have a Hamiltonian formalism for the light-front membrane theory. Unfortunately, this theory is still rather difficult to quantize. Unlike string theory, where the equations of motion are linear in this formalism, for the membrane the equations of motion (2.27) are nonlinear and difficult to solve.

2.4. MATRIX REGULARIZATION

In 1982 a remarkably clever regularization of the light-front membrane theory was found by Goldstone and Hoppe in the case where the surface Σ is a sphere S^2 [1]. According to this regularization procedure, functions on the membrane surface are mapped to finite size $N \times N$ matrices. Just as in the quantization of a classical mechanical system defined in terms of a Poisson brackets, the Poisson bracket appearing in the membrane theory is replaced in the matrix regularization of the theory by a matrix commutator.

The matrix regularization of the theory can be generalized to membranes of arbitrary topology, but is perhaps most easily understood by considering the case discussed in [1], where the membrane has the topology of a sphere S^2 for all values of τ. In this case the world-sheet of the membrane surface at fixed time can be described by a unit sphere with a rotationally invariant canonical symplectic form. Functions on this membrane can be described in terms of functions of the three Cartesian coordinates ξ_1, ξ_2, ξ_3 on the unit sphere satisfying

$$\xi_1^2 + \xi_2^2 + \xi_3^2 = 1. \tag{2.38}$$

The Poisson brackets of these functions are given by

$$\{\xi_A, \xi_B\} = \epsilon_{ABC}\xi_C.$$

This is essentially the same algebraic structure as that defined by the commutation relations of the generators of $SU(2)$. It is therefore natural to associate these coordinate functions on S^2 with the matrices generating $SU(2)$ in the N-dimensional representation. In terms of the conventions we are using here, when the normalization constant ν is integral, the correct correspondence is

$$\xi_A \to \frac{2}{N} J_A$$

where J_1, J_2, J_3 are generators of the N-dimensional representation of $SU(2)$ with $N = \nu$, satisfying the commutation relations

$$-i[J_A, J_B] = \epsilon_{ABC} J_C \, .$$

In general, any function on the membrane can be expanded as a sum of spherical harmonics

$$f(\xi_1, \xi_2, \xi_3) = \sum_{l,m} c_{lm} y_{lm}(\xi_1, \xi_2, \xi_3) \tag{2.39}$$

The spherical harmonics can in turn be written as a sum of monomials in the coordinate functions:

$$y_{lm}(\xi_1, \xi_2, \xi_3) = \sum_k t^{(lm)}_{A_1 \ldots A_l} \xi_{A_1} \cdots \xi_{A_l}$$

where the coefficients $t^{(lm)}_{A_1 \ldots A_l}$ are symmetric and traceless (because $\xi_A \xi_A = 1$). Using the above correspondence, a matrix approximation to each of the spherical harmonics with $l < N$ can be constructed, which we denote by Y.

$$Y_{lm} = \left(\frac{2}{N}\right)^l \sum t^{(lm)}_{A_1 \ldots A_l} J_{A_1} \cdots J_{A_l} \tag{2.40}$$

For a fixed value of N only the spherical harmonics with $l < N$ can be constructed because higher order monomials in the generators J_A do not generate linearly independent matrices. Note that the number of independent matrix entries is precisely equal to the number of independent spherical harmonic coefficients which can be determined for fixed N

$$N^2 = \sum_{l=0}^{N-1} (2l + 1) \tag{2.41}$$

The matrix approximations (2.40) of the spherical harmonics can be used to construct matrix approximations to an arbitrary function of the form (2.39)

$$F = \sum_{l<N,m} c_{lm} Y_{lm} \tag{2.42}$$

The Poisson bracket in the membrane theory is replaced in the matrix regularized theory with the matrix commutator according to the prescription

$$\{f, g\} \to \frac{-iN}{2} [F, G]. \tag{2.43}$$

Similarly, an integral over the membrane at fixed τ is replaced by a matrix trace through

$$\frac{1}{4\pi} \int d^2\sigma f \to \frac{1}{N} \mathrm{Tr}\, F \tag{2.44}$$

The Poisson bracket of a pair of spherical harmonics takes the form

$$\{y_{lm}, y_{l'm'}\} = g^{l''m''}_{lm,l'm'} y_{l''m''}. \tag{2.45}$$

The commutator of a pair of matrix spherical harmonics (2.40) can be written

$$[Y_{lm}, Y_{l'm'}] = G^{l''m''}_{lm,l'm'} Y_{l''m''}. \tag{2.46}$$

It can be verified that in the large N limit the structure constant of these algebras agree

$$\lim_{N\to\infty} \frac{-iN}{2} G^{l''m''}_{lm,l'm'} \to g^{l''m''}_{lm,l'm'} \tag{2.47}$$

As a result, it can be shown that for any smooth functions on the membrane f, g defined in terms of convergent sums of spherical harmonics, with Poisson bracket $\{f, g\} = h$ we have

$$\lim_{N\to\infty} \frac{1}{N} \mathrm{Tr}\, F = \frac{1}{4\pi}\int d^2\sigma f \tag{2.48}$$

and it is possible to show that

$$\lim_{N\to\infty} ((\frac{-iN}{2})[F, G] - H) = 0 \tag{2.49}$$

This last relation is really shorthand for the statement that

$$\lim_{N\to\infty} \frac{1}{N} \mathrm{Tr} \left(((\frac{-iN}{2})[F, G] - H)J \right) = 0 \tag{2.50}$$

where J is the matrix approximation to any smooth function j on the sphere.

We now have a dictionary for transforming between continuum and matrix-regularized quantities. The correspondence is given by

$$\xi_A \leftrightarrow \frac{2}{N} J_A \qquad \{\cdot,\cdot\} \leftrightarrow \frac{-iN}{2}[\cdot,\cdot] \qquad \frac{1}{4\pi}\int d^2\sigma \leftrightarrow \frac{1}{N}\mathrm{Tr} \tag{2.51}$$

The matrix regularized membrane Hamiltonian is therefore given by

$$\begin{aligned} H &= (2\pi l_p^3)\mathrm{Tr}\left(\frac{1}{2}\mathbf{P}^i\mathbf{P}^i\right) - \frac{1}{(2\pi l_p^3)}\mathrm{Tr}\left(\frac{1}{4}[\mathbf{X}^i, \mathbf{X}^j][\mathbf{X}^i, \mathbf{X}^j]\right) \\ &= \frac{1}{(2\pi l_p^3)}\mathrm{Tr}\left(\frac{1}{2}\dot{\mathbf{X}}^i\dot{\mathbf{X}}^i - \frac{1}{4}[\mathbf{X}^i, \mathbf{X}^j][\mathbf{X}^i, \mathbf{X}^j]\right). \end{aligned} \tag{2.52}$$

This Hamiltonian gives rise to the matrix equations of motion

$$\ddot{\mathbf{X}}^i + [[\mathbf{X}^i, \mathbf{X}^j], \mathbf{X}^j] = 0$$

which must be supplemented with the Gauss constraint

$$[\dot{\mathbf{X}}^i, \mathbf{X}^i] = 0\,. \tag{2.53}$$

This is a classical theory with a finite number of degrees of freedom. The quantization of such a system is straightforward, although solving the quantum theory can in practice be quite tricky. Thus, we have found a well-defined quantum theory describing the matrix regularization of the relativistic membrane theory in light-front coordinates.

There are a number of rather deep mathematical reasons why the matrix regularization of the membrane theory works. One way of looking at this regularization is in terms of the underlying symmetry of the theory. After gauge-fixing, the membrane theory has a residual invariance under the group of time-independent area-preserving diffeomorphisms of the membrane world-sheet. This diffeomorphism group can be described in a natural mathematical way as a limit of the matrix group $U(N)$ as $N \to \infty$. In the discrete theory the area-preserving diffeomorphism symmetry thus is replaced by the $U(N)$ matrix symmetry. The matrix regularization can also be viewed in terms of a geometrical quantization of the operators associated with functions on the membrane. From this point of view the matrix membrane is like a "fuzzy" membrane which is discrete yet preserves the $SU(2)$ rotational symmetry of the original smooth sphere. This point of view ties into recent developments in noncommutative geometry.

We will not pursue these points of view in any depth here. We will note, however, that from both points of view it is natural to generalize the construction to higher genus surfaces. We discuss the matrix torus explicitly in section 4.2.3.

2.5. THE BOSONIC MEMBRANE IN A GENERAL BACKGROUND

So far we have only considered the membrane in a flat background Minkowski geometry. Just as for strings, it is natural to generalize the discussion to a bosonic membrane moving in a general background metric $g_{\mu\nu}$ and 3-form field $A_{\mu\nu\rho}$. The introduction of a general background metric modifies the Nambu-Goto action by replacing $h_{\alpha\beta}$ in (2.13) with

$$h_{\alpha\beta} = \partial_\alpha X^\mu \partial_\beta X^\nu g_{\mu\nu}(X). \tag{2.54}$$

The membrane couples to the 3-form field as an electrically charged object, giving an additional term to the action of the form $\int A_{\alpha\beta\gamma}$ where $A_{\alpha\beta\gamma}$ is the pullback to the world-volume of the membrane of the 3-form field. This gives a total Nambu-Goto-type action for the membrane in a general background of the form

$$S = -T \int d^3\sigma \left(\sqrt{-\det h_{\alpha\beta}} + 6\dot{X}^\mu \partial_1 X^\nu \partial_2 X^\rho A_{\mu\nu\rho}(X) \right). \tag{2.55}$$

With an auxiliary world-volume metric, this action becomes

$$\begin{aligned} S \;=\; -\frac{T}{2} \int d^3\sigma \Big[& \sqrt{-\gamma} \left(\gamma^{\alpha\beta} \partial_\alpha X^\mu \partial_\beta X^\nu g_{\mu\nu}(X) - 1 \right) \\ & + 12 \dot{X}^\mu \partial_1 X^\nu \partial_2 X^\rho A_{\mu\nu\rho}(X) \Big] \end{aligned} \tag{2.56}$$

We can gauge fix the action (2.56) using the same gauge (2.21) as in the flat space case. We can then consider carrying out a similar procedure for quantizing the membrane in a general background as we described in the case of the flat background. We will return to this question in section 6.3 when we discuss in more detail the prospects for constructing matrix theory in a general background.

2.6. THE SUPERMEMBRANE

Now let us turn our attention to the supermembrane. In order to make contact with M-theory, and indeed to make the membrane theory well-behaved it is necessary to add supersymmetry to the theory. Supersymmetric membrane theories can be constructed classically in dimensions 4, 5, 7 and 11. These theories have different degrees of supersymmetry, with 2, 4, 8 and 16 independent supersymmetric generators respectively. It is believed that all the supermembrane theories other than the 11D maximally supersymmetric theory suffer from anomalies in the Lorentz algebra. Thus, just as $D = 10$ is the natural dimension for the superstring, $D = 11$ is the natural dimension for the supermembrane.

The formalism for describing the supermembrane is rather technically complicated. We will not use most of this formalism in the rest of these lectures, so we restrict ourselves here to a fairly concise discussion of the structure of the supersymmetric theory. The reader not interested in the details of how the supersymmetric form of matrix theory is derived may wish to skip directly to the result of this analysis, the supersymmetric matrix theory Hamiltonian (2.89), on first reading. In Section 2.6.1 we describe using superfield notation the supermembrane action in a general background and its symmetries. We discuss in particular the fact that the κ-symmetry of the theory at the classical level guarantees already that the background geometry satisfies the equations of motion of 11D supergravity. In 2.6.2 we describe in more explicit form the supermembrane action in a flat background. We describe the light-front form of the theory in 2.6.3, where we show how the regularized theory gives precisely the Hamiltonian of the supersymmetric matrix theory.

2.6.1. *The supermembrane action*

In this section we describe the supermembrane action in an arbitrary background and its symmetries. In particular, we describe the κ-symmetry of the theory, which implies that the background obeys the classical equations of 11D supergravity. For further details see the original paper of Bergshoeff, Sezgin and Townsend [12] or the review paper of Duff [13].

The standard NSR description of the superstring gives a theory which is fairly straightforward to quantize. This formalism can be used in a straightforward fashion to describe the spectra of the five superstring theories. One disadvantage of this formalism is that the target space supersymmetry of the theory is difficult to show explicitly. There is another formalism, known as the Green-Schwarz formalism ([14], reviewed in [15]), in which the target space supersymmetry of the theory is much more clear. In the Green-Schwarz formalism additional Grassmann degrees of freedom are introduced which transform as space-time fermions but as world-sheet vectors. These correspond to space-time superspace coordinates for the string. The Green-Schwarz superstring action does not have a standard world-sheet supersymmetry (it can't, since there are no world-sheet fermions). The theory does, however, have a novel type of supersymmetry known as a κ-symmetry. The existence of the κ-symmetry in the classical Green-Schwarz string theory already implies that the theory is restricted to $D = 3, 4, 6$ or 10. This is already a much stronger restriction than can be gleaned from classical superstring with world-sheet supersymmetry.

Unlike the superstring, there is no known way of formulating the supermembrane in a world-volume supersymmetric fashion (although there has been some recent progress in this direction, for further references see [13]). A Green-Schwarz formulation of the supermembrane in a general background was first found by Bergshoeff, Sezgin and Townsend [12]. We now review this construction.

We consider an 11-dimensional target space with a general metric $g_{\mu\nu}$ described by an elfbein e^a_μ, and an arbitrary background gravitino field ψ_μ and 3-form field $A_{\mu\nu\rho}$. In superspace notation we describe the space as having 11 bosonic coordinates X^μ and 32 anticommuting fermionic coordinates $\theta^{\dot{\alpha}}$. These coordinates are combined into a single superspace coordinate

$$Z^M = (X^\mu, \theta^{\dot{\alpha}}) \tag{2.57}$$

where M is an index with 43 possible values. (Space-time spinor indices $\dot{\alpha}, \dot{\beta}, \ldots$ will carry a dot in this section to distinguish them from world-volume coordinate indices $\alpha, \beta, \ldots$). In superspace the elfbein becomes a 43-bein E^A_M, with $A = (a, \phi)$. There is also an antisymmetric superspace 3-form field B_{MNP}. The superspace formulation of 11D supergravity is written in terms of these two fields. The identification of the superspace degrees of freedom with the component fields e^a_μ, ψ_μ and $A_{\mu\nu\rho}$ is quite subtle, and involves a careful analysis of the supersymmetry transformations in both formalisms as well as gauge choices. At leading order in θ the component fields are identified through

$$\begin{aligned} E^a_\mu &= e^a_\mu + \mathcal{O}(\theta) \\ E^\phi_\mu &= \psi^\phi_\mu + \mathcal{O}(\theta) \end{aligned} \tag{2.58}$$

$$B_{\mu\nu\rho} = A_{\mu\nu\rho} + \mathcal{O}(\theta)$$

The identifications of E_M^A and B_{MNP} in terms of component fields through order θ^2 has only recently been determined [16]. The identification beyond this order has not been determined explicitly.

In terms of these superspace fields, the supermembrane action in a general background is given by

$$S = -\frac{T}{2}\int d^3\sigma \left[\sqrt{-\gamma}\left(\gamma^{\alpha\beta}\Pi_\alpha^a\Pi_\beta^b\eta_{ab} - 1\right) + \epsilon^{\alpha\beta\gamma}\Pi_\alpha^A\Pi_\beta^B\Pi_\gamma^C B_{ABC}\right] \quad (2.59)$$

where Π_α^A are the components of the pullback of the 43-bein to the membrane world-volume

$$\Pi_\alpha^A = \partial_\alpha Z^M E_M^A \quad (2.60)$$

and B_{ABC} is defined implicitly through

$$B_{MNP} = E_M^A E_N^B E_P^C B_{ABC} \quad (2.61)$$

The action (2.59) is very closely related to the superspace formulation of the Green-Schwarz action. The superstring action differs in that it has no cosmological term and that the antisymmetric field is a superspace 2-form field.

Let us now review the symmetries of the action (2.59). This action has global symmetries corresponding to space-time super diffeomorphisms, gauge transformations and discrete symmetries, as well as the local symmetries of world-volume diffeomorphisms and κ symmetry.

Super diffeomorphisms: Under a super diffeomorphism of the target space generated by a super vector field ξ^M the coordinate fields, 43-bein and 3-form field transform under

$$\begin{aligned}
\delta Z^M &= \xi^M \\
\delta E_M^A &= \xi^N\partial_N E_M^A + \partial_M\xi^N E_N^A \qquad (2.62)\\
\delta B_{MNP} &= \xi^Q\partial_Q B_{MNP} + (\partial_M\xi^Q)B_{QNP} - (\partial_N\xi^Q)B_{MQP} + (\partial_P\xi^Q)B_{MNQ}
\end{aligned}$$

Super gauge transformations: This global symmetry transforms the 3-form superfield by

$$\delta B_{MNP} = \partial_M\Sigma_{NP} - \partial_N\Sigma_{MP} + \partial_P\Sigma_{MN}. \quad (2.63)$$

Discrete symmetries: There is also a discrete symmetry $\mathbb{Z}_2$ corresponding to taking

$$B_{MNP} \rightarrow -B_{MNP} \quad (2.64)$$

and performing a space-time reflection on a single coordinate.

World-volume diffeomorphisms: Under a world-volume diffeomorphism generated by the vector field η^α the fields transform by

$$\delta Z^M = \eta^\alpha \partial_\alpha Z^M \tag{2.65}$$

κ-symmetry: The most interesting symmetry of the theory is the fermionic κ-symmetry. The parameter κ^ψ is taken to be an anticommuting world-volume scalar which transforms as a space-time 32-component spinor. Under this symmetry the coordinate fields transform under

$$\begin{aligned} \delta Z^M E_M^a &= 0 \\ \delta Z^M E_M^\phi &= (1+\Gamma)^\phi_\psi \kappa^\psi \end{aligned} \tag{2.66}$$

where

$$\Gamma = \frac{1}{6\sqrt{-\gamma}} \epsilon^{\alpha\beta\gamma} \Pi_\alpha^a \Pi_\beta^b \Pi_\gamma^c \Gamma_{abc}. \tag{2.67}$$

The κ-symmetry of the theory has a number of interesting features. For one thing, it can be shown that (2.66) is only a symmetry of the theory when the background fields E_M^a, B_{MNP} obey the equations of motion of the classical 11D supergravity theory. Thus, 11D supergravity emerges from the membrane theory even at the classical level. For the details of this analysis, see [12]. This situation is similar to that which arises in the Green-Schwarz formulation of the superstring theories. In the Green-Schwarz formalism there is a local κ-symmetry on the string world-sheet only when the backgrounds satisfy the supergravity equations of motion.

Another interesting aspect of the κ-symmetry arises from the algebraic fact that

$$\Gamma^2 = 1. \tag{2.68}$$

This implies that $(1 + \Gamma)$ is a projection operator. We can thus use κ-symmetry to gauge away half of the fermionic degrees of freedom $\theta^{\dot\alpha}$. This reduces the number of propagating fermionic degrees of freedom to 8. This is also the number of propagating bosonic degrees of freedom, as can be seen by going to a static gauge where $X^{0,1,2}$ are identified with $\tau, \sigma_{1,2}$ so that only the 8 transverse directions appear as propagating degrees of freedom.

In general, gauge-fixing the κ-symmetry in any particular way will break the Lorentz invariance of the theory. This makes it quite difficult to find any way of quantizing the theory without breaking Lorentz symmetry. This situation is again analogous to the Green-Schwarz superstring theory, where fixing of κ-symmetry also breaks Lorentz invariance and no covariant quantization scheme is known.

2.6.2. *The supermembrane in flat space*

To make the connection with matrix theory, we now restrict attention to a flat Minkowski background space-time with vanishing 3-form field $A_{\mu\nu\rho}$. We will return to a discussion of general backgrounds in section 6.3.

In flat space the 43-bein becomes

$$\begin{aligned} E^a_M &= (\delta^a_\mu, (\Gamma^a)_{\dot\alpha\dot\beta}\theta^{\dot\beta}) \\ E^\phi_M &= (0, \delta^\phi_{\dot\alpha}) \end{aligned} \tag{2.69}$$

The super 4-form field strength H_{MNPQ} has as its only nonvanishing components

$$H_{ab\phi\psi} = \frac{1}{3}(\Gamma_{ab})_{\phi\psi}. \tag{2.70}$$

From this and the definition $H = dB$ it is possible to derive the components of the super 3-form field B_{MNP}

$$\begin{aligned} B_{\mu\nu\rho} &= 0 \\ B_{\mu\nu\dot\alpha} &= \frac{1}{6}(\Gamma_{\mu\nu}\theta)_{\dot\alpha} \\ B_{\mu\dot\alpha\dot\beta} &= \frac{1}{6}(\Gamma_{\mu\nu}\theta)_{(\dot\alpha}(\Gamma^\nu\theta)_{\dot\beta)} \\ B_{\dot\alpha\dot\beta\dot\gamma} &= \frac{1}{6}(\Gamma_{\mu\nu}\theta)_{(\dot\alpha}(\Gamma^\mu\theta)_{\dot\beta}(\Gamma^\nu\theta)_{\dot\gamma)} \end{aligned} \tag{2.71}$$

From (2.60) it follows that

$$\Pi^\mu_\alpha = \partial_\alpha X^\mu + \bar\theta\Gamma^\mu\partial_\alpha\theta. \tag{2.72}$$

The membrane action (2.59) reduces in flat space to

$$S = -\frac{T}{2}\int d^3\sigma \left\{\sqrt{-\gamma}\left(\gamma^{\alpha\beta}\Pi^\mu_\alpha\Pi^\nu_\beta\eta_{\mu\nu} - 1\right)\right. \tag{2.73}$$

$$-\epsilon^{\alpha\beta\gamma}\left[\frac{1}{2}\partial_\alpha X^\mu(\partial_\beta X^\nu + \bar\theta\Gamma^\nu\partial_\beta\theta)\right. \tag{2.74}$$

$$\left.\left.+\frac{1}{6}(\bar\theta\Gamma^\mu\partial_\alpha\theta)(\bar\theta\Gamma^\nu\partial_\beta\theta)\right]\bar\theta\Gamma_{\mu\nu}\partial_\gamma\theta\right\}$$

The extra Wess-Zumino type terms which appear in this action are rather non-obvious from the point of view of the flat space-time theory, although they have arisen naturally in the superspace formalism. These are analogous to terms in the Green-Schwarz superstring action which were originally found by imposing κ-symmetry on the theory. The equation of motion for γ as usual sets $\gamma_{\alpha\beta}$ to be the pullback of the metric

$$\gamma_{\alpha\beta} = \Pi^\mu_\alpha\Pi^\nu_\beta\eta_{\mu\nu} \tag{2.75}$$

The action (2.73) has the target space supersymmetry

$$\begin{aligned} \delta\theta &= \epsilon \\ \delta X^\mu &= -\bar{\epsilon}\Gamma^\mu\theta \end{aligned} \tag{2.76}$$

This transformation leaves Π^μ_α invariant. The fact that it leaves the action invariant follows from the identity

$$\bar{\psi}_{[1}\Gamma^\mu\psi_2\bar{\psi}_3\Gamma_{\mu\nu}\psi_{4]} = 0 \tag{2.77}$$

which holds in 11 dimensions (as well as in dimensions 4, 5 and 7). The relation (2.77) is also necessary to show that the action is κ-symmetric. This relation is analogous to the relation $\bar{\epsilon}\Gamma_\mu\psi_{[1}\bar{\psi}_2\Gamma^\mu\psi_{3]} = 0$ which holds in 3, 4, 6 and 10 dimensions and which is necessary for the supersymmetry and κ-symmetry of the Green-Schwarz superstring action.

2.6.3. *The quantum supermembrane and supersymmetric matrix theory*
We now go to light-front gauge. As usual we define

$$X^\pm = (X^0 \pm X^{D-1})/\sqrt{2}. \tag{2.78}$$

We write the 32 × 32 Γ matrices in the block forms

$$\begin{aligned} \Gamma^+ &= \begin{pmatrix} 0 & 0 \\ \sqrt{2}i\mathbb{1}_{16} & 0 \end{pmatrix} \\ \Gamma^- &= \begin{pmatrix} 0 & \sqrt{2}i\mathbb{1}_{16} \\ 0 & 0 \end{pmatrix} \\ \Gamma_i &= \begin{pmatrix} \gamma^i & 0 \\ 0 & -\gamma^i \end{pmatrix} \end{aligned} \tag{2.79}$$

where γ^i are 16 × 16 Euclidean $SO(9)$ gamma matrices.

In addition to setting the gauge

$$X^+ = \tau \tag{2.80}$$

We can also use κ-symmetry to fix

$$\Gamma^+\theta = 0 \tag{2.81}$$

From the above form of the matrices Γ^μ it is clear that this projects onto the 16 Grassmann degrees of freedom $(0, \theta)$, and that as a consequence all expressions of the forms

$$\bar{\theta}\Gamma^\mu\partial_\alpha\theta, \qquad \mu \neq - \tag{2.82}$$

and

$$\bar{\theta}\Gamma^{ij}\partial_\alpha\theta \quad \text{or} \quad \bar{\theta}\Gamma^{+\mu}\partial_\alpha\theta \tag{2.83}$$

must vanish in this gauge. This simplifies the theory in this gauge considerably. First, we have

$$\Pi^\mu_\alpha = \partial_\alpha X^\mu, \qquad \mu \neq - \tag{2.84}$$

Second, we find that the terms on the second line of (2.73) simplify to

$$-\bar{\theta}\Gamma_{+i}\{X^i, \theta\} \tag{2.85}$$

Solving for the derivatives $\partial_\gamma X^-$ as in (2.33) we get

$$\begin{aligned} \dot{X}^- &= \frac{1}{2}\dot{X}^i\dot{X}^i + \frac{1}{\nu^2}\{X^i, X^j\}\{X^i, X^j\} + \bar{\theta}\Gamma_+\dot{\theta} \\ &= \Pi_0^- + \bar{\theta}\Gamma_+\dot{\theta} \end{aligned} \tag{2.86}$$

$$\begin{aligned} \partial_a X^- &= \dot{X}^i\partial_a X^i + \bar{\theta}\Gamma_+\partial_a\theta \\ &= \Pi_a^- + \bar{\theta}\Gamma_+\partial_a\theta \end{aligned} \tag{2.87}$$

Combining these observations, we find that the light-front supermembrane Hamiltonian becomes

$$H = \frac{\nu T}{4}\int d^2\sigma \left(\dot{X}^i\dot{X}^i + \frac{2}{\nu^2}\{X^i, X^j\}\{X^i, X^j\} - \frac{2}{\nu}\theta^T\gamma_i\{X^i, \theta\}\right) \tag{2.88}$$

where θ is a 16-component Majorana spinor of $SO(9)$.

It is straightforward to apply the matrix regularization procedure discussed in section 2.4 to this Hamiltonian. This gives the supersymmetric form of matrix theory

$$H = \frac{1}{(2\pi l_p^3)}\text{Tr}\left(\frac{1}{2}\dot{\mathbf{X}}^i\dot{\mathbf{X}}^i - \frac{1}{4}[\mathbf{X}^i, \mathbf{X}^j][\mathbf{X}^i, \mathbf{X}^j] + \frac{1}{2}\theta^T\gamma_i[\mathbf{X}^i, \theta]\right). \tag{2.89}$$

This is the matrix quantum mechanics theory which will play a central role in these lectures. This theory was derived in [2] from the regularized supermembrane action, but had been previously found and studied as a particularly simple example of a supersymmetric theory with gauge symmetry [17, 18, 19].

2.7. COVARIANT MEMBRANE QUANTIZATION

It is natural to think of generalizing the matrix regularization approach to the covariant formulation of the bosonic and supersymmetric membrane theories (2.26) and (2.73). Some progress was made in this direction by

Fujikawa and Okuyama in [20]. For the bosonic membrane it is straightforward to implement the matrix regularization procedure. The only catch is that the BRST charge needed to implement the gauge-fixing procedure cannot be simply expressed in terms of the Poisson bracket on the membrane. For the supermembrane, there is a more serious complication related to the κ-symmetry of the theory. Essentially, as mentioned above, any gauge-fixing of the κ-symmetry will break the 11-dimensional Lorentz invariance of the theory. This is the same difficulty as one encounters when trying to construct a covariant quantization of the Green-Schwarz superstring. The approach taken in the second paper of [20] is to fix the κ-symmetry in a way which breaks the 32 of SO(10, 1) into $16_R + 16_L$ of SO(9, 1). Thus, they end up with a matrix formulation of a theory with explicit SO(9, 1) Lorentz symmetry. Although this theory does not have the desired complete SO(10, 1) Lorentz symmetry of M-theory, there are many questions which might be addressed by this theory with limited Lorentz invariance. It would be interesting to study the quantum mechanics of this alternative matrix formulation of M-theory in further detail.

3. The BFSS conjecture

As we have already discussed, the fact that the light-front supermembrane theory can be regularized and described as a supersymmetric quantum mechanics theory has been known for over a decade. At the time that this theory was first developed, however, it was believed that the quantum supermembrane theory suffered from instabilities which would make the low-energy interpretation as a theory of quantized gravity impossible. In 1996 the supersymmetric Yang-Mills quantum mechanics theory was brought back into currency as a candidate for a microscopic description of an 11-dimensional quantum mechanical theory containing gravity by Banks, Fischler, Shenker and Susskind (BFSS). The BFSS proposal, which quickly became known as the "Matrix Theory Conjecture" was motivated not by the quantum supermembrane theory, but by considering the low-energy theory of a system of many D0-branes as a partonic description of light-front M-theory.

In this section we discuss the apparent instability of the quantized membrane theory and the BFSS conjecture. We describe the membrane instability in subsection 3.1. We give a brief introduction to M-theory in section 3.2, and describe the BFSS conjecture in subsection 3.3. In subsection 3.4 we describe the resolution of the apparent instability of the membrane theory by an interpretation in terms of a second-quantized gravity theory. Finally, in subsection 3.5 we review an argument due to Seiberg and Sen which shows that matrix theory should be equivalent to a discrete light-front quantization of M-theory, even at finite N.

3.1. MEMBRANE "INSTABILITY"

At the time that de Wit, Hoppe and Nicolai wrote the paper [2] showing that the regularized supermembrane theory could be described in terms of supersymmetric matrix quantum mechanics, the general hope seemed to be that the quantized supermembrane theory would have a discrete spectrum of states. In string theory the spectrum of states in the Hilbert space of the string can be put into one-to-one correspondence with elementary particle-like states in the target space. The facts that the massless particle spectrum contains a graviton and that there is a mass gap separating the massless states from massive excitations are crucial for this interpretation. For the supermembrane theory, however, the spectrum does not seem to have these properties. This can be seen in both the classical and quantum membrane theories. In this section we discuss this apparent difficulty with the membrane theory, which was first described in detail in [21].

The simplest way to see the instability of the membrane theory is to consider a classical bosonic membrane whose energy is proportional to the area of the membrane times a constant tension. Such a membrane can have long narrow spikes at very low cost in energy (See Figure 1). If the spike is roughly cylindrical and has a radius r and length L then the energy is $2\pi TrL$. For a spike with very large L but a small radius $r \ll 1/TL$ the energy cost is small but the spike is very long. This indicates that a quantum membrane will tend to have many fluctuations of this type, making it difficult to think of the membrane as a single pointlike object. Note that the quantum string theory does not have this problem since a long spike in a string always has energy proportional to the length of the string.

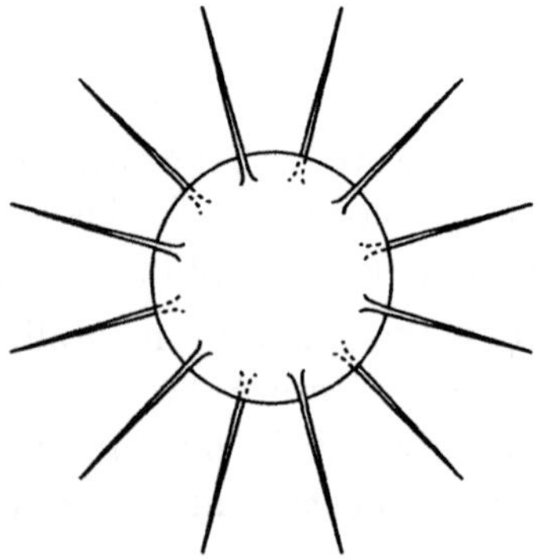

Figure 1. Classical membrane instability arises from spikes of infinitesimal area

In the matrix regularized version of the membrane theory, this instability appears as a set of flat directions in the classical theory. For example, if we have a pair of matrices with nonzero entries of the form

$$X^1 = \begin{pmatrix} x & 0 \\ 0 & 0 \end{pmatrix} \quad X^2 = \begin{pmatrix} 0 & y \\ y & 0 \end{pmatrix} \tag{3.1}$$

then a potential term $[X^1, X^2]^2$ corresponds to a term x^2y^2. If either $x = 0$ or $y = 0$ then the other variable is unconstrained, giving flat directions in the moduli space of solutions to the classical equations of motion. This corresponds classically to a marginal instability in the matrix theory with $N > 1$. (Note that in the previous section we distinguished matrices $\mathbf{X}^i$ from related functions X^i by using bold font for matrices. We will henceforth drop this font distinction as long as the difference can easily be distinguished from context.)

In the quantum bosonic membrane theory, the apparent instability from the flat directions is cured because of the 0-modes of off-diagonal degrees of freedom. In the above example, for instance, if x takes a large value then y corresponds to a harmonic oscillator degree of freedom with a large mass. The zero point energy of this oscillator becomes larger as x increases, giving an effective confining potential which removes the flat directions of the classical theory. This would seem to resolve the instability problem. Indeed, in the matrix regularized quantum bosonic membrane theory, there is a discrete spectrum of energy levels for the system of $N \times N$ matrices.

When we consider the supersymmetric theory, on the other hand, the problem reemerges. The zero point energies of the fermionic oscillators associated with the extra Grassmann degrees of freedom in the supersymmetric theory conspire to precisely cancel the zero point energies of the bosonic oscillators. This cancellation gives rise to a continuous spectrum in the supersymmetric matrix theory. This result was formally proven by de Wit, Lüscher and Nicolai in [21]. They showed that for any $\epsilon > 0$ and any energy $E \in [0, \infty)$ there exists a state ψ in the $N = 2$ maximally supersymmetric matrix model which is normalizable ($\int |\psi|^2 = ||\psi||^2 = 1$) and which has

$$||(H - E)\psi||^2 < \epsilon.$$

This implies that the spectrum of the supersymmetric matrix quantum mechanics theory is continuous[1]. This result indicated that it would not be possible to have a simple interpretation of the states of the theory in terms of a discrete particle spectrum. After this work there was little further development on the supersymmetric matrix quantum mechanics theory as a theory of membranes or gravity until almost a decade later.

[1]Note that [21] did not resolve the question of whether a state existed with identically vanishing energy $H = 0$. This question was not resolved until the much later work of Sethi and Stern [22] showed that such a marginally bound state does indeed exist in the maximally supersymmetric theory.

3.2. M-THEORY

The concept of M-theory has played a fairly central role in the development of the web of duality symmetries which relate the five string theories to each other and to supergravity [23, 11, 24, 25, 26]. M-theory is a conjectured eleven-dimensional theory whose low-energy limit corresponds to 11D supergravity. Although there are difficulties with constructing a quantum version of 11D supergravity, it is a well-defined classical theory with the following field content [27]:

e_I^a: a vielbein field (bosonic, with 44 components)
ψ_I: a Majorana fermion gravitino (fermionic, with 128 components)
A_{IJK}: a 3-form potential (bosonic, with 84 components).

In addition to being a local theory of gravity with an extra 3-form potential field, M-theory also contains extended objects. These consist of a two-dimensional supermembrane and a 5-brane, which couple electrically and magnetically to the 3-form field.

One way of defining M-theory is as the strong coupling limit of the type IIA string. The IIA string theory is taken to be equivalent to M-theory compactified on a circle S^1, where the radius of compactification R of the circle in direction 11 is related to the string coupling g through $R = g^{2/3} l_p = g l_s$, where l_p and $l_s = \sqrt{\alpha'}$ are the M-theory Planck length and the string length respectively. The decompactification limit $R \to \infty$ corresponds then to the strong coupling limit of the IIA string theory. (Note that we will always take the eleven dimensions of M-theory to be labeled $0, 1, \ldots, 8, 9, 11$; capitalized roman indices $I, J, \ldots$ denote 11-dimensional indices).

Given this relationship between compactified M-theory and IIA string theory, a correspondence can be constructed between various objects in the two theories. For example, the Kaluza-Klein photon associated with the components $g_{\mu 11}$ of the 11D metric tensor can be associated with the R-R gauge field A_μ in IIA string theory. The only object which is charged under this R-R gauge field in IIA string theory is the D0-brane; thus, the D0-brane can be associated with a supergraviton with momentum p_{11} in the compactified direction. The membrane and 5-brane of M-theory can be associated with different IIA objects depending on whether or not they are wrapped around the compactified direction; the correspondence between various M-theory and IIA objects is given in Table 1.

3.3. THE BFSS CONJECTURE

In 1996, motivated by recent work on D-branes and string dualities, Banks, Fischler, Shenker and Susskind (BFSS) proposed that the large N limit

TABLE 1. Correspondence between objects in M-theory and IIA string theory

M-theory	IIA
KK photon ($g_{\mu 11}$)	RR gauge field A_μ
supergraviton with $p_{11} = 1/R$	D0-brane
wrapped membrane	IIA string
unwrapped membrane	IIA D2-brane
wrapped 5-brane	IIA D4-brane
unwrapped 5-brane	IIA NS5-brane

of the supersymmetric matrix quantum mechanics model (2.89) should describe all of M-theory in a light-front coordinate system [3]. Although this conjecture fits neatly into the framework of the quantized membrane theory, the starting point of BFSS was to consider M-theory compactified on a circle S^1, with a large momentum in the compact direction. As we have just discussed, when M-theory is compactified on S^1 the corresponding theory in 10D is the type IIA string theory, and the quanta corresponding to momentum in the compact direction are the D0-branes of the IIA theory. In the limits where the radius of compactification R and the compact momentum p_{11} are both taken to be large, this correspondence relates M-theory in the "infinite momentum frame" (IMF) to the nonrelativistic theory of many D0-branes in type IIA string theory.

The low-energy Lagrangian for a system of many type IIA D0-branes is the matrix quantum mechanics Lagrangian arising from the dimensional reduction to $0 + 1$ dimensions of the 10D super Yang-Mills Lagrangian

$$\mathcal{L} = \frac{1}{2gl_s} \mathrm{Tr} \left[\dot{X}^a \dot{X}^a + \frac{1}{2}[X^a, X^b]^2 + \theta^T (i\dot{\theta} - \Gamma_a [X^a, \theta]) \right] \tag{3.2}$$

(the gauge has been fixed to $A_0 = 0$.) The corresponding Hamiltonian is

$$H = \frac{1}{2gl_s} \mathrm{Tr} \left(\dot{X}^i \dot{X}^i - \frac{1}{2}[X^i, X^j][X^i, X^j] + \theta^T \gamma_i [X^i, \theta] \right) . \tag{3.3}$$

Using the relations $R = g^{2/3} l_{11} = gl_s$, we see that in string units ($2\pi l_s^2 = 1$) we can replace $gl_s = R = 2\pi l_{11}^3$. So the Hamiltonian (3.3) arising in the matrix quantum mechanics picture is in fact precisely equivalent to the matrix membrane Hamiltonian (2.89). This connection and its possible physical significance was first pointed out by Townsend [28]. The matrix theory Hamiltonian is often written, following BFSS, in the form

$$H = \frac{R}{2}\mathrm{Tr}\left(P^i P^i - \frac{1}{2}[X^i, X^j][X^i, X^j] + \theta^T \gamma_i [X^i, \theta]\right) \tag{3.4}$$

where we have rescaled $X/g^{1/3} \to X$ and written the Hamiltonian in Planck units $l_{11} = 1$.

The original BFSS conjecture was made in the context of the large N theory. It was later argued by Susskind that the finite N matrix quantum mechanics theory should be equivalent to the discrete light-front quantized (DLCQ) sector of M-theory with N units of compact momentum [29]. We describe in section (3.5) below an argument due to Seiberg and Sen which makes this connection more precise and which justifies the use of the low-energy D0-brane action in the BFSS conjecture.

While the BFSS conjecture was based on a different framework from the matrix quantization of the supermembrane theory we have discussed above, the fact that the membrane naturally appears as a coherent state in the matrix quantum mechanics theory was a substantial piece of additional evidence given by BFSS for the validity of their conjecture. Two additional pieces of evidence were given by BFSS which extended their conjecture beyond the previous work on the matrix membrane theory.

One important point made by BFSS is that the Hilbert space of the matrix quantum mechanics theory naturally contains multiple particle states. This observation, which we discuss in more detail in the following subsection, resolves the problem of the continuous spectrum discussed above. Another piece of evidence given by BFSS for their conjecture is the fact that quantum effects in matrix theory give rise to long-range interactions between a pair of gravitational quanta (D0-branes) which have precisely the correct form expected from light-front supergravity. This result was first shown by a calculation of Douglas, Kabat, Pouliot and Shenker [30]; we will discuss this result and its generalization to more general matrix theory interactions in Section 5.

3.4. MATRIX THEORY AS A SECOND QUANTIZED THEORY

The classical equations of motion for a bosonic matrix configuration with the Hamiltonian (2.89) are (up to an overall constant)

$$\ddot{X}^i = -[[X^i, X^j], X^j]. \tag{3.5}$$

If we consider a block-diagonal set of matrices

$$X^i = \begin{pmatrix} \hat{X}^i & 0 \\ 0 & \tilde{X}^i \end{pmatrix}$$

with first time derivatives $\dot{X}^i$ which are also of block-diagonal form, then the classical equations of motion for the blocks are separable

$$\ddot{\hat{X}}^i = -[[\hat{X}^i, \hat{X}^j], \hat{X}^j]$$
$$\ddot{\tilde{X}}^i = -[[\tilde{X}^i, \tilde{X}^j], \tilde{X}^j]$$

If we think of each of these blocks as describing a matrix theory object with center of mass

$$\hat{x}^i = \frac{1}{\hat{N}} \mathrm{Tr}\, \hat{X}^i$$
$$\tilde{x}^i = \frac{1}{\tilde{N}} \mathrm{Tr}\, \tilde{X}^i$$

then we have two objects obeying classically independent equations of motion (See Figure 2). It is straightforward to generalize this construction to a block-diagonal matrix configuration describing k classically independent objects. This gives a simple indication of how matrix theory can encode, even in finite N matrices, a configuration of multiple objects. In this sense it is natural to think of matrix theory as a second quantized theory from the point of view of the target space.

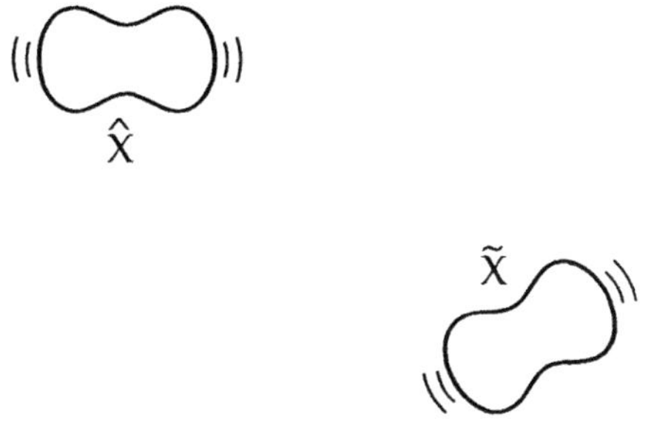

Figure 2. Two matrix theory objects described by block-diagonal matrices

Given the realization that matrix theory should describe a second quantized theory, the puzzle discussed above regarding the continuous spectrum of the theory is easily resolved. If there is a state in matrix theory corresponding to a single graviton of M-theory (as we will discuss in more detail in section 4.1) with $H = 0$ which is roughly a localized state, then by taking two such states to have a large separation and a small relative velocity

v it should be possible to construct a two-body state with an arbitrarily small total energy. Since the D0-branes of the IIA theory correspond to gravitons in M-theory with a single unit of longitudinal momentum, we would therefore naturally expect to have a continuous spectrum of energies even in the theory with $N = 2$. This resolves the puzzle found by de Wit, Lüscher and Nicolai in a very pleasing fashion, which suggests that matrix theory is perhaps even more powerful than string theory, which only gives a first-quantized theory in the target space.

The second quantized nature of matrix theory can also be seen naturally in the continuous membrane theory. Recall that the instability of membrane theory appears in the classical theory of a continuous membrane when we consider the possibility of long thin spikes of negligible energy, as discussed in section 3.1. In a similar fashion, it is possible for a classical smooth membrane of fixed topology to be mapped to a configuration in the target space which looks like a system of multiple distinct macroscopic membranes connected by infinitesimal tubes of negligible energy (See Figure 3). In the limit where the tubes become very small, their effect on the classical dynamics of the multiple membrane configuration disappears and we effectively have a system of multiple independent membranes moving in the target space. At the classical level, the sum of the genera of the membranes in the target space must be equal to or smaller than the genus of the single world-sheet membrane, but when quantum effects are included handles can be added to the membrane as well as removed [31]. These considerations seem to indicate that any consistent quantum theory which contains a continuous membrane in its effective low-energy theory must contain configurations with arbitrary membrane topology and must therefore be a "second quantized" theory from the point of view of the target space.

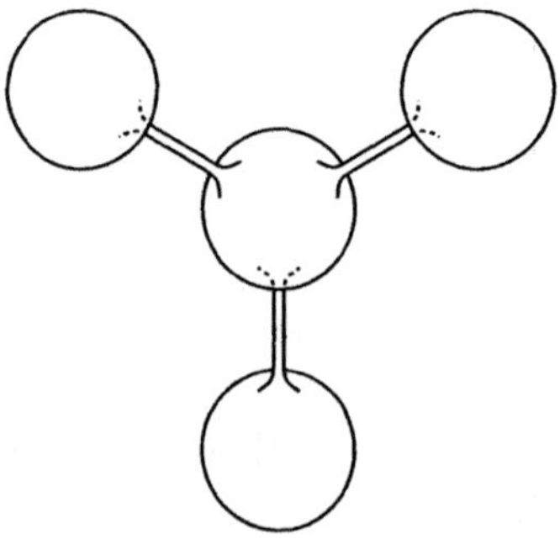

Figure 3. Membrane of fixed (spherical) topology mapped to multiple membranes connected by tubes in the target space

A theory which has been compactified on a lightlike circle can be viewed as a limit of a theory compactified on a spacelike circle where the size of the spacelike circle becomes vanishingly small in the limit. This point of view was used by Seiberg and Sen in [32, 33] to argue that light-front compactified M-theory is described through such a limiting process by the low-energy Lagrangian for many D0-branes, and hence by matrix theory. In this section we go through this argument in detail.

Consider a space-time which has been compactified on a lightlike circle by identifying

$$\begin{pmatrix} x \\ t \end{pmatrix} \sim \begin{pmatrix} x - R/\sqrt{2} \\ t + R/\sqrt{2} \end{pmatrix} \tag{3.6}$$

This theory has a quantized momentum in the compact direction

$$P^+ = \frac{N}{R} \tag{3.7}$$

The compactification (3.6) can be described as a limit of a family of spacelike compactifications

$$\begin{pmatrix} x \\ t \end{pmatrix} \sim \begin{pmatrix} x - \sqrt{R^2/2 + R_s^2} \\ t + R/\sqrt{2} \end{pmatrix} \tag{3.8}$$

parameterized by the size $R_s \to 0$ of the spacelike circle, which is taken to vanish in the limit.

The system satisfying (3.8) is related through a boost to a system with the identification

$$\begin{pmatrix} x' \\ t' \end{pmatrix} \sim \begin{pmatrix} x' - R_s \\ t' \end{pmatrix} \tag{3.9}$$

where

$$\begin{pmatrix} x' \\ t' \end{pmatrix} = \begin{pmatrix} \frac{1}{\sqrt{1-\beta^2}} & \frac{\beta}{\sqrt{1-\beta^2}} \\ \frac{\beta}{\sqrt{1-\beta^2}} & \frac{1}{\sqrt{1-\beta^2}} \end{pmatrix} \begin{pmatrix} x \\ t \end{pmatrix} \tag{3.10}$$

The boost parameter β is given by

$$\beta = \frac{1}{\sqrt{1 + \frac{2R_s^2}{R^2}}} \cong 1 - \frac{R_s^2}{R^2}. \tag{3.11}$$

In the context of matrix theory we are interested in understanding M-theory compactified on a lightlike circle. This is related through the above limiting process to a family of spacelike compactifications of M-theory, which we know can be identified with the IIA string theory. At first glance, it may seem that the limit we are considering here is difficult to analyze from the IIA point of view. The IIA string coupling and string length are related to the compactification radius and 11D Planck length as usual by

$$\begin{aligned} g &= (\frac{R_s}{l_{11}})^{3/2} \\ l_s^2 &= \frac{l_{11}^3}{R_s} \end{aligned} \tag{3.12}$$

Thus, in the limit $R_s \to \infty$ the string coupling g becomes small as desired; the string length l_s, however, goes to ∞. Since $l_s^2 = \alpha'$, this corresponds to a limit of vanishing string tension. Such a limiting theory is very complicated and would not seem to provide a useful alternative description of the theory.

Let us consider, however, how the energy of the states we are interested in behaves in the class of limiting theories with spacelike compactification. If we want to describe the behavior of a state which has light-front energy P^- and compact momentum $P^+ = N/R$ then the spatial momentum in the theory with spatial R_s compactification is $P' = N/R_s$. The energy in the spatially compactified theory is

$$E' = N/R_s + \Delta E, \tag{3.13}$$

where ΔE has the energy scale we are interested in understanding. The term N/R_s in the energy is simply the mass-energy of the N D0-branes which correspond to the momentum in the compactified M-theory direction. Relating back to the near lightlike compactified theory we have

$$\begin{pmatrix} P \\ E \end{pmatrix} = \begin{pmatrix} \frac{1}{\sqrt{1-\beta^2}} & -\frac{\beta}{\sqrt{1-\beta^2}} \\ -\frac{\beta}{\sqrt{1-\beta^2}} & \frac{1}{\sqrt{1-\beta^2}} \end{pmatrix} \begin{pmatrix} P' \\ E' \end{pmatrix} \tag{3.14}$$

so

$$P^- = \frac{1}{\sqrt{2}}(E - P) = \frac{1}{\sqrt{2}} \frac{1+\beta}{\sqrt{1-\beta^2}} \Delta E \approx \frac{R}{R_s} \Delta E \tag{3.15}$$

As a result we see that the energy ΔE of the IIA configuration needed to approximate the light-front energy P^- is given by

$$\Delta E \approx P^- \frac{R_s}{R} \tag{3.16}$$

We know that the string scale $1/l_s$ becomes small as $R_s \to 0$. We can compare the energy scale of interest to this string scale, however, and find

$$\frac{\Delta E}{(1/l_s)} = \frac{P^-}{R} R_s l_s = \frac{P^-}{R} \sqrt{R_s l_{11}^3} \tag{3.17}$$

This ratio vanishes in the limit $R_s \to 0$, which implies that although the string scale vanishes, the energy scale of interest is smaller still. Thus, it is reasonable to study the lightlike compactification through a limit of spatial compactifications in this fashion.

To make the correspondence between the light-front compactified theory and the spatially compactified limiting theories more transparent, we perform a change of units to a new Planck length $\tilde{l}_{11}$ in the spatially compactified theories in such a way that the energy of the states of interest is independent of R_s. For this condition to hold we must have

$$\Delta E \tilde{l}_{11} = P^- \frac{R_s l_{11}^2}{R \tilde{l}_{11}} \tag{3.18}$$

where E, R and P^- are independent of R_s and all units have been explicitly included. This requires us to keep the quantity

$$\frac{R_s}{\tilde{l}_{11}^2} \tag{3.19}$$

fixed in the limiting process. Thus, in the limit $\tilde{l}_{11} \to 0$.

We can now summarize the discussion with the following story: to describe the sector of M-theory corresponding to light-front compactification on a circle of radius R with light-front momentum $P^+ = N/R$ we may consider the limit $R_s \to 0$ of a family of IIA configurations with N D0-branes where the string coupling and string length

$$\begin{aligned} \tilde{g} &= (R_s/\tilde{l}_{11})^{3/2} \to 0 \\ \tilde{l}_s &= \sqrt{\tilde{l}_{11}^3/R_s} \to 0 \end{aligned} \tag{3.20}$$

are defined in terms of a Planck length $\tilde{l}_{11}$ and compactification length R_s which satisfy

$$R_s/\tilde{l}_{11}^2 = R/l_{11}^2 \tag{3.21}$$

All transverse directions scale normally through

$$\tilde{x}^i/\tilde{l}_{11} = x^i/l_{11} \tag{3.22}$$

To give a very concrete example of how this limiting process works, let us consider a system with a single unit of longitudinal momentum

$$P^+ = \frac{1}{R} \tag{3.23}$$

We know that in the corresponding IIA theory, we have a single D0-brane whose Lagrangian has the Born-Infeld form

$$\mathcal{L} = -\frac{1}{\tilde{g}\tilde{l}_s}\sqrt{1 - \dot{\tilde{x}}^i \dot{\tilde{x}}^i} \tag{3.24}$$

Expanding the square root we have

$$\mathcal{L} = -\frac{1}{\tilde{g}\tilde{l}_s}\left(1 - \frac{1}{2}\dot{\tilde{x}}^i \dot{\tilde{x}}^i + \mathcal{O}(\dot{\tilde{x}}^4)\right). \tag{3.25}$$

Replacing $\tilde{g}\tilde{l}_s \to R_s$ and $\tilde{x} \to x\tilde{l}_{11}/l_{11}$ gives

$$\mathcal{L} = -\frac{1}{R_s} + \frac{1}{2R}\dot{x}^i \dot{x}^i + \mathcal{O}(R_s/R). \tag{3.26}$$

Thus, we see that all the higher order terms in the Born-Infeld action vanish in the $R_s \to 0$ limit. The leading term is the D0-brane energy $1/R_s$ which we subtract to compare to the M-theory light-front energy P^-. Although we do not know the full form of the nonabelian Born-Infeld action describing N D0-branes in IIA, it is clear that an analogous argument shows that all terms in this action other than those in the nonrelativistic supersymmetric matrix theory action will vanish in the limit $R_s \to 0$.

This argument apparently demonstrates that matrix theory gives a complete description of the dynamics of DLCQ M-theory. There are several caveats which should be taken into account, however, with respect to this discussion. First, in order for this argument to be correct, it is necessary that there exists a well-defined theory with the properties expected of M-theory, and that there exist a well-defined IIA string theory which arises as the compactification of M-theory. Neither of these statements is at this point definitely established. Thus, this argument must be taken as contingent upon the definition of these theories. Second, although we know that 11D supergravity arises as the low-energy limit of M-theory, this argument does not necessarily indicate that matrix theory describes DLCQ supergravity in the low-energy limit. It may be that to make the connection to supergravity it is necessary to deal with subtleties of the large N limit.

In the remainder of these lectures we will discuss some more explicit approaches to connecting matrix theory with supergravity. In particular, we will see how far it is possible to go in demonstrating that 11D supergravity arises from calculations in the finite N version of matrix theory, which is a completely well-defined theory. In the last sections we will return to a more general discussion of the status of matrix theory.

4. M-theory objects from matrix theory

In this section we discuss how the matrix theory degrees of freedom can be used to construct the various objects of M-theory: the supergraviton, supermembrane and 5-brane. We discuss each of these objects in turn in subsections 4.1, 4.2, 4.3, after which we give a general discussion of the structure of extended objects and their charges in subsection 4.4

4.1. SUPERGRAVITONS

Since in DLCQ M-theory there should be a pointlike state corresponding to a longitudinal graviton with $p^+ = N/R$ and arbitrary transverse momentum p^i, we expect from the massless condition $m^2 = -p^I p_I = 0$ that such an object will have matrix theory energy

$$E = \frac{p_i^2}{2p^+}$$

We discuss such states first classically and then in the quantum theory.

4.1.1. *Classical supergravitons*

The classical matrix theory potential is $-[X^i, X^j]^2$, from which we have the classical equations of motion

$$\ddot{X}^i = -[[X^i, X^j], X^j].$$

One simple class of solutions to these equations of motion can be found when the matrices minimize the potential at all times and therefore all commute. Such solutions are of the form

$$X^i = \begin{pmatrix} x_1^i + v_1^i t & 0 & 0 & \ddots \\ 0 & x_2^i + v_2^i t & \ddots & 0 \\ 0 & \ddots & \ddots & 0 \\ \ddots & 0 & 0 & x_N^i + v_N^i t \end{pmatrix}$$

This corresponds to a classical N-graviton solution, where each graviton has

$$p_a^+ = 1/R \quad p_a^i = v_a^i/R \quad E_a = v_a^2/(2R) = (p_a^i)^2/2p^+$$

A single classical graviton with $p^+ = N/R$ can be formed by setting

$$x_1^i = \cdots = x_N^i, \quad v_1^i = \cdots = v_N^i$$

so that the trajectories of all the components are identical. Although this may seem like a very simple model for a graviton, it is precisely such matrix configurations which are used as a background in most computations of quantum effects in matrix theory corresponding to gravitational interactions, as will be discussed further in the following.

4.1.2. *Quantum supergravitons*

The picture of a supergraviton in quantum matrix theory is somewhat more subtle than the simple classical picture just discussed. Let us first consider the case of a single supergraviton with $p^+ = 1/R$. This corresponds to the U(1) case of the super Yang-Mills quantum mechanics theory. The Hamiltonian is simply

$$H = \frac{1}{2R}\dot{X}^2$$

since all commutators vanish in this theory. The bosonic part of the theory is simply a free nonrelativistic particle. In the fermionic sector there are 16 spinor variables with anticommutation relations

$$\{\theta_\alpha, \theta_\beta\} = \delta_{\alpha\beta}.$$

By using the standard trick of writing these as 8 fermion creation and annihilation operators

$$\theta_i^\pm = \frac{1}{\sqrt{2}}(\theta_i \pm \theta_{i+8}) \qquad 1 \leq i \leq 8$$

we see that the Hilbert space for the fermions is a standard fermion Fock space of dimension $2^8 = 256$. Indeed, this is precisely the number of states needed to represent all the polarization states of the graviton (44), the antisymmetric 3-tensor field (84) and the gravitino (128). For details of how the polarization states are represented in terms of the fermionic Fock space, see [2, 34].

The case when $N > 1$ is much more subtle. We can factor out the overall $U(1)$ so that every state in the $SU(N)$ quantum mechanics theory has 256 corresponding states in the full theory. For the matrix theory conjecture to be correct, as BFSS pointed out, it should then be the case that for every N there exists a unique threshold bound state in the $SU(N)$ theory with $H = 0$. As mentioned before, no definitive answer as to the existence of such a state was given in the early work on matrix theory. This result was finally proven by Sethi and Stern for $N = 2$ in [22]. Progress towards proving the result for arbitrary values of N was made in [35, 36, 37], and the result for a general gauge group was given in [38] (see also [39]).

4.2. MEMBRANES

In this section we discuss the description of M-theory membranes in terms of the matrix quantum mechanics degrees of freedom. It is clear from the derivation of matrix theory as a regularized supermembrane theory that there must be matrix configurations which in the large N limit give arbitrarily good descriptions of any membrane configuration. It is instructive, however, to study in detail the structure of such membrane configurations. In subsection 4.2.1 we discuss the significance of the matrix representation of membranes in the language of type IIA D0-branes. In subsection 4.2.2 we discuss in some detail how a spherical membrane can be very accurately described by matrices even with small values of N. In subsection 4.2.3 we discuss higher genus matrix membranes. In subsection 4.2.4 we discuss noncompact matrix membranes, and finally in subsection 4.2.5 we discuss M-theory membranes which are wrapped on the longitudinal direction and appear as strings in the IIA theory.

4.2.1. *D2-branes from D0-branes*

As we have mentioned, it is clear from inverting the matrix membrane regularization procedure that smooth membranes can be approximated by finite size matrices. This construction may seem less natural in the language of type IIA string theory, where it corresponds to a construction of a IIA D2-brane out of the degrees of freedom describing a system of N D0-branes. In fact, however, the fact that this construction is possible is simply the T-dual of the familiar statement that D0-branes are described by the magnetic flux of the gauge field living on a set of N D2-branes [40]. Both of these statements can in turn be seen by performing T-duality on a diagonally wrapped D1-brane on a 2-torus.

To see this explicitly, consider a set of N D2-branes on a torus T^2 with k units of magnetic flux

$$\frac{1}{2\pi}\int F = k \tag{4.1}$$

Under a T-duality transformation on one direction of the torus, the gauge field component A_2 is replaced by an infinite matrix

$$X^2 = i\partial_2 + A_2$$

representing a transverse scalar field for a set of N D1-branes living on the dual torus $(T^2)^*$. These matrices are infinite because they contain information about winding strings connecting the infinite number of copies of each brane which live on the infinite covering space of the dual torus. (This construction is described in more detail in section 6.1.) This T-dual configuration corresponds to a single D-string which is diagonally wound N

times around the x^1 direction and k times around the x^2 direction; this can be seen from the fact that the T-dual of (4.1) is $\partial_1 X^2 = (kL_2^*/NL_1)\mathbb{1}$. Since under T-duality in the x^2 direction a D1-brane wrapped in the x^2 direction becomes a D0-brane, we can identify the flux (4.1) with k D0-branes in the original theory.

Further T-dualizing in the direction x^1, we replace

$$X^1 = i\partial_1 + A_1.$$

where X^1, X^2 are now infinite matrices describing transverse fields of a system of N D0-branes on the dual torus $(T^2)^{**}$. When the normalization constants are treated carefully, the flux condition (4.1) now becomes the condition on the D0-brane matrices

$$\mathrm{Tr}\,[X^1, X^2] = \frac{iAk}{2\pi} \tag{4.2}$$

where A is the area of the dual torus. Since the T-dual in the x^1 direction of the D-string wrapped in the x^2 direction is a D2-brane, we interpret k in (4.2) as the D2-brane charge of a system of N D0-branes.

This construction can be interpreted more generally, so that in general a pair of matrices X^a, X^b describing a D0-brane configuration satisfying

$$\mathrm{Tr}\,[X^a, X^b] = \frac{iA}{2\pi} \tag{4.3}$$

should be interpreted as giving rise to a piece of a D2-brane of area A. Of course, for finite matrices the trace of the commutator must vanish. This is simply a consequence of the fact that the net D2-brane charge of any compact object must vanish. However, not only is it possible to have a nonzero membrane charge when the matrices are infinite, but it is also possible to treat (4.3) as a local expression by restricting the trace to a subset of the diagonal elements. We will see a specific example of this in the next subsection. The local relation (4.3) will also be useful in constructing higher moments of the membrane charge, which can be nonzero even for finite size configurations, as we shall discuss later.

4.2.2. *Spherical membranes*

One extremely simple example of a membrane configuration which can be approximated very well even at finite N by simple matrix configurations is the symmetric spherical membrane [41]. Imagine that we wish to construct a membrane embedded in an isotropic sphere

$$x_1^2 + x_2^2 + x_3^2 = r^2$$

in the first three dimensions of $\mathbb{R}^{11}$. The embedding functions for such a continuous membrane can be written as linear functions

$$X^i = r\xi^i \qquad 1 \leq i \leq 3$$

of the three Euclidean coordinates ξ^i on the spherical world-volume. Using the matrix-membrane correspondence (2.51) we see that the matrix approximation to this membrane will be given by the $N \times N$ matrices

$$\mathbf{X}^i = \frac{2r}{N} J^i \qquad 1 \leq i \leq 3 \tag{4.4}$$

where J^i are the generators of $SU(2)$ in the N-dimensional representation.

It is quite interesting to see how many of the geometrical and physical properties of the sphere can be extracted from the algebraic structure of these matrices, even for small values of N. We list here some of these properties.

i) Spherical locus: The matrices (4.4) satisfy

$$\mathbf{X}_1^2 + \mathbf{X}_2^2 + \mathbf{X}_3^2 = \frac{4r^2}{N^2} C_2(N) \mathbb{1} = r^2(1 - 1/N^2)\mathbb{1}$$

where $C_2(N) = (N^2 - 1)/4$ is the quadratic Casimir of $SU(2)$ in the N-dimensional representation. This shows that the D0-branes are in a non-commutative sense "localized" on a sphere of radius $r + \mathcal{O}(1/N^2)$.

ii) Rotational invariance: The matrices (4.4) satisfy

$$R_{ij}\mathbf{X}_j = U(R) \cdot \mathbf{X}_i \cdot U(R^{-1})$$

where $R \in SO(3)$ and $U(R)$ is the N-dimensional representation of R. Thus, the spherical matrix configuration is rotationally invariant up to a gauge transformation.

iii) Spectrum: The matrix $\mathbf{X}^3 = 2rJ_3/N$ (as well as the other matrices) has a spectrum of eigenvalues which are uniformly distributed in the interval $[-r, r]$. This is precisely the correct distribution if we imagine a perfectly symmetric sphere with D0-branes distributed uniformly on its surface and project this distribution onto a single axis.

iv) Local membrane charge: As discussed above, the expression (4.3) gives an area for a piece of a membrane described by a pair of matrices. We can use this formula to check the interpretation of the matrix sphere. We do this by computing the membrane charge in the 1-2 plane of the half of the configuration with eigenvalues $X^3 > 0$. This should correspond to the projected area of the "upper hemisphere" of the sphere. We compute

$$A_h = -2\pi i \mathrm{Tr}_{1/2} \left[\mathbf{X}^1, \mathbf{X}^2\right]$$

where the trace is restricted to the set of eigenvalues where $X^3 > 0$ in the standard representation. This is possible since $[\mathbf{X}^1, \mathbf{X}^2] \sim \mathbf{X}^3$. We find

$$A_h = 2\pi \frac{4}{N^2} r^2 \mathrm{Tr}_{1/2}\, J_3 = \pi r^2 (1 + \mathcal{O}(1/N^2))$$

thus, we find precisely the expected area of the projected hemisphere.

v) Energy: In M-theory we expect the tension energy of a (momentarily) stationary membrane sphere to be

$$e = \frac{4\pi r^2}{(2\pi)^2 l_{11}^3} = \frac{r^2}{\pi l_{11}^3}$$

Using $p^I p_I = -e^2$ we see that the light-front energy should be

$$E = \frac{e^2}{2p^+} \tag{4.5}$$

in 11D Planck units. Let us compute the matrix membrane energy. It is given by

$$E = -\frac{1}{4R}[\mathbf{X}^i, \mathbf{X}^j]^2 = \frac{2r^4}{NR} + \mathcal{O}(N^{-3})$$

in string units. This is easily seen to agree with (4.5).

It is also straightforward to verify that the equations of motion for the membrane are correctly reproduced in matrix theory.

Thus, we see that many of the geometrical and physical properties of the membrane can be extracted from algebraic information about the structure of the appropriate membrane configuration. The discussion we have carried out here has only applied to the simple case of the rotationally invariant spherically embedded membrane. It is straightforward to extend the discussion to a membrane of spherical topology and arbitrary shape, however, simply by using the matrix-membrane correspondence (2.51) to construct matrices approximating an arbitrary smooth spherical membrane. We now turn to the question of membranes with non-spherical topology.

4.2.3. *Higher genus membranes*

So far we have only discussed membranes of spherical topology. It is possible to describe compact membranes of arbitrary genus by generalizing this construction, although an explicit construction is only known for the sphere and torus. In this section we give a brief description of the matrix torus, following the work of Fairlie, Fletcher and Zachos [42, 43, 44].

We consider a torus defined by two coordinates $x_1, x_2 \in [0, 2\pi]$ with symplectic form $\omega_{ij} = \epsilon_{ij}/\pi$ corresponding to a total volume $\int d^2x\, \omega = 4\pi$ as in the case of the sphere discussed in section 2.4. As in the case of the sphere we wish to find a map from functions on the torus to matrices which is compatible with the correspondence

$$\{\cdot,\cdot\} \leftrightarrow \frac{-iN}{2}[\cdot,\cdot] \qquad \frac{1}{4\pi^2}\int d^2x \leftrightarrow \frac{1}{N}\mathrm{Tr} \tag{4.6}$$

A natural (complex) basis for the functions on T^2 is given by the Fourier modes

$$y_{nm}(x_1, x_2) = e^{inx_1 + imx_2} \tag{4.7}$$

The real functions on T^2 are given by the linear combinations

$$\frac{1}{2}\left(y_{nm} + y_{-n\,-m}\right), \qquad \frac{-i}{2}\left(y_{nm} - y_{-n\,-m}\right). \tag{4.8}$$

The Poisson bracket algebra of the functions y_{nm} is

$$\{y_{nm}, y_{n'm'}\} = -\pi(nm' - mn')y_{n+n',m+m'} \tag{4.9}$$

To describe the matrix approximations for these functions we use the 't Hooft matrices

$$U = \begin{pmatrix} 1 & & & & \\ & q & & & \\ & & q^2 & & \\ & & & \ddots & \\ & & & & q^{N-1} \end{pmatrix} \tag{4.10}$$

and

$$V = \begin{pmatrix} & 1 & & & \\ & & 1 & & \\ & & & \ddots & \\ & & & & 1 \\ 1 & & & & \end{pmatrix} \tag{4.11}$$

where

$$q = e^{\frac{2\pi i}{N}}. \tag{4.12}$$

The matrices U, V satisfy

$$UV = q^{-1}VU. \tag{4.13}$$

In terms of these matrices we can define

$$Y_{nm} = q^{nm/2}U^nV^m = q^{-nm/2}V^mU^n \tag{4.14}$$

so that the matrix approximation to an arbitrary function

$$f(x_1, x_2) = \sum_{n,m} c_{nm}y_{nm}(x_1, x_2) \tag{4.15}$$

is given by

$$F = \sum_{n,m} c_{nm} Y_{nm}. \tag{4.16}$$

By computing

$$\begin{aligned}[Y_{nm}, Y_{n'm'}] &= (q^{(mn'-nm')/2} - q^{(nm'-mn')/2})Y_{n+n',m+m'} \\ &\to \frac{2\pi i}{N}(mn' - nm')Y_{n+n',m+m'}\end{aligned}$$

We see that for fixed n, m, n', m' in the large N limit the matrix commutation relations correctly reproduce (4.9) just as in the case of the sphere.

As a concrete example let us consider embedding a torus into $\mathbb{R}^4 \subset \mathbb{R}^9$ so that the membrane fills the locus of points satisfying

$$X_1^2 + X_2^2 = r^2 \qquad X_3^2 + X_4^2 = s^2. \tag{4.17}$$

Such a membrane configuration can be realized through the following matrices

$$\begin{aligned}\mathbf{X}_1 &= \frac{r}{2}(U + U^\dagger) \\ \mathbf{X}_2 &= \frac{-ir}{2}(U - U^\dagger) \\ \mathbf{X}_3 &= \frac{s}{2}(V + V^\dagger) \\ \mathbf{X}_4 &= \frac{-is}{2}(V - V^\dagger)\end{aligned} \tag{4.18}$$

It is straightforward to check that this matrix configuration has geometrical properties analogous to those of the matrix membrane sphere discussed in the previous subsection. In particular, the equation (4.17) is satisfied identically as a matrix equation. Note, however that this configuration is not gauge invariant under $U(1)$ rotations in the 12 and 34 planes—only under a $\mathbb{Z}_N$ subgroup of each of these $U(1)$'s.

4.2.4. *Infinite membranes*

So far we have discussed compact membranes, which can be described in terms of finite-size $N \times N$ matrices. In the large N limit it is also possible to construct membranes with infinite spatial extent. The matrices X^i describing such configurations are infinite-dimensional matrices which correspond to operators on a Hilbert space. Infinite membranes are of particular interest because they can be BPS states which solve the classical equations of motion of matrix theory. Extended compact membranes cannot be static solutions of the equations of motion since their membrane tension always causes them to contract and oscillate, as in the case of the spherical membrane.

The simplest infinite membrane is the flat planar membrane corresponding in IIA theory to an infinite D2-brane. This solution can be found by looking at the limit of the spherical membrane at large radius. It is simpler, however, to simply directly construct the solution by regularizing the flat membrane of M-theory. As in the other cases we have studied, we wish to quantize the Poisson bracket algebra of functions on the brane. Functions on the infinite membrane can be described in terms of two coordinates x_1, x_2 with a symplectic form $\omega_{ij} = \epsilon_{ij}$ giving a Poisson bracket

$$\{f(x_1, x_2), g(x_1, x_2)\} = \partial_1 f \partial_2 g - \partial_1 g \partial_2 f. \tag{4.19}$$

This algebra of functions can be "quantized" to the algebra of operators generated by Q, P satisfying

$$[Q, P] = \frac{i\epsilon^2}{2\pi} \mathbb{1} \tag{4.20}$$

where ϵ is a constant parameter. As usual in the quantization process there are operator-ordering ambiguities which must be resolved in determining a general map from functions expressed as polynomials in x_1, x_2 to operators expressed as polynomials of Q, P.

This gives a map from functions on $\mathbb{R}^2$ to operators which allows us to describe fluctuations around a flat membrane geometry with a single unit of $P^+ = 1/R$ in each region of area ϵ^2 on the membrane. Configurations of this type were discussed in the original BFSS paper [3] and their existence used as additional evidence for the validity of their conjecture. Note that this configuration only makes sense in the large N limit.

In addition to the flat membrane solution there are other infinite membranes which are static solutions of M-theory in flat space. In particular, there are BPS solutions corresponding to membranes which are holomorphically embedded in $\mathbb{C}^4 = \mathbb{R}^8 \subset \mathbb{R}^9$. These are static solutions of the membrane equations of motion. Finding a matrix theory description of such membranes is possible but involves some somewhat subtle issues related to choosing a regularization which preserves the complex structure of the brane. The details of this construction for a general holomorphic membrane are discussed in [45].

4.2.5. *Wrapped membranes as matrix strings*

So far we have discussed M-theory membranes which are unwrapped in the longitudinal direction and which therefore appear as D2-branes in the IIA language of matrix theory. It is also possible to describe wrapped M-theory membranes which correspond to strings in the IIA picture. The charge in matrix theory which measures the number of strings present is proportional to

$$\frac{i}{R}\mathrm{Tr}\left([X^i, X^j]\dot{X}^j + [[X^i, \theta^{\dot{\alpha}}], \theta^{\dot{\alpha}}]\right) \tag{4.21}$$

This result can be understood in several ways. It was found in [46] as a central charge in the matrix theory SUSY algebra corresponding to string charge; we will discuss this algebra further in the subsection 4.4. An intuitive way of understanding why (4.21) measures string charge is by a T-duality argument analogous to that used in 4.2.1 to derive the D2-brane charge of a system of D0-branes. If we compactify on a 2-torus in the i and j directions, the D0-branes become D2-branes and the bosonic part of (4.21) becomes

$$\frac{1}{R}F^{ij}F_{j0}. \tag{4.22}$$

This is the part of the energy-momentum tensor usually referred to as the Poynting vector in the 4D theory, and which describes momentum in the i direction. Such momentum is of course T-dual to string winding in the original picture, so we understand the identification of the original charge (4.21) as counting fundamental IIA strings corresponding to wound M-theory membranes. Configurations with nonzero values of this charge were considered by Imamura in [47].

To realize a classical configuration in matrix theory which contains fundamental strings it is clear from the form of the charge that we need to construct configuration with local membrane charge extended in a pair of directions X^i, X^j and to give the D0-branes velocity in the X^j direction. For example, we could consider an infinite planar membrane (as discussed in the previous subsection) sliding along itself according to the equation

$$X^1 = Q + t\mathbb{1} \tag{4.23}$$

$$X^2 = P \tag{4.24}$$

This corresponds to an M-theory membrane which has a projection onto the X^1, X^2 plane and which wraps around the compact direction as a periodic function of X^1 so that the IIA system contains a D2-brane with infinite strings extended in the X^2 direction since

$$\dot{X}^1[X^1, X^2] \sim \mathbb{1}. \tag{4.25}$$

Another example of a matrix theory system containing fundamental strings can be constructed by spinning the torus from (4.18) in the 12 plane to stabilize it. This gives the system some fundamental strings wrapped around the 34 circle. By taking the radius r to be very small we can construct a configuration of a single fundamental string wrapped in a circle of radius s. As $s \to \infty$ this becomes an infinite fundamental string.

It is interesting to note that there is no classical matrix theory solution corresponding to a classical string which is truly 1-dimensional and has no local membrane charge. This follows from the appearance of the commutator $[X^i, X^j]$ in the string charge, which vanishes unless the matrices describe a configuration with at least two dimensions of spatial extent. We can come very close to a 1-dimensional classical string configuration by considering a one-dimensional array of D0-branes at equal intervals on the X^1 axis

$$X^1 = a \begin{pmatrix} \ddots & \ddots & \ddots & & \\ \ddots & 1 & 0 & \ddots & \\ \ddots & 0 & 0 & 0 & \ddots \\ & \ddots & 0 & -1 & \ddots \\ & & \ddots & \ddots & \ddots \end{pmatrix} \tag{4.26}$$

We can now construct an excitation of the off-diagonal elements of X^2 corresponding to a string threading through the line of D0-branes

$$X^2 = b \begin{pmatrix} \ddots & \ddots & \ddots & & \\ \ddots & 0 & e^{i\omega t} & \ddots & \\ \ddots & e^{-i\omega t} & 0 & e^{i\omega t} & \ddots \\ & \ddots & e^{-i\omega t} & 0 & \ddots \\ & & \ddots & \ddots & \ddots \end{pmatrix} \tag{4.27}$$

where $\omega = a$. In the classical theory, this configuration can have arbitrary string charge. If the mode (4.27) is quantized then the string charge is quantized in the correct units. This string is almost 1-dimensional but has a small additional extent in the X^2 direction corresponding to the extra dimension of the M-theory membrane. From the M-theory point of view this extra dimension must appear because the membrane cannot have momentum in a direction parallel to its direction of extension since it has no internal degrees of freedom. Thus, the momentum in the compact direction represented by the D0-branes must appear on the membrane as a fluctuation in some transverse direction.

4.3. 5-BRANES

The M-theory 5-brane can appear in two possible guises in type IIA string theory. If the 5-brane is wrapped around the compact direction it becomes a D4-brane in the IIA theory, while if it is unwrapped it appears as an

NS 5-brane. We will refer to these two configurations as "longitudinal" and "transverse" 5-branes in matrix theory. We begin by discussing the transverse 5-brane.

A priori, one might think that it should be possible to see both types of 5-branes in matrix theory. Several calculations, however, indicate that the transverse 5-brane does not carry a conserved charge which can be described in terms of the matrix degrees of freedom. In principle, if this charge existed we would expect it to appear both in the supersymmetry algebra of matrix theory (discussed in the next subsection) and in the set of supergravity currents whose interactions are described by perturbative matrix theory calculations (discussed in section 5.1.2). In fact, no charge or current with the proper tensor structure for a transverse 5-brane appears in either of these calculations.

One way of understanding this apparent puzzle is by comparing to the situation for D-branes in light-front string theory [46]. Due to the Virasoro constraints, strings in the light-front formalism must have Neumann boundary conditions in both the light-front directions X^+, X^-. Thus, in light-front string theory there are no transverse D-branes which can be used as boundary conditions for the string. A similar situation holds for membranes in M-theory, which can end on M5-branes. The boundary conditions on the bosonic membrane fields which can be derived from the action (2.22) state that

$$(\bar{h}h^{ab}\partial_b X^i)\delta X^i = 0 \tag{4.28}$$

Combined with the Virasoro-type constraint

$$\partial_a X^- = \dot{X}^i \partial_a X^i \tag{4.29}$$

we find that, just as in the string theory case, membranes must have Neumann boundary conditions in the light-front directions.

These considerations would seem to lead to the conclusion that transverse 5-branes simply cannot be constructed in matrix theory. On the other hand, it was argued in [48] that there may be a way to construct a transverse 5-brane using S-duality, at least when the theory has been compactified on a 3-torus. To construct an infinite extended transverse 5-brane in this fashion would require performing an S-duality on (3+1)-dimensional $\mathcal{N} = 4$ supersymmetric Yang-Mills theory with gauge group $U(\infty)$, which is a poorly understood procedure to say the least. In [49], however, a finite size transverse 5-brane with geometry $T^3 \times S^2$ was constructed using S-duality of the four-dimensional $U(N)$ with finite N. Furthermore, it was shown that this object couples correctly to the supergravity fields even in the absence of an explicit transverse 5-brane charge. This seems to indicate that transverse 5-branes in matrix theory can be constructed locally, but that they are essentially solitonic objects and do not carry independent conserved quantum numbers. It would be nice to have a more explicit construction of a general class of such finite size transverse 5-branes, particularly in the noncompact version of matrix theory.

We now turn to the wrapped, or "longitudinal", M5-brane which we will refer to as the "L5-brane". This object appears as a D4-brane in the IIA theory. An infinite D4-brane was considered as a matrix theory background in [50] by including extra fields corresponding to strings stretching between the D0-branes of matrix theory and the background D4-brane. As in the case of the membrane, however, we would like to find a way to explicitly describe a dynamical L5-brane using the matrix degrees of freedom. Just as for the D2-brane, it may be surprising that a D4-brane can be constructed from a configuration of D0-branes. This can be seen from the same type of T-duality argument we used for the D2-brane in 4.2.1. By putting D4-branes and D0-branes on a torus T^4 we find that the charge-volume relation analogous to (4.2) for a D4-brane is [48, 6]

$$\mathrm{Tr}\ \epsilon_{ijkl} X^i X^j X^k X^l = \frac{V}{2\pi^2} \tag{4.30}$$

This is the T-dual of the instanton number in a 4D gauge theory which measures D0-brane charge on D4-branes.

Unlike the case of the membrane, there is no general theory describing an arbitrary L5-brane geometry in matrix theory language. In fact, the only L5-brane configurations which have been explicitly constructed to date are those corresponding to the highly symmetric geometries $S^4, \mathbb{C}P^2$ and $\mathbb{R}^4$. We now make a few brief comments about these configurations.

The L5-brane with isotropic S^4 geometry is similar in many ways to the membrane with S^2 geometry discussed in section 4.2.2. There are a number of unusual features of the S^4 system, however, which deserve mention. For full details of the construction see [51].

A rotationally invariant spherical L5-brane can only be constructed for those values of N which are of the form

$$N = \frac{(n+1)(n+2)(n+3)}{3} \tag{4.31}$$

where n is integral. For N of this form we define the configuration by

$$X_i = \frac{r}{n} G_i, \quad i \in \{1, \ldots, 5\}. \tag{4.32}$$

where G_i are the generators of the n-fold symmetric tensor product representation of the five four-dimensional Euclidean gamma matrices Γ_i satisfying $\Gamma_i \Gamma_j + \Gamma_j \Gamma_i = 2\delta_{ij}$

$$G_i^{(n)} = (\Gamma_i \otimes \mathbb{1} \otimes \cdots \otimes \mathbb{1} + \mathbb{1} \otimes \Gamma_i \otimes \mathbb{1} \otimes \cdots \otimes \mathbb{1} + \cdots + \mathbb{1} \otimes \cdots \otimes \mathbb{1} \otimes \Gamma_i)_{\mathrm{S}}$$

where the subscript S indicates that only the completely symmetric representation is used. For any n this configuration has the geometrical properties expected of n superimposed L5-branes contained in the locus of points describing a 4-sphere. As for the spherical membrane discussed in 4.2.2 the configuration is confined to the appropriate spherical locus

$$X_1^2 + X_2^2 + X_3^2 + X_4^2 + X_5^2 \approx r^2 \mathbb{1}. \tag{4.33}$$

The configuration is symmetric under $SO(5)$ and has the correct spectrum and the local D4-brane charge of n spherical branes. The energy and equations of motion of this system agree with those expected from M-theory.

Although the system can only be defined in a completely symmetric fashion for certain values of n, N, this does not seem like a fundamental issue. This constraint is a consequence of the imposition of exact rotational symmetry on the system. It may be that for large and arbitrary N it is possible to construct a very good approximation to a spherical L5-brane which breaks rotational invariance to a very small degree. A more fundamental problem, however, is that there is no obvious way of including small fluctuations of the membrane geometry around the perfectly isotropic sphere in a systematic way. In the case of the membrane, we know that for any particular geometry the fluctuations around that geometry can be encoded into matrices which form an arbitrarily good approximation to a smooth fluctuation through the procedure of replacing functions described in terms of an orthonormal basis by appropriate matrix analogues. In the case of the L5-brane we have no such procedure. In fact, there seems to be an obstacle to including all degrees of freedom corresponding to local fluctuations of the brane. It is natural to speculate by analogy with the membrane case that arbitrary fluctuations should be encoded in symmetric polynomials in the matrices G_i. It can be shown, however, that this is not possible. This geometry has been discussed in a related context in the noncommutative geometry literature [52] as a noncommutative version of S^4. There also, it was found that not all functions on the sphere could be consistently quantized.

As for the infinite membrane, the infinite L5-brane with geometry of a flat $\mathbb{R}^4 \subset \mathbb{R}^9$ can be viewed as a local limit of a large spherical geometry or it can be constructed directly. We need to find a set of operators X^{1-4} on some Hilbert space satisfying

$$\epsilon_{ijkl} X^i X^j X^k X^l = \frac{\epsilon^4}{2\pi^2} \mathbb{1}. \tag{4.34}$$

Such a configuration can be constructed using matrices which are tensor products of the form $\mathbb{1} \otimes Q, P$ and $Q, P \otimes \mathbb{1}$. This gives a "stack of D2-branes" solution with D2-brane charge as well as D4-brane charge [46]. It is also possible to construct a configuration with no D2-brane charge by identifying X^a with the components of the covariant derivative operator for an instanton on S^4

$$X^i = i\partial^i + A_i. \tag{4.35}$$

This construction is known as the Banks-Casher instanton [53]. Just as for the spherical L5-brane, it is not known how to construct small fluctuations of the membrane geometry around any of these flat solutions.

The only other known configuration of an L5-brane in matrix theory corresponds to a brane with geometry $\mathbb{C}P^2$. This configuration was constructed by Nair and Randjbar-Daemi as a particular example of a coset space G/H with $G = SU(3)$ and $H = U(2)$ [54]. They choose the matrices

$$X_i = \frac{rt_i}{\sqrt{N}} \tag{4.36}$$

where t_i are generators spanning $\mathbf{g}/\mathbf{h}$ in a particular representation of $SU(3)$. The geometry defined in this fashion seems to be in some ways better behaved than the S^4 geometry. For one thing, configurations of a single brane with arbitrarily large N can be constructed. Furthermore, it seems to be possible to include all local fluctuations as symmetric functions of the matrices t_i. This configuration is also somewhat confusing, however, as it extends in only four spatial dimensions, which makes the geometrical interpretation somewhat unclear.

Clearly there are many aspects of the L5-brane in matrix theory which are not understood. The principal outstanding problem is to find a systematic way of describing an arbitrary L5-brane geometry including its fluctuations. One approach to this might be to find a way of regularizing the world-volume theory of an M5-brane in a fashion similar to the matrix regularization of the supermembrane. It is also possible that understanding the structure of noncommutative 4-manifolds might help clarify this question. This is one of many places where noncommutative geometry seems to tie in closely with matrix theory. We will discuss other such connections with noncommutative geometry later in these lectures.

4.4. EXTENDED OBJECTS FROM MATRICES

We have seen that not only pointlike graviton states, but also objects extended in one, two, and four transverse directions can be constructed from matrix degrees of freedom. In this subsection we make some general comments about the appearance of these extended objects and their structure.

One systematic way of understanding the conserved charges associated with the longitudinal and transverse membrane and the longitudinal 5-brane in matrix theory arises from considering the supersymmetry algebra of the theory. The 11-dimensional supersymmetry algebra takes the form

$$\{Q_\alpha, Q_\beta\} \sim P^I(\gamma_I)_{\alpha\beta} + Z^{I_1 I_2}(\gamma_{I_1 I_2})_{\alpha\beta} + Z^{I_1\ldots I_5}(\gamma_{I_1\ldots I_5})_{\alpha\beta} \tag{4.37}$$

where the central terms Z correspond to 2-brane and 5-brane charges. The supersymmetry algebra of Matrix theory was explicitly computed by Banks, Seiberg and Shenker [46]. Similar calculations had been performed previously [17, 2]; however, in these earlier analyses terms such as $\mathrm{Tr}\,[X^i, X^j]$ and $\mathrm{Tr}\, X^{[i}X^j X^k X^{l]}$ were dropped since they vanish for finite N. The full supersymmetry algebra of the theory takes the schematic form

$$\{Q, Q\} \sim P^I + z^i + z^{ij} + z^{ijkl}, \tag{4.38}$$

as we would expect for the light-front supersymmetry algebra corresponding to (4.37). The charge

$$z^i \sim i\mathrm{Tr}\,\left(\{P^j, [X^i, X^j]\} + [[X^i, \theta^\alpha], \theta^\alpha]\right) \tag{4.39}$$

corresponds to longitudinal membranes (strings), the charge

$$z^{ij} \sim -i\mathrm{Tr}\,[X^i, X^j] \tag{4.40}$$

corresponds to transverse membranes and

$$z^{ijkl} \sim \mathrm{Tr}\, X^{[i}X^j X^k X^{l]} \tag{4.41}$$

corresponds to longitudinal 5-brane charge. For all the extended objects we have described in the preceding subsections, these results agree with the charges we motivated by T-duality arguments.

Note that the charges of all the extended objects in the theory vanish when the matrix size N is finite. Physically, this corresponds to the fact that any finite-size configuration of strings, 2-branes and 4-branes must have net charges which vanish.

Another approach to understanding the charges associated with the extended objects of matrix theory arises from the study of the coupling of these objects to supergravity fields, which we will discuss in the next section. From this point of view, perturbative matrix theory calculations can be used to determine not only the conserved charges of the theory, but also the higher multipole moments of all the components of the supercurrent describing the matrix configuration. For example [55, 56], the multipole moments of the membrane charge $z^{ij} = -2\pi i\mathrm{Tr}\,[X^i, X^j]$ can be written in terms of the matrix moments

$$z^{ij(k_1\cdots k_n)} = -2\pi i\;\mathrm{STr}\,\left([X^i, X^j]X^{k_1}\cdots X^{k_n}\right) \tag{4.42}$$

which are the matrix analogues of the moments

$$\int d^2\sigma \, \{X^i, X^j\} X^{k_1} \cdots X^{k_n} \tag{4.43}$$

for the continuous membrane. The symbol STr indicates a symmetrized trace, wherein the trace is averaged over all possible orderings of the terms $[X^i, X^j]$ and X^{k_ν} appearing inside the trace. This corresponds to a particular ordering prescription in applying the matrix-membrane correspondence to (4.43). There is no *a priori* justification for this ordering prescription, but it is a consequence of explicit calculations of interactions between general matrix theory objects as described in the next section. The same prescription can be used to define the multipole moments of the longitudinal membrane and 5-brane charges.

Although as we have mentioned, the conserved charges in matrix theory corresponding to extended objects all vanish at finite N, the same is not true of the higher moments of these charges. For example, the isotropic spherical matrix membrane configuration discussed in section 4.2.2 has nonvanishing membrane dipole moments

$$\begin{aligned} z^{12(3)} = z^{23(1)} = z^{31(2)} &= -2\pi i \mathrm{Tr}\left([X^1, X^2] X^3\right) \\ &= \frac{4\pi r^3}{3}(1 - 1/N^2) \end{aligned} \tag{4.44}$$

which agrees with the membrane dipole moment $4\pi r^3/3$ of the smooth spherical membrane up to terms of order $1/N^2$. Using the multipole moments of a fixed matrix configuration we can essentially reproduce the complete spatial dependence of the matter configuration to which the matrices correspond. This higher moment structure describing higher-dimensional extended objects through lower-dimensional objects is very general, and has a precise analog in describing the supercurrents and charges of Dirichlet $(p + 2k)$-branes in terms of the world-volume theory of a system of Dp-branes [49]. This structure has many possible applications to D-brane physics as well as to matrix theory. For example, it was recently pointed out by Myers [57] that putting a system of Dp-branes in a constant background $(p+4)$-form flux will produce a dielectric effect in which spherical bubbles of D$(p+2)$-branes will be formed with dipole moments which screen the background field.

5. Interactions in matrix theory

In this section we discuss interactions in matrix theory between block matrices describing general time-dependent matrix theory configurations which may include gravitons, membranes and 5-branes. We begin by reviewing the perturbative Yang-Mills formalism in background field gauge. This formalism can be used to carry out loop calculations in matrix theory, giving results which can be related to supergravity interactions. We carry out two explicit examples of this calculation at one-loop order: first for a pair of 0-branes with relative velocity v, following [30], then for the leading order term in the interaction between an arbitrary pair of bosonic background configurations, following [56]. Following these examples, we summarize the extent to which perturbative Yang-Mills calculations of this kind have been shown to agree with classical supergravity. At the level of linearized supergravity, it has been found that there is an infinite series of terms in the one-loop matrix theory effective potential which precisely reproduce all tree-level supergravity interactions arising from the exchange of a single graviton, 3-form quantum or gravitino. There is limited information about the extent to which nonlinear supergravity effects are reproduced by higher-loop matrix theory calculations, however. While it has been shown that the nonlinear structure of 3-graviton scattering is correctly reproduced by a two-loop matrix theory calculation, there is not a clear picture of what should be expected beyond this. We discuss these results and how they are related to supersymmetric nonrenormalization theorems which protect some terms in the perturbative Yang-Mills expansion from higher-loop corrections.

In this section we primarily focus on the problem of deriving classical 11-dimensional supergravity from matrix theory. A very interesting, but more difficult, question is whether matrix theory can also successfully reproduce string/M-theory corrections to classical supergravity. The first such corrections would be $\mathcal{R}^4$ corrections to the Einstein-Hilbert action. Some work has been done investigating the question of whether these terms can be seen in matrix theory [58, 59, 60, 61, 62, 63]. While more work needs to be done in this direction, the results of [61, 62, 63] indicate that the perturbative loop expansion in matrix theory probably does not correctly reproduce quantum effects in M-theory. The most likely explanation for this discrepancy is that such terms are not subject to nonrenormalization theorems, and are only reproduced in the large N limit. We discuss these issues again briefly in the last section.

In subsection 5.1 we describe two-body interactions in matrix theory, and in subsection 5.2 we discuss interactions between more than two objects. Section 5.3 contains a brief discussion of interactions involving longitudinal momentum transfer, which correspond to nonperturbative processes in matrix theory.

The background field formalism [64] for describing matrix theory interactions between block matrices which are widely separated in eigenvalue space was first used by Douglas, Kabat, Pouliot and Shenker in [30] to describe interactions between a pair of D0-branes in type IIA string theory moving with relative velocity v. In this subsection we discuss their result and the generalization to general bosonic background configurations. The matrix theory Lagrangian is

$$\mathcal{L} = \frac{1}{2R}\mathrm{Tr}\left[D_0 X^i\, D_0 X^i + \frac{1}{2}[X^i, X^j]^2 + \theta^T(i\dot{\theta} - \gamma_i[X^i, \theta])\right] \tag{5.1}$$

where

$$D_0 X^i = \partial_t X^i - i[A, X^i]. \tag{5.2}$$

We wish to expand each of the matrix theory fields around a classical background. We will assume here for simplicity that the background has a vanishing gauge field and vanishing fermionic fields. For a discussion of the general situation with background fermionic fields as well as bosonic fields see [65]. We expand the bosonic field in terms of a background plus a fluctuation

$$X^i = B^i + Y^i.$$

We choose the background field gauge

$$D_\mu^{\mathrm{bg}} A^\mu = \partial_t A - i[B^i, X^i] = 0. \tag{5.3}$$

This gauge can be implemented by adding a term $-(D_\mu^{\mathrm{bg}} A^\mu)^2$ to the action and including the appropriate ghosts. The nice feature of this gauge is that the terms quadratic in the bosonic fluctuations simplify to the form

$$\dot{Y}^i \dot{Y}^i - [B^i, Y^j]^2 - [B^i, B^j][Y^i, Y^j] \tag{5.4}$$

The complete gauge-fixed action including ghosts is written in Euclidean time $\tau = it$ as

$$S = S_0 + S_2 + S_3 + S_4 \tag{5.5}$$

where

$$\begin{aligned} S_0 &= \frac{1}{2R}\int d\tau \mathrm{Tr}\left[\partial_\tau B^i \partial_\tau B^i + \frac{1}{2}[B^i, B^j]^2\right] \\ S_2 &= \frac{1}{2R}\int d\tau \mathrm{Tr}\Big[\partial_\tau Y^i \partial_\tau Y^i - [B^i, Y^j][B^i, Y^j] - [B^i, B^j][Y^i, Y^j] \\ &\quad + \partial_\tau A \partial_\tau A - [B^i, A][B^i, A] - 2i\dot{B}^i[A, Y^i] \\ &\quad + \partial_\tau \bar{C}\partial_\tau C - [B^i, \bar{C}][B^i, C] + \theta^T\dot{\theta} - \theta^T\gamma_i[B^i, \theta]\Big] \end{aligned} \tag{5.6}$$

and where S_3 and S_4 contain terms cubic and quartic in the fluctuations Y^i, A, C, θ. Note that we have taken $A \to -iA$ as appropriate for the Euclidean formulation.

We now wish to use this gauge-fixed action to compute the effective potential governing the interaction between a pair of matrix theory objects. In general, to calculate the interaction potential to arbitrary order it is necessary to include the terms S_3 and S_4 in the action. The propagators for each of the fields can be computed from the quadratic term S_2. A systematic diagrammatic expansion will then yield the effective potential to arbitrary high order. We begin our discussion of matrix theory interactions, however, with the simplest case: the interaction of two objects at leading order in the inverse separation distance. In 5.1.1 we discuss the simplest case of this situation, the scattering of a pair of gravitons. In 5.1.2 we discuss the situation of two general matrix theory objects, giving an explicit calculation for the leading term in the case where both objects are purely bosonic. After working out these explicit examples we review what is known about the scattering of a general pair of matrix theory objects to arbitrary order in section 5.1.3. We review the special case of a pair of gravitons in section 5.1.4. We discuss the N-body problem in 5.2.

5.1.1. *Two graviton interactions at leading order*

As we have discussed in 4.1.1, a classical background describing a pair of gravitons with relative velocity v and impact parameter b (and no polarization information) is given in the center of mass frame by

$$B^1 = \frac{-i}{2}\begin{pmatrix} v\tau & 0 \\ 0 & -v\tau \end{pmatrix} \tag{5.7}$$

$$B^2 = \frac{1}{2}\begin{pmatrix} b & 0 \\ 0 & -b \end{pmatrix} \tag{5.8}$$

$$B^i = 0, \qquad i > 2 \tag{5.9}$$

Inserting these backgrounds into (5.6) we see that at a fixed value of time the Lagrangian at quadratic order for the 10 complex bosonic off-diagonal components of A and Y^i is that of a system of 10 harmonic oscillators with mass matrix

$$(\Omega_b)^2 = \begin{pmatrix} r^2 & -2iv & 0 & \cdots & 0 \\ 2iv & r^2 & 0 & \ddots & 0 \\ 0 & 0 & r^2 & \ddots & \vdots \\ \vdots & \ddots & \ddots & \ddots & 0 \\ 0 & 0 & \cdots & 0 & r^2 \end{pmatrix} \tag{5.10}$$

where $r^2 = b^2 + (vt)^2$ is the instantaneous separation between the gravitons.

There are two complex off-diagonal ghosts with $\Omega^2 = r^2$.

There are 16 fermionic oscillators with a mass-squared matrix

$$(\Omega_f)^2 = r^2 \mathbb{1}_{16\times 16} + v\gamma_1 \tag{5.11}$$

This matrix can be found by writing

$$P^\dagger P = -\partial^2 + (\Omega_f)^2 \tag{5.12}$$

where

$$P = \partial - v\tau\gamma_1 - b\gamma_2 \tag{5.13}$$

To perform a completely general calculation of the two-body effective interaction potential to all orders in $1/r$ it is necessary to perform a diagrammatic expansion using the exact propagator for the bosonic and fermionic fields. For example, the bosonic propagator satisfying

$$(-\partial^2 + b^2 + v^2\tau^2)\Delta_B(\tau, \tau'|b^2 + v^2\tau^2) = \delta(\tau - \tau') \tag{5.14}$$

is given by the expression [66]

$$\begin{aligned} \Delta_B(\tau, \tau'|b^2 + v^2\tau^2) &= \int_0^\infty ds\, e^{-b^2 s} \sqrt{\frac{v}{2\pi \sinh 2sv}} \times \\ & \exp\left(-\frac{v}{2\sinh 2sv}((\tau^2 + \tau'^2)\cosh 2sv - 2\tau\tau')\right). \end{aligned} \tag{5.15}$$

In general, even for a simple 2-graviton calculation there is a fair amount of algebra involved in extracting the effective potential using propagators of the form (5.15). If, however, we are only interested in calculating the leading term in the long-range interaction potential we can simplify the calculation by making the quasi-static assumption that all the oscillator frequencies ω of interest are large compared to the ratio v/r of the relative velocity divided by the separation scale. In this regime, we can make the approximation that all the oscillators stay in their ground state over the interaction time-scale, so that the effective potential between the two objects is simply given by the sum of the ground-state energies of the boson, ghost and fermion oscillators

$$V_{\rm qs} = \sum_b \omega_b - \sum_g \omega_g - \frac{1}{2}\sum_f \omega_f. \tag{5.16}$$

Note that the bosonic and ghost oscillators are complex so that no factor of $1/2$ is included.

In the situation of two-graviton scattering we can therefore calculate the effective potential by diagonalizing the frequency matrices Ω_b, Ω_g and Ω_f. We find that the bosonic oscillators have frequencies

$$\begin{aligned} \omega_b &= r \quad \text{with multiplicity 8} \\ \omega_b &= \sqrt{r^2 \pm 2v} \quad \text{with multiplicity 1 each.} \end{aligned}$$

The 2 ghosts have frequencies

$$\omega_g = r, \tag{5.17}$$

and the 16 fermions have frequencies

$$\omega_f = \sqrt{r^2 \pm v} \quad \text{with multiplicity 8 each.} \tag{5.18}$$

The effective potential for a two-graviton system with instantaneous relative velocity v and separation r is thus given by the leading term in a $1/r$ expansion of the expression

$$V = \sqrt{r^2 + 2v} + \sqrt{r^2 - 2v} + 6r - 4\sqrt{r^2 + v} + 4\sqrt{r^2 - v}. \tag{5.19}$$

Expanding in v/r^2 we see that the terms of order $r, v/r, v^2/r^3$ and v^3/r^5 all cancel. The leading term is

$$V = -\frac{15}{16}\frac{v^4}{r^7} + \mathcal{O}(\frac{v^6}{r^{11}}) \tag{5.20}$$

This potential was first computed by Douglas, Kabat, Pouliot and Shenker [30]. This result agrees with the leading term in the effective potential between two gravitons with $P^+ = 1/R$ in light-front 11D supergravity. We will discuss the supergravity side of this calculation in more detail in the following section, where we generalize this calculation to an arbitrary pair of matrix theory objects.

5.1.2. *General 2-body systems and linearized supergravity at leading order*
We now generalize the discussion to an arbitrary pair of matrix theory objects, which are described by a block-diagonal background

$$B^i = \begin{pmatrix} \hat{X}^i & 0 \\ 0 & \tilde{X}^i \end{pmatrix} \tag{5.21}$$

where $\hat{X}^i$ and $\tilde{X}^i$ are $\hat{N} \times \hat{N}$ and $\tilde{N} \times \tilde{N}$ matrices describing the two objects. The separation distance between the objects, which we will use as an expansion parameter, is given by

$$r^i = \frac{1}{\hat{N}}\text{Tr}\,\hat{X}^i - \frac{1}{\tilde{N}}\text{Tr}\,\tilde{X}^i \tag{5.22}$$

There are $\hat{N}\tilde{N}$ independent complex off-diagonal components of the fluctuation matrices Y^i. We will find it useful to treat these components as an

$\hat{N}\tilde{N}$-component vector Z^i. We now construct a $\hat{N}\tilde{N} \times \hat{N}\tilde{N}$ matrix which acts on the Z^i vectors

$$K_i \equiv \hat{X}_i \otimes \mathbb{1}_{\tilde{N}\times\tilde{N}} - \mathbb{1}_{N\times N} \otimes \tilde{X}_i^T . \tag{5.23}$$

It is convenient to extract the centers of mass explicitly so that K^i can be rewritten as

$$K^i = r^i \mathbb{1} + \bar{K}^i \tag{5.24}$$

where $\bar{K}^i$ is of order 1 in terms of the separation scale r. The matrices K encode the adjoint action of the background B on the fluctuations Y so that we can extract the part of $[B, Y]$ depending on the off-diagonal fields Z through

$$[B^i, Y^j] \to K^i Z^j. \tag{5.25}$$

This formalism allows us to write the quadratic terms from (5.6) in the action for the off-diagonal fields in a simple form

$$\begin{aligned} &\dot{Y}^i\dot{Y}^i - [B^i, Y^j][B^i, Y^j] - [B^i, B^j][Y^i, Y^j] \\ &\quad \to \dot{Z}_i^\dagger \dot{Z}^i - Z_j^\dagger K^i K_i Z^j - 2Z_i^\dagger [K_i, K_j] Z^j \end{aligned} \tag{5.26}$$

Performing a similar operation for the terms quadratic in fluctuations of the A field, we find that the full frequency-squared matrices for the bosonic, ghost and fermionic fields can be written

$$\begin{aligned} \Omega_b^2 &= K^2 \mathbb{1}_{10\times 10} - 2iF_{\mu\nu} \\ \Omega_g^2 &= K^2 \mathbb{1}_{2\times 2} \\ \Omega_f^2 &= K^2 \mathbb{1}_{16\times 16} - iF_{\mu\nu}\gamma^\mu\gamma^\nu \end{aligned} \tag{5.27}$$

where $\gamma^0 = \mathbb{1}$ and the field strength matrix $F_{\mu\nu}$ is given by

$$\begin{aligned} F_{0i} &= \dot{K}^i \\ F_{ij} &= i[K^i, K^j] \end{aligned} \tag{5.28}$$

Note that each of the frequencies has a leading term r and subleading terms of order 1. Expanding the frequency matrices in powers of $1/r$ we find that for a completely arbitrary pair of objects described by the background matrices $\hat{X}^i$ and $\tilde{X}^i$ the potential vanishes to order $1/r^6$. At order $1/r^7$ we find that the potential is

$$V_{\text{leading}} = \text{Tr}\,(\Omega_b) - \frac{1}{2}\text{Tr}\,(\Omega_f) - 2\text{Tr}\,(\Omega_g) \tag{5.29}$$

$$= -\frac{5}{128 r^7}\text{STr}\,\mathcal{F} \tag{5.30}$$

where

$$\mathcal{F} = 24F^{\mu}{}_{\nu}F^{\nu}{}_{\lambda}F^{\lambda}{}_{\sigma}F^{\sigma}{}_{\mu} - 6F_{\mu\nu}F^{\mu\nu}F_{\lambda\sigma}F^{\lambda\sigma} \tag{5.31}$$

and STr indicates that the trace is symmetrized over all possible orderings of F's in the product F^4.

From the definition (5.23) it is clear that the field strength $F_{\mu\nu}$ decomposes into a piece from each of the two objects

$$F_{\mu\nu} = \hat{F}_{\mu\nu} - \tilde{F}_{\mu\nu} \tag{5.32}$$

where $\hat{F}_{\mu\nu}$ and $\tilde{F}_{\mu\nu}$ are defined through (5.28) in terms of $\hat{X}$ and $\tilde{X}$. We can therefore decompose the potential V_{leading} into a sum of terms which are written as products of a function of $\hat{X}$ and a function of $\tilde{X}$, where the terms can be grouped according to the number of Lorentz indices contracted between the two objects. With some algebra, we can write this potential as

$$V_{\text{leading}} = V_{\text{gravity}} + V_{\text{electric}} + V_{\text{magnetic}} \tag{5.33}$$

$$V_{\text{gravity}} = -\frac{15R^2}{4r^7}\left(\hat{\mathcal{T}}^{IJ}\tilde{\mathcal{T}}_{IJ} - \frac{1}{9}\hat{\mathcal{T}}^{I}{}_{I}\tilde{\mathcal{T}}^{J}{}_{J}\right) \tag{5.34}$$

$$V_{\text{electric}} = -\frac{45R^2}{r^7}\hat{\mathcal{J}}^{IJK}\tilde{\mathcal{J}}_{IJK} \tag{5.35}$$

$$V_{\text{magnetic}} = -\frac{45R^2}{r^7}\hat{\mathcal{M}}^{+-ijkl}\tilde{\mathcal{M}}^{-+ijkl} \tag{5.36}$$

This is, as we shall discuss shortly, precisely the form of the interactions we expect to see from 11D supergravity in light-front coordinates, where $\mathcal{T}, \mathcal{J}$ and $\mathcal{M}$ play the role of the (integrated) stress tensor, membrane current and 5-brane current of the two objects. The quantities appearing in this decomposition are defined as follows.

$\mathcal{T}^{IJ}$ is a symmetric tensor with components

$$\begin{aligned}
\mathcal{T}^{--} &= \frac{1}{R}\,\text{STr}\,\frac{\mathcal{F}}{96} \\
\mathcal{T}^{-i} &= \frac{1}{R}\,\text{STr}\left(\frac{1}{2}\dot{X}^i\dot{X}^j\dot{X}^j + \frac{1}{4}\dot{X}^iF^{jk}F^{jk} + F^{ij}F^{jk}\dot{X}^k\right) \\
\mathcal{T}^{+-} &= \frac{1}{R}\,\text{STr}\left(\frac{1}{2}\dot{X}^i\dot{X}^i + \frac{1}{4}F^{ij}F^{ij}\right) \\
\mathcal{T}^{ij} &= \frac{1}{R}\,\text{STr}\left(\dot{X}^i\dot{X}^j + F^{ik}F^{kj}\right) \\
\mathcal{T}^{+i} &= \frac{1}{R}\,\text{STr}\,\dot{X}^i \\
\mathcal{T}^{++} &= \frac{N}{R}
\end{aligned} \tag{5.37}$$

$\mathcal{J}^{IJK}$ is a totally antisymmetric tensor with components

$$\begin{aligned}
\mathcal{J}^{-ij} &= \frac{1}{6R}\mathrm{STr}\left(\dot{X}^i\dot{X}^kF^{kj} - \dot{X}^j\dot{X}^kF^{ki} - \frac{1}{2}\dot{X}^k\dot{X}^kF^{ij}\right. \qquad (5.38)\\
&\qquad \left.+\frac{1}{4}F^{ij}F^{kl}F^{kl} + F^{ik}F^{kl}F^{lj}\right)\\
\mathcal{J}^{+-i} &= \frac{1}{6R}\mathrm{STr}\left(F^{ij}\dot{X}^j\right)\\
\mathcal{J}^{ijk} &= -\frac{1}{6R}\mathrm{STr}\left(\dot{X}^iF^{jk} + \dot{X}^jF^{ki} + \dot{X}^kF^{ij}\right)\\
\mathcal{J}^{+ij} &= -\frac{1}{6R}\mathrm{STr}\,F^{ij}
\end{aligned}$$

Note that we retain some quantities — in particular $\mathcal{J}^{+-i}$ and $\mathcal{J}^{+ij}$ — which vanish at finite N (by the Gauss constraint and antisymmetry of F^{ij}, respectively). These terms represent BPS charges (for longitudinal and transverse membranes) which are only present in the large N limit. We define higher moments of these terms below which can be non-vanishing at finite N.

$\mathcal{M}^{IJKLMN}$ is a totally antisymmetric tensor with

$$\mathcal{M}^{+-ijkl} = \frac{1}{12R}\mathrm{STr}\left(F^{ij}F^{kl} + F^{ik}F^{lj} + F^{il}F^{jk}\right). \qquad (5.39)$$

At finite N this vanishes by the Jacobi identity, but we shall retain it because it represents the charge of a longitudinal 5-brane. The other components of $\mathcal{M}^{IJKLMN}$ do not appear in the Matrix potential. In principle, we expect another component of the 5-brane current, $\mathcal{M}^{-ijklm}$, to be well-defined. This term arises from a moving longitudinal 5-brane. This term does not appear in the 2-body interaction formula because it would couple to the transverse 5-brane charge $\mathcal{M}^{+ijklm}$ which, as we have discussed, vanishes in light-front coordinates. The component $\mathcal{M}^{-ijklm}$ can, however, be determined from the conservation of the 5-brane current, and is given by [67]

$$\mathcal{M}^{-ijklm} = \frac{5}{4R}\mathrm{STr}\left(\dot{X}^{[i}F^{jk}F^{lm]}\right). \qquad (5.40)$$

Let us compare the interaction potential (5.33) with the leading long-range interaction between two objects in 11D light-front compactified supergravity. The scalar propagator in 11D is

$$\Box^{-1}(x) = \frac{1}{2\pi R}\sum_n \int \frac{dk^- d^9k_\perp}{(2\pi)^{10}} \frac{e^{-i\frac{n}{R}x^- - ik^-x^+ + ik_\perp\cdot x_\perp}}{2\frac{n}{R}k^- - k_\perp^2} \qquad (5.41)$$

where n counts the number of units of longitudinal momentum k^+. To compare the leading term in the long-distance potential with matrix theory we only need to extract the term associated with $n = 0$, corresponding to interactions mediated by exchange of a supergraviton with no longitudinal momentum.

$$\Box^{-1}(x-y) = \frac{1}{2\pi R}\delta(x^+ - y^+)\frac{-15}{32\pi^4|x_\perp - y_\perp|^7} \tag{5.42}$$

Note that the exchange of quanta with zero longitudinal momentum gives rise to interactions that are instantaneous in light-front time, as recently emphasized in [68]. This is precisely the type of instantaneous interaction that we find at one loop in Matrix theory. Such action-at-a-distance potentials are allowed by the Galilean invariance manifest in the light-front formalism.

The graviton propagator can be written in terms of this scalar propagator as

$$D^{IJ,KL}_{\text{graviton}} = 2\kappa^2\left(\eta^{IK}\eta^{JL} + \eta^{IL}\eta^{JK} - \frac{2}{9}\eta^{IJ}\eta^{KL}\right)\Box^{-1}(x-y) \tag{5.43}$$

where $2\kappa^2 = (2\pi)^5 R^3$ in string units. The effective supergravity interaction between two objects having stress tensors $\hat{T}_{IJ}$ and $\tilde{T}_{KL}$ can then be expressed as

$$S = -\frac{1}{4}\int d^{11}x d^{11}y\, \hat{T}_{IJ}(x) D^{IJ,KL}_{\text{graviton}}(x-y)\tilde{T}_{KL}(y) \tag{5.44}$$

This interaction has a leading term of precisely the form (5.34) if we define $\mathcal{T}^{IJ}$ to be the integrated component of the stress tensor

$$\mathcal{T}^{IJ} \equiv \int dx^- d^9x_\perp T^{IJ}(x). \tag{5.45}$$

It is straightforward to show in a similar fashion that (5.35) and (5.36) are precisely the forms of the leading supergravity interaction between membrane currents and 5-brane currents of a pair of objects.

We can calculate the components of the source currents (5.37), (5.38) and (5.39) for all the matrix theory objects we have discussed: the supergraviton, the membrane and the L5-brane. For all these objects the currents have the expected values, at least to order $1/N^2$. For example, the stress tensor of a graviton can be written in the form

$$\mathcal{T}^{IJ} = \frac{p^I p^J}{p^+} \tag{5.46}$$

where

$$p^+ = N/R, \quad p^i = p^+\dot{x}^i, \quad p^- = p_\perp^2/2p^+ \tag{5.47}$$

The stress tensor and membrane current of the membrane can be computed in the continuum membrane theory from the action (2.56) for the bosonic membrane in a general background. Using the matrix-membrane correspondence (2.51) it is possible to show that the matrix definitions above are compatible with the expressions for the stress tensor and membrane current of the continuum membrane, although the matrix expressions are not uniquely determined by this correspondence.

We have thus shown that to leading order in the separation distance the interaction between any pair of objects in supergravity is precisely reproduced by one-loop quantum effects in matrix theory. We have only shown this explicitly in the case of a pair of bosonic backgrounds, following [56]. The more general case where fermionic background fields are included is discussed in [65]. In the following sections we discuss what is known about the extension of these results to higher order in $1/r$ and to interactions of more than two distinct objects.

5.1.3. *General 2-body interactions*

In the previous subsections we have considered only the leading $1/r^7$ terms in the 2-body interaction potential. If we consider all possible Feynman diagrams which might contribute to higher-order terms, it is straightforward to demonstrate by power counting that the complete potential can be written as a sum of terms of the form

$$V = \sum_{n,k,l,m,p} V_{n,k,l,m,p,\alpha} R^{n-1} \frac{X^l D^p F^k \psi^{2m}}{r^{3n+2k+l+3m+p-4}}. \tag{5.48}$$

where n counts the number of loops in the relevant diagrams and ψ describes the fermionic background fields. Each D either indicates a time derivative or a commutator with an X, as in $\psi[X, \psi]$. The summation over the index α indicates a sum over many possible index contractions for every combination of F's, X's and D's and Γ matrices between the ψ's.

For a completely general pair of objects, only terms in the one-loop effective action have been understood in terms of supergravity. At one-loop order, when the fields are taken on-shell by imposing the matrix theory equations of motion, all terms with $k+m+p < 4$ which have been calculated vanish. All terms with $k+m+p = 4$ which have been calculated have $m \geq p$ and can be written in the form

$$V_{1,4-m-p,l,m,p,\alpha} \frac{X^l F^{(4-m-p)} \psi^{2(m-p)} (\psi D \psi)^p}{r^{7+m-p+l}}. \tag{5.49}$$

In this expression, the grouping of ψ terms indicates the contraction of spinor indices—in general, the terms can be ordered in an arbitrary fashion when considered as $U(N)$ matrices. The terms (5.49) have been explicitly determined for $m < 2$ in [56, 65], where they were shown to precisely correspond to multipole interaction terms in linearized supergravity. We now briefly describe some of those terms which have been interpreted in this fashion

$m = p = 0,\ k = 4,\ l = 0$: These are the leading $1/r^7$ terms in the interaction potential between a pair of purely bosonic objects discussed above. They are precisely equivalent to the leading term in the supergravity potential between a pair of objects with appropriate integrated stress tensors, membrane currents and 5-brane currents.

$m = p = 0,\ k = 4,\ l > 0$: This infinite set of terms was shown in [56] to be equivalent to the higher-order terms in the linearized supergravity potential arising from higher moments of the bosonic parts of the stress tensor, membrane current and 5-brane current. The simplest example (discussed in [55]) is the term of the form $F^4 X/r^8$ which appears in the case of a graviton moving in the long-range gravitational field of a matrix theory object with angular momentum

$$J^{ij} = T^{+i(j)} - T^{+j(i)} \tag{5.50}$$

where the first moment of the matrix theory stress tensor component T^{+i} is defined through (as discussed in subsection 4.4)

$$T^{+i(j)} = \frac{1}{R}\mathrm{Tr}\,\left(\dot{X}^i X^j\right) \tag{5.51}$$

In [65] it was shown that terms of the general form $F^4 X^l/r^{7+l}$ can describe higher-moment membrane-5-brane and D0-brane-D6-brane interactions as well as membrane-membrane and 5-brane-5-brane interactions, generalizing previous results in [50, 69].

$m = 1,\ p = 1,\ k = 2,\ l \geq 0$: The terms of the form

$$\frac{F^2(\psi D\psi)X^l}{r^{7+l}} \tag{5.52}$$

correspond again to leading and higher-moment interactions in linearized supergravity, where now the components of the (integrated) gravity currents have contributions from the fermionic backgrounds as well as the bosonic backgrounds. These terms are also related to linearized supercurrent interactions arising from single gravitino exchange, as discussed in [70, 65].

$m = 2,\ p = 2,\ k = 0,\ l \geq 0$: The terms of the form

$$\frac{(\psi D\psi)(\psi D\psi)X^l}{r^{7+l}} \tag{5.53}$$

correspond, just like the terms (5.52), to fermionic contributions to the linearized supergravity interaction arising from fermion contributions to the integrated supergravity currents.

$m = 1,\ p = 0,\ k = 3,\ l \geq 0$: The terms of the form

$$\frac{F^3 \psi\psi X^l}{r^{8+l}} \tag{5.54}$$

have a similar interpretation to the terms (5.52). In these terms, however, the dipole moments of the currents have nontrivial fermionic contributions in which no derivatives act on the fermions [65]. The simplest example of this is the spin contribution to the matrix theory angular momentum

$$J^{ij}_{\text{fermion}} = \frac{1}{4R} \text{Tr}\left(\psi \gamma^{ij} \psi\right) \tag{5.55}$$

This contribution was first noted in the context of spinning gravitons in [71]. This angular momentum term couples to the component $T^{-i} \sim F^3$ of the matrix theory stress-energy tensor through terms of the form $\hat{J}^{ij}\tilde{T}^{-i}r^j/r^9$.

$m > 1,\ k = 4 - m,\ l \geq 0$: The terms of the form

$$\frac{F^2\psi^4 X^l}{r^{9+l}}, \quad \frac{F\psi^6 X^l}{r^{10+l}}, \quad \frac{\psi^8 X^l}{r^{11+l}}, \quad \frac{(\psi D\psi)F\psi^2 X^l}{r^{8+l}}, \quad \frac{(\psi D\psi)\psi^4 X^l}{r^{9+l}} \tag{5.56}$$

have not been completely calculated or related to supergravity interactions, although as we will discuss in the following section these terms are known and agree with supergravity interactions in the special case $N = 2$. From the structure which has already been understood it seems most likely that these terms arise from fermion contributions to the higher multipole moments of the supergravity currents, and that these terms will also agree with the corresponding supergravity interactions.

This is all that is known about the 2-body interaction for a completely general (and not necessarily supersymmetric) pair of matrix theory objects. To summarize, it has been shown that all terms of the form $F^k(\psi D\psi)^p\psi^{2(m-p)}$ with $k + p + m = 4$ correspond to supergravity interactions, at least for the terms with $m < 2$. It seems likely that this correspondence persists for the remaining values of $m > 1$, but the higher order fermionic contributions to the multipole moments of the supergravity currents have not yet been calculated for a general matrix theory object. It is likely that all these terms are protected by a supersymmetric nonrenormalization theorem of the type discussed in the following section. This has not yet been proven, but might follow from arguments similar to those in

[72]. The only other known results are for a pair of gravitons, which we now review.

5.1.4. *General two-graviton interactions*

In the case of a pair of gravitons, the general interaction potential (5.48) simplifies to

$$V = \sum_{n=0}^{\infty}\sum_{k=0}^{\infty}\sum_{m=0}^{4} V_{n,k,m} R^{n-1} \frac{v^k \psi^{2m}}{r^{3n+2k+3m-4}} \tag{5.57}$$

The sum over m is finite since in the $SU(2)$ theory all terms with fermions can be described in terms of a product of 2, 4, 6 or 8 ψ's. The leading terms for each value of m have been computed using the one-loop approach, and agree with supergravity. The sum of these terms is (see [73] and references therein for further details)

$$\begin{aligned} V_{(1)} \;=\; -\frac{15}{16}\Big[& v^4 + 2v^2 v_i D^{ij}\partial_j + 2v_i v_j D^{ik} D^{jl}\partial_k\partial_l \qquad (5.58)\\ &+\frac{4}{9} v_i D^{ij} D^{km} D^{lm}\partial_j\partial_k\partial_l + \frac{2}{63} D^{in} D^{jn} D^{km} D^{lm}\partial_i\partial_j\partial_k\partial_l\Big]\frac{1}{r^7} \end{aligned}$$

where

$$D^{ij} = \psi\gamma^{ij}\psi \tag{5.59}$$

The term with a single D proportional to $1/r^8$ arises from the spin angular momentum term described in (5.55).

No further checks have been made on the matrix theory/supergravity correspondence for terms with nontrivial fermion backgrounds. Simplifying to the spinless case, the complete effective potential (5.57) simplifies still further to

$$V = \sum_{n,k} V_{n,k} R^{n-1} \frac{v^k}{r^{3n+2k-4}}. \tag{5.60}$$

Following [74], we write these terms in matrix form

$$\begin{array}{rcccccccc} V & = & \frac{1}{R} V_{0,2}\, v^2 & & & & & & \\ & + & V_{1,4}\frac{v^4}{r^7} & + & V_{1,6}\frac{v^6}{r^{11}} & + & V_{1,8}\frac{v^8}{r^{15}} & + & \cdots \\ & + & R\, V_{2,4}\,\frac{v^4}{r^{10}} & + & R\, V_{2,6}\,\frac{v^6}{r^{14}} & + & R\, V_{2,8}\,\frac{v^8}{r^{18}} & + & \cdots \\ & + & R^2\, V_{3,4}\,\frac{v^4}{r^{13}} & + & R^2\, V_{3,6}\,\frac{v^6}{r^{17}} & + & R^2\, V_{3,8}\,\frac{v^8}{r^{21}} & + & \cdots \\ & + & \vdots & + & \vdots & + & \vdots & + & \ddots \end{array} \tag{5.61}$$

where each row gives the contribution at fixed loop order. We will now give a brief review of what is known about these coefficients. First, let us note that in Planck units this potential is (restoring factors of $\alpha' = l_{11}^3/R$ by dimensional analysis)

$$V = \sum_{n,k} V_{n,k} \frac{l_{11}^{3n+3k-6}}{R^{k-1}} \frac{v^k}{r^{3n+2k-4}}. \tag{5.62}$$

Since the gravitational coupling constant is $\kappa^2 = 2^7\pi^8 l_{11}^9$ we only expect terms with

$$n + k \equiv 2 \pmod 3 \tag{5.63}$$

to correspond with classical supergravity interactions, since all terms in the classical theory have integral powers of κ. Of the terms explicitly shown in (5.61) only the diagonal terms satisfy this criterion. By including factors of $\hat{N}$ and $\tilde{N}$ for semi-classical graviton states with finite momentum P^+ and comparing to supergravity, one can argue that the terms on the diagonal are precisely those which should correspond to classical supergravity. The terms beneath the diagonal should vanish for a naive agreement with supergravity at finite N, while the terms above the diagonal correspond to quantum gravity corrections. It was argued in [74] that the sum of diagonal terms corresponding to the effective classical supergravity potential between two gravitons should be given by

$$V_{\text{classical}} = \frac{2}{15R^2}\left(1 - \sqrt{1 - \frac{15R}{2}v^2}\right). \tag{5.64}$$

Now let us discuss the individual terms in (5.61). As we have discussed, the one-loop analysis gives a term

$$V_{1,4} = -\frac{15}{16} \tag{5.65}$$

which agrees with linearized supergravity. The analysis of DKPS [30] can be extended to the remaining one-loop terms. The next one-loop term vanishes

$$V_{1,6} = 0. \tag{5.66}$$

Some efforts have been made to relate the higher order terms $V_{1,8}, \ldots$ to quantum effects in 11D supergravity, but so far this interpretation is not clear. For some discussion of this issue see [61, 62] and references therein. The term

$$V_{2,4} = 0 \tag{5.67}$$

was computed by Becker and Becker [66]. As expected, this term vanishes. The term

$$V_{2,6} = \frac{225}{32} \tag{5.68}$$

was computed in [74]. This term agrees with the expansion of (5.64). A general expression for the two-loop effective potential given by the second line of (5.61) was given in [75].

It was shown in [72, 76] by Paban, Sethi and Stern that there can be no higher-loop corrections to the v^4 and v^6 terms on the diagonal. Their demonstration of these results follows from a consideration of the terms with the maximal number of fermions which are related to the v^4 and v^6 terms by supersymmetry. For example, this is the ψ^8/r^{11} term in the case v^4. They show that the fermionic terms are uniquely determined by supersymmetry, and that this in turn uniquely fixes the form of the bosonic terms proportional to v^4 and v^6 (see also [77] for more about the case of v^4). Thus, they have shown that that

$$V_{(n>1),4} = V_{(n>2),6} = 0. \tag{5.69}$$

This nonrenormalization theorem was originally conjectured by BFSS in analogy to similar known theorems for higher-dimensional theories [78].

This completes our summary of what is known about 2-body interactions in matrix theory. The complete set of known terms is given by

$$\begin{array}{rcl} V & = & \frac{1}{2R}v^2 \\ & & \begin{array}{ccccccccc} + & -\frac{15}{16}\frac{v^4}{r^7} & + & 0 & + & (\text{known}) & \rightarrow & & \\ + & 0 & + & \frac{225}{32}R\frac{v^6}{r^{14}} & + & (\text{known}) & \rightarrow & & \\ + & 0 & + & 0 & + & ? & + & \cdots & \\ & \downarrow & & \downarrow & + & \vdots & + & \ddots & \end{array} \end{array} \tag{5.70}$$

It has been proposed that for arbitrary N the analogues of the higher-loop diagonal terms should naturally take the form of a supersymmetric Born-Infeld type action [79, 80, 81, 82], which would give rise in the case $N = 2$ to a sum of the form (5.64). There is as yet, however, no proof of this statement beyond two loops. One particular obstacle to calculating the higher-loop terms in this series is that it is necessary to integrate over loops containing propagators of massless fields. These propagators can give rise to subtle infrared problems with the calculation. Some of these difficulties can be avoided by trying to reproduce higher-order supergravity interactions from interactions of more than two objects in matrix theory, the subject to which we will turn in section 5.2.

5.1.5. *The Equivalence Principle in matrix theory*

We have seen that the form of the linearized theory of 11D supergravity is precisely reproduced by a one-loop calculation in matrix theory. This equivalence follows provided that the expressions in (5.37-5.39), as well as the higher moments of these expressions and related expressions for the fermion components of the supercurrent are interpreted as definitions of the stress tensor, membrane current and other supercurrent components of a given matrix theory object. It is perhaps somewhat surprising given that this correspondence holds exactly at finite N to observe that Einstein's Equivalence Principle breaks down at finite N, even in the linearized theory [56].

The Equivalence Principle essentially states that given a background gravitational field produced by some source matter configuration, any two objects which are small compared to the scale of variation in the metric and which have the same initial space-time velocity vector $\dot{x}^I$ will follow identical trajectories through space-time. This follows from the fact that objects which are moving in the influence of a gravitational field follow geodesics in space-time. Of course, this result is only valid if the objects are not influenced by any other fields in the theory such as an electromagnetic 1-form or 3-form field.

To see a simple example of a case where the equivalence principle is violated in matrix theory, consider a source at the origin consisting of a single graviton with $p^+ = \tilde{N}/R$ and $\tilde{v}^i = 0$. This source produces a long-range gravitational field and no 3-form or gravitino field. Now consider a probe object at a large distance r. We take the probe to be a small membrane sphere, initially stationary, of radius r_0 and with longitudinal momentum $p^+ = N/R$. It is straightforward to calculate the energy p^- of the membrane; we find that the 11-momentum of the membrane has the light-front components

$$p^+ = N/R \qquad p^i = 0 \qquad p^- = \frac{8r_0^4}{RN^3}c_2 \,.$$

The initial velocity of the membrane is then

$$\dot{x}^+ = 1 \qquad \dot{x}^i = 0 \qquad \dot{x}^- = \frac{p^-}{p^+} = 8r_0^4\frac{c_2}{N^4} \,.$$

According to the equivalence principle, any two membrane spheres with different values of r_0 but the same value of $\dot{x}^- = r_0^4 c_2/N^4$ should experience precisely the same acceleration. Using the general formula for the 2-body interaction potential in matrix theory, however, it is straightforward to calculate

$$\ddot{x}^i = -\frac{R}{N}\frac{\partial V_{\text{matrix}}}{\partial x^i} = -1680R\tilde{N}\frac{x^i}{|x|^9}\frac{r_0^8}{N^8}\left(c_2^2 - \frac{1}{3}c_2\right) \,.$$

The leading term in an expansion in $1/N$ of this acceleration is indeed a function of $\dot{x}^-$. Thus, in the large N limit the equivalence principle is indeed satisfied. The subleading term, however, has a different dependence on r_0 and N. Thus, the equivalence principle is not satisfied at finite N.

This result implies that even if finite N matrix theory is equivalent to DLCQ M-theory, this theory does not seem to be related to a smooth theory of Einstein-Hilbert gravity, even on a compact space and with restrictions on longitudinal momentum. This is not a problem if one only takes seriously the large N version of the conjecture. If one wishes to make sense of the finite N theory in terms of some theory with a reasonable classical limit, however, it may be necessary to consider some new ideas for what this theory may be. It is tempting to think that the theory at finite N might be some sort of theory of classical gravity on a noncommutative space. Since the equivalence principle in the form we have been using it depends upon the geodesic equations, which are defined only on a smooth commutative space, it is natural to imagine that this principle might have to be corrected at finite N when the space has nontrivial noncommutative structure.

5.2. THE N-BODY PROBLEM

So far we have seen that in general the linearized theory of supergravity is correctly reproduced by an infinite series of terms arising from one-loop calculations in matrix theory. We have also discussed 2-loop calculations of two-graviton interactions which seem to agree with supergravity. If matrix theory is truly to reproduce all of classical supergravity, however, it must reproduce all the nonlinear effects of the fully covariant gravitational theory. The easiest way to study these nonlinearities is to consider N-body interaction processes. For example, following [83] let us consider a probe body at position r_3 in the long-range gravitational field produced by a pair of bodies at positions $r_1 = 0, r_2 \ll r_3$. We can consider a perturbative expansion of Einstein's equations. At leading order we have the linearized theory which gives a long-range field satisfying (schematically, dropping indices)

$$\partial^2 h \sim T$$

where T is a matter source. At the next order we have

$$\partial^2 h + h\partial^2 h + (\partial h)^2 \sim T + Th,$$

which we can rewrite in the form

$$\partial^2 h \sim T + Th + h\partial^2 h + (\partial h)^2 \tag{5.71}$$

The action of a probe object in the long-range field produced by objects 1 and 2 can be written in a double expansion in the inverse separations r_3

and r_2 as

$$T_3 h_{12} \sim \frac{T_3(T_1+T_2)}{r_3^7} + \frac{T_3 T_{(12)}}{r_3^7 r_2^7} + \cdots \tag{5.72}$$

where $T_{(12)}$ is an interaction term contributing through the quadratic terms on the RHS of (5.71). On the matrix theory side, an apparently analogous calculation can be performed by first doing the one-loop calculation we have already described to find the linearized interaction between the 3rd object and the 1-2 system, and then doing a further loop integration to evaluate the quantum corrections to the long-range field generated by the first two sources, giving an expression of the form

$$\frac{T_3 \langle T_{1+2} \rangle}{r_3^7}.$$

We expect quantum corrections to the expectation value of the schematic form

$$\langle T_{1+2} \rangle \sim T_1 + T_2 + \frac{T_{(12)}}{r_2^7} + \cdots$$

which roughly conforms to the structure expected from (5.72). Thus, in principle, it seems like it should be possible to make a correspondence between the double power series expansions computed in the two theories, given the results of the one-loop expansion for a completely general pair of objects such as was calculated in [65]. Indeed, a simple subset of terms were shown to correspond in this way in [84]. The terms considered in that paper were the terms in the 3-graviton interaction potential proportional to v_3^4/r_3^7. Considering the form discussed above for the components of the matrix stress tensor, it is clear that such terms only arise in the part of the interaction potential corresponding to

$$\frac{v_3^4 \langle T_{1+2}^{++} \rangle}{r_3^7}. \tag{5.73}$$

But the stress tensor component

$$\langle T_{1+2}^{++} \rangle = \frac{N_1 + N_2}{R}$$

is a constant which suffers no quantum corrections in matrix theory. This is a conserved charge: the total longitudinal momentum of the 1-2 system, and is responsible for the long-range component h^{++} of the metric. It is therefore easy to see that this term is correctly reproduced by matrix theory. The terms corresponding to other powers of v are more complicated, however, as the relevant components of T_{1+2} are corrected by quantum effects.

In addition to the practical complications of the calculation, there are several conceptual subtleties in using the approach we have just described to making a concrete correspondence between the matrix theory and supergravity descriptions of a general 3-body interaction process. The first subtlety arises, as was pointed out by Okawa and Yoneya in [85], from the fact that the complete gravity action is not simply the probe-source term (5.72), but also contains terms cubic in the gravitational field ($\int h^3$). These terms have a more complicated structure than the simple probe-source terms considered above, and it is more complicated to relate them to the results of the matrix theory calculation. The second subtlety which arises is that the precise choice of gauge made in the matrix theory calculation has a very strong impact on the form of the expressions found in the resulting effective action. Of course, for any physical quantity such as an S-matrix element, the result of a complete calculation will be independent of gauge choice. Nonetheless, to compare terms in the fashion we are suggested here will require a careful choice of gauge in matrix theory to match the appropriate gauge chosen in the supergravity theory. From this point of view, it is somewhat remarkable that in the calculation of the leading-order terms the natural gauge choices in the two theories (background field gauge in matrix theory and linearized gauge in supergravity) give rise to results which can be easily compared.

In any case, one might hope to navigate through these complications in the general 3-body problem, although this clearly would involve a substantial amount of work. In a very impressive pair of papers by Okawa and Yoneya [85, 86] (see also the more recent work [63]), the full S-matrix calculation was carried out for the interaction between 3 gravitons in both matrix theory and in supergravity, and it was shown that there was a precise agreement between all terms. Unlike other work on this problem, Okawa and Yoneya did not use the double expansion to simplify the problem but simply carried out the complete calculation.

One would naturally like to extend these results beyond the 3-body problem to the general N-body problem. The hierarchy of scales leading to the double expansion discussed above can be generalized, so that one has a different scale for each distance in the problem. This organizes the large number of terms in the N-body interaction into a more manageable structure. To date, however there has been very little work done on the problem of understanding higher order nonlinearities in the theory beyond those involved in the 3-body problem.

A very intriguing paper by Dine, Echols and Gray [87] attempts to find a matrix-supergravity correspondence for some special terms in the general N-body interaction potential. Although they find that some terms agree, they also find some terms which appear in the matrix theory potential which have the wrong scaling behavior to correspond to supergravity terms. We briefly describe these terms here in the language we have been using of stress tensor components.

For a 3-graviton system the term (5.73) is associated with an infinite series of higher-moment terms, as described in subsection 5.1.3 and in more detail in [55, 56]. The first of these higher moment terms is

$$v_3^4 \langle T^{++(ij)} \rangle_{12} \partial_i \partial_j \frac{1}{r_3^7} \tag{5.74}$$

This expectation value is given by

$$\langle X^i X^j \rangle \sim \frac{\delta_{ij}}{r_2} + \frac{v_2^i v_2^j + \delta^{ij} v_2^2}{r_2^5} + \cdots$$

The contribution to (5.74) from the first delta function vanishes since $\partial^2 r^{-7} = 0$ away from the origin in the 9-dimensional transverse space. The second term gives rise to a term in the 3-body potential of the form

$$V_a \sim \frac{v_3^4 (v_2 \cdot \partial)^2}{r_2^5} \frac{1}{r_3^7}$$

Dine, Echols and Gray argue that such a term should also be found in supergravity, giving an example of an agreement between two-loop matrix theory and tree level supergravity in the $U(3)$ theory at order v^6/r^{14}. This argument can be repeated by taking a higher moment of this term in a 4-body system

$$V_a \sim v_4^4 (v_3 \cdot \partial)^2 \frac{1}{r_4^7} \langle X^i X^j \rangle_{12} \partial_i \partial_j \frac{1}{r_3^5}$$

This time, however, the first term in the expectation value does not give 0, so that matrix theory predicts a term of the form

$$V_a \sim v_4^4 \left((v_3 \cdot \partial)^2 \frac{1}{r_4^7} \right) \left(\partial^2 \frac{1}{r_3^5} \right) \frac{1}{r_2}$$

As argued by Dine, Echols and Gray, this term has the wrong scaling to correspond to a classical supergravity interaction. Indeed, this term is of the form v^6/r^{17}, corresponding to a term "below the diagonal", which is expected to vanish.

The appearance of this term in the matrix theory perturbation series is troubling. It seems to indicate that there may be a breakdown of the correspondence between matrix theory and even classical supergravity. This is the first concrete calculation where the two perturbative expansions have been shown to contain terms which may disagree. On the other hand, there are subtleties in this calculation which may need be resolved. For one thing, there are the issues of gauge choices mentioned above. This calculation implicitly assumes a gauge which may not be appropriate for comparison to the 4-body interaction terms being considered in supergravity. There are also issues of infrared divergences which may lead to unexpected cancellations. In any case, clearly more work is needed to determine whether this indeed represents a breakdown of the relationship between matrix theory and classical supergravity which works so well for lower order terms.

5.3. LONGITUDINAL MOMENTUM TRANSFER

In this section we have so far concentrated on interactions in matrix theory and supergravity where no longitudinal momentum is transferred from one object to another. A supergravity process in which longitudinal momentum is transferred from one object to another is described in the IIA theory by a process where one or more D0-branes are exchanged between coherent states consisting of clumps of D0-branes. Such processes are exponentially suppressed since the D0-branes are massive, and thus are not relevant for the expansion of the effective potential in terms of $1/r$ which we have been discussing. In the matrix theory picture, this type of exponentially suppressed process can only appear from nonperturbative effects. Clearly, however, for a full understanding of interactions in Matrix theory it will be necessary to study processes with longitudinal momentum transfer in detail and to show that they also correspond correctly with processes in supergravity and M-theory. Some progress has been made in this direction. Polchinski and Pouliot have calculated the scattering amplitude for two 2-branes for processes in which a 0-brane is transferred from one 2-brane to the other [88]. In the Yang-Mills picture on the world-volume of the 2-branes, the incoming and outgoing configurations in this calculation are described in terms of an $U(2)$ gauge theory with a scalar field taking a VEV which separates the branes. The transfer of a 0-brane corresponds to an instanton-like process where a unit of flux is transferred from one brane to the other. The amplitude for this process was computed by Polchinski and Pouliot and shown to be in agreement with expectations from supergravity. This result suggests that processes involving longitudinal momentum transfer may be correctly described in Matrix theory. It should be noted, however, that the Polchinski-Pouliot calculation is not precisely a calculation of membrane scattering with longitudinal momentum transfer in Matrix theory since it is carried out in the 2-brane gauge theory language. In the T-dual Matrix theory picture the process in question corresponds to a scattering of 0-branes in a toroidally compactified space-time with the transfer of membrane charge. Processes with 0-brane transfer and the relationship between these processes and graviton scattering in matrix theory have been studied further in [89, 90, 91, 92].

6. Matrix theory in a general background

So far we have only discussed matrix theory as a description of M-theory in infinite flat space. In this section we consider the possibility of extending the theory to compact and curved spaces. As a preliminary to the discussion of compactification, we give an explicit description of T-duality in gauge theory language in subsection 6.1. We then discuss the compactification of the theory on tori in subsection 6.2. Following the discussion of matrix theory compactification, we turn in subsection 6.3 to the problem of using matrix theory methods to describe M-theory in a curved background space-time.

6.1. T-DUALITY

In this subsection we briefly review how T-duality may be understood from the point of view of super Yang-Mills theory. For more details see [93, 6].

In string theory, T-duality is a symmetry which relates the type IIA theory compactified on a circle of radius R_9 with type IIB theory compactified on a circle with dual radius $\hat{R}_9 = \alpha'/R_9$. In the perturbative type II string theory, T-duality exchanges winding and momentum modes of the closed string around the compact direction. For open strings, Dirichlet and Neumann boundary conditions are exchanged by T-duality, so that Dirichlet p-branes are mapped under T-duality to Dirichlet $(p \pm 1)$-branes [94].

It was argued by Witten [95] that the low-energy theory describing a system of N parallel Dp-branes in flat space is the dimensional reduction of $\mathcal{N} = 1$, $(9+1)$-dimensional super Yang-Mills theory to $p+1$ dimensions. In the case of N D0-branes, this gives the Lagrangian (3.2). To understand T-duality from the point of view of this low-energy field theory, we consider the simplest case of N D0-branes moving in a space which has been compactified in a single direction by identifying

$$x^9 \approx x^9 + 2\pi R^9. \tag{6.1}$$

To interpret this equivalence in terms of the matrix degrees of freedom of the D0-branes it is natural to pass to the covering space $\mathbb{R}^{9,1}$, where the N D0-branes are each represented by an infinite number of copies labeled by integers $n \in \mathbb{Z}$. We can thus describe the dynamics of N D0-branes on $\mathbb{R}^{8,1} \times S^1$ by a set of infinite matrices $M^i_{ma,nb}$ where $a, b \in \{1, \ldots, N\}$ are $U(N)$ indices and $m, n \in \mathbb{Z}$ index copies of each D0-brane which differ by translation in the covering space (See Figure 4). In terms of this set of

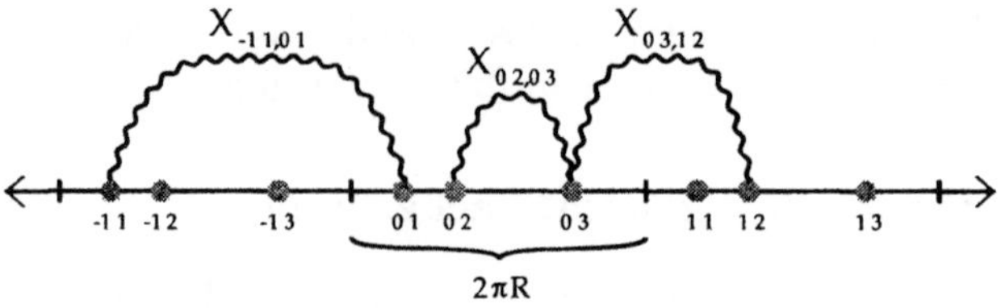

Figure 4. D0-branes on the cover of S^1 are indexed by two integers

infinite matrices, the quotient condition (6.1) becomes a set of constraints on the allowed matrices which can be written (dropping the $U(N)$ indices a, b) as

$$\begin{aligned} X^i_{mn} &= X^i_{(m-1)(n-1)}, \quad i < 9 \\ X^9_{mn} &= X^9_{(m-1)(n-1)}, \quad m \neq n \\ X^9_{nn} &= 2\pi R_9 \mathbb{1} + X^9_{(n-1)(n-1)} \ . \end{aligned} \tag{6.2}$$

From the structure of the constraints (6.2) it is natural to interpret the matrices X^i_{mn} in terms of the $(n-m)$th Fourier modes of a theory on the dual circle. The infinite matrix X^9 becomes a covariant derivative operator

$$X^9 \to (2\pi\alpha')(i\partial_9 + A_9) \tag{6.3}$$

in a $U(N)$ Yang-Mills theory on the dual torus, while X^i for $i < 9$ becomes an adjoint scalar field. The fermionic fields in the theory can be interpreted similarly.

This gives a precise equivalence between the low-energy world-volume theory of a system of N D0-branes on S^1 and a system of N D1-branes on the dual circle. The relationship between winding modes X^i_{mn} in the D0-brane theory and modes with $n - m$ units of momentum in the dual theory corresponds precisely to the mapping from winding to momentum modes in the closed string theory under T-duality.

This argument can easily be generalized to a system of multiple Dp-branes transverse to a torus T^d, which are equivalent to a system of wrapped D$(p+d)$-branes on the dual torus. When we compactify in multiple dimensions, the possibility arises of having a topologically nontrivial gauge field configuration on the dual torus. To discuss this possibility it is useful to use a slightly more abstract language to describe the T-duality.

The constraints (6.2) can be formulated by saying that there exists a translation operator U under which the infinite matrices X^a transform as

$$UX^aU^{-1} = X^a + \delta^{a9} 2\pi R_9 \mathbb{1}. \tag{6.4}$$

This relation is satisfied formally by the operators

$$X^9 = i\partial_9 + A_9, \qquad U = e^{2\pi i\hat{x}^9 R_9} \tag{6.5}$$

which correspond to the solutions discussed above. In this formulation of the quotient theory, the operator U generates the group $\Gamma = \mathbb{Z}$ of covering space transformations. Generally, when we take a quotient theory of this type, however, there is a more general constraint which can be satisfied. Namely, the translation operator only needs to preserve the state up to a gauge transformation. Thus, we can consider the more general constraint

$$UX^aU^{-1} = \Omega(X^a + \delta^{a9}2\pi R_9 \mathbb{1})\Omega^{-1}. \tag{6.6}$$

where $\Omega \in U(N)$ is an arbitrary element of the gauge group. This relation is satisfied formally by

$$X^9 = i\partial^9 + A_9, \qquad U = \Omega \cdot e^{2\pi i\hat{x}^9 R_9} \tag{6.7}$$

This is precisely the same type of solution as we have above; however, there is the additional feature that the translation operator now includes a nontrivial gauge transformation. On the dual circle $\hat{S}^1$ this corresponds to a gauge theory on a bundle with a nontrivial boundary condition in the compact direction 9.

A similar story occurs when several directions are compact. In this case, however, there is a constraint on the translation operators in the different compact directions. For example, if we have compactified on a 2-torus in dimensions 8 and 9, the generators U_8 and U_9 of a general twisted sector must generate a group isomorphic to $\mathbb{Z}^2$ and therefore must commute. The condition that these generators commute can be related to the condition that the boundary conditions in the dual gauge theory correspond to a well-defined $U(N)$ bundle over the dual torus. For compactifications in more than one dimension such boundary conditions can define a topologically nontrivial bundle. It is interesting to note that this construction can even be generalized to situations where the generators U_i do not commute. Physically, such a configuration is produced when there is a background NS-NS B field. This construction leads to a dual theory which is described by gauge theory on a noncommutative torus [96, 97, 98]. A description of this scenario along the lines of the preceding discussion is given in [99]. The connection between nontrivial background field configurations and noncommutative geometry has been a subject of much recent interest [100].

6.2. MATRIX THEORY ON TORI

From the discussion in the previous section, it follows that the matrix theory description of M-theory compactified on a torus T^d becomes $(d+1)$-dimensional super Yang-Mills theory. The argument of Seiberg and Sen in

[32, 33] is valid in this situation, so that $U(N)$ super Yang-Mills theory on $(T^d)^*$ should describe M-theory compactified on T^d. When $d \leq 3$ the quantum super Yang-Mills theory is renormalizable so this is a sensible way to approach the theory. As the dimension of the torus increases, however, the matrix description of the theory develops more and more complications. In general, the super Yang-Mills theory on the d-torus encodes the full U-duality symmetry group of M-theory on T^d in a rather nontrivial fashion.

Compactification of the theory on a two-torus was discussed by Sethi and Susskind [101]. They pointed out that as the T^2 shrinks, a new dimension appears whose quantized momentum modes correspond to magnetic flux on the T^2. In the limit where the area of the torus goes to 0, an $O(8)$ symmetry appears. This corresponds with the fact that IIB string theory appears as a limit of M-theory on a small 2-torus [102, 103].

Compactification of the theory on a three-torus was discussed in [104, 48]. In this case, M-theory on T^3 is equivalent to $(3+1)$-dimensional super Yang-Mills theory on a torus. This theory is conformal and finite. M-theory on T^3 has a special type of T-duality symmetry under which all three dimensions of the torus are inverted. In the matrix description this is encoded in the Montanen-Olive S-duality of the 4D super Yang-Mills theory.

When more than three dimensions are toroidally compactified, the theory undergoes even more remarkable transformations [105]. When compactified on T^4, the manifest symmetry group of the theory is $SL(4, Z)$. The expected U-duality group of M-theory compactified on T^4 is $SL(5, Z)$, however. It was pointed out by Rozali [106] that the U-duality group can be completed by interpreting instantons on T^4 as momentum states in a fifth compact dimension. This means that Matrix theory on T^4 is most naturally described in terms of a $(5 + 1)$-dimensional theory with a chiral $(2, 0)$ supersymmetry. This unusual $(2, 0)$ theory with 16 supersymmetries [107] appears to play a crucial role in numerous aspects of the physics of M-theory and 5-branes, and has been studied extensively in recent years.

Compactification on tori of higher dimensions than four continues to lead to more complicated situations, particularly when one gets to T^6, when the matrix theory description seems to be as complicated as the original M-theory. A significant amount of literature has been produced on this subject, to which the reader is referred to further details (see [7, 10] for reviews and further references). Despite the complexity of T^6 compactification, however, it was suggested by Kachru, Lawrence and Silverstein [108] that compactification of Matrix theory on a more general Calabi-Yau 3-fold might actually lead to a simpler theory than that resulting from compactification on T^6. If this speculation is correct and a more explicit description of the theory on a Calabi-Yau compactification could be found, it might make matrix theory a possible approach for studying realistic 4D phenomenology.

6.3. MATRIX THEORY IN CURVED BACKGROUNDS

We now consider matrix theory in a space which is infinite but may be curved or have other nontrivial background fields. We would like to generalize the matrix theory action to one which includes a general supergravity background given by a metric tensor, 3-form field, and gravitino field which together satisfy the equations of motion of 11D supergravity. This issue has been discussed in [109, 110, 111, 112, 32, 113, 114, 65]. In [32] it was argued that light-front M-theory on an arbitrary compact or non-compact manifold should be reproduced by the low-energy D0-brane action on the same compact manifold; no explicit description of this low-energy theory was given, however. In [112] an explicit prescription was given for the first few terms of a matrix theory action on a general Kähler 3-fold which agreed with a general set of axioms proposed in [111]. In [109] and [113], however, it was argued that no finite N matrix theory action could correctly reproduce physics on a large K3 surface. We review here an explicit proposal for a formulation of matrix theory in an arbitrary background geometry originally presented in [65].

If we assume that matrix theory is a correct description of M-theory around a flat background, then there is a large class of curved backgrounds for which we know it is possible to construct a matrix theory action for $N \times N$ matrices. This is the class of backgrounds which can be produced as long-range fields produced by some other supergravity matter configuration with a known description in matrix theory. Imagine that a background metric $g_{IJ} = \eta_{IJ} + h_{IJ}$, a 3-form field A_{IJK} and a gravitino field ψ_I of light-front compactified 11-dimensional supergravity can be produced by a matter configuration described in matrix theory by matrices $\tilde{X}^i$. Then the matrix theory action describing $N \times N$ matrices X^i in this background should be precisely the effective action found by considering the block-diagonal matrix configuration

$$X^i = \begin{bmatrix} X^i & 0 \\ 0 & \tilde{X}^i \end{bmatrix}$$

(and a similar fermion configuration) and integrating out the off-diagonal fields as well as fluctuations around the background $\tilde{X}$.

From the results found in [56, 65], we know that for weak background fields, the first few terms in an expansion of this effective action in the background metric are given by

$$\begin{aligned} S_{\text{eff}} &= S_{\text{matrix}} + \int dx\, T^{IJ}(x) h_{IJ}(x) + \cdots \\ &= S_{\text{matrix}} + \int dx^{+} \{ T^{IJ} h_{IJ}(0) + T^{IJ(i)} \partial_i h_{IJ}(0) + \cdots \} + \cdots \end{aligned} \tag{6.8}$$

where $T^{IJ(\cdots)}$ are the moments of the matrix theory stress-energy tensor, and there are analogous terms for the coupling of the membrane, 5-brane and fermionic components of the supercurrent to A_{IJK} and S_I. If the standard formulation of matrix theory in a flat background is correct, the absence of corrections to the long-range $1/r^7$ potential around an arbitrary matrix theory object up to at least order $1/r^{11}$ implies that this formulation must be correct at least up to terms of order $\partial^4 h$ and h^2.

As we have derived it, this formulation of the effective action is only valid for certain background geometries which can be produced by well-defined matrix theory configurations. It is natural, however, to suppose that this result can be generalized to an arbitrary background. Thus, it is proposed in [65] that up to nonlinear terms in the background, the general form of the matrix theory action in an arbitrary but weak background is given by

$$\begin{aligned} S_{\text{weak}} = \int d\tau \sum_{n=0}^{\infty} \sum_{i_1,\ldots,i_n} \frac{1}{n!} \Big(& T^{IJ(i_1\cdots i_n)} \partial_{i_1} \cdots \partial_{i_n} h_{IJ} \\ & + J^{IJK(i_1\cdots i_n)} \partial_{i_1} \cdots \partial_{i_n} A_{IJK} \\ & + M^{IJKLMN(i_1\cdots i_n)} \partial_{i_1} \cdots \partial_{i_n} A^{D}_{IJKLMN} \\ & + \text{fermion terms} \Big) \end{aligned} \tag{6.9}$$

Let us make several comments about this action. First, this formulation is only appropriate for backgrounds with no explicit x^- dependence, as we do not understand how to encode higher modes in the compact direction in the components of the supergravity currents. Second, note that the coupling to A^D is free of ambiguity since the net 5-brane charge must vanish for any finite matrices, so that only first and higher derivatives of A^D appear in the action. Third, note that though we only have explicit expressions for the fermion terms in the zeroeth and some of the first moments of T, we may in principle generalize the calculations of [56, 65] to determine all the fermionic contributions from higher order terms in the one loop matrix theory potential.

The linearized couplings in the action (6.9) are motivated by the results of one-loop calculations in matrix theory. In principle, it may be possible to extend the formulation of matrix theory in weak background fields to higher order by performing general higher-loop calculations in matrix theory. For example, a complete description of the 2-loop interaction in matrix theory between an arbitrary 3 background configurations would suggest the form of the coupling between one object considered as a probe and the quadratic terms in the background produced by the other pair of objects. Generally, knowing the full n-loop interaction between $n+1$ matrix theory objects would suggest the nth order coupling of the matrix degrees of freedom to the background fields. Unfortunately, as we have discussed such calculations are rather complicated. In addition to the technical difficulties of doing the general 2-loop calculation, there are subtleties related to the gauge choice and possible infrared divergences. Furthermore, finite N calculations will only help us to learn the higher-order couplings to the background if the results of these calculations are protected by supersymmetric nonrenormalization theorems, and as we have discussed there is no strong reason to believe that such nonrenormalization theorems hold for the general n-loop $SU(N)$ calculation. Thus, to write a completely general coupling of matrix theory to a nontrivial supergravity background, it is probably necessary to find a new general principle, such as a matrix version of the principle of coordinate invariance.

Another approach which one might take to define matrix theory in a general background geometry is to follow the original derivation of matrix theory as a regularized membrane theory, but to include a general background geometry instead of a flat background as was used in [1, 2]. The superspace formulation of a supermembrane theory in a general 11D supergravity background was given in [12]. In principle, it should be possible to simply apply the matrix regularization procedure to this theory to derive matrix theory in a general background geometry. Unfortunately, however, the connection between superspace fields and component fields is not well-understood in this theory. Until recently, in fact, the explicit expressions for the superspace fields were only known up to first order in the component fermion fields θ [115]. In [16], this analysis was extended to quadratic order in θ with the goal of finding an explicit formulation of the supermembrane in general backgrounds in terms of component fields, to which the matrix regulation procedure could be applied to generate a general background formulation of matrix theory. These results can be compared with the proposal just described for the linear couplings to the background. The two formulations seem to be completely compatible [116], although extra terms appear in the matrix theory action which cannot be predicted from the form of the continuous membrane theory.

In [111], Douglas proposed that any formulation of matrix theory in a curved background should satisfy a number of axioms. All these axioms are satisfied in a straightforward fashion by the proposal in [65], except one:

this exception is the axiom that states that a pair of D0-branes at points x^i and y^i should correspond to diagonal 2×2 matrices where the masses of the off-diagonal fields should be equal to the geodesic distance between the points x^i and y^i in the given background metric. In [117] it was shown that the linearized terms in the action (6.9) are consistent with this condition and that the linear variation in geodesic distance between a pair of D0-branes is correctly reproduced by coupling the matrix theory stress tensor to the background metric through a combinatorial identity which follows from the particular ordering implied by the symmetrized trace form of the multipole moments of the stress tensor. The fact that this condition can be satisfied at linear order provides hope that it might be possible to extend the action to all orders in a consistent way. In [112], it was indeed shown by Douglas, Kato and Ooguri that a set of some higher order terms for the action on a Ricci-flat Kähler manifold can be found which are consistent with the geodesic length condition, but these authors also found that this condition did not uniquely determine most of the terms in the action so that a more general principle is still needed to construct the action to all orders.

We synopsize the discussion in this section as follows: (6.9) seems to be a consistent proposal for the linearized couplings between matrix theory and weak supergravity background fields. The expressions for the higher moments of the supergravity currents which couple to the derivatives of the background fields are known up to terms quadratic in the fermions, and the remaining terms can be found from a one-loop matrix theory computation. This proposal can be generalized to mth order in the background fields, where matrix expressions are needed for quantities which can be determined from an m-loop matrix theory calculation. Whether these terms can be calculated and sensibly organized into higher-order couplings of matrix theory to background fields depends on whether higher-loop matrix theory results are protected by supersymmetric nonrenormalization theorems. It is worth emphasizing that the definitions of the matrix theory currents we have described here depend upon gauge choices for the propagating supergravity fields. For a given gauge choice, the theory is only defined for backgrounds compatible with the gauge condition. Making the appropriate gauge choices represents another obstacle to carrying out this analysis to higher order.

7. Outlook

We conclude with a brief review of the connection between matrix theory and M-theory, and a short discussion of the current state of affairs and the outlook for further developments in matrix theory.

We have discussed two complementary ways of thinking about matrix theory: first as a quantized regularized theory of a supermembrane, which naturally describes a second-quantized theory of objects moving in an 11-dimensional target space, and second as the DLCQ of M-theory which is equivalent to a simple limit of type IIA string theory through the Seiberg-Sen limiting argument.

Using matrix degrees of freedom, it is possible to describe pointlike objects which have many of the physical properties of supergravitons. It is also possible to use the matrix degrees of freedom to describe extended objects which behave like the supermembrane and 5-brane of M-theory. For supergravitons and membranes this story seems fairly complete; for 5-branes, only a few very special geometries have been described in matrix language, and a complete description of dynamical (longitudinal) 5-branes, even at the classical level, is still lacking.

As we have discussed, to date all perturbative calculations except the 3-loop calculation of Dine, Echols and Gray indicate that matrix theory correctly reproduces classical 11D supergravity. It has been suggested that the agreement between the theories at 1-loop and 2-loop orders is essentially an accident of supersymmetry, however there is little understanding of how to interpret or organize higher-loop terms. There is also very little understanding at this point of how quantum corrections to the supergravity theory can be understood in terms of matrix theory, although there is evidence [61, 62, 63] that quantum gravity effects are not reproduced by perturbative calculations in matrix theory but will require a better understanding of the large N limit of the theory.

At this point there are essentially 4 possible scenarios for the validity of the matrix theory conjecture:

i) Matrix theory is correct, and DLCQ supergravity is reproduced at finite N by perturbative matrix theory calculations.

ii) Matrix theory is correct in the large N limit, and noncompact supergravity is reproduced by a naive large N limit of the standard perturbative matrix theory calculations.

iii) Matrix theory is correct in the large N limit, but to connect it with supergravity, even at the classical level, it is necessary to deal with subtleties in the large N limit. (i.e., there are problems with the standard perturbative analysis at higher order)

iv) Matrix theory is simply wrong, and further terms need to be added to the dimensionally reduced super Yang-Mills action to find agreement with M-theory even in the large N limit.

Now let us examine the evidence:

• The breakdown of the Equivalence Principle seems incompatible with (i), but compatible with all other possibilities.

• If the result of Dine, Echols and Gray in [87] is correct, and has been correctly interpreted, clearly (i) and (ii) are not possible. The fact that the methods of Paban, Sethi and Stern for proving nonrenormalization theorems in the $SU(2)$ theory break down for $SU(3)$ at two loops and at higher loop order [118] also hints that (ii) may not be correct.

• The analysis of Seiberg and Sen seems to indicate that one of the possibilities (i)-(iii) should hold.

It seems that (iii) is the most likely possibility, given this limited evidence. There are several issues which are extremely important in understanding how this problem will be resolved. The first is the issue of Lorentz invariance. If a theory contains linearized gravity and is Lorentz invariant, then it is well known that it must be either the complete generally covariant gravity theory or just the pure linearized theory. Since we know that matrix theory has some nontrivial nonlinear structure which reproduces part of the nonlinearity of supergravity, it would seem that the conjecture must be valid if and only if the theory is Lorentz invariant. Unfortunately, so far there is no complete understanding of whether the quantum theory is Lorentz invariant (classical Lorentz invariance was demonstrated in [119]). It was suggested by Lowe in [120] that the problems found in [87] might be related to a breakdown of Lorentz invariance and that in fact extra terms must be added to the theory to restore this invariance; this would lead to possibility (iv) above.

Another critical issue in understanding how the perturbative matrix theory calculations should be interpreted is the issue of the order of limits. In the perturbative calculations discussed here we have assumed that the longitudinal momentum parameter N is fixed for each of the objects we are taking as a background, and we have then taken the limit of large separations between each of the objects. Since the size of the wavefunction describing a given matrix theory object will depend on N but not on the separation from a distant object, this gives a systematic approximation scheme in which the bound state and wavefunction effects for each of the bodies can be ignored in the perturbative analysis. If we really are interested in the large N theory, however, the correct order of limits to take is the opposite. We should fix a separation distance r and then take the large N limit. Unfortunately, in this limit we have no systematic approximation scheme. The wavefunctions for each of the objects overlap significantly as the size of the objects grows. Indeed, it was argued recently by Polchinski [121] that the size of the bound state wavefunction of N D0-branes will grow at least as fast as $N^{1/3}$. As emphasized by Susskind in [122], this overlap of wavefunctions makes the theory very difficult to analyze. Indeed, if possibility (iii) above is correct, it may be very difficult to use matrix theory to reproduce all the nonlinear structure of classical supergravity, let alone to derive new results about quantum supergravity. On the other hand, it may be that whatever mechanism allows the one-loop and two-loop matrix theory results to correctly reproduce the first few terms in supergravity and to evade the problem of wavefunction overlap may persist at higher orders. Indeed, one of the must important outstanding questions regarding matrix theory is to understand precisely which terms in the naive perturbative expansion of the quantum mechanics will agree with classical supergravity, and more importantly, *why* these terms agree. As mentioned in the last section, one of the other main outstanding problems in matrix theory is understanding how the matrix quantum mechanics theory behaves when M-theory is compactified on a curved manifold. In order to use matrix theory to make new statements about corrections to classical supergravity in phenomenologically interesting models such as M-theory on compact 7-manifolds or orbifolds, it will be necessary to solve both of these problems. In each case, a certain amount of luck will be needed for it to be possible to probe physically interesting questions using existing computational techniques.

In these lectures we have focused on understanding some basic aspects of matrix theory: the definitions of the theory in terms of the membrane and DLCQ of M-theory, and the construction of the objects and supergravity interactions of M-theory using matrix degrees of freedom. We conclude with a few brief words about some of the topics we have not discussed.

In addition to the matrix model of M-theory, there have been numerous related models suggested in the literature in the last few years. Some of these which have received particular attention are the $(0+0)$-dimensional matrix model of IIB string theory suggested by Ishibashi, Kawai, Kitazawa and Tsuchiya [123], the $(1+1)$-dimensional matrix string theory of Dijkgraaf, Verlinde and Verlinde [124] and the family of AdS/CFT conjectures proposed by Maldacena [125]. All these proposals relate a particular limit of string theory or M-theory in a fixed background to a field theory. Many connections between these models have been made, and in fact most of these proposals are related by a duality symmetry to the matrix theory we have discussed here. A fundamental question at this point, however, is how we may move away from a fixed background and discuss questions of cosmological significance.

Even within the framework of the matrix model of M-theory we have discussed in these lectures, there are many very interesting directions and particular applications which have been pursued which we did not have time to review here in any detail. These include questions about black holes in matrix theory (see, *e.g.*, [126, 127] and references therein), higher dimensional compactifications and the matrix model of the (2, 0) theory which arises upon compactification on T^4 ([106], see [10] for a review and further references), the detailed structure of the $N = 2$ bound state (see *e.g.*, [128, 129] and references therein), and many other directions of recent research.

In closing, it seems that matrix theory has achieved something which just a few years ago would have been deemed virtually impossible to accomplish in such a simple fashion: it gives a well-defined framework for M-theory and quantum gravity which reduces any problem, at least in light-front coordinates, to a computation which can in principle be defined and fed into a computer. Thus, in some sense this may be the first concrete answer to the problem of finding a consistent theory of quantum gravity. Unfortunately, even though this theory is a simple quantum mechanics theory, and not even a field theory, it is computationally intractable at this point to ask many of the really interesting questions about M-theory using this model. It is clearly a very interesting problem to try to find better ways of doing interesting M-theory calculations using the matrix model. But even if matrix theory is never able to give us a computational handle on some of the subtle aspects of M-theory, it certainly has given us a new perspective on how to think about a microscopic theory of quantum gravity. One of the most interesting aspects of the matrix picture is the appearance of dynamical higher-dimensional extended objects from a system of ostensibly pointlike degrees of freedom, as discussed in Section 4. It seems likely that this feature of matrix theory may play a key role in future attempts to describe a more covariant or background-independent microscopic model for M-theory, string theory or quantum gravity.

Acknowledgments

I would like to thank L. Thorlacius and the other organizers of the "Quantum Geometry" NATO Advanced Study Institute for putting together an excellent summer school and for inviting me to participate. Thanks to the students and other lecturers for providing a stimulating atmosphere and for many interesting discussions and questions. I would also like to thank N. Prezas and M. Van Raamsdonk for pointing out errors in a preliminary version of these lecture notes and suggesting constructive changes.

This work is supported in part by the A. P. Sloan Foundation and in part by the DOE through contract #DE-FC02-94ER40818.

Some of the material in these lectures was previously presented as part of a course at MIT: physics 8.871, Fall 1998. Some of this material was also previously presented in lectures at the Korean Institute for Advanced Study in October 1998 and at the Komaba '99 workshop in Tokyo, 1999.

References

1. J. Goldstone, unpublished; J. Hoppe, MIT Ph.D. thesis (1982); J. Hoppe, in proc. Int. Workshop on Constraint's Theory and Relativistic Dynamics; eds. G. Longhi and L. Lusanna (World Scientific, 1987).
2. B. de Wit, J. Hoppe and H. Nicolai, *Nucl. Phys.* **B305** [FS 23] (1988) 545.
3. T. Banks, W. Fischler, S. Shenker, and L. Susskind, "M Theory as a Matrix Model: A Conjecture," *Phys. Rev.* **D55** (1997) 5112, hep-th/9610043.
4. T. Banks, "Matrix Theory," *Nucl. Phys. Proc. Suppl.* **67** (1998) 180, hep-th/9710231.
5. D. Bigatti and L. Susskind, "Review of matrix theory," hep-th/9712072.
6. W. Taylor, "Lectures on D-branes, gauge theory and M(atrices)," Proceedings of Trieste summer school 1997, to appear; hep-th/9801182.
7. N. Obers and B. Pioline, "U-duality and M-theory," *Phys. Rep.* **318** (1999) 113, hep-th/9809039.
8. H. Nicolai and R. Helling, "Supermembranes and M(atrix) theory," hep-th/9809103.
9. B. de Wit, "Supermembranes and super matrix models," hep-th/9902051.
10. T. Banks, "TASI lectures on Matrix Theory," hep-th/9911068.
11. E. Witten, "String Theory Dynamics in Various Dimensions," *Nucl. Phys.* **B443** (1995) 85, hep-th/9503124.
12. E. Bergshoeff, E. Sezgin and P. K. Townsend, *Ann. of Phys.* **185** (1988) 330.
13. M. J. Duff, "Supermembranes," hep-th/9611203.
14. M. B. Green and J. H. Schwarz, *Phys. Lett.* **B136** (1984) 367; *Nucl. Phys.* **B243** (1984) 285.
15. M. B. Green, J. H. Schwarz and E. Witten, *Superstring theory* (Cambridge University Press, Cambridge, 1987).
16. B. de Wit, K. Peeters and J. Plefka "Superspace geometry for supermembrane backgrounds," hep-th/9803209.
17. M. Claudson and M. Halpern, *Nucl. Phys.* **B250** (1985) 689.
18. R. Flume, *Ann. of Phys.* **164** (1985) 189.
19. M. Baake, P. Reinicke and V. Rittenberg, *J. Math. Phys.* **26** (1985) 1070.
20. K. Fujikawa and K. Okuyama, "On a Lorentz covariant matrix regularization of membrane theories," hep-th/9706027; "SO(9, 1) invariant matrix formulation of supermembrane," hep-th/9709044;.
21. B. de Wit, M. Luscher and H. Nicolai, *Nucl. Phys.* **B320** (1989) 135.
22. S. Sethi and M. Stern, "D-brane Bound States Redux," hep-th/9705046.
23. C. M. Hull and P. K. Townsend, *Nucl. Phys.* **B438** (1995) 109, hep-th/9410167.
24. M. J. Duff, J. T. Liu and R. Minasian, *Nucl. Phys.* **B452** (1995) 261, hep-th/9506126.
25. J. H. Schwarz , *Phys. Lett.* **367** (1996) 97, hep-th/9510086.
26. P. Horava and E. Witten, *Nucl. Phys.* **B460** (1996) 506, hep-th/9510209.
27. E. Cremmer, B. Julia and J. Scherk, *Phys. Lett.* **B76** (1978) 409.
28. P. K. Townsend, *Phys. Lett.* **B373** (1996) 68, hep-th/9512062.
29. L. Susskind, "Another Conjecture about M(atrix) Theory," hep-th/9704080.
30. M. R. Douglas, D. Kabat, P. Pouliot, and S. Shenker, "D-Branes and Short Distances in String Theory," *Nucl. Phys.* **B485** (1997) 85, hep-th/9608024.
31. N. Prezas and W. Taylor, in preparation.
32. N. Seiberg, "Why is the Matrix Model Correct?," *Phys. Rev. Lett.* **79** (1997) 3577, hep-th/9710009.
33. A. Sen, "D0 Branes on T^n and Matrix Theory," *Adv. Theor. Math. Phys.* **2** (1998) 51, hep-th/9709220.
34. J. Plefka and A. Waldron, "On the quantum mechanics of M(atrix) theory," *Nucl. Phys.* **B512** (1998) 460, hep-th/9710104.
35. M. Porrati and A. Rozenberg, "Bound states at threshold in supersymmetric quan-

tum mechanics," hep-th/9708119.
36. M. Green and M. Gutperle, "D-particle bound states and the D-instanton measure," hep-th/9711107.
37. G. Moore, N. Nekrasov and S. Shatashvili "D-particle bound states and generalized instantons," hep-th/9803265.
38. V. G. Kac and A. V. Smilga, "Normalized vacuum states in $N = 4$ supersymmetric quantum mechanics with any gauge group," hep-th/9908096.
39. A. Hanany, B. Kol and A. Rajaraman, "Orientifold points in M-theory," *JHEP* **9910** (1999) 027, hep-th/9909028.
40. M. R. Douglas, "Branes within Branes," in Cargese 97: *Strings, branes and dualities* p. 267, hep-th/9512077.
41. D. Kabat and W. Taylor, "Spherical membranes in Matrix theory," *Adv. Theor. Math. Phys.* **2**, 181-206, hep-th/9711078.
42. D. B. Fairlie, P. Fletcher and C. K. Zachos, "Trigonometric Structure Constants For New Infinite Algebras," *Phys. Lett.* **B218**, (1989) 203.
43. D. B. Fairlie and C. K. Zachos, "Infinite Dimensional Algebras, Sine Brackets And SU(Infinity)," *Phys. Lett.* **B224**, (1989) 101.
44. D. B. Fairlie, P. Fletcher and C. K. Zachos, "Infinite Dimensional Algebras And A Trigonometric Basis For The Classical Lie Algebras," *J. Math. Phys.* **31**, (1990) 1088.
45. L. Cornalba and W. Taylor, "Holomorphic curves from matrices," *Nucl. Phys.* **B536** (1998) 513-552, hep-th/9807060.
46. T. Banks, N. Seiberg, and S. Shenker, "Branes from Matrices," *Nucl. Phys.* **B497** (1997) 41, hep-th/9612157.
47. Y. Imamura, "A comment on fundamental strings in M(atrix) theory," *Prog. Theor. Phys.* **98** (1997) 677, hep-th/9703077.
48. O. J. Ganor, S. Ramgoolam and W. Taylor, "Branes, Fluxes and Duality in M(atrix)-Theory," *Nucl. Phys.* **B492** (1997) 191-204; hep-th/9611202.
49. W. Taylor and M. Van Raamsdonk, "Multiple Dp-branes in weak background fields," hep-th/9910052.
50. M. Berkooz and M. R. Douglas, "Five-branes in M(atrix) Theory," hep-th/9610236.
51. J. Castelino, S. Lee and W. Taylor, "Longitudinal 5-branes as 4-spheres in Matrix theory," *Nucl. Phys.* **B526** (1998) 334-350, hep-th/9712105.
52. H. Grosse, C. Klimčík, P. Prešnajder , "On finite 4D quantum field theory in non-commutative geometry," hep-th/9602115.
53. T. Banks and A. Casher, *Nucl. Phys.* **B167** (1980) 215.
54. V. P. Nair and S. Randjbar-Daemi, "On brane solutions in M(atrix) theory," hep-th/9802187.
55. W. Taylor and M. Van Raamsdonk, "Angular momentum and long-range gravitational interactions in Matrix theory," *Nucl. Phys.* **B532** (1998) 227-244, hep-th/9712159.
56. D. Kabat and W. Taylor, "Linearized supergravity from Matrix theory," *Phys. Lett.* **B426** (1998) 297-305, hep-th/9712185.
57. R. C. Myers, "Dielectric-branes," hep-th/9910053.
58. L. Susskind, Talk at Strings '97.
59. P. Berglund and D. Minic, "A note on effective Lagrangians in matrix theory," *Phys. Lett.* **B415** (1997) 122, hep-th/9708063.
60. M. Serone, "A Comment on the R^4 Coupling in M(atrix) Theory," *Phys. Lett.* **B422** (1998) 88, hep-th/9711031.
61. E. Keski-Vakkuri and P. Kraus, "Short distance contributions to graviton-graviton scattering: Matrix theory versus supergravity," hep-th/9712013.
62. K. Becker and M. Becker, "On Graviton Scattering Amplitudes in M-theory," *Nucl. Phys.* **B506** (1997) 48, hep-th/9712238.

63. R. Helling, J. Plefka, M. Serone and A. Waldron, "Three graviton scattering in M-theory," *Nucl. Phys.* **B559** (1999) 184; hep-th/9905183.
64. L. F. Abbott, "Introduction to the background field method," *Acta Phys. Polon.* **B13** (1982) 33; "The background field method beyond one loop," *Nucl. Phys.* **B185** (1981) 189.
65. W. Taylor and M. Van Raamsdonk, "Supergravity currents and linearized interactions for matrix theory configurations with fermion backgrounds," *JHEP* **9904** (1999) 013, hep-th/9812239.
66. K. Becker and M. Becker, "A Two-Loop Test of M(atrix) Theory," *Nucl. Phys.* **B506** (1997) 48-60, hep-th/9705091.
67. M. Van Raamsdonk, "Conservation of supergravity currents from matrix theory," *Nucl. Phys.* **B542** (1999) 262, hep-th/9803003.
68. S. Hellerman and J. Polchinski, "Compactification in the Lightlike Limit," hep-th/9711037.
69. W. Taylor, "Adhering 0-branes to 6-branes and 8-branes," *Nucl. Phys.* **B508** (1997) 122-132; hep-th/9705116.
70. V. Balasubramanian, D. Kastor, J. Traschen and K. Z. Win, "The spin of the M2-brane and spin-spin interactions via probe techniques," hep-th/9811037.
71. P. Kraus, "Spin-Orbit interaction from Matrix theory," hep-th/9709199.
72. S. Paban, S. Sethi and M. Stern, "Constraints from extended supersymmetry in quantum mechanics," *Nucl. Phys.* **B534** (1998) 137, hep-th/9805018.
73. J. Plefka, M. Serone and A. Waldron, "D = 11 SUGRA as the low energy effective action of Matrix Theory: three form scattering," hep-th/9809070.
74. K. Becker, M. Becker, J. Polchinski, and A. Tseytlin, "Higher Order Graviton Scattering in M(atrix) Theory," *Phys. Rev.* **D56** (1997) 3174, hep-th/9706072.
75. K. Becker and M. Becker, "Complete solution for M(atrix) Theory at two loops," *JHEP* **9809**:019 (1998), hep-th/9807182.
76. S. Paban, S. Sethi and M. Stern, "Supersymmetry and higher-derivative terms in the effective action of Yang-Mills theories," *JHEP* **9806**:012 (1998), hep-th/9806028.
77. S. Hyun, Y. Kiem and H. Shin, "Supersymmetric completion of supersymmetric quantum mechanics," hep-th/9903022.
78. M. Dine and N. Seiberg, "Comments on higher derivative operators in some SUSY field theories," *Phys. Lett.* **B409** (1997) 239, hep-th/9705057.
79. I. Chepelev and A. Tseytlin, "Long-distance interactions of branes: correspondence between supergravity and super Yang-Mills descriptions," hep-th/9709087.
80. E. Keski-Vakkuri and P. Kraus, "Born-Infeld actions from Matrix theory," hep-th/9709122.
81. V. Balasubramanian, R. Gopakumar and F. Larsen, "Gauge theory, geometry and the large N limit," hep-th/9712077.
82. I. Chepelev and A. Tseytlin, "On Membrane Interaction in Matrix Theory," hep-th/9801120.
83. M. Dine and A. Rajaraman, "Multigraviton Scattering in the Matrix Model," hep-th/9710174.
84. W. Taylor and M. Van Raamsdonk, "Three-graviton scattering in Matrix theory revisited," *Phys. Lett.* **B438** (1998), 248-254, hep-th/9806066.
85. Y. Okawa and T. Yoneya, "Multibody interactions of D-particles in supergravity and matrix theory," *Nucl. Phys.* **B538** (1999) 67, hep-th/9806108.
86. Y. Okawa and T. Yoneya, "Equations of motion and Galilei invariance in D-particle dynamics," hep-th/9808188.
87. M. Dine, R. Echols and J. P. Gray, "Tree level supergravity and the matrix model," hep-th/9810021.
88. J. Polchinski and P. Pouliot, "Membrane Scattering with M-Momentum Transfer," hep-th/9704029.
89. N. Dorey, V. V. Khoze and M. P. Mattis, *Nucl. Phys.* **B502** (1997) 94-106,

hep-th/9704197.
90. T. Banks, W. Fischler, N. Seiberg, and L. Susskind, *Phys. Lett.* **B408** (1997) 111-116, hep-th/9610043.
91. S. Hyun, Y. Kiem and H. Shin, "Effective action for membrane dynamics in DLCQ M-theory on a two torus," *Phys. Rev.* **D59** (1999) 021901,hep-th/9808183.
92. E. Keski-Vakkuri and P. Kraus, "M-momentum transfer between gravitons, membranes, and 5-branes as perturbative gauge theory processes," hep-th/9804067.
93. W. Taylor, "D-brane Field Theory on Compact Spaces," *Phys. Lett.* **B394** (1997) 283; hep-th/9611042.
94. J. Dai, R. G. Leigh, and J. Polchinski, *Mod. Phys. Lett.* **A4** (1989) 2073.
95. E. Witten, "Bound States of Strings and p-Branes," *Nucl. Phys.* **B460** (1996) 335, hep-th/9510135.
96. A. Connes, M. R. Douglas and A. Schwarz, "Noncommutative geometry and Matrix theory: compactification on tori," hep-th/9711162.
97. M. R. Douglas and C. Hull, "D-branes and the noncommutative torus," hep-th/9711165.
98. P.-M. Ho, Y.-Y. Wu and Y.-S. Wu, "Towards a noncommutative approach to matrix compactification," hep-th/9712201.
99. Y.-K. E. Cheung and M. Krogh, "Noncommutative geometry from 0-branes in a background B field," *Nucl. Phys.* **B528**, 185 (1998), hep-th/9803031.
100. N. Seiberg and E. Witten, "String theory and noncommutative geometry," *JHEP* **9909**:032 (1999), hep-th/9908142.
101. S. Sethi and L. Susskind, *Phys. Lett.* **B400** (1997) 265-268, hep-th/9702101.
102. P. Aspinwall, "Some relationships between dualities in string theory," hep-th/9508154.
103. J. H. Schwarz, *Phys. Lett.* **B360** (1995) 13, Erratum-*ibid.* **B364** (1995) 252, hep-th/9508143.
104. L. Susskind, "T-duality in M(atrix)-theory and S-duality in Field Theory," hep-th/9611164.
105. W. Fischler, E. Halyo, A. Rajaraman and L. Susskind, "The Incredible Shrinking Torus," hep-th/9703102.
106. M. Rozali, "Matrix theory and U-duality in seven dimensions," hep-th/9702136.
107. N. Seiberg, "Notes on Theories with 16 Supercharges," hep-th/9705117.
108. S. Kachru, A. Lawrence and E. Silverstein, "On the Matrix Description of Calabi-Yau Compactifications," hep-th/9712223.
109. M. R. Douglas, H. Ooguri and S. Shenker, "Issues in M(atrix) Theory Compactification," *Phys. Lett.* **B402** (1997) 36, hep-th/9702203.
110. M. R. Douglas, "D-branes in curved space," *Adv. Theor. Math. Phys.* **1** (1998) 198, hep-th/9703056.
111. M. R. Douglas, "D-branes and matrix theory in curved space," hep-th/9707228.
112. M. R. Douglas, A. Kato and H. Ooguri, "D-brane actions on Kaehler manifolds," *Adv. Theor. Math. Phys.* **1** (1998) 237, hep-th/9708012.
113. M. R. Douglas and H. Ooguri, "Why Matrix Theory is Hard," *Phys. Lett.* **B425** (1998) 71, hep-th/9710178.
114. G. Lifschytz, "DLCQ-M(atrix) Description of String Theory, and Supergravity," hep-th/9803191.
115. E. Cremmer and S. Ferrara, *Phys. Lett.* **B91** (1980) 61.
116. K. Millar and W. Taylor, in preparation.
117. W. Taylor and M. Van Raamsdonk, "Multiple D0-branes in weakly curved backgrounds," hep-th/9904095.
118. S. Sethi and M. Stern, "Supersymmetry and the Yang-Mills effective action at finite N," hep-th/9903049.
119. B. de Wit, V. Marquard and H. Nicolai, *Comm. Math. Phys.* **128** (1990) 39.
120. D. Lowe, "Constraints on higher derivative operators in the matrix theory effective Lagrangian," hep-th/9810075.

121. J. Polchinski, "M-theory and the light cone," *Prog. Theor. Phys. Suppl.* **134** (1999) 158, hep-th/9903165.
122. L. Susskind, "Holography in the flat space limit," hep-th/9901079.
123. N. Ishibashi, H. Kawai, Y. Kitazawa and A. Tsuchiya, "A Large-N Reduced Model as Superstring," hep-th/9612115.
124. R. Dijkgraaf, E. Verlinde, H. Verlinde, *Nucl. Phys.* **B500** (1997) 43-61, hep-th/9703030.
125. J. Maldacena, "The large N limit of superconformal field theories and supergravity," *Adv. Theor. Math. Phys.* **2** (1998) 231, hep-th/9711200.
126. T. Banks, W. Fischler, I. R. Klebanov and L. Susskind, "Schwarzschild black holes from Matrix theory," hep-th/9709091.
127. D. Kabat and G. Lifschytz, "Approximations for strongly coupled supersymmetric quantum mechanics," , hep-th/9910001.
128. M. B. Halpern and C. Schwartz, "Asymptotic search for ground states of SU(2) matrix theory," *Int. J. Mod. Phys.* **A13** (1998) 4367, hep-th/9712133.
129. J. Frohlich, G. M. Graf, D. Hasler and J. Hoppe, "Asymptotic form of zero energy wave functions in supersymmetric matrix models," hep-th/9904182.

20 8'99

THE HOLOGRAPHIC PRINCIPLE

DANIELA BIGATTI
Institute of Theoretical Physics, KU Leuven,
B-3001 Heverlee, Belgium
email: daniela.bigatti@fys.kuleuven.ac.be

AND

LEONARD SUSSKIND
Department of Physics, Stanford University,
Stanford CA 94305-4060
email: susskind@sewerrat.stanford.edu

Abstract. These lectures review the Holographic principle. The first lecture describes the puzzle of black hole information loss that led to the idea of Black Hole Complementarity and subsequently to the Holographic Principle itself. The second lecture discusses the holographic entropy bound in general space-times. The final two lectures are devoted to the ADS/CFT duality as a special case of the principle. The presentation is self contained and emphasizes the physical principles. Very little technical knowledge of string theory or supergravity is assumed.

1. Black Hole Complementarity

New scientific ideas are usually characterized by simple organizing principles that can be expressed in a phrase or two. The invariance of the speed of light, the equivalence principle, the uncertainty principle and survival of the fittest are famous examples. Is there a comparable simple summary of the new principles which our science is now uncovering? Some people think it is supersymmetry, others think it is duality. But the real world is not supersymmetric, nor is it known to have dual descriptions in any deep sense. Our own view is that the lasting idea will be the *holographic principle* [1][2], the assertion that the number of possible states of a region of space is the same as that of a system of binary degrees of freedom distributed on

L. Thorlacius and T. Jonsson (eds.), M-Theory and Quantum Geometry, 179-225.

the boundary of the region. The number of such degrees of freedom is not indefinitely large but is bounded by the area of the region in Planck units. These lectures are about the motivations and evidence for this principle.

The holographic principle grew out of the deep insights of Bekenstein [3] and Hawking [4] in the 70's. In particular Hawking raised a very profound question concerning the consistency of gravitation and the usual operational principles of quantum mechanics [5]. To state the paradox clearly it is useful to think of a black hole as an intermediate state in a scattering process. Particles, perhaps in the form of stars, galaxies or just ordinary quanta come together in an initial state $|in\rangle$. A black hole forms and evaporates leaving outgoing photons, gravitons neutrinos and other quanta. No energy is lost in the process so there are no unaccounted for degrees of freedom in the final state. According to the usual rules, such a process is described by a unitary scattering matrix S.

$$|out\rangle = S|in\rangle \tag{1.1}$$

Since S is unitary we can also write

$$|in\rangle = S^{\dagger}|out\rangle \tag{1.2}$$

In other words it must be possible to recover the initial quantum state from the final state in a unique way. However, Hawking gave arguments, that appeared to many as completely persuasive, that information is irretrievably lost when matter falls behind the horizon of the black hole. Thus, from an operational point of view, the rules of quantum mechanics as set out by Dirac would have to be modified as collision energies approach and exceed the Planck energy. In particular the conventional S matrix would not exist.

Not everyone believed Hawking's arguments [6] [7]. *Black hole complementarity* [8] and the holographic principle [1] [2] are counter-proposals that preserve intact the general principles of quantum mechanics but question some of the naive beliefs about locality and the objectivity and invariance of space-time events.

1.1. THE SCHWARZSCHILD BLACK HOLE

To understand the issues we will need to review the geometry of black holes. There are many kinds of black holes in string theory but we will confine our attention to the usual $3+1$ dimensional Schwarzschild case.

The ordinary Schwarzschild black hole is described by the metric

$$ds^2 = \left(1 - \frac{2MG}{r}\right) dt^2 - \left(1 - \frac{2MG}{r}\right)^{-1} dr^2 - r^2 d\Omega^2 \tag{1.3}$$

M, G and $d\Omega^2$ are the black hole mass, the gravitational constant and the metric of a unit 2-sphere. The curvature singularity at $r = 0$ will not concern us but the coordinate singularity at the Schwarzschild radius $r = 2MG$ defines the all important horizon. Despite its singular importance, the horizon is not a mathematical singularity of the geometry, at least in the usual sense. To see that let us concentrate on the "near horizon limit". We consider a small angular region near a point on the horizon. Define

$$y = r - 2MG \tag{1.4}$$

For $y << 2MG$ the metric has the form

$$ds^2 = \frac{y}{2MG}dt^2 - \frac{2MG}{y}dy^2 - dx^i dx^i \tag{1.5}$$

where dx^i is an element of length in the two dimensional plane tangent to the horizon. Now define

$$\begin{aligned} \rho &= \sqrt{8MGy} \\ \omega &= \frac{t}{4MG} \end{aligned} \tag{1.6}$$

and the metric takes the form

$$ds^2 = \rho^2 d\omega^2 - d\rho^2 - dx^i dx^i \tag{1.7}$$

Expression (1.7) is the metric of ordinary Minkowski space in hyperbolic polar coordinates. If we define

$$\begin{aligned} X^+ &= \rho e^{\omega} \\ X^- &= -\rho e^{-\omega} \end{aligned} \tag{1.8}$$

the metric becomes

$$ds^2 = dX^+ dX^- - dx^i dx^i \tag{1.9}$$

which is the standard light cone form of the Minkowski metric. From this fact it is apparent that the horizon is not singular.

The relation between the flat minkowski coordinates $X^{\pm}$ and the Schwarzschild coordinates r, t is shown in figure(1) for the region outside the horizon.

The entire horizon $r = 2MG$ is mapped to the point (2D-surface) $X^+ = X^- = 0$. The extended horizon is defined by the 3 dimensional surface $X^- = 0$. Notice that a signal originating from a point behind the horizon, $X^- > 0$ can never escape to the outside, $X^- < 0$. For the region $X^+ > 0$, the extended horizon coincides with the asymptotic limiting value of

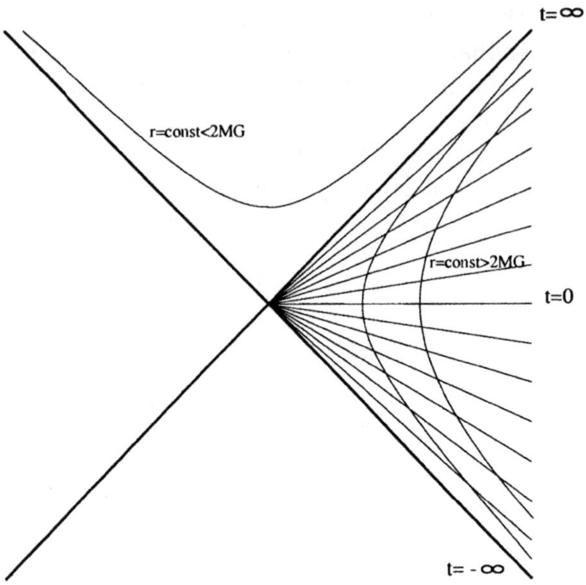

Fig. 1

Schwarzschild time $t = \infty$. Although the flat Minkowski coordinates only describe the near horizon region, a generalization to Kruskal-Szekeres (KS) coordinates covers the whole black hole space-time. Suppressing the angular coordinates Ω the KS metric has the form

$$ds^2 = F(X^+X^-)dX^+dX^- \tag{1.10}$$

where $F \to 1$ for $X^+X^- \to 0$ and

$$F \to \frac{16M^2G^2}{\rho^2} \tag{1.11}$$

for $X^+X^- \to \infty$. Equation (1.11) insures that the metric far from the black hole tends to flat space

$$ds^2 \to dt^2 - dr^2 - r^2 d\Omega^2 \tag{1.12}$$

In KS coordinates the singularity at $r = 0$ is defined by the *space-like* surface

$$X^+X^- = M^2G^2 \tag{1.13}$$

In figure (2) the geometry of the black hole is shown for the region $X^+ > 0$.

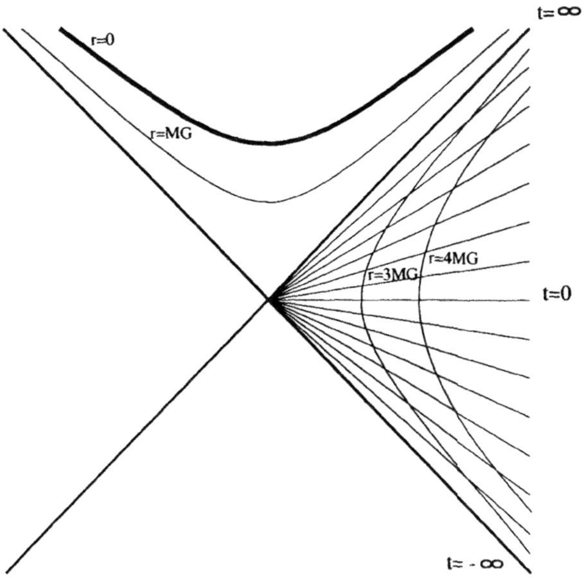

Fig. 2

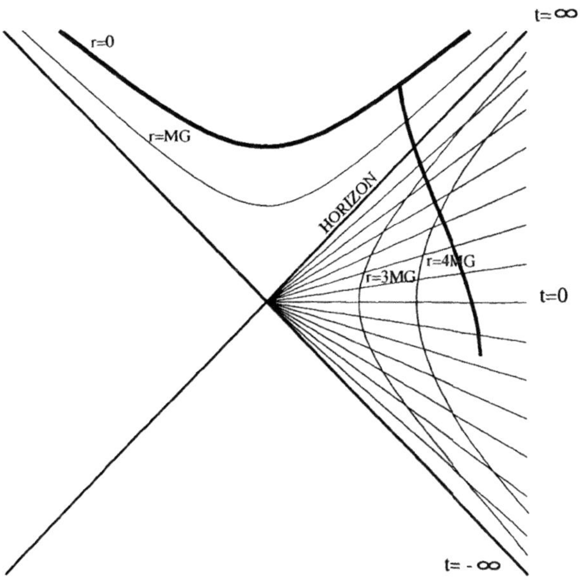

Fig. 3

Now consider a particle trajectory which begins outside the black hole, falls through the horizon and eventually hits the singularity as shown in figure (3). In Schwarzschild coordinates the particle does not cross the horizon until infinite time has elapsed. Thus from the viewpoint of an observer outside the black hole, the particle asymptotically approaches the horizon, but never crosses it. Indeed, all the matter which collapsed to form the black hole never crosses the horizon in finite Schwarzschild time. Classically it forms progressively thinner layers which asymptotically approach the horizon.

On the other hand, from the point of view of a freely falling observer accompanying the infalling particle the horizon is crossed after a finite time. In fact from figure 3 it is obvious that nothing special happens to the infalling matter at the horizon. This discrepancy is the first instance of an under-appreciated complementarity or relativity between the descriptions of matter by external and infalling observers.

1.2. PENROSE DIAGRAMS

Penrose diagrams provide an intuitively clear way to visualize the global geometry of black holes. They are especially useful for spherically symmetric geometries. The Penrose diagram describes the r, t plane. Here are the rules for a Penrose diagram.

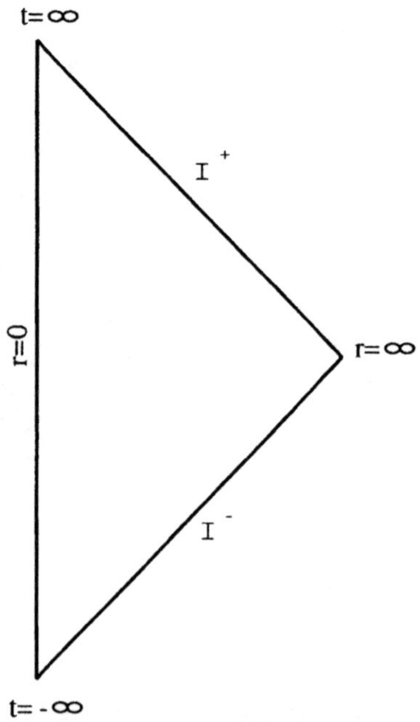

Fig. 4

1. Use coordinates which map the entire geometry to a finite portion of the plane.

2. The coordinates should be chosen so that radial light rays correspond to line oriented at ±45 degrees to the vertical.

As an example the Penrose diagram for flat space is shown in figure (4). The vertical axis is the spatial origin at $r = 0$ and the point labeled $r = \infty$ represents the asymptotic endpoints of space-like lines. The points $t = \pm\infty$ are the points where time-like trajectories begin and end. Light rays enter from past null infinity, $\Im^-$ and exit at future null infinity, $\Im^+$.

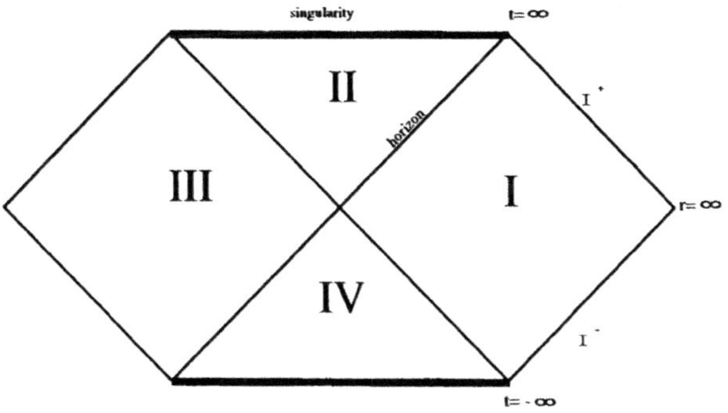

Fig. 5

The Penrose diagram for the Schwarzschild geometry is shown in figure (5). As we will see the regions III and IV are unphysical. Region I is the outside of the black hole and like flat Minkowski space it has space-like, time-like and null infinities. Obviously future directed time-like or light-like trajectory that begins in region II will collide with the singularity. Thus region II is identified as being behind the horizon. The extended horizon (from now on called the horizon) is the light-like line $t = \infty$.

A real black hole must be formed in a collapse. Thus in the remote past there is no black hole and the geometry should resemble the lower portion of figure (4). At late times the black hole has formed and the geometry should resemble figure (5). Thus the Penrose diagram for the collapse looks is shown in figure (6).

1.3. BLACK HOLE THERMODYNAMICS

It is well known that black holes are thermodynamic objects [3] [4] [9]. In addition to their energy, M they have a temperature and entropy. To understand this we need to study the behavior of quantum fields in the near horizon geometry. We will see later that quantum field theory can not

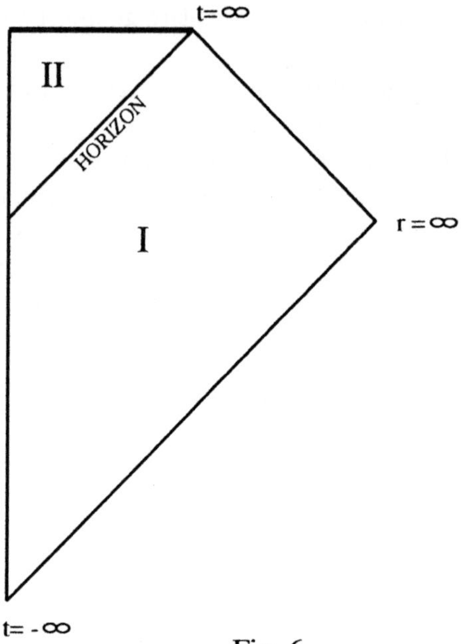

Fig. 6

really be an adequate description of a world including gravity but it is a starting point which will allow us to abstract some important lessons.

As we have seen, the near horizon geometry is just Minkowski space described in hyperbolic polar coordinates. In particular the portion of the near horizon region ($X^+X^- < 0$) outside the black hole is called Rindler space.

The usual time coordinate of Minkowski space is $x^0 = \frac{X^+ + X^-}{2}$ and conjugate to it is the momentum component p_0. However, p_0 is not the energy or Hamiltonian appropriate to the study of black holes by distant observers. For such observers the natural time is the Schwarzschild time $t = 4MG\omega$. The conjugate Hamiltonian which represents the energy or Mass of the black hole is

$$H_t = \frac{1}{4MG} H_\omega = \frac{i}{4MG} \partial_\omega \tag{1.14}$$

where H_ω is a dimensionless Hamiltonian conjugate to the dimensionless Rindler time ω.

An observer outside the horizon has no access to the degrees of freedom behind the horizon. For this reason all observations can be described in terms of a density matrix $\Re$ obtained by tracing over the degrees of freedom behind the horizon [9]. To derive the form of the density matrix for external observations we begin with the Minkowski space vacuum. The coordinates of Minkowski space are

$$\begin{aligned} x^0 &= (X^+ + X^-)/2 \\ x^3 &= (X^+ - X^-)/2 \end{aligned} \tag{1.15}$$

and the horizon coordinates x^i. The instant of Rindler time $\omega = 0$ coincides with the half-surface

$$\begin{aligned} x^0 &= 0 \\ x^3 &> 0 \end{aligned} \tag{1.16}$$

The other half of the surface $x^3 < 0$ is behind the horizon and is to be traced over.

Let us consider a set of quantum fields labeled ϕ. To specify the field configuration at $x^0 = 0$ we need to give the values of ϕ on both half-surfaces. Let ϕ_I and ϕ_F represent the field configurations for $x^3 > 0$ and $x^3 < 0$ respectively. A quantum state is represented by a wave functional

$$\Psi(\phi) = \Psi(\phi_I, \phi_F) \tag{1.17}$$

We use the standard Euclidean Feynman path integral formula to compute Ψ.

$$\Psi(\phi_I, \phi_F) = \int d\phi \exp -S \tag{1.18}$$

where the path integral is over all fields in the future half space $ix^0 > 0$ with boundary condition $\phi = (\phi_I, \phi_F)$ at $x^0 = 0$.

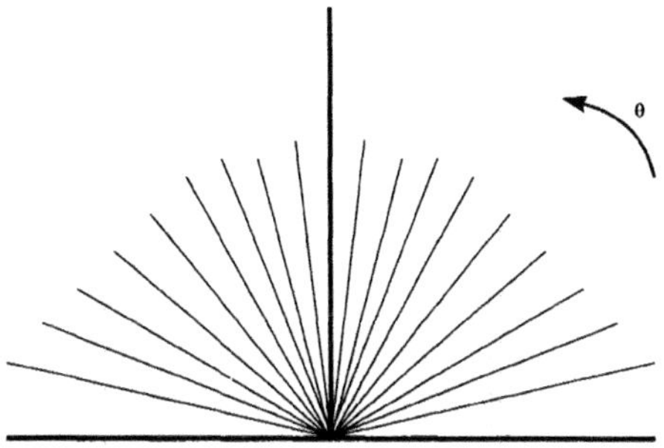

Fig. 7

The trick to compute the density matrix $\Re$ is to divide the upper half plane $ix^0 > 0$ into infinitesimal angular wedges as in figure (7). The path integral can then be evaluated in terms of a generator of angular rotations. This generator is nothing but iH_ω. Thus the expression for the Minkowski vacuum is

$$\Psi(\phi_F, \phi_I) = \langle\phi_F|\exp(-H_\omega\pi)|\phi_I\rangle \tag{1.19}$$

In other words the Minkowski vacuum wave functional is a transition amplitude for elapsed Euclidean time π.

Now consider the density matrix given by

$$\Re = \int d\phi_F \Psi^*(\phi_F, \phi_I')\Psi(\phi_F, \phi_I) \tag{1.20}$$

Using eq.(1.19)and the completeness of the states $\langle\phi_F|$ gives

$$\Re = \langle\phi_I'|\exp(-2\pi H_\omega)|\phi_I\rangle \tag{1.21}$$

or more concisely

$$\Re = \exp(-H_\omega/T_\omega) \tag{1.22}$$

with $T_\omega = 1/2\pi$.

Equation (1.22) is has the remarkable property of being a thermal density matrix for temperature T_ω. Notice that the derivation is exact and in no way relies on the free field approximation. It is valid for any quantum field theory for any strength of coupling.

The temperature $T_\omega = 1/2\pi$ does not have the usual dimensions of energy. This is because the Rindler time and therefore the Rindler Hamiltonian is dimensionless. To convert to a proper temperature with dimensions of energy we consider the proper time interval corresponding to an interval $d\omega$. From eq.(1.7)

$$ds = \rho d\omega \tag{1.23}$$

Thus an observer at distance ρ from the horizon converts from dimensionless quantities using the conversion factor ρ. The proper temperature at distance ρ is given by

$$T(\rho) = \frac{1}{\rho}T_\omega = \frac{1}{2\pi\rho} \tag{1.24}$$

In this way we arrive at the important conclusion that an observer outside a black hole but in the near horizon region will detect a temperature that varies as the inverse distance from the horizon [9].

Next consider the temperature as measured by a distant observer asymptotically far from the black hole. The proper time variable for such an observer is the Schwarzschild time $t = 4MG\omega$. Thus such distant observers measure temperature

$$T_H = \frac{T_\omega}{4\pi MG} \tag{1.25}$$

This is the Hawking temperature [4] of the black hole. It represents the true thermodynamic temperature of an isolated black hole.

The thermodynamic relation between temperature and mass (energy) allow us to compute an entropy for the black hole. Using

$$dM = TdS \tag{1.26}$$

we find

$$S = 4\pi MG \tag{1.27}$$

or in terms of the horizon area A

$$S = \frac{A}{4G} \tag{1.28}$$

Equation (1.28) is far more general than the derivation given here. It applies to every kind of black hole, be it rotating, charged or in arbitrary dimensions. In the general $(d+1)$ dimensional case the concept of two dimensional area only needs to be replaced by the $(d-1)$ dimensional measure of the horizon which we continue to call area.

1.4. THE THERMAL ATMOSPHERE

Because the region above the horizon has a non-vanishing temperature, it has a kind of thermal atmosphere [10] consisting of thermally excited quanta. In regions where the field theory is weakly coupled the thermal atmosphere consists of ordinary black body radiation. Some of these quanta have sufficient energy to escape the gravitational pull of the black hole and appear as Hawking radiation. However, for a large black hole, this process is very slow. The equilibrium approximation for the thermal atmosphere of the near horizon region is a very good one.

The thermal atmosphere contributes to the entropy of the black hole [11]. Let us consider the ordinary quantum fields of the standard model or its suitable generalization. For simplicity lets ignore the interactions as well as masses. The entropy stored in the shell between ρ and $\rho + d\rho$ for free massless fields is given by

$$\frac{dS}{d\rho d^2 x^i} = cT^3 \tag{1.29}$$

where c is a constant proportional to the effective number of massless fields at that temperature. Using $T = 1/2\pi\rho$ we find

$$S \sim A \int \frac{d\rho}{\rho^3} \tag{1.30}$$

Evidently if this formula made sense all the way to $\rho = 0$ the entropy of the black hole would be infinite. But since we know that the entropy is

$A/4G$ the field theory description must break down at some small ρ_0. In this case the entropy in the thermal atmosphere of ordinary quanta will be

$$S \sim Ac/\rho_0^2 \tag{1.31}$$

Since the total black hole entropy is $A/4G$ the contribution from the thermal atmosphere must be less than this. Accordingly [11] ρ_0 can not be smaller than $\sim G^{1/2}$.

Perhaps a more illuminating way to express this is to say that the number of effective degrees of freedom must tend to zero as the Planck temperature is approached [12]. In conventional quantum field theory the number of effective degrees of freedom is a non-decreasing function of temperature. The finiteness of black hole entropy is the first evidence that quantum field theory overestimates the number of independent degrees of freedom.

It is not too surprising that quantum field theory has too many degrees of freedom at short distances to describe a world with gravity. The non-renormalizability of quantum gravity has led to many suggestions of a Planck scale cutoff over the years. Roughly speaking, the idea was that there should be about 1 binary degree of freedom per Planck volume. What we will see in the following is that this idea still vastly overestimates the number of degrees of freedom. The correct reduction in the number of degrees of freedom is that there is no more than $1/R$ degrees of freedom per Planck volume where R is infrared cutoff radius, that is, the size of the spatial region being studied.

1.5. THE QUANTUM XEROX PRINCIPLE

The Holographic Principle represents a radical departure from the principles of local quantum field theory. In order to understand why we are driven to it we need to follow Hawking's original arguments about the loss of quantum coherence in black hole processes. The argument as I will present it is based on a principle that I call the *quantum Xerox principle.* It prohibits the existence of a machine which can duplicate the information in a quantum system and in so doing, produce two copies of the original information. To illustrate an example, consider a two-state system with states $|u\rangle$ and $|d\rangle$. We will call the system a q-bit. The general state of the q-bit is

$$|\psi\rangle = a|u\rangle + b|d\rangle \tag{1.32}$$

Now assume we had a machine which could clone the q-bit and duplicate a second q-bit in the same state. We can express this by

$$|\psi\rangle \to |\psi\rangle|\psi\rangle \tag{1.33}$$

For example

$$\begin{aligned} |u\rangle &\rightarrow |u\rangle|u\rangle \\ |d\rangle &\rightarrow |d\rangle|d\rangle \end{aligned} \tag{1.34}$$

Suppose a q-bit in the quantum state $|u\rangle + |d\rangle$ is fed into the machine. The output of the machine is

$$\begin{aligned} \{|u\rangle + |d\rangle\} \otimes \{|u\rangle + |d\rangle\} = |u\rangle &\otimes |u\rangle \\ +|d\rangle &\otimes |d\rangle \\ +|d\rangle &\otimes |u\rangle \\ +|u\rangle &\otimes |d\rangle \end{aligned} \tag{1.35}$$

However this is inconsistent with the most basic principle of quantum mechanics, the linearity of the evolution of state vectors. Linearity together with eq. (1.34) requires

$$|u\rangle + |d\rangle \rightarrow |u\rangle \otimes |u\rangle + |d\rangle \otimes |d\rangle \tag{1.36}$$

In this way we see that the principles of quantum mechanics forbid the duplication of quantum information. What has all this to do with black holes?

Consider the following thought experiment [13]. A black hole is formed as in figure (6). Before the black hole has a chance to evaporate a q-bit is thrown in. According to the observer who falls with the q-bit, the information at a later time will be localized behind the horizon at point (a) in figure (8). On the other hand an observer outside the horizon eventually sees all of the energy returned in the form of Hawking radiation. In order that the usual laws of quantum mechanics are satisfied for the outside observer, the q-bit of information must be found in the state of the outgoing evaporation products localized at point (b) in figure (8). Since there can not be two copies of the same information it would seem that either the infalling observer or the outside observer must experience a violation of the known laws of nature. Either the horizon is not such a benign place as we thought ($\sim$ Minkowski space) and infalling matter is severely disrupted or else the outside observer experiences a loss of information in contradiction with quantum principles!

The principle of Black Hole Complementarity flatly denies that either of these undesirable things happens. According to this principle no real observer ever detects a violation of the usual laws of nature. External observations are assumed to be consistent with a description in which infalling information is absorbed, thermalized near the hot horizon and returned in the form of subtle correlations in the Hawking radiation. Furthermore, infalling observers detect nothing unusual at the almost flat horizon and only

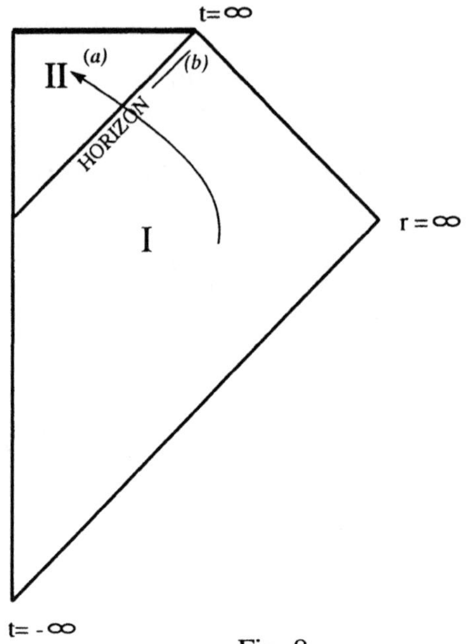

Fig. 8

experience violent effects as the singularity is approached. Reconciliation of these two facts will require that we radically modify our naive ideas of locality so that the space-time location of an event loses its invariant significance and becomes a relative concept.

As we have seen, quantum mechanics forbids information cloning. Let us take that to mean that no real observer is ever allowed to detect duplicate information. The outside observer has no problem with this since she can never detect signals from behind the horizon. However, it is more subtle to argue that observers behind the horizon can never detect duplicate information. Here is how it might happen:

An observer, O, stationed outside the horizon in figure (9) collects information stored in the Hawking radiation. After some time she has collected the information stored in the infalling q-bit. At that time, she jumps into the black hole, carrying the information to point (c) behind the horizon . Now there are two copies of the q-bit behind the horizon, one at (a) and one at (c). A signal from (a) to (c) can reveal that information has been duplicated. In fact we will argue that there is a *quantum Xerox censorship* mechanism which always prevents this from happening. To understand it we need one more concept.

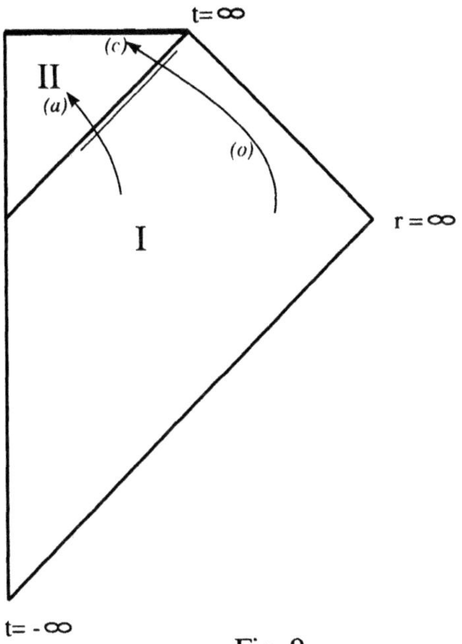

Fig. 9

1.6. INFORMATION RETENTION TIME

Consider a conventional complex system such as a piece of coal. Suppose the coal begins in its ground state and is heated by shining a laser beam on it. As its temperature rises it begins to glow and emit thermal radiation. Assume the laser beam is modulated so that it can convey information and that it sends in a bit.

Let S be the maximum entropy that the coal is heated to before being allowed to cool back to its ground state. By the time it does cool, all the information in the laser beam has been returned in the almost thermal radiation. An interesting question is how many photons are involved in carrying out the single bit. The answer has been given in a paper by Don Page [14]. The number of photons that have to be measured in order to collect a single bit is of order $S/2$. This is roughly half the photons that will be emitted. Another way to say it is that no information can be retrieved until the coal has cooled to the point where its entropy is about half its maximum value.

Given the luminosity, this restriction on collecting information from thermal radiation can be translated to a time scale for the coal to retain the original bit. This time is called the information retention time. How long is it for a massive black hole? The answer can easily be deduced from the known luminosity of black holes. In $(3+1)$ dimensions one finds

$$t_R \sim G^2 M^3 \tag{1.37}$$

For times much shorter than t_R we can expect that information which has been absorbed by the thermal horizon to be inaccessible.

1.7. QUANTUM XEROX CENSORSHIP

Let us return to the thought experiment in figure (9) designed to detect information duplication behind the horizon. The resolution of the dilemma is as follows. The point (c) must occur before the trajectory of O intersects the singularity. On the other hand O may not cross the horizon until the information retention time has elapsed. The implication of these two constraints is most easily seen using KS coordinates

$$\begin{aligned} X^+ &= \rho e^{\omega} \\ X^- &= -\rho e^{-\omega} \\ \omega &= \frac{t}{4MG} \end{aligned} \tag{1.38}$$

An observer outside the horizon must wait a time $t \sim M^3 G^2$ to collect a bit from the Hawking radiation. Thus she may not jump into the black hole until $(X^+ \sim e^{M^2 G})$. On the other hand the singularity is at $X^+ X^- = M^2 G^2$. This means that O will hit the singularity at a point satisfying

$$X^- < \exp -M^2 G \tag{1.39}$$

Thus for the original infalling system to send a signal which will reach O before she hits the singularity, the message must be sent within a time interval δt of the same order of magnitude, an incredibly short time.

Classically there is no obstruction to sending as much information as you like in as small a time as you like using as little energy as you like. Quantum physics changes this. A bit of information requires at least one quantum to transmit it. The uncertainty principle requires that the quantum have energy of order $(\delta t)^{-1}$. Thus the message requires a photon of energy

$$E_{signal} \sim \exp M^2 G \tag{1.40}$$

This is completely inconsistent with the assumption that the entire black hole, including the q-bit had energy M. If the observer at (a) had that much energy available, the black hole would have been much heavier and its horizon would have been at a very different place. Thus we see that quantum mechanics and gravity conspire to prevent O from detecting duplicate information.

We can now see that there is something wrong with the usual ideas of local quantum field theory in black hole backgrounds. The points (a) and (b) can be widely separated by a large space-like separation. Quantum field theory would say that the fields at these two points are independent commuting variables and it would predict correlations between them. But as we have seen, these correlations are unmeasurable by any real observer subject to the usual limitations of relativity and quantum mechanics. If you share the belief that a theory should not predict things which are in principle unobservable then you must conclude that local quantum field theory in a black hole background is the wrong starting point.

1.8. BARYON VIOLATION AND BLACK HOLE HORIZONS

It is generally conceded that there are no additive conserved quantities in a consistent quantum theory that includes gravity except for those that couple to long range fields. If nothing else, black hole evaporation will lead to violations of global conservation laws such as baryon conservation. An interesting question is where in the black hole geometry does the violation take place? Does it happen at or near the almost flat horizon or only at the violently curved singularity [15] or, is it more subtile as suggested by black hole complementarity [13]?

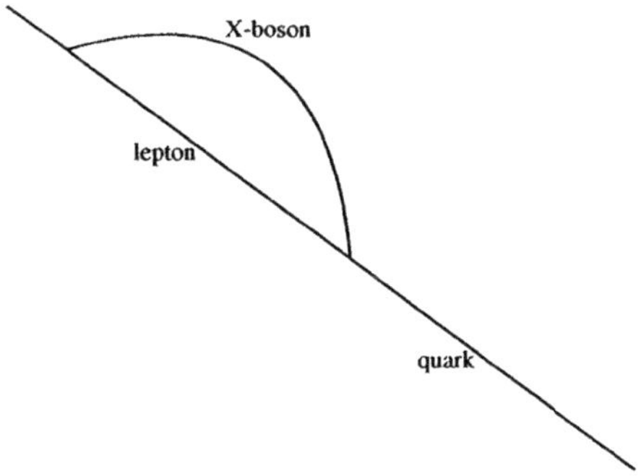

Fig. 10

For definiteness lets assume that baryon violation takes place in a conventional Grand Unification scheme such as $SU(5)$. Begin with a system of baryons and an observer all falling freely through the horizon of a very large black hole. Since the near horizon limit is nearly flat it is certain that the freely falling observer will detect negligible baryon violating effects in this region. However as time elapses the system will enter regions in which the curvature becomes of order the GUT scale. At that point there is every reason to think that baryon violating effects will be observed if the observer is in any shape to observe them.

The observer outside the black hole has a very different story to tell. According to him, the baryonic system entered the near horizon region where it was subjected to increasing proper temperature. When the temperature becomes of order M_{gut} the baryons are exposed to a flux of high energy particles in the thermal atmosphere and baryon violating processes must occur. Who is correct?

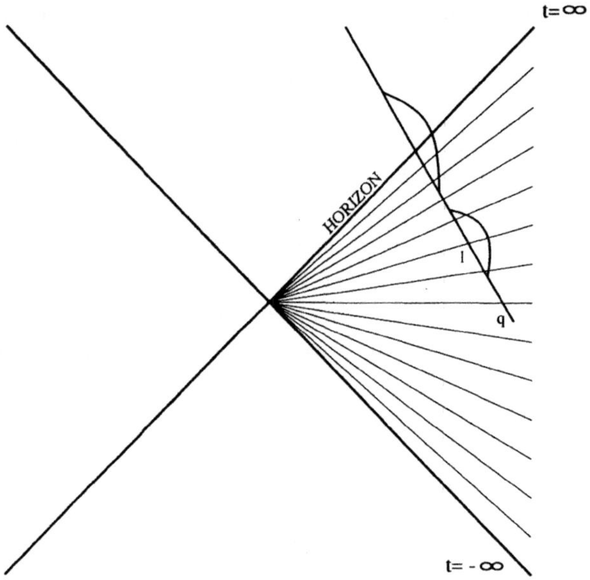

Fig. 11

In order to answer this question consider the propagation of a quark through empty space. Virtual baryon violating processes of the kind shown in the Feynman Diagram in figure (10) are continuously taking place. In other words the quark spends part of the time in a virtual state with the wrong baryon number even in empty flat space. What percentage of the time is the baryon number wrong? One might think the answer is incredibly small given the stability of the proton. But it is not. An explicit calculation gives a probability of order g^2 where g is the gauge coupling constant. Thus the quark has the wrong baryon number about 1 percent of the time. The reason we don't see this as baryon violation is that the lifetime of the intermediate states is of order the gut scale. The baryon number is constantly undergoing very rapid quantum fluctuations. The usual approximately conserved quantum number is the time averaged baryon number normalized to 1 for the nucleon. Now consider a quark falling through the horizon as in figure (11). It is evident from the figure that there is a significant probability that when the quark passes the horizon at $t = \infty$ it has the wrong baryon number. From the viewpoint of the infalling observer doing ordinary low energy experiments on the baryon the fluctuation is too fast to see. However, from the outside the rapid fluctuations slow down and the quark is caught frozen with the wrong baryon number. Of course this description fails to take gravitation into account but it nevertheless shows that understanding the apparent contradictory descriptions involves analyzing the behavior of matter at extremely short time scales and high frequencies.

Another thought experiment can illuminate the interplay between gravity and quantum mechanics. Suppose an observer O falls through the horizon just before the baryon as in figure (12). This observer sends out a signal (photon) which interacts with the infalling baryon and measures its baryon number. The signal is then received by a distant observer. Let us suppose that the experiment is arranged so that the signal-photon encounters the baryon in a region where the temperature is at least M_{gut}. In the rest frame of the infalling quark, it has a time of order M_{gut}^{-1} before it crosses the horizon. Thus the photon must be concentrated in a wave packet of size less that or equal to M_{gut}^{-1}. Its energy must be so high that it will resolve the baryon violating virtual state and will therefore have a finite probability of reporting baryon violation at the horizon. Complementarity works!

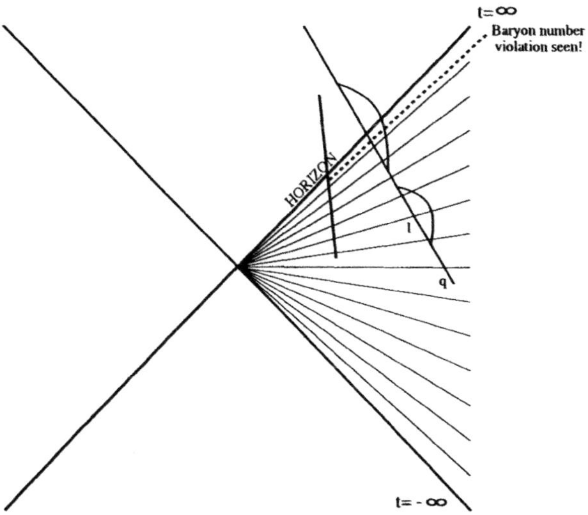

Fig. 12

Ordinary quantum field theory can not resolve the paradoxes of black holes. We have already seen that Q.F.T. drastically overestimates the number of ultraviolet degrees of freedom in the near horizon region and leads to a divergent entropy in the thermal atmosphere. String theory is widely believed to be a consistent quantum mechanical framework that includes gravitation. If so it must differ from Q.F.T. in very non-trivial ways at short times.

Although we are far from achieving a definitive understanding of black hole complementarity in string theory, there are some simple and suggestive ways to see that string theory is very different from Q.F.T. at high frequency [16].

Let us consider a string falling through a horizon. For our purposes we can approximate the horizon by the light-like surface $X^- = 0$. To study the string as it falls we use light cone coordinates. It is conventional to use X^+ for the light cone time variable. We are going to be unconventional and use X^-. Thus we choose the string theory gauge

$$\tau = X^- \tag{1.41}$$

The string starts out at negative X^- and reaches the horizon at $X^- = 0$.

Suppose the string falls through the horizon near $X^+ = 1$. Using

$$\begin{aligned} X^- &= -\rho e^{-\omega} \\ X^+ &= \rho e^{\omega} = 1 \end{aligned} \tag{1.42}$$

we find that near the string

$$X^- = \exp\left(-2\omega\right) = -\exp\left(-t/2MG\right) \tag{1.43}$$

The unusual properties of strings can already be seen at the level of free string theory. In light cone gauge a free string is described by a set of transverse coordinates $x^m(\sigma)$ where $0 \leq \sigma < 2\pi$. The coordinates are expressed in terms of harmonic oscillator variables $\alpha(n)$ and $\tilde{\alpha}(n)$. In string units

$$x(\sigma) = x_{cm} + \sum_n \left(\frac{\alpha(n)}{n} e^{in(\tau-\sigma)} + \frac{\tilde{\alpha}(n)}{n} e^{in(\tau+\sigma)} \right) \tag{1.44}$$

The question that will interest us has to do with the spatial size of the string. For simplicity we will consider the ground state of the string which classically has zero size. We usually envision the quantum fluctuations to spread the string over a size of order l_s, the string scale. However explicit calculation gives a very different result. The spatial size R will be defined in an obvious way.

$$R^2 = \langle 0|(x - x_{cm})^2|0\rangle \tag{1.45}$$

Using the standard commutation rules for the $\alpha's$ we find

$$R^2 = \sum_n \frac{1}{n} = \log(\infty) \tag{1.46}$$

Evidently the spatial size of the string is dependent of the frequency cutoff. If the frequency cutoff for a given observation is n_{max} then the apparent size of the string is

$$R^2 = \log n_{max} \tag{1.47}$$

We see a small string only if we average over sufficient time (τ) to eliminate the very high frequencies. This lesson is an important one and it will be repeated later in the form of the *ultraviolet infrared connection* in lecture III.

Consider the outside observer's description of the infalling string as it approaches the horizon. At any given point the string has a light cone time $|\tau|$ before it crosses the horizon at $\tau = 0$. Thus it makes no sense for the outside observer to average modes of frequency smaller than $|\tau|^{-1}$. In other words the frequency cutoff appropriate for an outside observer increases as the horizon is approached. Using eq.(1.47) and setting $n_{max} = |\tau|^{-1}$ we find

$$R^2 = \log \tau = t/2MG \tag{1.48}$$

Free string theory predicts that as a string falls toward the horizon it grows and appears to become an increasing tangled mass of string but only to the external observer. The infalling observer, depending on how she interacts with the string has a fixed time resolution and sees no growth.

1.10. THE SPACE TIME UNCERTAINTY RELATION

Even more revealing are the fluctuations of the longitudinal [17] coordinate X^+ (usually called X^-). First consider a classical point particle. It crosses the horizon, ($X^- = 0$), at a finite value of X^+. At that point the radial space-like distance from the horizon vanishes.

$$\rho^2 = -X^+X^- = 0 \tag{1.49}$$

Now consider the falling string. The coordinate $X^+(\sigma)$ is not an independent variable in string theory. To find out how it behaves we use the constraint equation

$$\partial_\sigma X^+ = \partial_\sigma x^i \partial_\tau x^i \tag{1.50}$$

The fact that the string does not require an independent degree of freedom for fluctuations in the X^- direction was one of the early indications of the large reduction in the number of degrees of freedom expected in a holographic theory. Using eq.(1.50) we can express $X^+(\sigma)$ in terms of harmonic oscillators. An explicit calculation gives

$$\begin{aligned} (\Delta X^+)^2 &\equiv \langle 0|(X^+ - X^+_{cm})^2|0\rangle \\ &= l_s^2 \sum_n \frac{1}{n^3} \\ &= l_s^2 n_{max}^2 \end{aligned} \tag{1.51}$$

This is a special case of a fundamental new uncertainty relation [17] [18] which occurs throughout string theory and which we will return to. To write it in a more suggestive form we write $n_{max} = (\Delta\tau)^{-1}$ or equivalently $n_{max} = (\delta X^-)^{-1}$. Equation (1.51) then takes the symmetrical form

$$\Delta X^+ \Delta X^- = l_s^2 \tag{1.52}$$

This is the *string uncertainty principle.* It implies that there is a fundamental unit of area in the X^+, X^- plane. It is reminiscent of uncertainty principles which occur in non-commutative geometry but it is not put in by hand.

To appreciate the implications of the space time uncertainty relation, let us consider an infalling massless string whose center of mass moves along the trajectory $X^+ = 1$. As X^- tends to zero the fluctuation in X^+, as seen by an outside observer, increases like l_s^2/X^-. Thus the stringy matter will be spread over region $X^+X^- \leq l_s^2$. From the point of view of Schwarzschild coordinates, instead of asymptotically approaching the horizon, the stringy matter can not be localized more precisely than to say that it is within a proper distance l_s from the horizon.

What we are seeing is a new relativity principle. According to the usual relativity principles, two observers in relative motion will disagree about the length of rods and the rate of clocks. But there is an invariant concept, the event, which occurs at a well defined space-time location. Even this is eliminated by black hole complementarity. External and freely falling observers will radically disagree about where and when events such as baryon violation take place or where the energy and momentum of a string is located. As we have seen, quantum mechanics and relativity conspire to insure that no observer ever sees a violation of the laws of quantum mechanics.

We have also seen that the origin of this relativity of descriptions is the behavior of the very high frequency fluctuations which are invisible to the freely falling observer but which dominate the description of the outside observer.

How can it be that the usual ideas of local quantum field theory fail so badly? In the remaining lectures we will see that conventional ideas of locality badly overestimate the number of independent degrees of freedom of a system. The key to black hole complementarity is the vast reduction implied by the holographic principle.

2. Entropy Bounds

2.1. MAXIMUM ENTROPY

The Holographic Principle is about the counting of quantum states of a system. We begin by considering a large region of space Γ. For simplicity we take the region to be a sphere. Now consider the space of states that describe arbitrary systems that can fit into Γ such that the region outside Γ is empty space. Our goal is to determine the dimensionality of that state-space. Lets consider some preliminary examples. Suppose we are dealing with a lattice of spins. Let the lattice spacing be a and the volume of Γ be V. The number of spins is V/a^3 and the number of orthogonal states supported in Γ is

$$N_{states} = 2^{\frac{V}{a^3}} \tag{2.1}$$

A second example is a continuum quantum field theory. In this case the number of quantum states will diverge for obvious reasons. We can limit the states, for example by requiring the energy density to be no larger than some bound ρ_{max}. In this case the states can be counted using some concepts from thermodynamics. One begins by computing the thermodynamic entropy density s as a function of the energy density ρ. The total entropy is

$$S = s(\rho)V \tag{2.2}$$

The total number of states is of order

$$N_{states} \sim \exp S = \exp s(\rho_{max})V \tag{2.3}$$

In each case the number of distinct states is exponential in the volume V. This is a very general property of conventional local systems and represents the fact that the number of independent degrees of freedom is additive in the volume.

In counting the states of a system the entropy plays a central role. In general entropy is not really a property of a given system but also involves ones state of knowledge of the system. To define entropy we begin with some restrictions that express what we know, for example, the energy within certain limits, the angular momentum and whatever else we may know. The entropy is by definition the logarithm of the number of quantum states that satisfy the given restrictions.

There is another concept that we will call the *maximum entropy*. This is a property of the system. It is the logarithm of the total number of states.

In other words it is the entropy given that we know nothing about the state of the system. For the spin system the maximum entropy is

$$S_{max} = \frac{V}{a^3} \log 2 \tag{2.4}$$

This is typical of the maximum entropy. Whenever it exists it is proportional to the volume. More precisely it is proportional to the number of simple degrees of freedom that it takes to describe the system.

Let us now consider a system that includes gravity. Again we focus on a spherical region of space Γ with a boundary $\partial\Gamma$. The area of the boundary is A. Suppose we have a thermodynamic system with entropy S that is completely contained within Γ. The total mass of this system can not exceed the mass of a black hole of area A or else it will be bigger than the region.

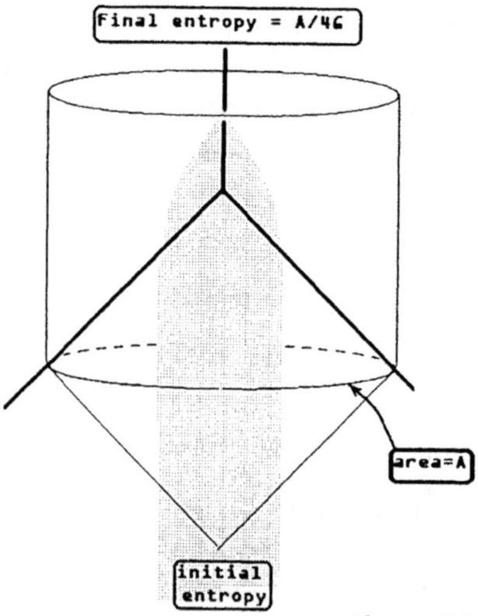

Figure (13)

Now imagine collapsing a spherically symmetric shell of matter with just the right amount of energy so that together with the original mass it forms a black hole which just fills the region. In other words the area of the horizon of the black hole is A. This is shown in figure (13). The result of this process is a system of known entropy, $S = A/4G$. But now we can use the second law of thermodynamics to tell us that the original entropy inside Γ had to be less than or equal to $A/4G$. In other words the maximum entropy of a region of space is proportional to its area measured in Planck units. Such bounds are called *holographic*.

2.2. ENTROPY ON LIGHT-LIKE SURFACES

We will see that it is most natural to define holographic entropy bounds on light-like surfaces [2] as opposed to space-like surfaces. Under certain circumstances the bounds can be translated to space-like surfaces but not always.

Let us start with an example in asymptotically flat space-time. We assume that flat Minkowski coordinates X^+, X^-, x^i can be defined at asymptotic distances. In this lecture we will revert to the usual convention in which X^+ is used as a light cone time variable. We will now define a "light sheet". Consider the set of all light rays which lie in the surface $X^+ = X_0^+$ in the limit $X^- \to +\infty$. In ordinary flat space this congruence of rays define a flat 3-dimensional light-like surface. In general they define a light like surface called a light sheet. The light sheet will typically have singular caustic lines but can be defined in a unique way [19]. When we vary X_0^+ the light sheets fill all space-time except for those points that lie behind black hole horizons.

Now consider a space-time point p. We will assign it light-cone coordinates as follows. If it lies on the light sheet X_0^+ we assign it the value $X^+ = X_0^+$. Also if it lies on the light ray which asymptotically has transverse coordinate x_0^i we assign it $x^i = x_0^i$. The value of X^- that we assign will not matter. The two dimensional x^i plane is called the Screen.

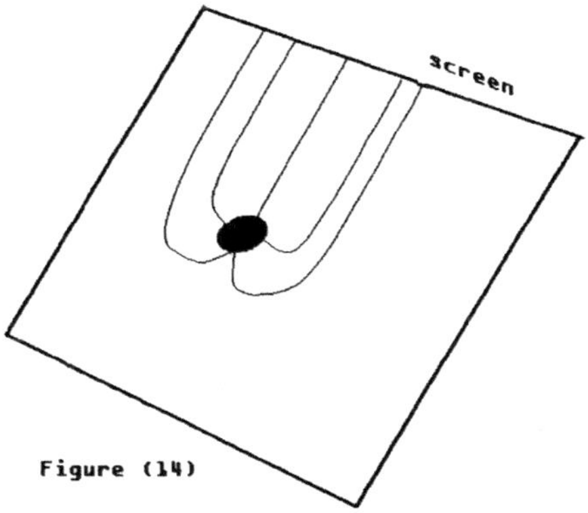

Figure (14)

Next assume a black hole passes through the light sheet X_0^+. The stretched horizon [1] of the black hole describes a two dimensional surface in the 3 dimensional light sheet as shown in figure (14). Each point on the stretched horizon has unique coordinates X^+, x^i. More generally if there are several black holes passing through the light sheet we can map each of their stretched horizons to screen in a single valued manner.

Since the entropy of the black hole is equal to $1/4G$ times the area of the horizon we can define an entropy density of $1/4G$ on the stretched horizon. The mapping to the screen then defines an entropy density in the x^i plane, $\sigma(x)$. It is a remarkable fact that $\sigma(x)$ is always less than or equal to $1/4G$.

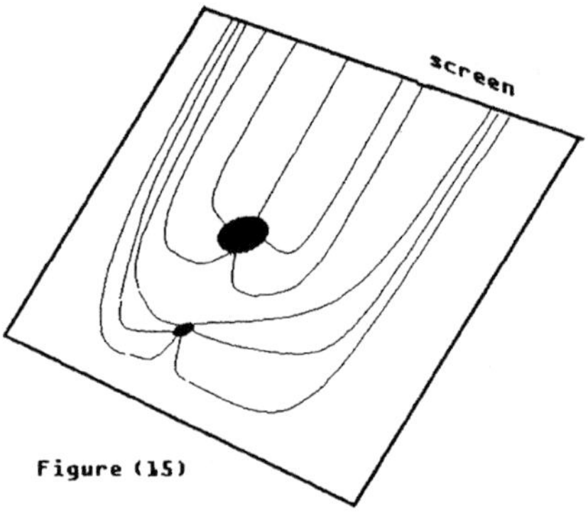

Figure (15)

To prove that $\sigma(x) \le \frac{1}{4G}$ we make use of the focusing theorem of general relativity. The focusing theorem depends on the positivity of energy and is based on the tendency for light to bend around regions of non-zero energy. Consider bundle of light rays with cross sectional area α. The light rays are parameterized by an affine parameter λ. The focusing theorem says that

$$\frac{d^2\alpha}{d\lambda^2} \le 0 \tag{2.5}$$

Consider a bundle of light rays in the light sheet which begin on the stretched horizon and go off to $X^- = \infty$. Since the light rays defining the light sheet are parallel in the asymptotic region $d\alpha/d\lambda \to 0$. The focusing theorem tells us that as we work back toward the horizon, the area of the bundle decreases. It follows that the image of a patch of horizon on the screen is larger than the patch itself. The holographic bound immediately follows.

$$\sigma(x) \le \frac{1}{4G} \tag{2.6}$$

[1]The stretched horizon is a time-like surface just outside the mathematical light-like surface. Its precise definition is not important here.

This is a surprising conclusion. No matter how we distribute the black holes in 3 dimensional space, the image of the entropy on the screen always satisfies the entropy bound (2.6). An example which helps clarify how this happens involves two black holes. Suppose we try to hide one of them behind the other along the X^- axis, thus doubling the entropy density in the x plane. The bending and focusing of light always acts as in figure (15) to prevent $\sigma(x)$ from exceeding the bound. These considerations lead us to the more general conjecture that for any system, when it is mapped to the screen the entropy density obeys the bound (2.6).

2.3. ROBERTSON WALKER GEOMETRY

This kind of bound has been generalized to *flat* Robertson Walker geometries by Fischler and Susskind [20] and to more general geometries by Bousso [21] [22]. First review the RW case. We will consider the general case of $d+1$ dimensions. The metric has the form

$$ds^2 = dt^2 - a(t)^2 dx^m dx^m \tag{2.7}$$

where the index m runs over the d spatial directions. The function $a(t)$ is assumed to grow as a power of t.

$$a(t) = a_0 t^p \tag{2.8}$$

Lets also make the usual simplifying cosmological assumptions of homogeneity. In particular we assume that the spatial entropy density (per unit d volume) is homogeneous. Later, following Bousso, we will relax these assumptions.

At time t we consider a spherical region Γ of volume V and area A. The boundary $(d-1)$-sphere, $\partial\Gamma$, will play the same role as the screen in the previous discussion. The light-sheet is now defined by the backward light cone formed by light rays that propagate from $\partial\Gamma$ into the past.

As in the previous case the holographic bound applies to the entropy passing through the light sheet. The bound states that the total entropy passing through the light sheet does not exceed $A/4G$. The key to a proof is again the focusing theorem. We observe that at the screen the area of the outgoing bundle of light rays is increasing as we go to later times. In other words the light sheet has positive expansion into the future and negative expansion into the past. The focusing theorem again tells us that if we map the entropy of black holes passing through the light sheet to the screen, the resulting density satisfies the holographic bound.

It is now easy to see why we concentrate on light sheets instead of space like surfaces. Obviously if the spatial entropy density is uniform and we choose Γ big enough, the entropy will exceed the area. However if Γ is larger than the particle horizon at time t the light sheet is not a cone but rather a truncated cone which is cut off by the big bang at $t = 0$. Thus a portion of the entropy present at time t never passed through the light sheet. If we only count that portion of the entropy which did pass through the light sheet it will scale like the area A. We will return to the question of space-like bounds after discussing Bousso's generalization [21] of the FS bound.

2.4. BOUSSO'S GENERALIZATION

Consider an arbitrary cosmology. Take a space-like region Γ bounded by the space-like boundary $\partial\Gamma$. Following Bousso [21], at any point on the boundary we can construct four light rays that are perpendicular to the boundary. We will call these the four *branches.* Two branches go toward the future. One of them is composed of outgoing rays and the other is ingoing. Similarly two branches go to the past. On any of these branches a light ray, together with its neighbors define a positive or negative expansion as we move away from the boundary. In ordinary flat space-time if $\partial\Gamma$ is convex the outgoing (ingoing) rays have positive (negative) expansion. However in non-static universes other combinations are possible. For example in a rapidly contracting universe the outgoing future branch may have negative expansion.

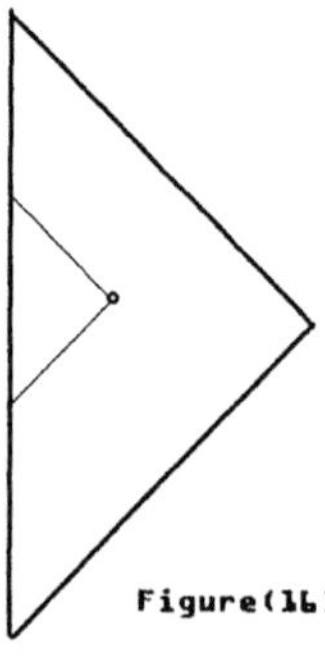

Figure(16)

If we consider general boundaries the sign of the expansion of a given branch may vary as we move over the surface. For simplicity we restrict attention to those regions for which a given branch has a unique sign. We can now state Bousso's rule:

From the boundary $\partial\Gamma$ construct all light sheets which have negative expansion as we move away. These light sheets may terminate at the tip of a cone or a caustic or even a boundary of the geometry. Bousso's bound states that the entropy passing through these light sheets is less that $A/4G$ where A is the boundary of $\partial\Gamma$.

To help visualize how Bousso's construction works we will consider spherically symmetric geometries and use Penrose diagrams to describe them. The Penrose diagram represents the radial and time directions. Each point of such a diagram really stands for a 2-sphere (more generally a $(d-1)$-sphere). The four branches at a given point on the Penrose diagram are represented by a pair of 45 degree lines passing through that point. However we are only interested in the branches of negative expansion. For example in figure(16) we illustrate flat space-time and the negative expansion branches of a typical local 2-sphere.

In general as we move around in the Penrose diagram the particular branches which have negative expansion may change. For example if the cosmology initially expands and then collapses, the outgoing future branch will go from positive to negative expansion. Bousso introduced a notation to indicate this. The Penrose diagram is divided into a number of regions depending on which branches have negative expansion. In each region the negative expansion branches are indicated by their directions at a typical point. Thus in figure (17) we draw the Penrose-Bousso (PB) diagram for a positive curvature, matter dominated universe that begins with a bang and ends with a crunch. It consists of four distinct regions.

In region I of figure (17) the expansion of the universe causes both past branches to have negative expansion. Thus we draw light surfaces into the past. These light surfaces terminate on the initial boundary of the geometry and are similar to the truncated cones that we discussed in the flat RW case. The holographic bound in this case says that the entropy passing through either backward light surface is bounded by the area of the 2-sphere at point p. Bousso's rule tells us nothing in this case about the entropy on space like surfaces bounded by p.

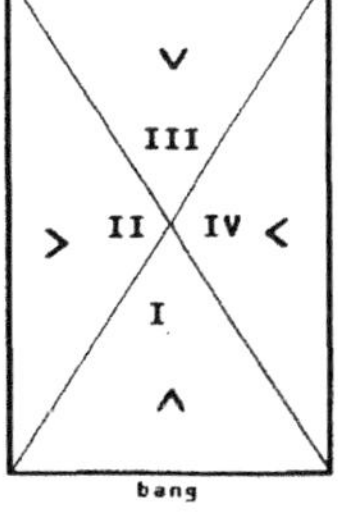

Figure (17)

Now move on to region II. The relevant light sheets in this region begin on the 2-sphere q and both terminate at the spatial origin. These are untruncated cones and the entropy on both of them is holographically bounded. There is something new in this case. We find that the entropy is bounded on a future light sheet. Now consider a space like surface bounded by q and extending to the spatial origin. It is evident that any matter which passes through the space-like surface must also pass through the future light sheet. By the second law of thermodynamics the entropy on the space-like surface can not exceed the entropy on the future light sheet. Thus in this case the entropy in a space-like region can be holographically bounded. Thus, one condition for a space-like bound is that the entropy is bounded by a corresponding future light sheet. With this in mind we return to region I. For region I there is no future bound and therefore the entropy is not bounded on space-like regions with boundary p.

In region III the entropy bounds are both on future light sheets. Nevertheless there is no space-like bound. The reason is that not all matter which passes through space-like surfaces is forced to pass through the future light sheets.

Region IV is identical to region II with the spatial origin being replaced by the diametrically opposed antipode. Thus we see that there are light-like bounds in all four regions but only in II and IV are there holographic bounds on space-like regions.

Another example of interest is inflationary cosmology. The PB diagram for de-Sitter space is shown in figure (18a). This time region I has both light sheets pointing to the future. This is due to the fact that de-Sitter space is initially contracting. In order to describe inflationary cosmology we must terminate the de Sitter space at some late time and attach it to a conventional RW space. This is shown in figure (18b). The dotted line where the two geometries are joined is the reheating surface where the entropy of the universe is created.

Let us focus on the point p in figure (18b). It is easy to see that in an ordinary inflationary cosmology p can be chosen so that the entropy on the space-like surface $p - q$ is bigger than the area of p. However Bousso's rule applied to point (p) only bounds the entropy on the past light sheet. In this case most of the newly formed entropy on the reheating surface is not counted since it never passed through the past light sheet. Typical inflationary cosmologies can be studied to see that the past light sheet bound is not violated.

As a final example we consider anti-de Sitter (AdS) space. The PB diagram consists of an infinite strip bounded on the left by the spatial origin and of the right by the AdS boundary. The PB diagram consists of a single region in which both negative expansion light sheets point toward

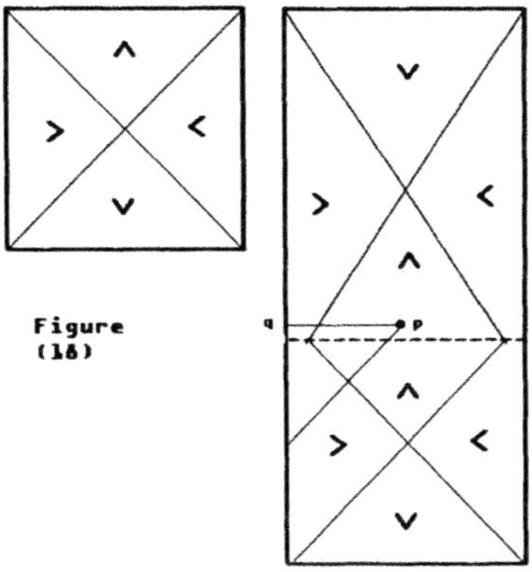

Figure (18)

the origin. Let us consider a static surface of large area A far from the spatial origin. The surface is denoted by the dotted vertical line L in figure (19). We will think of L as an infrared cutoff.

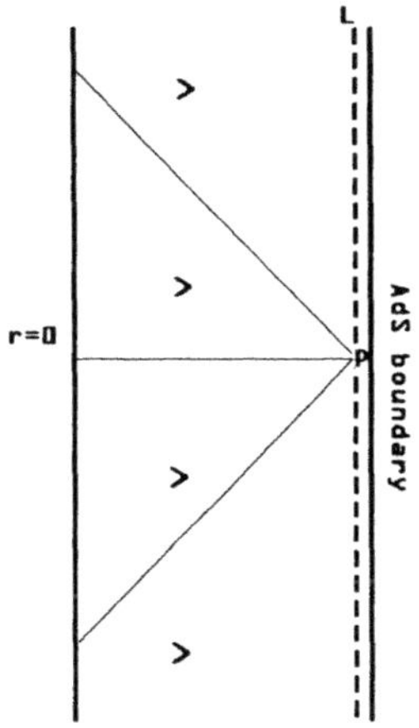

Figure (19)

Consider an arbitrary point p on L. Evidently Bousso's rules bound the entropy on past and future light sheets bounded by p. Therefore the entropy on any space-like surface bounded by p and including the origin is also holographically bounded. In other words the entire region to the left of L can be foliated with space-like surfaces such that the maximum entropy on each surface is $A/4G$.

AdS space is an example of a special class of geometries which have time-like killing vectors and which can be foliated by surfaces that satisfy the Holographic bound. These two properties imply a very far reaching conclusion. All physics taking place in such backgrounds (in the interior of the infrared cutoff L) must be described in terms of a Hamiltonian that acts in a Hilbert space of dimensionality

$$N_{states} = \exp(A/4G) \tag{2.9}$$

The holographic description of AdS space is the subject of the next lecture.

3. The AdS/CFT Correspondence and the Holographic Principle

3.1. ADS SPACE

As we saw in Lecture II, AdS space enjoys certain properties which make it a natural candidate for a holographic Hamiltonian description. In this lecture we will review the holographic description of $AdS(5) \otimes S(5)$ [23] [24][25]. This space arises in type IIb string theory, either as the near horizon geometry of a stack of D3-branes or as a solution of ten dimensional supergravity. We will begin with a brief review of AdS geometry.

For our purposes 5 dimensional AdS space may be considered to be a solid 4 dimensional spatial ball times the infinite time axis. The geometry can be described by dimensionless coordinates t, r, Ω where t is time, r is the radial coordinate ($0 \leq r < 1$) and Ω parametrizes the unit 3-sphere. The geometry has uniform curvature R^{-2} where R is the radius of curvature. The metric we will use is

$$ds^2 = \frac{R^2}{(1-r^2)^2}\left\{(1+r^2)^2 dt^2 - 4dr^2 - 4r^2 d\Omega^2\right\} \tag{3.1}$$

There is another form of the metric which is in common use,

$$ds^2 = \frac{R^2}{y^2}\left\{dt^2 - dx^i dx^i - dy^2\right\} \tag{3.2}$$

where i runs from 1 to 3.

The metric (3.2) is related to (3.1) in two different ways. First of all it is an approximation to (3.1) in the vicinity of a point on the boundary at $r = 1$. The 3 sphere is replaced by the flat tangent plane parameterized by x^i and the radial coordinate is replaced by y with $y = (1 - r)$.

The second way that (3.1) and (3.2) are related is that (3.2) is the exact metric of an incomplete patch of AdS space. A time-like geodesic can get to $y = \infty$ in a finite proper time so that the space in eq. (3.2) is not geodesically complete. The metric (3.2) describes the near horizon geometry of a stack of D3-branes located at the horizon $y = \infty$. The metric (3.2) may be expressed in terms of the coordinate $z = 1/y$.

$$ds^2 = R^2\left\{z^2(dt^2 - dx^i dx^i) - \frac{1}{z^2}dz^2\right\} \tag{3.3}$$

In this form the horizon is at $z = 0$ and the boundary is at $z = \infty$.

To construct the space $AdS(5) \otimes S(5)$ all we have to do is define 5 more coordinates ω_5 describing the unit 5 sphere and add a term to the metric

$$ds_5^2 = R^2 d\omega_5^2 \tag{3.4}$$

Although the boundary of AdS is an infinite proper distance from any point in the interior of the ball, light can travel to the boundary and back in a finite time. For example, it takes a total amount of (dimensionless) time $t = \pi$ for light to make a round trip from the origin at $r = 0$ to the boundary at $r = 1$ and back. For all practical purposes AdS space behaves like a finite cavity with reflecting walls. The size of the cavity is of order R. In what follows we will think of the cavity size R as being much larger than any microscopic scale such as the Planck or string scale.

3.2. HOLOGRAPHY IN ADS SPACE

In order to have a benchmark for the counting of degrees of freedom in $AdS(5) \otimes S(5)$ imagine constructing a cutoff field theory in the interior of the ball. A conventional cutoff would involve a microscopic length scale such as the 10 dimensional Planck length l_p. One way to do this would be to introduce a spatial lattice in nine dimensional space . It is not generally possible to make a regular lattice but a random lattice with an average spacing l_p is possible. We can then define a simple theory such as a Hamiltonian lattice theory on the space. In order to count degrees of freedom we also need to regulate area of the boundary of AdS which is infinite. The way to do that was hinted at in lecture II. We introduce a surface L at $r = 1 - \delta$. The total 9 dimensional spatial volume in the interior of L is easily computed using the metric (3.1).

$$V(\delta) \sim \frac{R^9}{\delta^3} \tag{3.5}$$

and the number of lattice sites and therefore the number of degrees of freedom is

$$\frac{V}{l_p^9} \sim \frac{1}{\delta^3} \frac{R^9}{l_p^9} \tag{3.6}$$

In such a theory we also will find that the maximum entropy is of the same order of magnitude.

On the other hand the holographic bound discussed in lecture II requires the maximum entropy and the number of degrees of freedom to be of order

$$S_{max} \sim \frac{A}{l_p^8} \tag{3.7}$$

where A is the 8 dimensional area of the boundary L. This is also easily computed. We find

$$S_{max} \sim \frac{1}{\delta^3} \frac{R^8}{l_p^8} \tag{3.8}$$

In other words when R/l_p becomes large the holographic description requires a reduction in the number of independent degrees of freedom by a factor l_p/R. To say it slightly differently, the holographic principle implies a complete description of all physics in the bulk of a very large AdS space in terms of only l_p/R degrees of freedom per spatial Planck volume.

3.3. THE ADS/CFT CORRESPONDENCE

The correspondence between string theory in $AdS(5) \otimes S(5)$ and Super Yang Mills (SYM) theory on the boundary has been discussed in other lectures in this school and we will only review some of the salient features. The correspondence states that there is a complete equivalence between superstring theory in the bulk of $AdS(5) \otimes S(5)$ and maximally supersymmetric (16 real supercharges), $3+1$ dimensional, $SU(N)$, SYM theory on the boundary of the AdS space [23][24][25]. In these lectures SYM theory will always refer to this particular version of supersymmetric gauge theory.

It is well known that SYM is conformally invariant and therefore has no dimensional parameters. It will be convenient to define the theory to live on the boundary parametrized by the dimensionless coordinates t, Ω or t, x. The corresponding momenta are also dimensionless. In fact we will use the convention that all SYM quantities are dimensionless. On the other hand the bulk gravity theory quantities such as mass, length and temperature carry their usual dimensions. To convert from SYM to bulk variables the conversion factor is R. Thus if E_{sym} and M represent the energy in the SYM and bulk theories

$$E_{sym} = MR$$

Similarly bulk time intervals are given by multiplying the t interval by R.

One might think that the boundary of $AdS(5) \otimes S(5)$ is $(8+1)$ dimensional but there is an important sense in which it is $3+1$ dimensional. To see this let us Weyl rescale the metric by a factor $\frac{R^2}{(1-r^2)^2}$ so that the rescaled metric at the boundary is finite. The new metric is

$$dS^2 = \left\{(1+r^2)^2 dt^2 - 4dr^2 - 4r^2 d\Omega^2\right\} + \left\{(1-r^2)^2 d\omega_5^2\right\} \tag{3.9}$$

Notice that the size of the 5-sphere shrinks to zero as the boundary at $r=1$ is approached. The boundary of the geometry is therefore $3+1$ dimensional.

Let us return to the correspondence between the bulk and bounday theories. The ten dimensional bulk theory has two dimensionless parameters. These are:

1. The radius of curvature of the AdS space measured in string units R/l_s

2. The dimensionless string coupling constant g.

The string coupling constant and length scale are are related to the ten dimensional Planck length and Newton constant by

$$l_p^8 = g^2 l_s^8 = G \tag{3.10}$$

On the other side of the correspondence, the gauge theory also has two constants. They are

1. The rank of the gauge group N

2. The gauge coupling g_{ym}

The relation between the string and gauge parameters was given by Maldacena [23]. It is

$$\begin{aligned} \frac{R}{l_s} &= (N g_{ym}^2)^{\frac{1}{4}} \\ g &= g_{ym}^2 \end{aligned} \tag{3.11}$$

We can also write ten dimensional Newton constant in the form

$$G = R^8/N^2 \tag{3.12}$$

There are two distinct limits that are especially interesting, depending on one's motivation. The AdS/CFT correspondence has been widely studied as a tool for learning about the behavior of gauge theories in the strongly coupled 't Hooft limit. From the gauge theory point of view the 't Hooft is defined by

$$\begin{aligned} g_{ym} &\to 0 \\ N &\to \infty \\ g_{ym}^2 N &= constant \end{aligned} \tag{3.13}$$

From the bulk string point of view the limit is

$$\begin{aligned} g &\to 0 \\ \frac{R}{l_s} &= constant \end{aligned} \tag{3.14}$$

Thus the strongly coupled 't Hooft limit is also the classical string theory limit in a fixed and large AdS space. This limit is dominated by classical supergravity theory.

The interesting limit from the viewpoint of the holographic principle is a different one. We will be interested in the behavior of the theory as the AdS radius increases but with the parameters that govern the microscopic physics in the bulk kept fixed. This means we want the limit

$$\begin{aligned} g &= constant \\ R/l_s &\rightarrow \infty \end{aligned} \tag{3.15}$$

On the gauge theory side this is

$$\begin{aligned} g_{ym} &= constant \\ N &\rightarrow \infty \end{aligned} \tag{3.16}$$

Our goal will be to show that the number of quantum degrees of freedom in the gauge theory description satisfies the holographic behavior in eq. (3.8).

3.4. THE INFRARED ULTRAVIOLET CONNECTION

In either of the metrics (3.1) or (3.2) the proper area of any finite coordinate patch tends to ∞ as the boundary of AdS is approached. Thus we expect that the number of degrees of freedom associated with such a patch should diverge. This is consistent with the fact that a continuum quantum field theory such as SYM has an infinity of modes in any finite three dimensional patch. In order to do a more refined counting [26] we need to regulate both the area of the AdS boundary and the number of ultraviolet degrees of freedom in the SYM. As we will see, these apparently different regulators are really two sides of the same coin. We have already discussed infrared (IR) regulating the area of AdS by introducing a surrogate boundary L at $r = 1 - \delta$ or similarly at $y = \delta$.

That the the IR regulator of the bulk theory is equivalent to an ultraviolet (UV) regulator in the SYM theory is called the IR/UV connection [26]. It can be motivated in a number of ways. In this lecture we give an argument based on the quantum fluctuations of the positions of the D3-branes which are nominally located at the origin of the coordinate z in eq. (3.3). The location of a point on a 3 brane is defined by six coordinates z, ω_5. We may also choose the six coordinates to be cartesian coordinates $(z^1, ..., z^6)$. The original coordinate z is defined by

$$z^2 = (z^1)^2 + ... + (z^6)^2 \tag{3.17}$$

The coordinates z^m are represented in the SYM theory by six scalar fields on the world volume of the branes. If the six scalar fields ϕ^n are canonically normalized then the precise connection between the $z's$ and $\phi's$ is

$$z = \frac{g_{ym} l_s^2}{R^2} \phi \tag{3.18}$$

Strictly speaking eq.(3.18) does not make sense because the fields ϕ are $N \times N$ matrices. The situation is the same as in matrix theory where we identify the N eigenvalues of the matrices in eq.(3.18) to be the coordinates z^m of the N D3-branes. As in matrix theory the geometry is noncommutative and only configurations in which the six matrix valued fields commute have a classical interpretation. However the radial coordinate $z = \sqrt{z^m z^m}$ can be defined by

$$z^2 = \left(\frac{g_{ym} l_s^2}{R^2}\right)^2 \frac{1}{N} Tr\phi^2 \tag{3.19}$$

A question which is often asked is; Where are the D3-branes located in the AdS space? The usual answer is that they are at the horizon $z = 0$. However our experiences in lecture I with similar questions should warn us that the answer may be more subtle. In lecture I (see the discussion from eq(1.45) to eq.(1.52)) a question was asked about the location of a string. What we found is that the answer depends on what frequency range it is probed with. High frequency or short time probes see the string widely spread in space while low frequency probes see a well localized string.

To answer the corresponding question about D3-branes we need to study the quantum fluctuations of their position. The fields ϕ are scalar quantum fields whose scaling dimensions are known to be exactly $(length)^{-1}$. From this it follows that any of the N^2 components of ϕ satisfies

$$\langle \phi_{ab}^2 \rangle \sim \delta^{-2} \tag{3.20}$$

where δ is the ultraviolet regulator of the field theory. It follows from eq(3.20) that the average value of z satisfies

$$< z >^2 \sim \left(\frac{g_{ym} l_s^2}{R^2}\right)^2 \frac{N}{\delta^2} \tag{3.21}$$

or, using eq's(3.12)

$$< z >^2 \sim \delta^{-2} \tag{3.22}$$

In terms of the coordinate y which vanishes at the boundary of AdS

$$< y >^2 \sim \delta^2 \tag{3.23}$$

Evidently low frequency probes see the branes at $z = 0$ but as the frequency of the probe increases the brane appears to move toward the boundary at $z = \infty$. The precise connection between the UV SYM cutoff and the bulk-theory IR cutoff is given by eq.(3.23).

3.5. COUNTING DEGREES OF FREEDOM

Let us now turn to the problem of counting the number of degrees of freedom needed to describe the region $y > \delta$ [26]. The UV/IR connection implies that this region can be described in terms of an ultraviolet regulated theory with a cutoff length δ. Consider a patch of the boundary with unit coordinate area. Within that patch there are $1/\delta^3$ cutoff cells of size δ. Within each such cell the fields are constant in a cutoff theory. Thus each cell has of order N^2 degrees of freedom corresponding to the $N \otimes N$ components of the adjoint representation of $U(N)$. Thus the number of degrees of freedom on the unit area is

$$N_{dof} = \frac{N^2}{\delta^3} \tag{3.24}$$

On the other hand the 8-dimensional area of the regulated patch is

$$A = \frac{R^3}{\delta^3} \times R^5 = \frac{R^8}{\delta^3} \tag{3.25}$$

and the number of degrees of freedom per unit area is

$$\frac{N_{dof}}{A} \sim \frac{N^2}{R^8} \tag{3.26}$$

Finally we may use eq.(3.12)

$$\frac{N_{dof}}{A} \sim \frac{1}{G} \tag{3.27}$$

This is exactly what is required by the holographic principle.

3.6. ADS BLACK HOLES

The apparently irreconcilable demands of black hole thermodynamics and the principles of quantum mechanics have led us to a very strange view of the world as a hologram. Now we will return, full circle, to see how the holographic description of $AdS(5) \otimes S(5)$ provides a description of black holes. What would be most interesting would be to give a holographic description of 10-dimensional black hole formation and evaporation in an $AdS(5) \otimes S(5)$ space which is much larger than the black hole. Unfortunately we will see that this is far beyond our present ability. There are however, black hole solutions in $AdS(5) \otimes S(5)$ which are within our current understanding. These are the black holes which have Schwarzschild radii as large or larger than the radius of curvature R. Such black holes are stable against decay and do not evaporate. In fact these black holes homogeneously fill the

5-sphere. They are solutions of the dimensionally reduced 5-dimensional Einstein equations with a negative cosmological constant. The thermodynamics can be derived from the black hole solutions by first computing the area of the horizon and then using the Bekenstein Hawking formula

One finds that the entropy is related to their mass by

$$S = c\left(M^3 R^{11} G^{-1}\right)^{\frac{1}{4}} \tag{3.28}$$

Where G is the ten dimensional Newton constant and c is a numerical constant. Using the thermodynamic relation $dM = TdS$ we can compute the relation between mass and temperature.

$$M = c\frac{R^{11}T^4}{G} \tag{3.29}$$

or in terms of dimensionless SYM quantities

$$\begin{aligned} E_{sym} &= c\frac{R^8}{G}T^4_{sym} \\ &= cN^2 T^4_{sym} \end{aligned} \tag{3.30}$$

Eq.(3.30) has a surprisingly simple interpretation. Recall that in $3+1$ dimensions the Stephan-Boltzmann law for the energy density of radiation is

$$E = T^4 V \tag{3.31}$$

where V is the volume. In the present case the relevant volume is the dimensionless 3-area of the unit boundary sphere. Furthermore there are $\sim N^2$ quantum fields in the $U(N)$ gauge theory so that apart from a numerical constant eq.(3.30) is nothing but the Stephan-Bolzmann law for black body radiation. Evidently the holographic description of the AdS black holes is a simple as it could be; a black body thermal gas of N^2 species of quanta propagating on the boundary hologram.

3.7. THE HORIZON

The high frequency quantum fluctuation of the location of the D3-branes are invisible to a low frequency probe. Roughly speaking this is insured by the renormalization group as applied to the SYM description of the branes. The renormalization group is what insures that our bodies are not severely damaged by constant exposure to high frequency vacuum fluctuations. We are not protected in the same way from classical fluctuations. An example is the thermal fluctuations of fields at high temperature. All probes sense thermal fluctuations of the brane locations. Let us return to eq.(3.20) but now, instead of using eq.(3.21) we use the thermal field fluctuations of ϕ. For each of the N^2 components the thermal fluctuations have the form

$$< \phi^2 >= T_{sym}^2 \tag{3.32}$$

and we find eqs.(3.22) and (3.23) replaced by

$$\begin{aligned} < z >^2 &\sim T_{sym}^2 \\ < y >^2 &\sim T_{sym}^{-2} \end{aligned} \tag{3.33}$$

It is clear that the thermal fluctuations will be strongly felt out to a coordinate distance $z = T_{sym}$. In terms of r the corresponding position is

$$1 - r \sim 1/T_{sym} \tag{3.34}$$

In fact this coincides with the location of the horizon of the AdS black hole.

A more precise definition of the horizon was given by Kabat and Lifschytz [27]. In the D-brane description the zero temperature stack of branes can be thought of as an extreme black brane with the horizon at $z = 0$. We would like to find something special about the corresponding point $\phi = 0$ in the SYM description. Let us displace one of the branes of the stack to a classical location z. At zero temperature supersymmetry insures the stability of this configuration. From the gauge theory point of view we have shifted a scalar field and broken the gauge symmetry to $U(1) \otimes U(N-1)$. The effect is to give the "W-bosons" a mass $g\phi$. From the brane point of view we have given a mass to the strings which extend between the displaced brane at z and the others at $z = 0$. Now we see what is special about $z = 0$. If we place a brane probe at a distance from the horizon there are massive modes of the brane. These modes become massless at the horizon. Presumably if we went even further these modes would become tachyonic and lead to an instability involving the irreversible production of strings connecting the probe and stack.

Kabat and Lifschytz [27] conjecture that this is the general feature of horizons in both the AdS/CFT theory and Matrix theory. In the AdS case we begin with a spontaneously broken SYM at finite temperature. It is well known that the mass of the W boson is corrected by finite temperature effects. Kabat and Lifschytz argue that at finite temperature the tachyonic instability occurs at a non-zero value of ϕ. This value corresponds to the position of the horizon.

The string theory correspondence gives a fairly convincing picture of the thermal effects on the W mass [27]. Let the probe brane be at z. The thermal effects are represented by a black hole or black brane with a horizon at z_H. We assume $z > z_H$. Now the string connecting the probe to the stack is terminated at the black hole horizon and its mass is

$$M = (z - z_H)/l_s \tag{3.35}$$

As $z \to z_H$ the string becomes massless and then tachyonic.

4. The Flat Space Limit

4.1. THE FLAT SPACE LIMIT

Gauge theory, gravity correspondences are especially interesting because they provide nonperturbative definitions of some quantum-gravity systems. The first example was matrix theory which uses SYM theory to define 11 dimensional supergravity in the DLCQ framework. To effectively decompactify the light cone direction we must pass to the large N limit keeping the gauge coupling fixed.

It has also been proposed that the AdS/CFT correspondence can be used to give a non-perturbative definition of type IIb string theory [28]. For this purpose we regard AdS space in the form of eq.(3.1) as a finite cavity with reflecting walls. It provides an ideal "box" for the purpose of infrared regulating a theory. Although the actual metric distance from any point in the bulk geometry to the boundary is infinite, it nevertheless closely resembles an ordinary finite box of size R. For example the time for light to propagate from $r = 0$ to the boundary and back is finite πR. Another indication of the finiteness of the box is that the energy eigenvalues of a particle moving in the metric (3.1) are discrete with the scale of energy being $1/R$.

To define the infinite volume limit we want to let $R \to \infty$ while keeping fixed the microscopic parameters of the theory such as g and l_s. We also want to keep fixed the energy and length scales in string units. Let us see what this means in terms of SYM quantities. From eq's(3.11) we see that we must allow $N \to \infty$ while keeping g_{ym} fixed just as in matrix theory. Furthermore the SYM energy is related to the mass M by $E_{sym} = MR = Ml_s(Ng_{ym}^2)^{\frac{1}{4}}$ Accordingly, to keep M fixed we must allow E_{sym} to grow like $N^{\frac{1}{4}}$ while time intervals must scale like $t \to N^{-\frac{1}{4}}$. Matrix theory also requires a scaling of energy with N but it is different. Instead of eq.(4.1) matrix theory involves energy of order $1/N$.

The next question is what quantities make sense in the limit

$$\begin{array}{ccc} N & \to & \infty \\ g_{ym} & = & constant \\ E_{sym} & \to N^{\frac{1}{4}} & \end{array} \tag{4.1}$$

The answer must be that any quantity that has a well defined flat space limit in ten dimensional IIb string theory should correspond to a quantity with a good limit under (4.1). The most obvious quantities are the spectrum and scattering matrix of stable particles. The only such particles are the massless supergravity multiplet. This includes Kaluza-Klein particles with non-zero momentum on the 5-sphere. From the point of view of the 5 dimensional AdS space these objects have non zero mass but they are stable. The 5-dimensional AdS mass of a particle with momentum k on the 5-sphere is

$$M = |k| \tag{4.2}$$

or in terms of the $S(5)$ angular momentum J

$$M = J/R \tag{4.3}$$

The existence and stability of these ten-dimensionally massless particles has been established beyond doubt from properties of the SYM theory. The existence and properties of an S-matrix have also been studied [29][28] but much less can be rigorously established. The idea for constructing scattering amplitudes is to use appropriate local gauge-invariant operators in the boundary theory as sources of the bulk particles. The particles can be aimed from the boundary toward the origin ($r = 0$) of the cavity and by carefully controlling the sources they can be made to interact in a small enough region that the curvature of the space is irrelevant. All kinds of interesting phenomena could occur during the collision. This includes the formation and evaporation of 10 dimensional black holes. You can look up the details of this kind of construction in the papers by Polchinski and Susskind [28] [29]. In this lecture we will concentrate on a couple of the poorly understood issues connected with the holographic description of in the interior of AdS.

4.2. HIGH ENERGY GRAVITONS DEEP IN THE BULK

The first issue has to do with the description of high energy particles far from the boundary. Let us consider a massless graviton emitted from the boundary with vanishing $S(5)$ momentum. The creation operator for emitting the graviton is made out of the energy-momentum tensor of the boundary theory by integrating T_{ij} with a test function whose frequency spectrum is concentrated around some value ω.

$$\omega = pR = pl_s(g_{ym}^2 N)^{\frac{1}{4}} \tag{4.4}$$

Acting with the resulting operator creates a graviton of bulk momentum p propagating from the boundary toward the origin.

Once the particle has entered the bulk and passed the surrogate boundary at $y = \delta$, the holographic principle requires that it has a description in the regulated SYM theory with momentum cutoff $1/\delta$. Let us first consider the case of low graviton momentum by which we mean $pR = \omega < 1/\delta$. In this case the source function is slowly varying on the cutoff scale and the ordinary renormalization group strategy applies. Integrating out the modes beyond the cutoff results in a renormalized theory. Because the SYM theory is scale invariant, the cutoff theory has the same form as the original theory and the graviton description is the same as in the continuum theory.

However, the renormalization group does not apply to situations in which the field theoretic source functions vary more rapidly than the cutoff scale. Thus if $(p > \delta/R)$ there is no guarantee that the cutoff theory can describe the graviton correctly. The problem is that the holographic principle demands that we be able to describe all the physical states in the region $y > \delta$ by states of the cutoff theory even if they contain high energy gravitons.

To phrase the paradox differently, note that a massless particle with momentum p moving in the y direction can be localized in the x plane with an uncertainty

$$R\Delta x \sim \frac{1}{p} \tag{4.5}$$

Thus it should be possible to distinguish two such particles if their separation x is of order $1/pR$ or bigger. On the other hand the largest momenta in the cutoff SYM theory is $1/\delta << pR$. How is it possible to construct such well localized objects out of the low momentum modes of the SYM fields? We will argue that the only possible answer is that the high energy graviton is created by operators that involve many SYM quanta. In other words the effective operator which creates the high energy graviton in the cutoff theory must be high order in the fundamental SYM fields.

The order can be estimated by taking the total dimensionless energy ω of the graviton and dividing up among gauge quanta of energy $1/\delta$.

$$n = \omega\delta = pR\delta \tag{4.6}$$

To illustrate the point consider an n-particle wave function (as long as $n << N$ the SYM quanta can be treated as non-identical Boltzmann particles). As an example we choose a product wave function

$$\psi(x_1, x_2, \ldots, x_n) = \psi(x_1)\psi(x_2)\ldots\psi(x_n) \tag{4.7}$$

with

$$\psi(x) = \exp - \left(\frac{|x|}{\delta}\right) \tag{4.8}$$

Note that wave functions of this type are composed of momenta of order $1/\delta$ and make sense in the cutoff theory.

Suppose we have two such states which are identical except one of them is displaced a distance a in the x direction. The inner product of these states is given by

$$\left\{\int \psi^*(x)\psi(x-a)\right\}^n \sim \exp -na/\delta \tag{4.9}$$

The function $exp{-}na/\delta$ in eq.(4.9) is narrowly peaked on the cutoff scale if n is large. In other words these states are distinguishable when they are displaced by distance δ/n even though the largest individual momentum is only $1/\delta$.

Thus we see that fine details can be distinguished in the coarse grained theory but only if the gravitons and other bulk particles are identified as an increasingly large number of gauge quanta as the UV cutoff of the SYM is lowered and/or the momentum is increased. This is very similar to matrix theory in which a graviton of momentum P_- is represented by a number of partons which grow with P_-.

4.3. KALUZA KLEIN MODES

So far we have considered particles which are massless in the 5 dimensional sense. Now let us consider a graviton with non-vanishing 5-momentum k. We want to hold k fixed as we let $R \to \infty$. The 5 dimensional mass is k. Let us also assume p, the momentum in the y direction is also kept fixed. The dimensionless SYM energy of the state is

$$\omega = R\sqrt{k^2 + p^2} \tag{4.10}$$

Once again it is known how to create such particles by introducing a source at the boundary. The source in this case is a local gauge invariant SYM operator of the form

$$S_n = Tr(\phi)^n \tag{4.11}$$

This expression stands for an nth order monomial in the scalar SYM fields ϕ. The integer n is equal to the $S(5)$ angular momentum kR.

$$n = kR \tag{4.12}$$

To construct a creation operator for a particle of momentum p, k we integrate S_n with a test function of frequency ω given in eq.(4.10).

The puzzling feature of this prescription is that it injects the particle into the system with a local boundary operator. But a massive particle with energy $\sqrt{k^2 + p^2}$ can never get near the boundary. This can be seen from the motion of a massive classical particle in AdS space. If a particle of mass M moves along the y axis with total bulk energy $E = \omega/R$ then the closest it comes to the boundary is

$$y^* = M/E \tag{4.13}$$

where y^* is the classical turning point of the trajectory. It is also true that the solution of the classical wave equation for such a particle has its largest value at this point. For $y < y*$ the wave function quickly goes to zero. Somehow the local boundary field S_n must be creating bulk particles far from the boundary.

This behavior can be qualitatively be understood in an elementary way from the SYM theory. The operator S_n in eq.(4.11) describes the creation of n quanta. Suppose that the SYM energy ω is divided among the quanta so that each carries ω/n. Equivalently the quanta have wave length n/ω. According to the UV/IR connection quanta of this wave length correspond to bulk phenomena at $y = n/\omega$. Using eq's.(4.10) and (4.12) we see that this corresponds to the position $y*$. In this way we see that the local operator constructed from S_n by projecting out given frequency components actually corresponds to a bulk particle at its classical turning point.

Before concluding this final lecture the are some negative features of holographic descriptions which need to be mentioned. These negative features become apparent when we begin to ask how ordinary phenomena near the origin of a very large AdS space are described in SYM theory [30] [31]. Suppose we have some object which may be macroscopic in size but which is very much smaller than the radius of curvature R. According to the UV/IR connection if the object is near the origin only the longest wavelength modes of the SYM fields should be important for their description. On the 3-sphere this means the almost homogeneous modes. The number of such homogeneous modes is of order N^2 and these must be the degrees of freedom which describe entire physics within a region of size R near the origin. In other words all the physics within a region small enough to be considered flat must be described by the matrix degrees of freedom of the SYM and not by the spatial variations of the fields. There is nothing wrong with this except that we have no idea how to translate ordinary physics into the holographic description. For example we would have no idea how to determine if a given SYM state were describing a small ten dimensional black hole, a rock or an elephant of the same mass.

I would like to suggest that there is a way to do physics which is complementary to the holographic way but in which bulk phenomena are much easier to recognize. I would expect that this new way would be in terms of local bulk fields which would either include the gravitational field or would allow its construction in some simple way. What would be unusual about this theory is that it would be extremely rich in gauge redundancies, so rich in fact that when the gauge is completely fixed and the non-redundant degrees of freedom are counted their number would be proportional to the area in Planck units. By some particular gauge fixing this would be made manifest. But after insuring ourselves that the counting is holographic other gauge choices might be much better for recognizing ordinary local physics. The kind of theory I have in mind is some generalization of Chern Simons theory which does have the property that the real states live on the boundary. Unfortunately this is just a speculation at the moment.

References

1. G. 't Hooft, "Dimensional Reduction in Quantum Gravity," gr-qc/9310026.
2. L. Susskind, "The World as a Hologram," hep-th/9409089.
3. Bekenstein," Black Holes And The Second Law of Thermodynamics", *Lett. Nuovo Cim.* **4** (1972) 737
4. S. Hawking, "Particle Creation By Black Holes..." , *Commun. Math. Phys.* **43** (1975) 199
5. S. Hawking, "Breakdown Of Predictability In Gravitational Collapse...", *Phys.Rev. D* **14** (1976) 2460
6. C. R. Stephens, G. 't Hooft, B. F. Whiting, "Black Hole Evaporation without Information Loss", *Class. Quant. Grav.* **11** (1994) 621-648, gr-qc/9310006
7. T. Banks, L. Susskind, M. E. Peskin, Difficulties for the evolution of pure states into mixed states, SLAC-PUB-3258, Dec 1983; *Nucl. Phys. B* **244** (1984) 125
8. L. Susskind, L. Thorlacius, and J. Uglum, The stretched horizon and black hole complementarity, *Phys. Rev. D* **48** (1993) 3743-3761, hep-th/9306069
9. Unruh, "Notes On Black Hole Evaporation...," *Phys. Rev. D* **14** (1976) 870
10. "The Membrane Paradigm", edited by K. S. Thorne, R. H. Price, and D.A. Macdonald, Yale University Press, New Haven
11. G. 't Hooft, "On The Quantum Structure Of A Black Hole...," *Nucl. Phys. B* **256** (1985) 727
12. J. J. Atick, E. Witten, "The Hagedorn transition and the number of degrees of freedom of string theory", *Nucl. Phys. B* **310** (1988) 291
13. L. Susskind and L. Thorlacius, Gedanken experiments involving black holes, *Phys. Rev. D* **49** (1994) 966, hep-th/9308100
14. D. N. Page, "Expected entropy of a subsystem," *Phys. Rev. Lett.* **71** (1993) 1291, gr-qc/9305007
15. S. Coleman and S. Hughes, "Black Holes, Wormholes, and the Disappearance of Global Charge," *Phys. Lett.* **B309** (1993) 246, hep-th/9305123
16. L. Susskind, "String theory and the principle of black hole complementarity", *Phys. Rev. Lett.* **71** (1993) 2367, hep-th/9307168
17. L. Susskind, "Strings, black holes and Lorentz contraction", *Phys.Rev. D* **49** (1994) 6606, hep-th/9308139
18. T. Yoneya, "Schild Action and Space-Time Uncertainty Principle in String Theory", *Prog. Theor. Phys.* **97** (1997) 949, hep-th/9703078,
19. S. Corley, T. Jacobson, "Focusing and the Holographic Hypothesis," *Phys. Rev. D* **53** (1996) 6720, gr-qc/9602043
20. W. Fischler, L. Susskind, "Holography and Cosmology," hep-th/9806039
21. R. Bousso, "The Holographic Principle for General Backgrounds," hep-th/9911002 R. Bousso, "Holography in General Space-times," hep-th/9906022, JHEP 9906 (1999) 028

R. Bousso, A Covariant Entropy Conjecture, hep-th/9905177, JHEP 9907 (1999) 004

22. E. E. Flanagan, D. Marolf, R M. Wald, "Proof of Classical Versions of the Bousso Entropy Bound and of the Generalized Second Law," hep-th/9908070
23. J. M. Maldacena, "The Large N Limit of Superconformal Field Theories and Supergravity," hep-th/9711 200
24. S. S. Gubser, I. R. Klebanov, A. M. Polyakov, "Gauge Theory Correlators from Non-Critical String Theory," hep-th/9802109
25. E. Witten, "Anti De Sitter Space And Holography," hep-th/9802150.
26. L. Susskind, E. Witten, "The Holographic Bound in Anti-de Sitter Space", hep-th/9805114
27. D. Kabat, G. Lifschytz, Gauge theory origins of supergravity causal structure, hep-th/9902073, JHEP 9905 (1999) 005
    D. Kabat, G. Lifschytz, Tachyons and Black Hole Horizons in Gauge Theory, hep-th/9806214, JHEP 9812 (1998) 002
28. L. Susskind, "Holography in the Flat Space Limit," hep-th/9901079
29. J. Polchinski, "S-Matrices from AdS Spacetime," hep-th/9901076
30. Joseph Polchinski, Leonard Susskind, Nicolaos Toumbas, Negative Energy, Superluminosity and Holography, hep-th/9903228, *Phys. Rev. D* **60** (1999) 084006
31. L. Susskind, N. Toumbas, "Wilson Loops as Precursors," *Phys. Rev. D* **61** (2000) 044001 hep-th/9909013

17 8'99

17 8'99

BORN-INFELD ACTIONS AND D-BRANE PHYSICS

CURTIS G. CALLAN, JR.
Department of Physics
Joseph Henry Laboratories
Princeton University
Princeton, NJ 08540,
U.S.A.
email: callan@feynman.princeton.edu

Abstract. Some very interesting things can be learned about the physics of D-branes by studying the solutions of the effective action for low-energy fluctuations of the D-brane about its ground state. Many pieces of evidence indicate that the correct action is essentially the Born-Infeld action for non-linear electrodynamics (with some modifications to adapt it to the D-brane context). We motivate this description of the dynamics of branes and then explore a series of applications, starting with the simple case of branes in flat space and working our way up to branes in the curved backgrounds of large numbers of other branes. Some of these applications make contact with the AdS/CFT conjecture of Maldacena in interesting ways and these are developed in some detail.

1. D-Brane Solitons and the Born-Infeld Action

The solitons which will be the subject of these lectures were first found in the conventional way, as classical solutions of the low-energy supergravity effective action of string theory. As is by now standard lore, the bosonic effective action of type-IIB string theory is

$$S_{IIB} \propto \int d^{10}x\sqrt{-g}[e^{-2\phi}(R+4(\nabla\phi)^2-\frac{1}{3}H^2)-({G_{(1)}}^2+{G_{(3)}}^2+{G_{(5)}}^2)]\ . \quad (1)$$

This is an action for the metric $g_{\mu\nu}$, the scalar dilaton ϕ and a collection of $U(1)$ gauge field strengths associated with distinct conserved charges.

L. Thorlacius and T. Jonsson (eds.), M-Theory and Quantum Geometry, 227-253.

All string theories have the three-form field strength $H = H_{\mu\nu\lambda}$ whose primary source is the fundamental string itself. The $G_{(r)} = G_{\mu_1 \ldots \mu_r}$ are field strengths whose sources are extended objects of various dimensions, conventionally called D-branes. The basic D-brane solution of the equations of motion is easy to write down (it was discovered nearly a decade ago [1]):

$$
\begin{aligned}
ds^2 &= \frac{1}{\sqrt{H(r)}}[-f(r)dt^2 + \sum_1^p dx_i^2] + \quad \sqrt{H(r)}[f(r)^{-1}dr^2 + r^2 d\Omega_{8-p}^2] \\
H(r) &= 1 + R^{7-p}/r^{7-p} \qquad\qquad f(r) = 1 - r_0^{7-p}/r^{7-p} \\
e^{\phi} &= H(r)^{\frac{3-p}{4}} \qquad\qquad G_{8-p} \neq 0 \qquad H_3 = 0
\end{aligned}
\tag{2}
$$

This describes an object with p dimensions of extension (described by the coordinates x_i) invariant under the $p+1$-dimensional Poincaré group. Note that in general ϕ, and therefore the effective string coupling constant g_s, vary with distance r from the brane. R is an arbitrary parameter that fixes the mass density of the soliton and r_0 is the horizon radius. In order to describe the underlying soliton, as opposed to some thermally excited object, we simply eliminate the horizon by setting $r_0 \to 0$. This is the extremal limit and, if we examine the fermionic degrees of freedom that are part of the underlying supergravity, we discover that the extremal soliton has 16 unbroken supersymmetries out of the the 32 that are unbroken in the type-II vacuum. The unbroken supersymmetries can be used to prove exact mass formulas and other useful properties of the soliton.

The p-brane charge is manifested in the flux of the G_{8-p}-form field through any $(8-p)$-sphere surrounding $r = 0$: $\int_{S_{8-p}} G_{8-p} \propto R^{7-p}$. This charge is actually quantized (in units involving the string coupling $g_s$and the string tension α') and the extremal classical solution can be regarded as a superposition of N 'elementary' solitons. The relation between classical and quantum parameters is such that

$$
R^{7-p} \propto g_s(\alpha')^{\frac{7-p}{2}} N \qquad T_p \propto \frac{1}{g_s}(\alpha')^{-\frac{p+1}{2}} N \tag{3}
$$

where T_p is the tension, or energy per unit volume, of the soliton. Note that the tension blows up as $g_s \to 0$, as is normal for a semiclassical soliton solution. A further fact that will be exploited later is that in the limit of many superposed elementary solitons ($N \to \infty$), the local curvature of space time goes to zero ($\mathcal{R} \sim R^{-2} \sim N^{-\frac{2}{7-p}}$) and the approximation of dropping string theory corrections to the description of spacetime becomes good.

But, solving the classical equations is not enough to assure us that these solitons are *bona fide* string theory objects: the supergravity description has singularities and quantizing the fluctuating zero modes of the soliton typically runs into nonrenormalizability problems. Polchinski gave a direct string theory description of the soliton that reassures us on all these points [2]. He starts with a geometrical specification of the soliton (p-brane from now on) as a p-dimensional hyperplane ($X^{p+1}, \ldots, X^9 = 0$) in flat ten-dimensional space. This hyperplane is given dynamical meaning by declaring that (despite the fact that we start from a closed string theory) the theory supports *open* strings whose ends must lie on the p-brane but are otherwise free to move. This is summarized by the following mixture of Dirichlet and Neumann boundary conditions for the string worldsheet field $X^a(\sigma, \tau)$ (with $0 < \sigma < \pi$):

$$\begin{aligned} \partial_\tau X^a(0,\tau) = \partial_\tau X^a(\pi,\tau) = 0 \qquad & a = p+1, \ldots, 9 \\ \partial_\sigma X^a(0,\tau) = \partial_\sigma X^a(\pi,\tau) = 0 \qquad & a = 0, \ldots, p \end{aligned} \tag{4}$$

The usual open string has only Neumann boundary conditions and the appearance of Dirichlet conditions is why these objects are sometimes called 'Dirichlet branes'.

The standard rules for first quantization of the string yield excited states which behave like particles of various masses in $p+1$-dimensions (the string zero mode is constrained by the Dirichlet boundary conditions to move in the hyperplane defining the soliton). The massless states are of special importance because they define the collective coordinates for small excitations of the soliton. The massless multiplet is simple, but a bit surprising: it contains one $U(1)$ gauge boson, $9-p$ scalars and some gauginos in the fermionic sector. The scalars are expected, in the sense that they are the fields that implement small displacements of the soliton transverse to itself. The gauge field is a bit surprising, but essential. A closer look at supersymmetry issues shows that the massless fields and their interactions match what one gets from dimensional reduction (to the dimension appropriate to the soliton) of the 10-dimensional $\mathcal{N} = 1$ supersymmetric $U(1)$ gauge theory. In the case that will be of the most interest to us, the threebrane, the dimensional reduction yields four gauginos and therefore an $\mathcal{N} = 4$ supersymmetric $U(1)$ gauge theory from the $3+1$-dimensional point of view. The key point to remember is that the string theory approach to the solitons shows that their small fluctuations are actually described by a gauge theory. This is rather surprising, since we started with a theory of gravity that knows nothing about gauge fields!

Many independent arguments point to the conclusion that the dynamics of the worldvolume gauge fields are governed by a nonlinear generalization of the Maxwell action known as the Born-Infeld action. Pure Born-Infeld

$U(1)$ electromagnetism in ten dimensions is described by the action

$$L_{10} = -\frac{1}{(2\pi)^9 \alpha'^5 g} \int d^{10}x \sqrt{det(1 - 2\pi\alpha' F)} \tag{5}$$

where $F_{\mu\nu}$ is the gauge field strength. This was derived in the old days from string sigma models as the low-energy effective action of type-I string theory [3, 4]. It reduces to the Maxwell action in the weak-field limit but has a very specific nonlinear structure in the strong field limit that eventually turned out to have physical import (related to duality). Nowadays, it would be thought of as the action for a nine-brane (a brane that extends over all spacetime). It was ultimately realized that essentially the same sigma model calculation was involved in the derivation of the D-brane worldvolume action [5] and that the result was just the dimensional reduction of (5). To dimensionally reduce to the dynamics of a p-brane, one identifies $A_i(\xi_j)$ for $i, j = 0, 1, \ldots, p$ with a worldbrane $U(1)$ gauge field and identifies the A_a for $a = p + 1, \ldots, 9$ as worldvolume scalar fields $X^a(\xi_j)$ that describe the transverse displacements of the brane. The reduction of (5) would read

$$L_p \propto \int d^{p+1}\xi \sqrt{det(\eta_{ab}\partial_i X^a \partial_j X^b - F_{ij})} \tag{6}$$

but we need to generalize it substantially before we are ready to discuss its properties.

The action (5) (and its dimensional reductions) are of course appropriate to a flat background spacetime with no closed string fields (dilaton, antisymmetric tensor) excited. We will, however, be interested in the dynamics of D-branes in non-trivial backgrounds and will need to know the appropriate worldvolume action. The direct deduction of this action from string theory arguments is very laborious, although this tack was taken in the early days with some success [4]. An important fact that we have not emphasized so far is that the D-brane action possesses certain supersymmetries inherited from the ten-dimensional string theory in which it is embedded. Imposing the requisite degree of supersymmetry turns out to be very difficult and it seems that a proper understanding of this issue pretty much uniquely determines the curved space generalization of the Born-Infeld action. We don't have the space to go into detail on this point, but the results of careful investigations are simple enough to state and use. According to [6] the appropriately supersymmetric generalization of (5) to arbitrary backgrounds of type-IIA(B) fields is a sum of two terms $I_p = I_{DBI} + I_{WZ}$, both functionals of the embedding $X^m(\xi)$ of the p-brane (with coordinates ξ^i) into the background space (with coordinates X^m) and of the worldvolume gauge field A_i. To simplify things, we will explicitly state only the bosonic content of these actions. The first piece involves only the NS background fields of the type-II string theory (dilaton, graviton, antisymmetric tensor):

$$\begin{aligned} I_{DBI} &= -T_p \int d^{p+1}\xi e^{-\phi}\sqrt{-det(g_{ij} - \mathcal{F}_{ij})} \\ \mathcal{F}_{ij} &= F_{ij} - B_{ij},\ F = dA,\ B_{ij} = \text{pullback of } B_{mn} \\ g_{ij} &= G_{mn}(X(\xi))\partial_i X^m \partial_j X^n \ = \text{pullback of } G_{mn}\ . \end{aligned} \tag{7}$$

One notable feature of this is the way the worldvolume gauge field and the stringy antisymmetric tensor potential combine together: this is a consequence of an anomaly noticed in the early days of superstrings and is quite simply reproduced by the sigma model treatment of string dynamics [4]. The second term, the so-called Wess-Zumino action, involves the RR form fields of the type-II string theory and is compactly written

$$I_{WZ} = -T_p \int_{p+1} C \wedge e^{\mathcal{F}},\ C = \sum_{r=0}^{10} C^{(r)} \tag{8}$$

where $C^{(r)}$ is the pullback of the r-form RR potential to the brane. To evaluate the integral, one expands the exponential and picks out those terms which give a $p+1$-form and can be legitimately integrated over the p-brane. This action is in fact rather complicated, but we will see that interesting and non-trivial solutions can be found in selected cases.

2. Born-Infeld Dynamics of Branes in Flat Space

One of the more characteristic features of a $U(1)$ gauge field is that it has Coulomb solutions tied to a localized source charge. It is interesting to look for such solutions of the p-brane worldvolume gauge theory and to understand what they mean. It is easiest to explore this question for branes embedded in a flat background with no string fields of any kind excited, so that is where we will begin [7, 8].

For simplicity, we consider the case where a purely electric worldbrane gauge field is excited. It turns out that to achieve the lowest possible energy, transverse coordinates must be excited as well. Let us assume for the moment that only one such coordinate (call it X) is excited. We describe the embedding of the fluctuating brane by setting $X^m = (\xi^0, \vec{\xi}, X(\xi), 0, \dots, 0)$. The pullback metric is then $g_{ij} = \eta_{ij} + \partial_i X \partial_j X$ and it is an easy calculation to show that the worldbrane action (7) reduces to (defining $E_i = F_{0i}$)

$$L = -\frac{1}{g_p}\int d^p x \sqrt{(1-\vec{E}^2)(1+\vec{\nabla}X^2) + (\vec{E}\cdot\vec{\nabla}X)^2 - \dot{X}^2}\ . \tag{9}$$

The normalization corresponds to the background tension of a Dp-brane in its ground state (hence the factor of $1/g_p$, where we have defined $g_p \equiv g(2\pi)^p \alpha'^{\frac{p+1}{2}}$). Since we are going to want the energy function anyway, it is best to pass to the Hamiltonian formalism right away. The canonical momenta associated with $\vec{A}_i$ and X, respectively, are

$$\begin{aligned} \vec{D} &= g_p \frac{\delta L}{\delta \vec{E}} = \frac{\vec{E}(1+\vec{\nabla}X^2)-\vec{\nabla}X(\vec{E}\cdot\vec{\nabla}X)}{\sqrt{(1-\vec{E}^2)(1+\vec{\nabla}X^2)+(\vec{E}\cdot\vec{\nabla}X)^2-\dot{X}^2}} \\ P &= g_p \frac{\delta L}{\delta \dot{X}} = \frac{\dot{X}}{\sqrt{(1-\vec{E}^2)(1+\vec{\nabla}X^2)+(\vec{E}\cdot\vec{\nabla}X)^2-\dot{X}^2}} \end{aligned} \tag{10}$$

The corresponding Hamiltonian for D-brane dynamics turns out to be:

$$H = \frac{1}{g_p}\int d^p x \sqrt{(1+\vec{\nabla}X^2)(1+P^2)+\vec{D}^2+(\vec{D}\cdot\vec{\nabla}X)^2}\,. \tag{11}$$

The canonical momentum $\vec{D}$ (conjugate to the electric field $\vec{E}$) is of course subject to the constraint $\vec{\nabla}\cdot\vec{D} = 0$ (due to the absence of a canonical momentum for A_0). In some cases, we will be able to use this to eliminate the gauge field entirely in favor of some source terms and turn (11) into an action for the transverse displacement alone (i.e. for the shape of the surface).

We will first try to find static solutions of (11), i.e. solutions where $P = 0$ with $\vec{D} \neq 0$ and time-independent. Upon setting $P = 0$ in (11), we can rearrange the Hamiltonian to show that it obeys an interesting lower bound:

$$\begin{aligned} H_{P=0} &= \frac{1}{g_p}\int d^p\xi \sqrt{(1\pm\vec{\nabla}X\cdot\vec{D})^2+(\vec{\nabla}X\pm\vec{D})^2} \\ &\geq \frac{1}{g_p}\int d^p\xi (1\pm\vec{\nabla}X\cdot\vec{D})\,. \end{aligned} \tag{12}$$

This bound is saturated when $\vec{\nabla}X = \pm\vec{D}$ and can be shown to be a BPS condition, i.e. the condition that a non-vacuum state be annihilated by some global supersymmetry charges. We will touch on the supersymmetry aspect of our solutions only very briefly in these lectures, but it is a very important subject.

To find explicit solutions, we begin by solving the gauge constraint. We can always set $\vec{D} = \vec{\nabla}\Lambda$ so that

$$\vec{\nabla}\cdot\vec{D} = 0 \;\Rightarrow\; \nabla^2\Lambda = 0 \;\Rightarrow\; \Lambda = \Sigma_i \frac{c_i^p}{|\vec{r}-\vec{r}_i|^{p-2}}\,. \tag{13}$$

This is a multi-center Coulomb solution for the displacement field $\vec{D}$ with a finite number of source charges c_i^p for the constraint equation at locations

$\vec{r}_i$ on the brane. Now we observe that the lower bound functional in (12) can be expressed in terms of boundary values of the fields by integrating by parts and using the constraint equation on $\vec{D}$

$$H_{bound} = \frac{1}{g_p}\int d^p\xi \pm \frac{1}{g_p}\Sigma_i X(\xi_i)\Phi(\xi_i) \tag{14}$$

$$\Phi = \int_{r=\epsilon} d\Omega_{p-1}\partial_r\Lambda(r) = (p-2)\Omega_{p-1}c^p \ . \tag{15}$$

In this expression, Φ is just the integral of the flux of $\vec{D}$ through a small sphere around a Coulomb singularity (and Ω_{p-1} is just the area of the unit sphere in p dimensions). Note that the bound does not depend on the values of X away from the singularities and is in that sense 'topological' [9, 10]. To make use of these facts, we will need to know the allowed values of the source charges c^p. In [8], it was shown how to take the rather simple quantization conditions on the the worldvolume gauge field on the D1-brane and use T-duality to promote them to quantization conditions for the general Dp-brane. These conditions are equivalent in our language to the very simple result

$$g_p^{-1}\Phi = n/2\pi\alpha' = nT_f, \quad n = \text{integer} \tag{16}$$

where T_f is the *fundamental* string tension (independent of g_s, unlike T_p).

If we set $X = \Lambda$ (and choose the plus sign in (14)), the BPS condition $\vec{\nabla}X = \vec{D}$ is satisfied and the lower bound on the energy automatically saturated. A dramatic feature of this solution is that the transverse displacement is singular at the Coulomb singularities in the gauge field. Specifically, near a singularity at $r = 0$, the quantities of interest are singular as follows:

$$\Lambda \sim \frac{c^p}{r^{p-2}}, \quad X \sim \frac{c^p}{r^{p-2}}, \quad D_r = \partial_r\Lambda \sim \frac{(p-2)c^p}{r^{p-1}} \ . \tag{17}$$

Now we can interpret the energy functional (14). The first term is the Dp-brane area times the Dp-brane tension: this is the energy of the bare or unexcited Dp-brane. From (17), each singularity corresponds to a 'spike' in the transverse coordinate that runs off to infinity. Because of (16), this spike has an energy per unit transverse distance of nT_f: i.e. it is energetically identical to a number n of fundamental strings that come in from infinity and attach themselves to the Dp-brane. As a result of the BPS status of this solution the location on the brane of these terminating strings is arbitrary: they exert no force on each other. The situation is as described in the figure: This is a very explicit realization of the Polchinski picture of D-branes as loci where fundamental strings may terminate. The essential phenomenon is that, at the location of a Coulomb singularity, the Dp-brane

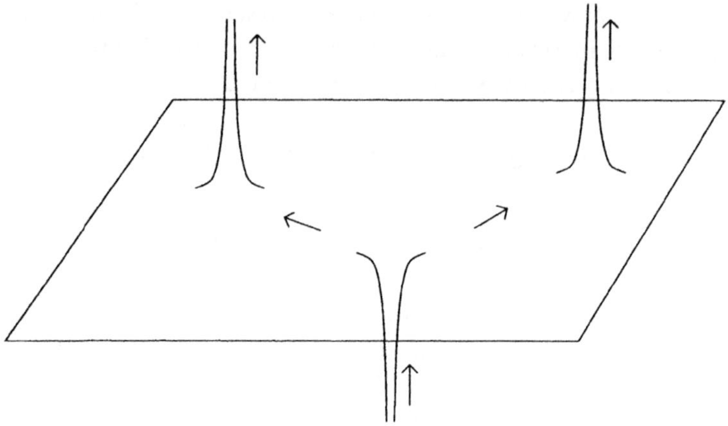

Figure 1. Brane with multiple strings attached on both sides. The arrows represent the direction of the electric field.

wraps itself on an S_{p-1} to become a one-brane (string) whose tension, by virtue of nonlinear features of the Born-Infeld action, equals that of the fundamental string. Without the specific nonlinearities of the BI action, the proper relation between fundamental string and D-brane tensions would not hold.

3. Branes in Curved Space and the Gauge Theory Connection

Polchinksi's open string theory approach to string solitons reveals that a $U(1)$ gauge field lives on the elementary soliton. The classical supergravity soliton solution indicates that it is possible to construct objects corresponding to multiple unit solitons. This poses the question of what happens to the gauge degrees of freedom on the individual solitons when N solitons are superposed. The short answer is that the $U(1)$ theory gets promoted to a $U(N)$ gauge theory. This is very interesting because the nonabelian theory has nontrivial interactions and very rich physics.

The way this works is a fairly simple extension of Polchinski's open string approach [11]. Suppose we superpose N solitons, indexed by $i = 1, \ldots, N$. Since open strings can terminate on any of the individual solitons, we must label the possible open strings by an index pair (i, j) to indicate where the two ends of the string terminate. Since the labels don't affect the dynamics, when we quantize, we will get N^2 copies of the same massless vector and scalar states as before: N copies arise from the (i, i) strings that begin and end on the same soliton and $N(N-1)$ copies arise from $(i \neq j)$ strings that begin and end on different solitons. This is just the structure of a $U(N)$ gauge theory with all the matter fields (scalars and gauginos in this case) lying in the adjoint representation. The self-interactions between the gauge fields and the scalars are quite non-trivial and are basically legislated by the gauge invariance and the extended supersymmetry of the theory. In fact, the simplest way to get an idea of the overall theory describing the collective coordinate dynamics is by dimensional reduction, to the dimension of the soliton worldsheet, of the ten-dimensional pure super Yang-Mills theory. A schematic version of the bosonic effective action of the soliton collective coordinates can easily be obtained in this way:

$$S_{YM}^{p} = T_p \int d^{p+1}x \; tr[F^2 + \sum_{a=p+1}^{9} |D\phi^a|^2 + \sum_{a,b=p+1}^{9} [\phi^a, \phi^b]^2 + \cdots] \quad (18)$$

The ellipses represent terms with more derivatives that are unimportant in a low-energy limit (and are hard to evaluate). This is a $U(N)$ gauge theory in $p+1$ dimensions with $9-p$ adjoint scalars endowed with a special potential term. As an aside, note that this potential has flat directions that correspond to spontaneously breaking the $U(N)$ gauge symmetry by introducing a transverse separation between some of the solitons: the gauge bosons arising from strings that run between separated branes get a mass proportional to the brane separation in a way that gives a geometrical realization of mass generation by the Higgs phenomenon. If a single brane moves away from the others, an isolated $U(1)$ (the gauge field living on that brane) separates off, which is why this direction in moduli space is called the 'Coulomb branch'. The action is, in general, nonrenormalizable and is ultimately defined by the full string theory. The soliton tension T_p normalizes the action and sets the energy scale. Since $T_p \sim g_s^{-1}$, this means that the effective coupling strength of the worldvolume Yang-Mills theory is $g_{YM}^2 = g_s$. This will be an important fact in our eventual translation table between gauge theory and quantum gravity results.

All of the above is true for a general p-brane. However, the $p = 3$ case has a number of special properties which make it possible to carry the gauge/string connection to much higher level of precision and usefulness than for other dimensions. These special properties are as follows: First, the solution displayed in (2) has the property for $p = 3$ that the dilaton field is independent of position. Since the dilaton determines the string coupling g_s, and g_s fixes the coupling of the Yang-Mills theory associated with multiple branes, g_{YM}^2 is unambiguous. This is not the case for the $p \neq 3$ branes for which these coupling constants vary from point to point (and sometimes blow up). Another reason why we like the three-brane is that it defines a four-dimensional gauge theory and this, of course, is the gauge theory that most interests us. The four-dimensional gauge theory is a very special one however: it is not too hard to show that it has $\mathcal{N} = 4$ extended supersymmetry (the spectrum contains four gauginos, for instance). This maximally supersymmetric gauge theory has uniquely specified particle content and interactions and it has been known for a long time that it has a vanishing beta function. This means that a) the theory is scale and conformal invariant and b) the coupling constant does not run and all quantities are functions of a coupling constant that can be chosen at will. The non-running of the coupling constant matches the constancy of the dilaton in the supergravity solution very nicely. We will shortly see that, despite preliminary indications to the contrary, the scale and conformal invariance are also properties of the associated supergravity three-brane solution.

The final salient point is that the geometry of the three-brane solution (at least the extremal version) is perfectly smooth and singularity-free. This is not the case for the other branes. To make further progress in understanding the supergravity approach to gauge theory, it is important to understand some of the special features of the three-brane supergravity solution. Specializing (2) to $p = 3$ and $r_0 = 0$ gives the following nontrivial metric and Ramond-Ramond form fields:

$$ds^2 = \frac{1}{\sqrt{H}}(-dt^2 + d\vec{x}^2) + \sqrt{H}(dr^2 + r^2 d\Omega_5^2)$$

$$C_{0123} = -H^{-1}, \quad G_{0123r} = \frac{H'}{2H^2}, \quad H = a + R^4/r^4, \tag{19}$$

where we will normally set $a = 1$, but will occasionally set $a = 0$ to focus on the $r \to 0$ limit of the geometry. It is also important that specializing (3) the $p = 3$ gives a crucial relation between stringy and gauge parameters:

$$R^4 \ = \ 4\pi N g_s \alpha'^4 \ . \tag{20}$$

It almost goes without saying, but the metric reduces to flat ten-dimensional spacetime in the $r \gg R$ limit (far from the brane). What is more interesting is the geometry in the $r \ll R$ limit:

$$ds^2 \to (\frac{r^2}{R^2}(-dt^2 + d\vec{x}^2) + \frac{R^2}{r^2}dr^2) + R^2 d\Omega_5^2$$

$$= \frac{R^2}{z^2}(-dt^2 + dz^2 + d\vec{x}^2) + R^2 d\Omega_5^2 \qquad (z = R^2/r) \tag{21}$$

In the second line, we have changed the coordinate system to bring out the fact that this is a very simple homogeneous space: the coordinates $(t, z, \vec{x})$ cover a space of constant negative curvature (known as AdS_5) while the remaining coordinates describe a five-sphere of constant radius R. Note that $r = 0$, the location of the threebrane itself, is an infinite proper distance from any finite point and that, since the space is homogeneous, the curvature is everywhere constant. This smooth inner region of the geometry, far inside $r = R$, is often called the 'throat'.

If the supergravity solution really describes the same thing as multiple superimposed D3-branes, then there must be a four-dimensional $SU(N)$ gauge theory hidden 'inside' the geometry we have just described. This is surprising because closed string theory lives in ten dimensions and has no intrinsic gauge degrees of freedom. Of course, any such gauge/gravitational equivalence has limits of validity and we should make them explicit before we go any further. In order to do explicit calculations, we will approximate string theory configurations (multiple threebanes mainly) by the metric and other fields of the semiclassical supergravity approximation. This implies two limits: we must take $g_{st} \ll 1$ to suppress string loop effects; we must take $R^2 \gg \alpha'$ to suppress higher derivative stringy corrections to the supergravity effective action (in other words, the scale of variation of the threebrane metric must be large compared to the string length). The latter condition can be expressed in purely gauge theory terms using (20):

$$\frac{R^4}{\alpha'^2} = 4\pi g_s N = 4\pi g_{YM}^2 N \gg 1 \quad (\text{with } g_{YM}^2 \ll 1,\ N \gg 1) \tag{22}$$

Note that the combination $g_{'tHooft}^2 = g_{YM}^2 N$ is the natural expansion parameter of large-N QCD. The bottom line is that semiclassical supergravity should tell us about the strong coupling limit of large-N gauge theory and not, unfortunately, about its most general behavior. The strong coupling limit has physical meaning since the $\mathcal{N} = 4$ supersymmetric $SU(N)$ gauge theory has a vanishing beta function and hence a non-running coupling that can be set to any value we like, weak or strong. Weak coupling behavior can be addressed using standard field theoretic perturbation theory. We are now learning that gravitational physics can tell us about the otherwise inaccessible strong-coupling behavior of gauge theories! This proposition was first stated unambiguously by Maldacena [12] who built on very important work of others, especially [13]. Our goal in what follows is not to explain or develop this 'AdS/CFT correspondence' for its own sake, but to use it as a context in which to probe features of the Born-Infeld action for D-brane dynamics. This will be a delicate matter, because we do not have the $SU(N)$ generalization of the Born-Infeld gauge theory in a form that can be used very explicitly.

With this in mind, it will be instructive to redo the analysis given in Sec.(2) of Coulomb solutions of the worldvolume gauge theory, but this time for a single D3-brane in the non-flat spacetime background of a large number of other D3-branes (19). To be explicit, the probe D3-brane is taken parallel to the N superposed D3-branes that set up the non-flat background, but displaced from them by a distance r in the space transverse to the branes. Using the fields of (19) in (7) and (8), we can write the effective action for fluctuations in the probe brane in the form

$$S_{D3} = -\int d^{3+1}\xi \frac{1}{H(r)}(\sqrt{-det\tilde{G}} - 1) \tag{23}$$

$$\tilde{G}_{ij} = \eta_{ij} + H(r)\partial_i X^a_\perp \cdot \partial_j X^a_\perp + \sqrt{H(r)} F_{ij} \tag{24}$$

where the embedding of the probe brane in the transverse space, $X^a(\xi)$, and the gauge field $F_{ij}(\xi)$ are taken as fields on the brane worldvolume and the transverse radial coordinate is given by $r = \sqrt{\sum_a (X^a)^2}$. The first term is the standard Nambu-Goto term and the second term is the Wess-Zumino term that arises from the existence of a background four-form potential according to the rules explained after (8). Note that for the case of an unexcited brane ($F_{ij} = 0$) sitting at a fixed transverse position ($X^a = cst$), the two terms in the action cancel. This corresponds to the well-known fact that parallel D3-branes exert no force on each other and hence that an unexcited probe brane must be able to sit at rest at any transverse location in this background.

Even though the action looks hopelessly nonlinear, it is tempting to follow the line of argument of Sec.(2) and look for Coulomb excitations of the gauge field on the probe brane. Just as in flat space, it will be necessary to excite both the Coulomb gauge field and a transverse displacement coordinate X. The appropriate generalization of (9) for the static case is easily found to be

$$L = -\frac{1}{g_p}\int d^p\xi \frac{1}{H}\{\sqrt{(1 - H\vec{E}^2)(1 + H\vec{\nabla}X^2) + (H\vec{E}\cdot\vec{\nabla}X)^2} - 1\}. \tag{25}$$

Because of the extra, rather nasty, non-linearity introduced by the function $H(r)$, one might think that it would not be possible to find exact solutions of this action. Quite remarkably, when the equations of motion that follow from (25) are worked out, it is easy to show that the flat-background Coulomb solution found in Sec.(2)

$$\vec{\nabla}E = \vec{\nabla}X, \quad \nabla^2 X = 0, \quad X = X^0 + \sum_i \frac{c_3^i}{|\vec{\xi} - \vec{\xi}_i|} \tag{26}$$

is also a solution in this specific curved background [14]. In exactly the same way as before, the singular transverse displacements of the D3-brane in the neighborhood of the Coulomb singularities describe the attachment of fundamental strings to the probe D3-brane sitting at $X = X^0$ in the transverse space.

Note that, depending on the (arbitrary) sign of the charge c_i, the fundamental string attached at the location of a Coulomb singularity either runs off to infinity ($|X| \to \infty$) or runs into $X = 0$, which is the location of the stack of N D3-branes that creates the background in the first place. The latter situation has the string theory interpretation of a finite-length fundamental string running between a single isolated D3-brane and a stack of N superposed D3-branes a distance X^0 away (the extra energy due to the string turns out to be $M = T_f X^0 = X^0/2\pi\alpha'$) [14, 15]. The gauge theory interpretation is equally simple: as described at the beginning of this section, $N+1$ superposed branes support an $SU(N+1)$ gauge theory; displacing one of the branes transverse to the remaining N superposed branes corresponds to the breaking of $SU(N+1)$ to $SU(N) \times U(1)$ by giving the appropriate v.e.v. to an adjoint scalar; the fundamental string stretched between the displaced brane and the stack of N superposed branes describes a gauge boson that has acquired a mass by the Higgs mechanism in the spontaneous breaking of $SU(N+1)$ (with mass, as required, proportional to the symmetry-breaking v.e.v.). The gauge theory analysis of this symmetry breaking would put the Higgsed gauge boson in the N-dimensional fundamental representation of the unbroken $SU(N)$: this is reproduced in the stringy analysis by the fact that the string can terminate on any one of the N superposed branes.

While the above-described connection between strings and broken gauge theory states is, in general, only approximate, the point of Maldcena's argument is that it should become exact in the special limit (22). More specifically, this means that fundamental strings extended in the anti-de Sitter geometry of (21) should be dynamically identical to the massive Higgsed gauge boson that arises in $3+1$-dimensional $\mathcal{N} = 4$ supersymmetric gauge theory when $SU(N+1)$ is spontaneously broken down to $SU(N) \times U(1)$. From the point of view of the $SU(N)$ gauge theory, this is just a massive external particle in the gauge fundamental, and we will occasionally call it a 'quark' for short. Although we will not be able to develop this point very far, the $\mathcal{N} = 4$ supersymmetric gauge theory also has an $SU(4)$ R-charge which is geometrically encoded in string theory in the orientation of the extended fundamental string in the six-dimensional space transverse to the D3-branes.

A simpler way to explore this dynamical equivalence than to look for solutions of (25) is to study the Nambu-Goto action for the fundamental string in the anti-de Sitter background of (21). The realisation of the broken symmetry situation we have been discussing should be provided by a static string running between $r = 0$ ($z = \infty$), the location of the N D3-branes that are the source of the background and are responsible for the $SU(N)$ gauge symmetry, and $r = r_c$ ($z_c = R^2/r_c$), the location of the single displaced

D3-brane that breaks the original $SU(N+1)$ gauge symmetry. The static Nambu-Goto action of such a string worldsheet in this geometry is easily calculated:

$$Action = -T_f \int dtdz \frac{R^2}{z^2} = -T_f \frac{R^2}{z_c} \Delta t = -T_f r_c \Delta t \ . \tag{27}$$

From the relation between static action and energy, we read off that the mass of such an extended string is $M = T_f r_c$, just as found in the more elaborate analysis of (25). We will use the notion that an extended string in anti-de Sitter space represents a heavy external 'quark' in a gauge theory as a tool for thinking about some interesting gauge theory questions.

4. Born-Infeld Analysis of the Baryon Vertex

There are some basic group-theoretic aspects of gauge theory which would appear to be hard to capture in Maldacena's geometrical representation of gauge theory dynamics. Consider, for instance, a configuration of n parallel strings in the anti-de Sitter background. According to what we have said above, each string corresponds to a heavy quark in the fundamental representation of the $SU(N)$ gauge group. From the gauge theory point of view, something special should happen when $n = N$ and it becomes possible to combine the quarks into a gauge singlet state (remember that the $\mathcal{N} = 4$ supersymmetric gauge theory is not confining so that non-singlet states are physically realizable). From the geometric (string theory) point of view, however, it is not so obvious why $n = N$ strings should be very different from $n \neq N$ strings. In what follows, we will construct solutions of the Born-Infeld action for a D5-brane in the background of a stack of N D3-branes. As we will soon see, this is a good way to describe the 'baryon vertex' of $n = N$ fundamental strings (or quarks) in the anti-de Sitter background [16, 17]. It will also provide a revealing geometrical explanation of the difference between gauge singlet and gauge non-singlet collections of quarks.

By building on work of Imamura [18], we can construct a variety of BPS-saturated solutions of the D5-brane worldvolume equations of motion [19]. To begin, we set up the equations for the Born-Infeld D5-brane in the background geometry of a stack of N D3-branes. The background fields are as displayed in (19) (and we write $H(r) = a + R^4/r^4$ so that we may treat the asymptotically flat D3-brane ($a = 1$) and the $AdS_5 \times \mathbf{S}^5$ ($a = 0$) geometries in parallel). The D5-brane worldvolume action is the Born-Infeld action calculated using the induced metric (including the worldvolume gauge field)

$$g^{ind}_{\alpha\beta} = g_{MN} \partial_\alpha X^M \partial_\beta X^N + \mathcal{F}_{\alpha\beta},$$

plus a WZW term constructed out of the five-form field strength and the worldvolume gauge field according to (8). Explicitly,

$$S = -T_5 \int d^6\xi \sqrt{-det(g^{ind}_{\alpha\beta})} + T_5 \int d^6\xi A_\alpha \partial_\beta X^{M_1} \wedge \ldots \wedge \partial_\gamma X^{M_5} G_{M_1 \ldots M_5} \ ,$$

where T_5 is the D5-brane tension. Note that, because the worldvolume is six-dimensional, the WZW action amounts to a linear source term, induced by the background Ramond-Ramond field, for the worldvolume gauge field A. In analogy to the situation discussed in (2), we expect that this charge must ultimately arise from the coupling of some number of fundamental strings to the D5-brane. The rather nontrivial geometrical details of how this happens will be worked out in the rest of this section.

We use the target-space time and the $\mathbf{S}^5$ spherical coordinates as coordinates on the D5-brane, $\xi_\alpha = (t, \theta_\alpha)$. For simplicity, we look for solutions where the D5-brane lies at a fixed point $\vec{x}$ on the threebrane stack and wraps over some time-independent closed five-dimensional surface in the six-dimensional space transverse to the threebranes. We will take a simple ansatz (with $SO(5)$ symmetry) in which the transverse space embedding is given by a function $r(\theta)$ and the gauge field is given by a function $A_0(\theta)$ (where θ is the polar angle in the $\mathbf{S}^5$ spherical coordinates). Less symmetric solutions and time-varying solutions are of interest too, but we will not deal with them here. Substituting this ansatz into the action and using (19) for the background fields, the action simplifies to

$$S = T_5 \Omega_4 \int dt d\theta \sin^4\theta \{ -r^4 H(r) \sqrt{r^2 + (r')^2 - F_{0\theta}^2} + 4 A_0 R^4 \}, \qquad (28)$$

where $\Omega_4 = 8\pi^2/3$ denotes the volume of the unit four-sphere and the A_0 term comes from the the WZW action.

The gauge field equation of motion following from this action reads

$$\partial_\theta \left[-\sin^4\theta \frac{(ar^4 + R^4)E}{\sqrt{r^2 + r'^2 - E^2}} \right] = 4R^4 \ (\sin\theta)^4,$$

where we have set $E = F_{0\theta}$ and the right-hand side is the source term coming from the WZW piece of the action. It is helpful to repackage this in terms of the displacement field D (the variation of the action with respect to E):

$$D = \frac{\sin^4\theta (ar^4 + R^4)E}{\sqrt{r^2 + r'^2 - E^2}} \quad \Rightarrow \quad \partial_\theta D(\theta) = -4R^4 \sin^4\theta. \qquad (29)$$

No matter what the embedding function $r(\theta)$, we can integrate the equation for D to find it as an explicit function of θ. The result is

$$D(\theta) = R^4 \left[\frac{3}{2}(\nu\pi - \theta) + \frac{3}{2} \sin\theta\cos\theta + \sin^3\theta\cos\theta \right], \qquad (30)$$

where the integration constant has been written in terms of a parameter $0 \leq \nu \leq 1$, whose meaning will be elucidated below. Since D, unlike E, is completely unaffected by the form of the function $r(\theta)$, it makes sense to express the action in terms of D and regard the result as a functional for $r(\theta)$. It is best to do this by a Legendre transformation, rewriting the original Lagrangian as

$$U = T_5\Omega_4 \int d\theta \{D \cdot E + \sin^4\theta(ar^4 + R^4)\sqrt{r^2 + (r')^2 - E^2}\} .$$

Integrating the DE term by parts using $E = -\partial_\theta A_0$, one reproduces (with a sign switch) the original Lagrangian (28). Using (29) we can eliminate E in favor of D to get the desired functional of $r(\theta)$ alone:

$$U = T_5\Omega_4 \int d\theta \sqrt{r^2 + (r')^2}\sqrt{D^2 + r^8 H(r)^2 \sin^8\theta}. \tag{31}$$

This functional is reasonably simple, but complicated by the fact that there is explicit dependence on θ. Hence there is no simple energy-conservation first integral that we can use to solve the equations (or at least analyse possible solutions).

However, in analogy with the discussion in Sec.(2) of the Born-Infeld action in flat space, we can show that this action has a nontrivial topological lower bound and we will find that the solutions which saturate the lower bound are simple and instructive. The basic observation is that the action (28) may be written in a 'sum of squares' form as follows [10]:

$$(r^2 + r'^2)(D^2 + \Delta^2) = \mathcal{Z}_{el}^2 + r^2(\Delta\cos\theta - D\sin\theta)^2(\frac{r'}{r} - f)^2 \tag{32}$$

where $\Delta = Hr^4\sin^4\theta$ and

$$\mathcal{Z}_{el} = r(\Delta\cos\theta - D\sin\theta)(1 + \frac{r'}{r}f)$$
$$f = \frac{\Delta\sin\theta + D\cos\theta}{\Delta\cos\theta - D\sin\theta} . \tag{33}$$

Even more remarkably, $\mathcal{Z}_{el}$ is identically a total derivative (identically in the sense that we only need to use the constraint identity (29) for $D(\theta)$ to show it):

$$\mathcal{Z}_{el} \equiv \frac{d}{d\theta}(D(\theta)r\cos\theta + (\frac{a}{5} + \frac{R^4}{r^4})(r\sin\theta)^5) \tag{34}$$

The resulting topological bound on the energy of the wrapped D5-brane configuration is

$$U \geq \int d\theta|\mathcal{Z}_{el}| = [D(\theta)r\cos\theta + (\frac{a}{5} + \frac{R^4}{r^4})(r\sin\theta)^5]_{\theta_i}^{\theta_f} \tag{35}$$

where the limits of integration are chosen so that $\mathcal{Z}_{el}$ doesn't change sign (and this may *not* be the full geometrical range $0 < \theta < \pi$). Explicit solutions will make this point clear. The condition for saturating this bound is

$$\frac{r'}{r} = f = \frac{\Delta \sin\theta + D\cos\theta}{\Delta\cos\theta - D\sin\theta} \tag{36}$$

Although we won't go very deeply into this point, this is a BPS condition, *i.e.* a condition that this non-vacuum soliton configuration preserve some residual supersymmetry [18, 19, 9]. As is typical for BPS conditions, it is a first-order differential equation for $r(\theta)$ (while the Euler-Lagrange equation for a general solution would be second-order).

At this point, it is probably most instructive to examine the behavior of explicit solutions of the BPS equations. Solutions can be found for the full, asymptotically flat, D3-brane metric ($a = 1$ in our expression for $H(r)$)) but they become particularly simple in the AdS throat limit ($a = 0$) and we will limit our attention to that case. Since the throat geometry is dual, in Maldacena's sense, to the $\mathcal{N} = 4$ gauge theory, this is of great physical interest anyway. As explained in [19], the general solution of the $a = 0$ BPS condition is

$$\begin{aligned} r(\theta) &= \frac{A}{\sin\theta}\left[\frac{\eta(\theta)}{\pi(1-\nu)}\right], \\ \eta(\theta) &= \theta - \pi\nu - \sin\theta\cos\theta\ . \end{aligned} \tag{37}$$

A is an arbitrary constant of integration: the fact that it simply rescales the solution is related to the scale invariance of the underlying gauge theory. The solution only makes sense in the range of θ where $\eta \geq 0$. An examination of η shows that it is positive (we take $0 \leq \nu \leq 1$) in some range $\theta_0(\nu) < \theta < \pi$. One can also show that $\theta_0(\nu)$ increases monotonically from zero to π as nu increases from zero to one.

Let us now examine the detailed behavior of the solution for different choices of ν, the mysterious constant of integration of the gauge constraint. First observe that, for any ν, $r \sim A/(\pi - \theta) \to \infty$ as $\theta \to \pi$: we shall see shortly that this 'spike' represents a bundle of fundamental strings in the manner described in [8, 7] and explained in Sec.(2). The way the surface behaves for small r depends on ν: for $\nu = 0$, the surface starts at a finite radius at $\theta = 0$; for $\nu > 0$, the surface starts at $r = 0$ and a finite value, $\theta_0(\nu)$, of θ. In the first case, the D5-brane wraps the entire S_5 and captures the entire flux of the five-form field strength. In the second case, the brane actually wraps only a fraction of the S_5 and captures only a fraction of the flux. The different behaviors are clearly displayed in Fig.(2). We can evaluate the energy of these solutions using (35). The θ integration limits are π and $\theta_0(\nu)$ and it is easy to see that the contribution of the lower limit

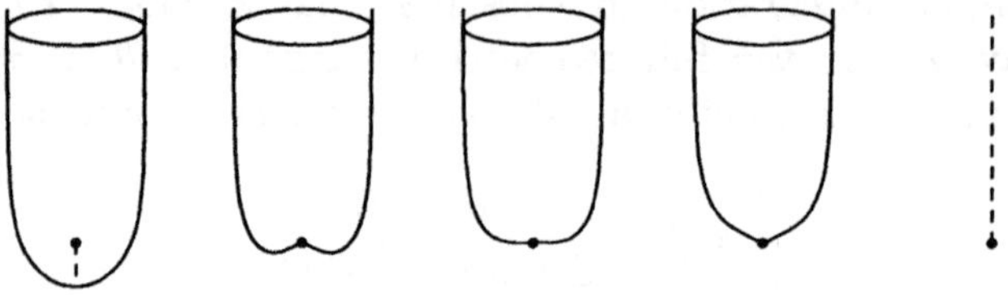

Figure 2. Polar plots of $r(\theta)$ for AdS 'tube' solutions. The cross-section of each 'tube' is an $\mathbf{S}^4$.

in (35) is always zero. Since $r \to \infty$ at the upper limit, the contribution of the upper limit simplifies to

$$U \to T_5 \Omega_4 D(\pi) r_{max} = T_5 \Omega_4 R^4 \frac{3\pi}{2}(1-\nu) r_{max} = (1-\nu) N T_f r_{max} \quad (38)$$

where 1) we have used the expressions for T_5 and R^4 in terms of fundamental string parameters to simplify things and 2) since $r(\pi)$ blows up, we have cutoff the radius at r_{max}. The energy blows up because the surface has unbounded length. What is relevant (and finite) is the energy per unit length (U/r_{max}), or tension, of the surface. It follows from (38) that the the spike has the tension of $n = (1-\nu)N$ fundamental strings.

We can now give a simple interpretation of the solutions in the light of the AdS/CFT correspondence: Since a fundamental string in anti-de Sitter space represents a heavy particle in the fundamental representation of the $SU(N)$ gauge group (heavy because it has gained a large mass from the spontaneous breaking of $SU(N+1)$ to $SU(N)$), the D5-brane solution with parameter ν represents a collection of $(1-\nu)N$ such heavy particles (or 'quarks'). The energy is linear in the number of particles because they satisfy a BPS condition and neither attract nor repel each other. The solution has at least one modulus (the scale factor A) and presumably has many others; to fully understand the state counting, one would have to quantize motions on the (as yet unexplored) moduli space. The way the solution is embedded in the transverse S_5 has to do with the representation of the $SO(6)$ ($SU(4)$, really) global symmetry inherited by the 'quarks' from the $\mathcal{N} = 4$ supersymmetry R-charge. Note that the solution for $\nu = 0$, the case where exactly N quarks are present and can be coupled to an $SU(N)$ singlet, is indeed special in the sense that the D5-brane maintains a finite distance from the singular point at $r = 0$, the location of the multiple D3-branes that support the $SU(N)$ gauge theory. On the other hand, when $\nu > 0$, the surface runs into $r = 0$ and it is not clear that the isolated D5-brane really decouples from the many degrees of freedom on the

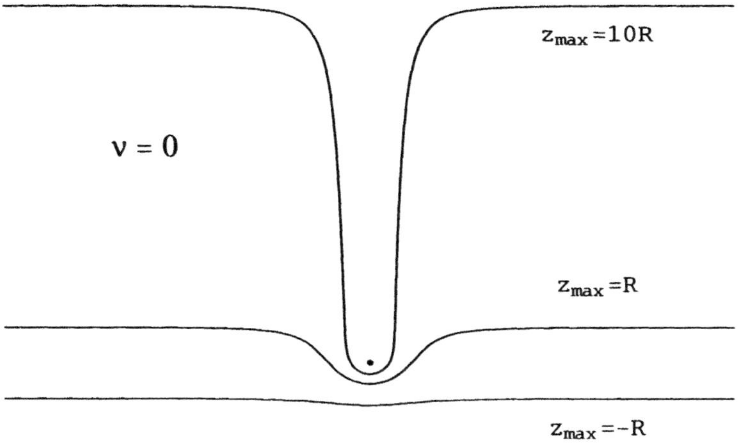

Figure 3. Solutions describing the creation of N fundamental strings as a D5-brane is dragged across a stack of D3-branes.

multi-D3-brane worldsheet. We therefore talk about a 'baryon vertex' in this context, even though there is no confinement and it is perfectly legal to construct color non-neutral ($\nu > 0$) collections of quarks.

For completeness, we mention that solutions of the type just discussed can also be found in the asymptotically flat geometry obtained by setting $a = 1$ in (19). This setup describes what happens when a D5-brane is 'dragged across' a stack of N D3-branes when the dimensions of extension of the two types of brane are totally orthogonal to each other. The solutions show that the D5-brane is trapped on the D3-brane singularity with the result that a 'tube' having the interpretation of a bundle of fundamental strings joining the two types of brane is created. The process is described graphically in Fig.(3). String creation was first understood algebraically [20, 21] and the worldbrane approach of [19] provides a very intuitive picture of what is otherwise a rather mysterious process.

5. Applications of the AdS/CFT Correspondence

As explained at the end of Sec.(3), in Maldacena's correspondence between gauge theories and gravity [12], external charges in the gauge theory are dual to macroscopic strings which trace out curves in the bulk of anti-de Sitter (AdS) space (the geometry of (19) for $a = 0$, to which we restrict our attention for the rest of this section). For simplicity, we consider the classic case of duality between $D = 3+1$ $\mathcal{N} = 4$ $SU(N)$ super-Yang-Mills (SYM) and Type IIB string theory on $AdS_5 \times S^5$. A solitary static quark (transforming in the fundamental of $SU(N)$) corresponds to a Type IIB string extended in the radial direction and sitting at a fixed angle on the S^5 (S^5 angles encode the global $SU(4)$ flavor R-symmetry of this theory); a string of opposite orientation represents an antiquark, transforming in the anti-fundamental of $SU(N)$). A single continuous fundamental string that starts on the boundary at one point in the $D = 3+1$ base space, arcs through the bulk of the AdS space and returns to the boundary at another point, corresponds to a quark-antiquark pair at a finite spatial separation in the dual gauge theory. That the energetics of the dual interpretations of these states correspond correctly has been shown in [22, 23].

There is much more to the AdS/CFT correspondence than energetics, however. As was shown in [24, 25], there is a very precise connection between the supergravity fields, evaluated at the boundary of AdS space, and expectation values of corresponding operators in the gauge theory. In Sec.(4), we showed that D5-brane surfaces, appropriately laid out in AdS space, correspond to multi-quark states of various kinds. These surfaces act as sources of the supergravity fields (graviton, dilaton and so on) and, by looking at the boundary values of the induced fields, we can infer the expectation values of gauge theory operators in the corresponding multi-quark states of the gauge theory. In what follows, we explore this line of thought, with a view to extracting more detailed information about the gauge theory significance of our baryon vertex construction. We refer the reader to [26] for a fuller development of this material.

To understand the basic ideas it will suffice to study how the correspondence works for the dilaton field. Its boundary values are known to be related to the expectation value of the operator

$$\mathcal{O}_{F^2} = \frac{1}{4g_{YM}^2} \operatorname{Tr}\left\{F^2 + [X_I, X_J][X^I, X^J] + \text{fermions}\right\} \tag{39}$$

in the dual gauge theory [13, 30]. In the above equation X^I, $I = 1, \ldots, 6$, denote the scalar fields of the $\mathcal{N} = 4$ SYM theory (fields which live in the vector of $SO(6)$ R-symmetry). This operator is sensitive, through the $\operatorname{Tr} F^2$ term, to the long-range gauge fields which are set up by the external quarks which our D-brane construct represents. It would be handier for the interpretation if the operator were 'pure' $\operatorname{Tr} F^2$, but we have to take what the duality gives us. Other supergravity fields are of interest as well: the graviton is dual to the gauge theory energy-momentum tensor, but this duality is technically more difficult to work out and we will not pursue it here.

Let us now briefly explain the GKPW recipe for extracting gauge theory expectation values from bulk supergravity fields [24, 25]. As promised, we will concentrate our attention on the dilaton field and its dual gauge theory operator $\mathcal{O}_{F^2}$. Consider, then, the bulk dilaton field $\phi(t, \vec{x}, z, \Omega_5)$ (we are using the coordinate system of (21)). The essence of the GKPW approach is the observation that the supergravity fields in the bulk of $AdS_5 \times S^5$ are uniquely determined by their values at the $z \to 0$ boundary of the space and that, as a consequence, the supergravity action S_{bulk} for the full space can be treated as a functional of the boundary values of the fields. The GKPW recipe, as applied to the dilaton and its dual operator, is the statement that

$$\langle \mathcal{O}_{F^2}(t, \vec{x}) \rangle = \frac{\delta}{\delta \phi_0(t, \vec{x}, z)} S_{bulk}|_{z \to 0} \ . \tag{40}$$

In ϕ_0, the subscript signifies projection of the full field $\phi(t, \vec{x}, z, \Omega_5)$ onto the constant S^5 spherical harmonic. The higher spherical harmonics are dual to operators that generalize $\mathcal{O}_{F^2}$ by including polynomials in X^I that lie in higher representations of $SO(6)$ (the detailed structure of the higher operators has been carefully worked out in [30]). In the applications of interest to us, the weak coupling limit may be taken and S_{bulk} may be approximated by the dilaton kinetic term alone. This leads to a great simplification of (40):

$$\langle \mathcal{O}_{F^2}(t, \vec{x}) \rangle = \frac{\Omega_5 R^8}{2\kappa^2} z^{-3} \frac{\partial}{\partial z} \phi_0(t, \vec{x}, z)|_{z \to 0} \tag{41}$$

where Ω_5 is the volume of S_5 and κ is related to the ten-dimensional Newton constant in the usual way. In other words, the gauge theory expectation value of $\mathcal{O}_{F^2}$ is just an appropriately normalized limit of the normal derivative of ϕ_0 at the boundary of the AdS space. We will see in a minute that the projection on S^5 spherical harmonics is rather trivial in the case of interest to us.

Our principal task is to calculate the dilaton field generated by a particular D-brane soliton and evaluate its behavior near $z = 0$. The soliton acts as a localized source for the ten-dimensional dilaton, as can be expressed by the source equation

$$\nabla^2 \phi(x^M) = 2\kappa^2 \int d^{p+1}\xi \, \delta^{(10)}(X^M - Y^M(\xi)) \Sigma_p(\xi) \tag{42}$$

where the source density Σ_p is the variation of the p-brane worldvolume action density with respect to the dilaton (holding the Einstein-frame metric, rather than the string-frame metric, constant) and $Y^M(\xi)$ is the embedding of the soliton in ten-dimensional spacetime. The dilaton dependence of the general Dp-brane action (and also of the fundamental string action) is easy to work out by making the well-known substitution $g^{string} = e^{\phi/2} g^{Einstein}$ on (7) (and on the simple Nambu-Goto action for the fundamental string):

$$S_p = \int d^{p+1}\xi e^{\frac{(p-3)}{4}\phi} \sqrt{\det(g^E_{ij} - e^{-\phi/2} \mathcal{F}_{ij})}$$

$$S_{fund} = \int d^2\xi e^{\phi/2} \sqrt{\det(g^E_{ij})} \ . \tag{43}$$

(Note that, after varying with respect to ϕ, we should set $\phi = 0$: the dilaton is constant in the AdS background and we can normalize its value to zero). We can read off from (43) the important facts that 1) the D3-brane is a source for the dilaton only if the worldvolume gauge field is turned on and 2) that if the gauge field is off, the D5-brane and the fundamental string dilaton sources are essentially identical.

The standard Green's function solution of (42) is

$$\phi(x^M) = 2\kappa^2 \int d^{p+1}\xi \ K(X - Y(\xi))\Sigma_p(\xi) \ , \tag{44}$$

where K is the scalar Green function in the $AdS_5 \times S^5$ geometry (*i.e.* (19) with $a = 0$). It is easily shown to be

$$K(X, X') = \frac{(zz')^4}{[(t-t')^2 + (\vec{x} - \vec{x}')^2 + (z\hat{n} - z'\hat{n}')^2]^4} \frac{1}{R^8} \tag{45}$$

where the $\hat{n}$ are unit vectors on the unit five-sphere which encode the S^5 coordinates of the arguments of K.

To give an end-to-end demonstration of how this works, we now carry out the calculation of the $\mathcal{O}_{F^2}$ expectation value for a single massive quark, represented by a static fundamental string extended in the AdS space. From (43) we conclude that the dilaton source density of the fundamental string is just the Nambu-Goto action density, which evaluates to $\sqrt{\det(g)} = R^2/z^2$ for a string extended in z. The solution for the dilaton field induced by a static string (located at $\vec{x} = \vec{x}_s$ on the boundary) is easily computed from (44) and (45):

$$\phi(z, \vec{x}, \hat{n}) = \frac{2\kappa^2 T_f}{R^6} z^4 \int dt dz' \frac{z'^2}{[t^2 + (\vec{x} - \vec{x}_s)^2 + (z\hat{n} - z'\hat{n}')^2]^4} \tag{46}$$

Note that, as $z \to 0$, the field vanishes as z^4 (exactly what is needed to give a finite result in (41)) and becomes independent of $\hat{n}$. In other words, the leading behavior of ϕ in the boundary limit is independent of S^5 angles and therefore automatically projected on the lowest S^5 spherical harmonic. The upshot of all this is

$$\langle \mathcal{O}_{F^2}(t, \vec{x}) \rangle = 4\Omega_5 R^2 T_f \int dt dz' \frac{z'^2}{\left[t^2 + (\vec{x} - \vec{x}_s)^2 + z'^2\right]^4}$$

$$\sim R^2 T_f \int_0^\infty dz \frac{z^2}{[(\vec{x} - \vec{x}_s)^2 + z^2]^{7/2}} \sim \frac{R^2 T_f}{|\vec{x} - \vec{x}_s|^4} \sim \frac{\sqrt{g^2_{YM} N}}{|\vec{x} - \vec{x}_s|^4} \ . \tag{47}$$

This makes perfect sense: the string corresponds to a quark carrying gauge charge; in this unconfined and conformal theory the gauge field strength generated by a charge will fall off like r^{-2} and an operator involving $\operatorname{Tr} F^2$ will fall off at least as fast as r^{-4}. The $\sqrt{g_{YM}^2 N}$ normalization of the expectation value is characteristic of all calculations involving strings in the AdS/CFT correspondence and does not have a simple intuitive explanation. For future reference, we remark that if we had done the same calculation for a quark-antiquark string configuration, we would have found that the $\operatorname{Tr} F^2$ vacuum expectation value falls off as r^{-7} for large distance. This does not correspond very neatly to our weak coupling intuition about how dipole field strengths fall off with distance from the source, so this calculation tells us something new (and hopefully true) about the behavior of dipole fields in the strong coupling limit of this type of gauge theory.

We will now carry out the analogous calculation for the D5-brane soliton corresponding to a bundle of M fundamental strings. As we have explained in Sec.(4), the D5-brane geometries vary with M, with the $M = N$ case corresponding to a particularly smooth geometry. From the gauge theory point of view, $M = N$ is special because the quarks can combine to a color-neutral fundamental representation of $SU(N)$. This fact should be visible in the expectation values of $\mathcal{O}_{F^2}$ and we would like to verify that this fact about group representations is realized in a credible way in the GKPW duality approach.

For a soliton of the type described in (37), for which the D5-brane embedding is given by a function $r(\theta)$ of the S^5 polar angle only, it is straightforward to work out from (43) the functional form of the dependence of the worldvolume action on a varying dilaton field:

$$S_{D5} = T_5 R^4 \int dt d\theta \sin^4\theta d\Omega_4 \{ -\sqrt{e^{\phi}\left[r^2 + r'^2\right] - E^2} + 4A_0 \}. \qquad (48)$$

The electric field $E = F_{0\theta}$ may be eliminated in favor of the displacement field

$$D = \frac{\sin^4\theta E}{\sqrt{r^2 + r'^2 - E^2}}. \qquad (49)$$

This is useful because D is a known explicit function of θ (30). After this replacement, (48) implies a linearized dilaton source term which can be written in the form

$$\delta S_{D5\phi} = -T_5 R^4 \int dt\, d\theta\, d\Omega_4 \phi \sqrt{r^2 + r'^2} \sqrt{D^2 + \sin^8\theta}. \qquad (50)$$

Inserting the implied value of Σ_{D5} into (44), we get an expression for the dilaton field induced by the baryon soliton solution (37):

$$\phi(z,\vec{x},\hat{n}) = \frac{2\kappa^2}{R^8}\int dt d\theta' d\Omega_4' \times$$
$$\frac{(zz')^4}{[t^2+(\vec{x}-\vec{x}_s)^2+(z\hat{n}-z'\hat{n}')^2]^4}\left[T_5 R^4\sqrt{r^2+r'^2}\sqrt{D^2+\sin^8\theta}\right] . \qquad (51)$$

To evaluate integrals like this, we will use the fact that $z = R^2/r$ is a known function of θ (37) and is independent of the S^4 angles. We now take the $z \to 0$ limit of ϕ and apply (41), to get the gauge theory vacuum expectation value:

$$\langle \mathcal{O}_{F^2}(t,\vec{x})\rangle = 4\Omega_5\Omega_4 R^4 T_5 \int dt d\theta \times$$
$$\frac{z^4}{[t^2+(\vec{x}-\vec{x}_s)^2+z^2]^4}\sqrt{r^2+r'^2}\sqrt{D^2+\sin^8\theta} . \qquad (52)$$

To get this expression we have integrated out the S^4 angles (hence the factor of Ω_4). For purposes of comparison with the fundamental string case (47), it is helpful to replace the integration over θ with an integration over z. With the help of the BPS condition we can recast (52) in the form

$$\langle \mathcal{O}_{F^2}(t,\vec{x})\rangle = (4\Omega_5\Omega_4)R^2(R^4T_5)\int dt dz \frac{z^2}{[t^2+(\vec{x}-\vec{x}_s)^2+z^2]^4}F(z) \qquad (53)$$

where

$$F(z) = \sqrt{1+r^2/r'^2}\sqrt{D^2+\sin^8\theta} = \frac{D^2+\sin^8\theta}{\sin^5\theta + D\cos\theta} \qquad (54)$$

and $F(z)$ is implicitly defined as a function of θ by (37) which defines a map between r (or z) and θ. Up to numerical factors of order one, (53) is exactly the same as the fundamental string answer (47) with the replacements $R^4T_5 \to T_f$ and $F(z) \to 1$.

To proceed with the evaluation of (53), we do the t integration explicitly and rescale the z integration to obtain

$$\langle \mathcal{O}_{F^2}(t,\vec{x})\rangle \sim R^2(R^4T_5)\int dz \frac{z^2}{[(\vec{x}-\vec{x}_s)^2+z^2]^{7/2}}F(z)$$
$$\sim \frac{R^2(R^4T_5)}{|\vec{x}-\vec{x}_s|^4}\int_0^\infty du \frac{u^2}{(1+u^2)^{7/2}}F(u|\vec{x}-\vec{x}_s|) . \qquad (55)$$

In writing this, we have dropped dimensionless constants of order one and we have also assumed that the soliton embedding runs over the entire range

$0 < z < \infty$. Referring back to the discussion of the soliton solution (37), we see that this implies that we are considering the 'fractional' baryon solutions with $\nu \neq 0$ or N. A fact that will shortly become important is that such solutions cover only a fraction of the five-sphere, starting at $\theta = \theta_0(\nu)$ at $r = 0$ ($z = \infty$) and running off to $r = \infty$ at $\theta = \pi$. We will say a few words at the end of this section about what happens in the case of the color-neutral baryon ($\nu = 0$), in which case there is a maximum value of z.

Our main interest is the behavior of this object as the spatial distance $|\vec{x} - \vec{x}_s|$ between the soliton source and the observation point gets large. The convergence of the integral is such that we may take the limit inside (55) to obtain

$$\langle \mathcal{O}_{F^2}(t, \vec{x}) \rangle \sim \frac{R^2(R^4 T_5)}{|\vec{x} - \vec{x}_s|^4} F(z \to \infty) \; . \tag{56}$$

This is the same scaling with separation as was found in (47) for the operator expectation value in the presence of a single fundamental string. Any differences are contained in constant normalization factors such as $F(z \to \infty)$. The latter factor can be simplified, using the definition (54) of F, as follows

$$F(z = \infty) \equiv F(\theta = \theta_0(\nu)) \equiv \frac{D(\theta_0)^2 + \sin^8 \theta_0}{\sin^5 \theta_0 + D(\theta_0) \cos \theta_0} \equiv \sin^3 \theta_0 \; . \tag{57}$$

The last equality comes from using the definition of θ_0, $\eta(\theta_0) = 0$ (see (37)) in the expression for D (see (30)) to show that $D(\theta_0) = \sin^3 \theta_0 \cos \theta_0$ and then simplifying the expression for F.

Putting all of this together and expressing R and T_5 in terms of fundamental constants, we finally obtain

$$\langle \mathcal{O}_{F^2}(t, \vec{x}) \rangle \sim \frac{\sqrt{g_{YM}^2 N}\; N \sin^3 \theta_0(\nu)}{|\vec{x} - \vec{x}_s|^4} \; . \tag{58}$$

This scales with distance in exactly the same way as (47) as is expected since the underlying gauge theory is conformally invariant. The factor $N \sin^3 \theta_0(\nu)$ is the only reflection of the fact that the soliton describes a collection of νN quarks in the $SU(N)$ gauge theory. From the definition of θ_0 as the zero of η (37), it is possible to show that this factor vanishes as ν takes on its limiting values $\nu = 0$ or $\nu = 1$ (corresponding to N or 0 quarks, respectively):

$$\sin^3 \theta_0 \sim \tfrac{3\pi}{2} \nu \qquad \text{as } \nu \to 0 \tag{59}$$

$$\sin^3 \theta_0 \sim \tfrac{3\pi}{2} (1 - \nu) \quad \text{as } \nu \to 1 \; . \tag{60}$$

Consider first the $\nu \to 1$ limit: $n = N(1-\nu) \ll N$ is precisely the number of quarks in the soliton and, by keeping careful track of constants, one can show that the result of (58) is exactly $N(1-\nu)$ times the result of (47) for a single quark. In other words, the $\mathrm{Tr}\, F^2$ expectation value is additive in the number n of quarks as long as that number is small compared to N. Now consider the $\nu \to 0$ limit (where the soliton describes exactly N quarks): (59) shows that the $\mathrm{Tr}\, F^2$ expectation value vanishes in the limit, consistent with the group theory expectation that N quarks will combine to form a gauge singlet state. What this really shows is that the $|\vec{x} - \vec{x}_s|^{-4}$ term in the large distance expansion of (55) vanishes: a closer examination shows that the $\nu = 0$ surface, which covers a finite range of z, leads to a $\mathrm{Tr}\, F^2$ expectation value which falls off as $|\vec{x} - \vec{x}_s|^{-7}$ (like a fundamental quark-antiquark dipole)[26].

The point of all this is simply that the geometrical approach to gauge theory that is the essence of the AdS/CFT duality correctly reproduces some of the non-trivial qualitative aspects of the representation theory of the $SU(N)$ gauge group. The most important such fact is that N quarks can combine to the identity, or gauge neutral, representation. In our approach, this distinction is represented geometrically by the fact that the soliton for N quarks stays away from the horizon at $r = 0$, while the soliton for $n \neq N$ quarks runs into the horizon. This sort of issue can and should be explored in greater depth.

6. Summary

In these lectures we have tried to give an impression of the range of D-brane dynamics that can be accounted for using the Born-Infeld style of worldvolume action. Our aim was not to be comprehensive or systematic, but rather to explore the richness of D-brane phenomena that can be understood through a semiclassical treatment of this action. Much was left out, and supersymmetry issues were given disgracefully short shrift in order to fit the the exposition into the confines of a few lectures. Here, as elsewhere in D-brane physics, it is amazing how much can be extracted from the D3-brane solution of type-IIB supergravity. Most of the interesting results reported here relied on special features of solutions of the worldvolume action of single branes (D3-branes or D5-branes) in the supergravity background of many superposed D3-branes. They illuminated phenomena as diverse as string creation in brane crossing events and (via the AdS/CFT correspondence) the long-range gauge fields generated by collections of quarks. Much remains to be done in this area and we hope to have provided the novice an accessible introduction to the basic ideas and techniques.

Acknowledgements

Much of the work on which these lectures is based was done in various collaborations with J. Maldacena, A. Guijosa and K. Savvidy and I am grateful to them for their help. I would also like to thank the organizers of the 1999 NATO Advanced Study Institute on Quantum Geometry for giving me the opportunity to present these lectures in such a pleasant and fascinating venue. This work was supported in part by the US Department of Energy under grant DE-FG02-91ER40671.

References

1. G.T. Horowitz and A. Strominger, Nucl. Phys. **B360**, 197 (1991).
2. J. Polchinski, Phys. Rev. Lett. **75**, 4724 (1995).
3. E.S. Fradkin and A.A. Tseytlin, Phys. Lett. **163B**, 123 (1985).
4. C.G. Callan, C. Lovelace, C.R. Nappi and S.A. Yost, Nucl. Phys. **B288**, 525 (1987).
5. R.G. Leigh, Mod. Phys. Lett. **A4**, 2767 (1989).
6. E. Bergshoeff and P.K. Townsend, Nucl. Phys. **B490**, 145 (1997).
7. G.W. Gibbons, Nucl. Phys. **B514**, 603 (1998).
8. C.G. Callan and J.M. Maldacena, Nucl. Phys. **B513**, 198 (1998).
9. J. Gomis, A.V. Ramallo, J. Simon and P.K. Townsend, "Supersymmetric baryonic branes," hep-th/9907022.
10. B. Craps, J. Gomis, D. Mateos and A. Van Proeyen, JHEP **04**, 004 (1999).
11. E. Witten, Nucl. Phys. **B460**, 335 (1996).
12. J. Maldacena, Adv. Theor. Math. Phys. **2**, 231 (1998).
13. I.R. Klebanov, Nucl. Phys. **B496**, 231 (1997).
14. J.P. Gauntlett, C. Koehl, D. Mateos, P.K. Townsend and M. Zamaklar, Phys. Rev. **D60**, 045004 (1999).
15. K. Ghoroku and K. Kaneko, "Born-Infeld strings between D-branes," hep-th/9908154.
16. D.J. Gross and H. Ooguri, Phys. Rev. **D58**, 106002 (1998).
17. E. Witten, JHEP **07**, 006 (1998).
18. Y. Imamura, Nucl. Phys. **B537**, 184 (1999).
19. C.G. Callan, A. Guijosa and K.G. Savvidy, Nucl. Phys. **B547** (1999) 127.
20. U. Danielsson, G. Ferretti and I.R. Klebanov, Phys. Rev. Lett. **79**, 1984 (1997).
21. O. Bergman, M.R. Gaberdiel and G. Lifschytz, Nucl. Phys. **B509**, 194 (1998).
22. S. Rey and J. Yee, "Macroscopic strings as heavy quarks in large N gauge theory and anti-de Sitter supergravity," hep-th/9803001.
23. J. Maldacena, Phys. Rev. Lett. **80**, 4859 (1998).
24. S.S. Gubser, I.R. Klebanov and A.M. Polyakov, Phys. Lett. **B428**, 105 (1998).
25. E. Witten, Adv. Theor. Math. Phys. **2**, 253 (1998).
26. C.G. Callan and A. Guijosa, "Undulating strings and gauge theory waves," hep-th/9906153.
27. U.H. Danielsson, E. Keski-Vakkuri and M. Kruczenski, JHEP **01**, 002 (1999).
28. T. Banks, M.R. Douglas, G.T. Horowitz and E. Martinec, "AdS dynamics from conformal field theory," hep-th/9808016.
29. V. Balasubramanian, P. Kraus, A. Lawrence and S.P. Trivedi, Phys. Rev. **D59**, 104021 (1999).
30. I. Klebanov, W.I. Taylor and M. Van Raamsdonk, "Absorption of dilaton partial waves by D3-branes," hep-th/9905174.
31. L. Susskind and E. Witten, "The holographic bound in anti-de Sitter space," hep-th/9805114.
32. A.W. Peet and J. Polchinski, Phys. Rev. **D59**, 065011 (1999).

12 8'99

13

LECTURES ON SUPERCONFORMAL QUANTUM MECHANICS AND MULTI-BLACK HOLE MODULI SPACES

RUTH BRITTO-PACUMIO
Jefferson Physical Laboratories
Harvard University
Cambridge, MA 02138, USA

JEREMY MICHELSON
New High Energy Theory Center
Rutgers University
126 Frelinghuysen Road
Piscataway, NJ 08854, USA

ANDREW STROMINGER
Jefferson Physical Laboratories
Harvard University
Cambridge, MA 02138, USA

AND

ANASTASIA VOLOVICH
Jefferson Physical Laboratories
Harvard University
Cambridge, MA 02138, USA
and
L.D. Landau Institute for Theoretical Physics,
Kosigina 2, Moscow, Russia

Abstract. This contribution to the proceedings of the 1999 NATO ASI on Quantum Geometry at Akureyri, Iceland, is based on notes of lectures given by A. Strominger. Topics include N-particle conformal quantum mechanics, extended superconformal quantum mechanics and multi-black hole moduli spaces.

L. Thorlacius and T. Jonsson (eds.), M-Theory and Quantum Geometry, 255-283.

1. Introduction

The problem of unifying quantum mechanics and gravity is one of the great unsolved problems in twentieth century physics. Progress has been slowed by our inability to carry out relevant physical experiments. Some progress has nevertheless been possible, largely through the use of gedanken experiments.

The quantum mechanical black hole has been a key ingredient of these gedanken experiments, beginning with [1, 2]. It provides an arena in which quantum mechanics and gravity meet head on. Such gedanken experiments have led to an astonishing depth and variety of insights, not only about the black holes themselves, but about string theory and quantum field theory in general. Nevertheless many aspects of quantum black holes remain enigmatic, and we expect they will continue to be a source of new insights.

Studies of quantum black holes have largely focused on the problem of quantum fields or strings interacting (by scattering or evaporation) with a single black hole. In these lectures we will address a different, less studied, type of gedanken experiment, involving an arbitrary number N of supersymmetric black holes. Configurations of N static black holes parametrize a moduli space $\mathcal{M}_{\mathcal{N}}$ [3, 4, 5]. The low-lying quantum states of the system are governed by quantum mechanics on $\mathcal{M}_{\mathcal{N}}$. As we shall see the problem of describing these states has a number of interesting and puzzling features. In particular $\mathcal{M}_{\mathcal{N}}$ has noncompact, infinite-volume regions corresponding to *near-coincident* black holes. These regions lead to infrared divergences and presents a challenge for obtaining a unitary description of multi-black hole scattering.

The main goal of these lectures is to describe the recent discovery of a superconformal structure [6, 7, 8, 9] in multi-black hole quantum mechanics. While the appearance of scale invariance at low energies follows simply from dimensional analysis, the appearance of the full conformal invariance requires particular values of the various couplings and is not *a priori* guaranteed. This structure is relevant both to the infrared divergences and the scattering, which however remain to be fully understood. We begin these lectures by developing the subject of conformal and superconformal quantum mechanics with N particles. Section 2 describes the simplest example [10] of single-particle conformally invariant quantum mechanics. The infrared problems endemic to conformal quantum mechanics as well as their generic cure are discussed in this context. Section 3 contains a discussion of conformally invariant N-particle quantum mechanics. Superconformal quantum mechanics is described in section 4. In section 5 the case of a test particle moving in a black hole geometry is discussed (following [11]) as a warm-up to the multi-black hole problem. The related issues of conformal invariance, infrared divergences and choices of time coordinate appear and are discussed in this simple context. In section 6 the five dimensional multi-black hole moduli space as well as its supersymmetric structure are described. It is shown that at low energies the supersymmetries are doubled and the $D(2,1;0)$ superconformal group makes an appearance. We close with a conjecture in section 7 on the possible relation to an M-brane description of the black hole and AdS_2/CFT_1 duality [12].

Many of the results described herein appeared recently in [13, 14].

2. A Simple Example of Conformal Quantum Mechanics

Let us consider the following Hamiltonian [10]:

$$H = \frac{p^2}{2} + \frac{g}{2x^2}. \tag{2.1}$$

In order to have an energy spectrum that is bounded from below, it turns out that we need to take $g \geq -1/4$, but otherwise g is an arbitrary coupling constant, though, following [10], we will consider only $g > 0$. Next introduce the operators

$$D = \tfrac{1}{2}(px + xp) \qquad\qquad K = \tfrac{1}{2}x^2. \tag{2.2}$$

D is known as the generator of dilations — it generates rescalings $x \to \lambda x$ and $p \to p/\lambda$ — and K is the generator of special conformal transformations. These operators obey the $SL(2,\mathbb{R})$ algebra

$$[D, H] = 2iH, \tag{2.3a}$$
$$[D, K] = -2iK, \tag{2.3b}$$
$$[H, K] = -iD. \tag{2.3c}$$

Since D and K do not commute with the Hamiltonian, they do not generate symmetries in the usual sense of relating degenerate states. Rather they can be used to to relate states with different eigenvalues of H [6, 7, 8, 9, 10].

Exercise 1 *Show that for any quantum mechanics with operators obeying the $SL(2,\mathbb{R})$ algebra (2.3), that if $|E\rangle$ is a state of energy E, then $e^{i\alpha D}|E\rangle$ is a state of energy $e^{2\alpha}E$. Thus, if there is a state of nonzero energy, then the spectrum is continuous.*

It follows from exercise 1 that the spectrum of the Hamiltonian (2.1) is continuous, and its eigenstates are not normalizable. Hence it is awkward to describe the theory in terms of H eigenstates.

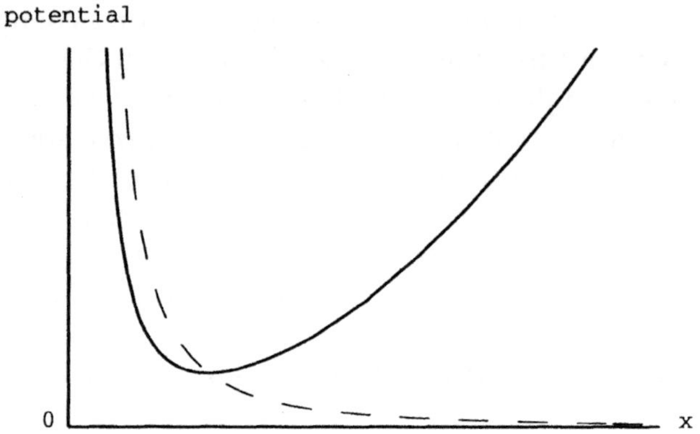

Figure 1. A comparison between the potentials for H and L_0. The dashed line is the potential energy part of H and the solid line is that for L_0. Note that the former has no minimum while the latter is a well.

This problem is easily rectified. Consider the linear combinations

$$L_{\pm 1} = \tfrac{1}{2}(aH - \frac{K}{a} \mp iD) \tag{2.4a}$$

$$L_0 = \tfrac{1}{2}(aH + \frac{K}{a}), \tag{2.4b}$$

where a is a parameter with dimensions of length-squared. These obey the $SL(2,\mathbb{R})$ algebra in the Virasoro form,

$$\left[L_1, L_{-1}\right] = 2L_0 \qquad \left[L_0, L_{\pm 1}\right] = \mp L_{\pm 1}. \tag{2.5}$$

In the following, we choose our units such that $a = 1$.

With the definitions (2.4b), (2.1) and (2.2), we have

$$L_0 = \frac{p^2}{4} + \frac{g}{4x^2} + \frac{x^2}{4}. \tag{2.6}$$

The potential energy part of this operator achieves its minimum and asymptotes to ∞ (see figure 1) and thus has a discrete spectrum with normalizable eigenstates.

Exercise 2 *Show that*

$$L^2 = L_0(L_0 - 1) - L_{-1}L_1 \tag{2.7}$$

is the $SL(2,\mathbb{R})$ Casimir operator. Thus show that, of the eigenstates of L_0, that with the smallest value of L_0 is annihilated by L_1. Also show that the eigenvalues of L_0 form an infinite tower above the "ground state", in integer steps.

∗Exercise 3 *Show that for the DFF model, the Casimir operator* (2.7) *takes the value*

$$L^2 = \frac{g}{4} - \frac{3}{16} \tag{2.8}$$

and thus that the "ground state" has $L_0 = \frac{1}{2}(1 \pm \sqrt{g + \frac{1}{4}})$. *(It turns out that the positive root is that for which the state is normalizable.)*

From exercise 2, we learn that the spectrum of L_0 is well defined, and thus has normalizable eigenstates. This motivated DFF to trade H for L_0, and use L_0 to generate the dynamics. We then have a well defined theory; this also justifies our use of the term "ground state" in exercises 2 and 3.

At this point it is a free world and one has the right to describe the theory in terms of L_0 rather than H eigenstates. Later on this issue will reappear in the context of black hole physics, and the trade of H for L_0 will take on a deeper significance.

3. Conformally Invariant N-Particle Quantum Mechanics

In this section, we find the conditions under which a general N-particle quantum mechanics admits an $SL(2,\mathbb{R})$ symmetry. Specifically, we derive the conditions for the existence of operators D and K obeying the algebra (2.3). N-particle quantum mechanics can be described as a sigma model with an N-dimensional target space. The general Hamiltonian is[1]

$$H = \tfrac{1}{2} P_a^\dagger g^{ab} P_b + V(X), \tag{3.1}$$

where $a, b = 1, \ldots, N$ and the metric g is a function of X. The canonical momentum P_a obeys $[P_a, X^b] = -i\delta_a^b$ and $[P_a, P_b] = 0$, and is given by

$$P_a = g_{ab}\dot{X}^b = -i\partial_a. \tag{3.2}$$

Exercise 4 *Given the norm* $(f_1, f_2) = \int d^N X \sqrt{g} f_1^* f_2$, *show that*

$$P_a^\dagger = \frac{1}{\sqrt{g}} P_a \sqrt{g} = P_a - i\Gamma^b_{ba}, \tag{3.3}$$

where Γ^c_{ab} *is the Christoffel symbol built from the metric* g_{ab}, *and the dagger denotes Hermitian conjugation. Thus,* $H\Psi = (-\nabla^2 + V)\Psi$, *for all (scalar) functions* $\Psi(X)$.

[1]In this and all subsequent expressions, the operator ordering is as indicated.

We first determine the conditions under which the theory, defined by equation (3.1), admits a dilational symmetry of the general form

$$\delta_D X^a = D^a(X). \tag{3.4}$$

This symmetry is generated by an operator

$$D = \tfrac{1}{2} D^a P_a + \text{h.c.} \tag{3.5}$$

which should obey equation (2.3a),

$$\left[D, H\right] = 2iH. \tag{2.3a}$$

From the definitions (3.5) and (3.1), one finds

$$\left[D, H\right] = -\frac{i}{2} P_a^\dagger (\pounds_D g^{ab}) P_b - i\pounds_D V - \frac{i}{4} \nabla^2 \nabla_a D^a, \tag{3.6}$$

where $\pounds_D$ is the usual Lie derivative obeying

$$\pounds_D g_{ab} = D^c g_{ab,c} + D^c{}_{,a} g_{cb} + D^c{}_{,b} g_{ac}. \tag{3.7}$$

Comparing equations (3.7) and (2.3a) reveals that a dilational symmetry exists if and only if there exists a conformal Killing vector D obeying

$$\pounds_D g_{ab} = 2 g_{ab} \tag{3.8a}$$

and

$$\pounds_D V = -2V. \tag{3.8b}$$

Note that equation (3.8a) implies the vanishing of the last term of equation (3.6). A vector field D obeying (3.8a) is known as a *homothetic* vector field, and the action of D is known as a *homothety* (pronounced h'MAWthitee).

Next we look for a special conformal symmetry generated by an operator $K = K(X)$ obeying equations (2.3b) and (2.3c):

$$\left[D, K\right] = -2iK, \tag{2.3b}$$

$$\left[H, K\right] = -iD. \tag{2.3c}$$

With equation (3.5), equation (2.3b) is equivalent to

$$\pounds_D K = 2K, \tag{3.9}$$

while equation (2.3c) can be written

$$D_a dX^a = dK. \tag{3.10}$$

Hence the one-form D is exact. One can solve for K as the norm of D^a,

$$K = \tfrac{1}{2} g_{ab} D^a D^b, \tag{3.11}$$

which is globally well defined. We shall adopt the phrase "closed homothety" to refer to a homothety whose associated one-form is closed and exact.

Exercise 5 *Show that conversely, given a vector field D^a obeying equation (3.8a) and $dD = 0$, that $D_a dX^a = dK$ where K is* defined *by equation (3.11). Thus, every "closed homothety" is an "exact homothety", and there is no significance in our choice of phrase. (We have chosen to use the phrase "closed homothety" in order to avoid confusion with a discussion of, say, quantum corrections.)*

$*$**Exercise 6** *Show that if a manifold admits a homothety (not necessarily closed), then the manifold is noncompact.*

We should emphasize that the existence of D did *not* guarantee the existence of K. It is not hard [14] to find examples of quantum mechanics with a D for which the corresponding unique candidate for K (by equation (3.11)) obeys neither equations (2.3c) nor (3.10). Indeed a generic homothety is not closed.[2]

4. Superconformal Quantum Mechanics

This section considers supersymmetric quantum mechanics with up to four supersymmetries and superconformal extensions with up to eight supersymmetries. In lower dimensions the Poincaré groups are smaller and hence so are the supergroups. This implies a richer class of supersymmetric structures for a given number of supercharges. In particular, in one dimension we shall encounter structures which cannot be obtained by reduction from higher dimensions.

4.1. A BRIEF DIVERSION ON SUPERGROUPS

Roughly, a supergroup is a group of matrices that take the block form

$$\left(\begin{array}{c|c} A & F_1 \\ \hline F_2 & B \end{array}\right), \tag{4.1}$$

[2] One can find even four dimensional theories that are dilationally, but not conformally, invariant by including higher derivative terms; for a scalar field $\phi(x^\mu)$, the Lagrangian $\mathcal{L} = f(\frac{\partial^\mu \phi \partial_\mu \phi}{\phi^4})\phi^4$ is dilationally invariant for any function f, but it is conformally invariant only for $f(y) = -\frac{1}{2}y - \frac{\lambda}{4!}$.[15]

Superalgebra	Dimension (#b,#f)	R-symmetry
$Osp(1\|2)$	(3,2)	1
$SU(1,1\|1)$	(4,4)	$U(1)$
$Osp(3\|2)$	(6,6)	$SU(2)$
$SU(1,1\|2)$	(6,8)	$SU(2)$
$D(2,1;\alpha), \alpha \neq -1, 0, \infty$	(9,8)	$SU(2) \times SU(2)$
$Osp(5\|2)$	(13,10)	$SO(5)$
$SU(1,1\|3)$	(12,12)	$SU(3) \times U(1)$
$Osp(6\|2)$	(18,12)	$SO(6)$
$G(3)$	(17,14)	G_2
$Osp(7\|2)$	(24,14)	$SO(7)$
$Osp(4^*\|4)$	(16,16)	$SU(2) \times SO(5)$
$SU(1,1\|4)$	(19,16)	$SU(4) \times U(1)$
$F(4)$	(24,16)	$SO(7)$
$Osp(8\|2)$	(31,16)	$SO(8)$
$Osp(4^*\|2n), n > 2$	$(2n^2 + n + 6,\ 8n)$	$SU(2) \times Sp(2n)$
$SU(1,1\|n), n > 4$	$(n^2 + 3,\ 4n)$	$SU(n) \times U(1)$
$Osp(n\|2), n > 8$	$(\frac{1}{2}n^2 - \frac{1}{2}n + 3,\ 2n)$	$SO(n)$

TABLE 1. The simple supergroups that contain an $SL(2,\mathbb{R})$ subgroup (see also [16]). The table is divided into those which have eight or fewer (ordinary) supersymmetries (including the exceptional supergroups) and those which have more than eight (ordinary) supersymmetries (for which there are no exceptional supergroups). The algebra of $Osp(4^*|2m)$ has bosonic part $SO^*(4) \times Usp(2m)$, where $SO^*(4) \cong SL(2,\mathbb{R}) \times SU(2)$ is a noncompact form of the $SO(4)$ algebra.

where A, B are ordinary matrices, and $F_{1,2}$ are fermionic matrices. We are interested in quantum mechanics with a supersymmetry whose supergroup includes $SL(2,\mathbb{R})$; that is, supergroups of the form

$$\left(\begin{array}{c|c} SL(2,\mathbb{R}) & \text{fermionic} \\ \hline \text{fermionic} & \text{R-symmetry} \end{array} \right). \tag{4.2}$$

There are many such supergroups; these have been tabulated in table 1.

One simple series of supergroups is the $Osp(m|n)$ series; the elements of $Osp(m|n)$ have the form

$$\left(\begin{array}{c|c} Sp(n) & \text{fermionic} \\ \hline \text{fermionic} & SO(m) \end{array}\right). \tag{4.3}$$

Since $Sp(2) \cong SL(2,\mathbb{R})$ [3] we are interested in $Osp(m|2)$. The simplest of these is $Osp(1|2)$, which is a subgroup of the others. We will describe the models with this symmetry group, for the supermultiplet defined in section 4.2, in section 4.3. We will skip $Osp(2|2) \cong SU(1,1|1)$ [4] — these models were described in [13] — and go directly to $Osp(4|2)$. In fact, it will turn out that, for the supermultiplet we consider, we will naturally obtain $D(2,1;\alpha)$ as the symmetry group, where α is a parameter that depends on the target space geometry. $Osp(4|2)$ is the special case of $\alpha = -2$, and appears, for example, when the target space is flat. The black hole system described in section 6 will turn out to have $D(2,1;0)$ superconformal symmetry.[5] We will explain this statement, and describe $D(2,1;\alpha)$ in more detail, in section 4.4. First, we should describe the supermultiplet under consideration.

4.2. QUANTUM MECHANICAL SUPERMULTIPLETS

There are many supermultiplets that one can construct in one dimension. In particular, unlike in higher dimensions, the smaller supersymmetry group does not require a matching of the numbers of bosonic and fermionic fields. Much of the literature — see, *e.g.* [17, 18, 19, 20, 21, 22] — concerns the so-called type A multiplet, with a real boson and complex fermion (X^a, ψ^a), which can be obtained by dimensional reduction of the $1+1$ dimensional $\mathcal{N} = (1,1)$ multiplet.

This is not the multiplet we will consider here. For the black hole physics that we will eventually consider, each black hole will have four bosonic (translational) degrees of freedom, as well as four fermionic degrees of freedom from the breaking of one half of the minimal (8 supercharge) super-

[3]The notation is such that only $Sp(2n)$ exist.

[4]The supergroup $U(m,n|p)$ is generated by matrices of the form (4.1), with $A \in U(m,n)$ and $B \in U(p)$. The subalgebra in which the matrices also obey $\mathrm{Tr}\, A = \mathrm{Tr}\, B$ generates $SU(m,n|p)$. However, with this definition, $SU(m,n|p = m+n)$ is not even semisimple, for the identity matrix obeys $\mathrm{Tr}\, A = \mathrm{Tr}\, B$ and generates a $U(1)$ factor. The quotient $PSU(m,n|m+n) \cong SU(m,n|m+n)/U(1)$ is simple, and is often denoted just $SU(m,n|m+n)$, as we have done for $SU(1,1|2)$.

[5]$D(2,1;0)$ (and $D(2,1;\infty)$) is omitted from Table 1 because it is the semidirect product $SU(1,1|2) \rtimes SU(2)$ and is therefore not simple.

symmetry in five dimensions.[6] Thus, we will consider the type B multiplet, consisting of a real boson and a *real* fermion $(X^a, \lambda^a = \lambda^{a\dagger})$. The supersymmetry transformation, parametrized by a real Grassman parameter ϵ, is given by

$$\delta_\epsilon X^a = -i\epsilon\lambda^a \qquad\qquad \delta_\epsilon \lambda^a = \epsilon \dot{X}^a, \tag{4.4}$$

where the overdot denotes a time derivative.

Exercise 7 *In an $\mathcal{N} = 1$ superspace formalism, the type B multiplet is given by a real supermultiplet $X^a(t,\theta) = X^{a\dagger}(t,\theta)$, where θ is the (real) fermionic coordinate, and we use the standard convention in which the lowest component of the superfield is notationally almost indistinguishable from the superfield itself. In components, we write*

$$X^a(t,\theta) = X^a(t) - i\theta\lambda^a(t). \tag{4.5}$$

The generator of supersymmetry transformations, Q (which obeys $Q^2 = H = i\frac{d}{dt}$) is given by

$$Q = \frac{\partial}{\partial\theta} + i\theta\frac{d}{dt}. \tag{4.6}$$

Show that

$$\delta_\epsilon X^a = \left[\epsilon Q, X^a\right], \tag{4.7}$$

as expected. Note also that $Q = Q^\dagger$, and thus both sides of equation (4.7) are, indeed, real. For completeness, we define the superderivative

$$D = \frac{\partial}{\partial\theta} - i\theta\frac{d}{dt}, \tag{4.8}$$

which obeys $D^2 = -i\frac{d}{dt}$ and $\{D, Q\} = 0$.

As we have already mentioned, there are many more multiplets than just the type B one; see *e.g.* [24, 23].

4.3. *Osp*(1|2)–INVARIANT QUANTUM MECHANICS

We now proceed to the simplest superconformal quantum mechanics for the Type B supermultiplet defined in the previous subsection. As in section 3, we use a Hamiltonian formalism.

[6] Recently, four-dimensional black holes have been described using a multiplet with 3 bosons and 4 fermions [23].

In general, the supercharge takes the form

$$Q = \lambda^a \Pi_a - \frac{i}{3} c_{abc} \lambda^a \lambda^b \lambda^c, \tag{4.9}$$

where we define

$$\Pi_a \equiv P_a - \frac{i}{2}\omega_{abc}\lambda^b\lambda^c + \frac{i}{2}c_{abc}\lambda^b\lambda^c \equiv P_a - \frac{i}{2}\Omega^+_{abc}\lambda^b\lambda^c, \tag{4.10}$$

where ω_{abc} is the spin connection with the last two indices contracted with the vielbein, and c_{abc} is a (so-far) general 3-form. The Hamiltonian is then given by

$$H = \tfrac{1}{2}\{Q, Q\}. \tag{4.11}$$

We remark that the bosonic part of this Hamiltonian is the special case of equation (3.1) with $V = 0$.

Exercise 8 *Show that the most general, renormalizable superspace action [24]*

$$S = i \int dt d\theta \left\{ \tfrac{1}{2} g_{ab} DX^a \dot{X}^b + \frac{i}{6} c_{abc} DX^a DX^b DX^c \right\}, \tag{4.12}$$

is given in terms of the component fields by

$$S = \int dt \left\{ \tfrac{1}{2} g_{ab} \dot{X}^a \dot{X}^b + \frac{i}{2}\lambda^a \left(g_{ab}\frac{D\lambda^b}{dt} - \dot{X}^c c_{abc}\lambda^b \right) - \frac{1}{6}\partial_d c_{abc}\lambda^d\lambda^a\lambda^b\lambda^c \right\}, \tag{4.13}$$

where

$$\frac{D\lambda^a}{dt} \equiv \dot{\lambda}^a + \dot{X}^b \Gamma^a_{bc}\lambda^c, \tag{4.14}$$

is the covariant time-derivative. (Note that $g_{ab} = g_{(ab)}$ and $c_{abc} = c_{[abc]}$ are arbitrary (though g_{ab} should be positive definite for positivity of the kinetic energy) functions of the superfield; e.g. $g_{ab} = g_{ab}(X(t,\theta))$.*) In terms of $\lambda^\alpha \equiv \lambda^a e_a^\alpha$, show that the action (4.13) is*

$$S = \int dt \left\{ \tfrac{1}{2} g_{ab}\dot{X}^a \dot{X}^b + \frac{i}{2}\delta_{\alpha\beta}\lambda^\alpha \frac{D\lambda^\beta}{dt} - \frac{i}{2}\dot{X}^c c_{c\alpha\beta}\lambda^\alpha\lambda^\beta \right.$$
$$\left. - \frac{1}{6} e^d_\delta \partial_d c_{abc} e^a_\alpha e^b_\beta e^c_\gamma \lambda^\delta\lambda^\alpha\lambda^\beta\lambda^\gamma \right\}, \tag{4.15}$$

where

$$\frac{D\lambda^\alpha}{dt} = \dot{\lambda}^\alpha + \dot{X}^a \omega_a{}^\alpha{}_\beta \lambda^\beta. \tag{4.16}$$

∗*Finally, show that equation (4.9) follows from equation (4.4) (or (4.7)).*

We note that, from equation (4.15), the fermions λ^α obey the canonical anticommutation relation

$$\{\lambda^\alpha, \lambda^\beta\} = \delta^{\alpha\beta}, \tag{4.17}$$

and commute with X^a and P_a.[7] It follows from equation (4.17), that the fermions can be represented on the Hilbert space by $\lambda^\alpha = \gamma^\alpha/\sqrt{2}$, where γ^α are the $SO(n)$ γ-matrices (n is the dimension of the target space), and that the wavefunction is an $SO(n)$ spinor. Thus Π_a is just the covariant derivative (with torsion c — see appendix A for a brief summary of calculus with torsion) on the Hilbert space.[8]

So far we have only discussed $\mathcal{N}=1$ super*symmetric* quantum mechanics, whereas we would like to discuss super*conformal* quantum mechanics. We have already shown in section 3 that in order to have conformal quantum mechanics, the metric g_{ab} must admit a closed homothety D^a, out of which were built the operators D and K. The supersymmetric extensions of the expressions (3.5) and (3.11) for D and K — that is including fermions — are given by replacing the P_a in equation (3.5) (which is a covariant derivative on the scalar wavefunction of the bosonic theory) with the covariant derivative Π_a:[9]

$$D = \tfrac{1}{2} D^a \Pi_a + \text{h.c.} \tag{4.18}$$

and

$$K = \tfrac{1}{2} D^a D_a. \tag{4.19}$$

However, it turns out that closure of the superalgebra places two constraints on the torsion c_{abc},

$$D^a c_{abc} = 0 \tag{4.20a}$$

and

$$\pounds_D c_{abc} = 2c_{abc}. \tag{4.20b}$$

The final operator that appears in an $Osp(1|2)$-invariant theory is

$$S = i\big[Q, K\big] = \lambda^a D_a. \tag{4.21}$$

[7]Note that this implies that (generically) λ^a does *not* commute with P_b, but rather, $[P_a, \lambda^b] = -i(\omega_a{}^b{}_c - \Gamma^b_{ac})\lambda^c$.

[8]It also follows [25, 26] that, for these theories, the Witten index, $\mathrm{Tr}(-1)^F$ [17, 18], is equivalent to the Atiyah-Singer index.

[9]But the reader should not extrapolate too far, for $H \neq \frac{1}{2}\Pi_a^\dagger g^{ab}\Pi_b$.

∗**Exercise 9** *Verify that (with equations* (4.20)*) the operators H, D, K, Q and S, defined by equations* (4.11), (4.18), (4.19), (4.9) *and* (4.21) *satisfy the $Osp(1|2)$ algebra*

$$\begin{aligned}
&[H,K]=iD, && [H,D]=-2iH, && [K,D]=2iK,\\
&\{Q,Q\}=2H, && [Q,D]=-iQ, && [Q,K]=-iS,\\
&\{S,S\}=2K, && [S,D]=iS, && [S,H]=iQ,\\
&\{S,Q\}=D, && [Q,H]=0, && [S,K]=0.
\end{aligned} \tag{4.22}$$

4.4. $D(2,1;\alpha)$–INVARIANT QUANTUM MECHANICS

The $D(2,1;\alpha)$ algebra is an $\mathcal{N}=4$ (actually $\mathcal{N}=4$B, since we use the type B supermultiplet) superconformal algebra, and thus contains four supercharges Q^m, $m=1,\dots,4$, and their superconformal partners S^m. Of course, for fixed m, Q^m, S^m, H, K, D should satisfy the $OSp(1|2)$ algebra (4.22). In addition, as is evident from table 1, there are two (commuting) sets of $SU(2)$ R-symmetry generators $R^r_\pm$, $r=1,2,3$, under which the supercharges Q^m and S^m transform as (2,2). There are no other generators, and the complete set of (anti)commutation relations, which define the algebra, are [27]

$$\begin{aligned}
&[H,K]=iD, && [H,D]=-2iH, && [K,D]=2iK,\\
&\{Q^m,Q^n\}=2H\delta^{mn}, && [Q^m,D]=-iQ^m, && [Q^m,K]=-iS^m,\\
&\{S^m,S^n\}=2K\delta^{mn}, && [S^m,D]=iS^m, && [S^m,H]=iQ^m,\\
&[R^r_\pm,Q^m]=it^{\pm r}_{mn}Q^n, && [R^r_\pm,S^m]=it^{\pm r}_{mn}S^n, && [R^r_\pm,R^s_{\pm'}]=i\delta_{\pm\pm'}\epsilon^{rst}R^t_\pm\\
&[R^r_\pm,H]=0, && [R^r_\pm,D]=0, && [R^r_\pm,K]=0,\\
&[Q^m,H]=0, && [S^m,K]=0, &&
\end{aligned} \tag{4.23}$$
$$\{S^m,Q^n\}=D\delta^{mn}-\frac{4\alpha}{1+\alpha}t^{+r}_{mn}R^r_+-\frac{4}{1+\alpha}t^{-r}_{mn}R^r_-,$$

where

$$t^{\pm r}_{mn}\equiv\mp\delta^r_{[m}\delta^4_{n]}+\tfrac{1}{2}\epsilon_{rmn}. \tag{4.24}$$

Clearly the $D(2,1;\alpha)$ algebra is not defined for $\alpha=-1$; for $\alpha=0$ (∞), R^r_+ (R^r_-) does not appear on the right-hand side of the commutation relations (4.23), and thus the group is the semidirect product of $SU(1,1|2)$ [10] (the unique group in table 1 with the correct number of generators and bosonic subalgebra) and $SU(2)$.

[10]See footnote 4 (page 9) for the definition of $SU(m,n|p)$.

Before we discuss the conditions under which the action (4.15) admits a $D(2,1;\alpha)$ superconformal symmetry, we should first discuss the conditions for $\mathcal{N} = 4$B supersymmetry.

4.4.1. *$\mathcal{N} = 4B$ Supersymmetric Quantum Mechanics.*
The conditions on the geometry for an $\mathcal{N} = 4$B theory have been given in [24, 28]. We will repeat them in the simplified form given in [13]. The object is to find supercharges Q^m, such that

$$\{Q^m, Q^n\} = 2\delta^{mn} H. \tag{4.25}$$

We take Q^4 to be the Q of equation (4.9) — *i.e.* the Noether charge associated with the symmetry generated by equation (4.4). We now look for three more symmetry transformations such that

$$[\delta_\epsilon^{(m)}, \delta_\eta^{(n)}] = -2i\delta_{mn}\eta\epsilon\frac{d}{dt}, \tag{4.26}$$

where $\delta_\epsilon^{(m)}$ is the $m^{\underline{\text{th}}}$ supersymmetry transformation, generated by the Grassmann variable ϵ. It is standard (see *e.g.* [29, 24, 28]) to give these transformations according to the following rather tedious exercise.

Exercise 10 *Define*

$$\delta_\epsilon^{(r)} X^a(t,\theta) = \epsilon I_b^{r\,a} D X^b, \tag{4.27}$$

where $I_b^{r\,a}(X(t,\theta))$ is some tensor-valued function on superspace. Then, show that the supersymmetry algebra (4.26) is obeyed iff $I_b^{r\,a}$ are almost complex structures obeying

$$I_a^{r\,c} I_c^{s\,b} + I_a^{s\,c} I_c^{r\,b} = -2\delta^{rs}\delta_a^b, \tag{4.28a}$$

with vanishing Nijenhuis concomitants,

$$N(r,s)_{ab}{}^c \equiv \left\{2I^{r\,d}_{[a}\partial_{|d|}I^{s\,c}_{b]} - 2I^{rc}_{d}\partial_{[a}I^{s\,d}_{b]}\right\} + (r \leftrightarrow s) = 0. \tag{4.28b}$$

From exercise 10, we learn that supersymmetry requires a complex target space, with three anticommuting complex structures. $\mathcal{N} = 4B$ supersymmetry is defined to have a quaternionic target space,

$$I_a^{r\,c} I_c^{s\,b} = -\delta^{rs}\delta_a^b + \epsilon^{rst} I_a^{t\,b}. \tag{4.29}$$

This provides a natural $SU(2)$ structure, which will give rise to self-dual rotations in the black hole context. Given such a manifold, it is convenient to define the three exterior derivatives d^r by

$$d^r\omega = (-1)^{p+1} I^r d I^r \omega ; I^r\omega \equiv \frac{(-1)^p}{p!} I^r_{a_1}{}^{b_1} \cdots I^r_{a_p}{}^{b_p} \omega_{b_1 \ldots b_p} dX^{a_1} \wedge \cdots \wedge dX^{a_p},$$
$$= \frac{1}{p!}\left[I^{rb}_a \partial_b \omega_{c_1 \ldots c_p} - p(\partial_a I^r_{c_1}{}^b)\omega_{bc_2 \ldots c_p}\right] dX^a \wedge dX^{c_1} \wedge \cdots \wedge dX^{c_p} \tag{4.30}$$

where $\omega = \frac{1}{p!}\omega_{a_1 \ldots a_p} dX^{a_1} \wedge \cdots \wedge dX^{a_p}$ is a p-form.

Exercise 11 *Show that, in complex coordinates adapted to* I^{rb}_a, $d^r = i(\partial - \bar{\partial})$.

Having defined the supersymmetry transformations using the quaternionic structure of the manifold, we should now check that the action is invariant. We will simply quote the result [13]. The action is invariant provided that[11]

$$\{d^r, d^s\} = 0, \tag{4.31a}$$
$$I^{rc}_a I^{sb}_c = -\delta^{rs}\delta^b_a + \epsilon^{rst} I^{tb}_a, \tag{4.31b}$$
$$g_{ab} = I^{rc}_a g_{cd} I^{rd}_b \ (\forall\, r) \tag{4.31c}$$
$$c = \frac{1}{6} c_{abc} dX^a \wedge dX^b \wedge dX^c = \tfrac{1}{2} d^r J^r \ (\forall\, r); J^r \equiv \tfrac{1}{2} I^{rc}_a g_{cb}\, dX^a \wedge dX^b. \tag{4.31d}$$

Equation (4.31a) is secretly a restatement of equation (4.28b) and equation (4.31b) was our demand (4.29). The new conditions are equations (4.31c) and (4.31d). Equation (4.31c) states that the metric is Hermitian with respect to each complex structure. Equation (4.31d) is a highly nontrivial differential constraint between the complex structures, which generalizes the hyperkähler condition, and from which the torsion c_{abc} is uniquely determined. It is equivalent to the condition that the quaternionic structure be covariantly constant:

$$\nabla^+_a I^{rc}_b \equiv \nabla_a I^{rc}_b + c^c{}_{ad} I^{rd}_b - c^d{}_{ab} I^{rc}_d = 0. \tag{4.32}$$

A manifold that satisfies the conditions (4.31) is known as a hyperkähler with torsion (HKT) (or sometimes weak HKT) manifold.

4.4.2. $\mathcal{N} = 4B$ *Superconformal Quantum Mechanics*

We must now find the further restrictions to a $D(2,1;\alpha)$-invariant superconformal quantum mechanics. Clearly, this will include equations (3.8a),

[11]For the more general case, with a Clifford, but not a quaternionic, structure see [24, 28, 30].

(3.10) and (4.20). The additional restrictions are obtained by demanding the proper behaviour of the R-symmetries, and are most easily phrased by defining the vector fields

$$D^{rb} \equiv D^a I_a^{rb}. \tag{4.33}$$

$D(2,1;\alpha)$-invariance then forces these to be Killing vectors

$$\pounds_{D^r} g_{ab} = 0, \tag{4.34}$$

which also obey the $SU(2)$ algebra for some normalization

$$\left[\pounds_{D^r}, \pounds_{D^s}\right] = -\frac{2}{\alpha+1}\epsilon^{rst}\pounds_{D^t}. \tag{4.35}$$

Equation (4.35) gives a geometric definition of α. Note that because the normalization of D^{ra} is specified by equations (3.8a) and (4.33), α is unambiguous. In fact, equation (4.35) is not a sufficiently strong condition for the proper closure of the algebra; we must have

$$\pounds_{D^r} I_a^{sb} = -\frac{2}{\alpha+1}\epsilon^{rst} I_a^{tb}. \tag{4.36}$$

Equation (4.36) implies equation (4.35).

These are the necessary and sufficient conditions for the quantum mechanics defined by (4.13) to be $D(2,1;\alpha)$ superconformal. They imply that

$$\alpha J^r = (\alpha+1)(d^r dK - \tfrac{1}{2}\epsilon^{rst} d^s d^t K); \tag{4.37}$$

i.e. that (at least for $\alpha \neq 0$) the HKT metric is described by a potential which is proportional to K. Actually, as discussed in more detail in [13], when the three complex structures are *simultaneously* integrable, there is always a potential, but a general HKT manifold admits a potential only under the conditions given in [31].

A general (but not most general!) set of models can be obtained from a function $L(X)$, where X^a are coordinates on $\mathbb{R}^{4n}$, and the I_a^{rb} are given by the self-dual complex structures on $\mathbb{R}^4$ tensored with the n-dimensional identity matrix.[12] If (but not iff — in particular, this is not true of the system described in section 6.3) $L(X)$ also obeys

$$X^a \partial_a L(X) = hL(X), \qquad X^a I_a^{rb} \partial_b L(X) = 0, \tag{4.38}$$

[12]This implies that the three complex structures are simultaneously integrable. Hellerman and Polchinski [32] have recently shown how to relax this limitation by generalizing the $\mathcal{N}=2$ superfield constraints of [29, 33].

then we obtain a $D(2,1;\alpha = -\frac{h+2}{2})$-invariant model, with

$$g_{ab} = \left(\delta_a^c \delta_b^d + I_a^{rc} I_b^{rd}\right) \partial_c \partial_d L(X), \tag{4.39a}$$

$$D^a = \frac{2}{h} X^a, \tag{4.39b}$$

$$K = \frac{h+2}{2h} L, \tag{4.39c}$$

and c_{abc} given by equation (4.31d). In an $\mathcal{N} = 2$ superspace formalism, X^a is a superfield obeying certain constraints [29, 33] and the potential L is the superspace integrand [13, 32, 30].

5. The Quantum Mechanics of a Test Particle in a Reissner-Nordström Background

Our goal is to apply the results on superconformal quantum mechanics to the quantum mechanics of a collection of supersymmetric black holes. As a warm-up in this section we consider the problem of a quantum test particle moving in the black hole geometry. The four-dimensional case was treated in [11], which will be followed and adapted to five dimensions in this section.

Consider a five-dimensional extremal Reissner-Nordström black hole of charge Q. The geometry of such a black hole is described by the metric

$$ds^2 = -\frac{dt^2}{\psi^2} + \psi d\vec{x}^2, \tag{5.1a}$$

and the gauge field

$$A = \psi^{-1} dt, \tag{5.1b}$$

where $\vec{x}$ is the $\mathbb{R}^4$ coordinate, and $\psi = 1 + \frac{Q}{|\vec{x}^2|}$. We have set $M_p = L_p = 1$. The horizon in these coordinates is at $|\vec{x}| = 0$.

Introduce a test particle with mass m and charge q. The particle action is

$$S = -m \int d\tau + q \int A. \tag{5.2}$$

Parametrize the particle's trajectory as $\vec{x} = \vec{x}(t)$. Eventually we will require the test particle to be supersymmetric (by imposing $q = m$). A supersymmetric test particle at rest at a fixed distance from the black hole, remains at rest, so it is sensible to consider a test particle that moves slowly. Accordingly we shall assume $|\dot{\vec{x}}| \ll 1$. In this parametrization, we can make the following substitution:

$$d\vec{x} = \dot{\vec{x}} dt, \tag{5.3}$$

which allows us to rewrite (5.1) to obtain the metric

$$ds^2 = -\frac{dt^2}{\psi^2} + \psi|\dot{\vec{x}}|^2 dt^2. \tag{5.4}$$

Now we solve the equation $ds^2 = -d\tau^2$ and find

$$d\tau = \frac{dt}{\psi} - \frac{1}{2}\psi^2|\dot{\vec{x}}|^2 dt + \mathcal{O}(\dot{x}^4), \tag{5.5}$$

which is substituted into (5.2) to obtain the action,

$$S = -m\int (\frac{dt}{\psi} - \frac{1}{2}\psi^2|\dot{\vec{x}}|^2 dt) + q\int \frac{dt}{\psi}. \tag{5.6}$$

For a supersymmetric test particle, $m = q$, this action reduces to

$$S = \frac{m}{2}\int \psi^2|\dot{\vec{x}}|^2 dt. \tag{5.7}$$

If the particle is near the horizon, at distances $r \ll \sqrt{Q}$, then we can approximate $\psi = \frac{Q}{|\vec{x}|^2}$, so that

$$S_p = \frac{mQ^2}{2}\int dt \frac{|\dot{\vec{x}}|^2}{|\vec{x}|^4}, \tag{5.8}$$

or, if we define a new quantity $\vec{y} = \frac{\vec{x}}{|\vec{x}|^2}$, then we see that we are actually in flat space:

$$S_p = \frac{mQ^2}{2}\int dt|\dot{\vec{y}}|^2. \tag{5.9}$$

Far from the black hole, spacetime and the moduli space look flat once again. Thus the moduli space can be described as two asymptotically flat regions connected by a wormhole whose radius scales as $\sqrt{Q}$. At low energies (relative to $M_p/\sqrt{Q}$) the wavefunctions spread out and do not fit into the wormhole. Hence the quantum mechanics is described by near and far superselection sectors that decouple completely at low energies.

This geometry leads to a problem. Consider the near horizon quantum theory. Given any fixed energy level E, there are infinitely many states of energy less than E. This suggests that there are infinitely many states of a test particle localized near the horizon of a black hole, which appears problematic for black hole thermodynamics. The possibility of such states arises from the large redshift factors near the horizon of a black hole. Similar problems have been encountered in studies of ordinary quantum fields in a black hole geometry.

The new observation of [11] is that this problem is in fact equivalent to the problem encountered by DFF [10] in their analysis of conformal quantum mechanics. To see this equivalence let ρ denote the radial coordinate $|\vec{y}|$. The Hamiltonian corresponding to (5.9) is

$$H = \frac{1}{2mQ^2}(p_\rho^2 + \frac{4}{\rho^2}J^2). \tag{5.10}$$

This is the DFF Hamiltonian of (2.1) with $g = \frac{4}{mQ^2}J^2$. The coordinate ρ grows infinite at the horizon. Thus this potential pushes a particle to the horizon whenever J^2 is nonzero. Our problem of infinitely many states at low energies is just the problem discussed by DFF.

Applying the DFF trick, as discussed in section 2, provides the solution to this problem. We work in terms of $H+K$ rather than H, since the former has a discrete spectrum of normalizable eigenstates. There is an $SL(2,\mathbb{R})$ symmetry generated by H, D and K, where D and K are defined to be

$$D = \tfrac{1}{2}(\rho p_\rho + p_\rho \rho); \tag{5.11a}$$
$$K = \tfrac{1}{2}mQ^2\rho^2. \tag{5.11b}$$

These generators satisfy equations (2.3).

The appearance of the $SL(2,\mathbb{R})$ symmetry was not an accident. It arises from the geometry of our spacetime. Near the horizon, we find that

$$ds^2 \to -\frac{r^4}{Q^2}dt^2 + \frac{Q}{r^2}dr^2 + Qd\Omega_3^2. \tag{5.12}$$

We recognize this metric as that of $AdS_2 \times S^3$. Introduce new coordinates $t^\pm = t \pm \frac{Q}{4r^2}$ on AdS_2. Now the metric can be written in the form

$$ds_2^2 = -\frac{Qdt^+dt^-}{(t^+ - t^-)^2}. \tag{5.13}$$

The $SL(2,\mathbb{R})$ isometry generators are then

$$h = \frac{\partial}{\partial t^+} + \frac{\partial}{\partial t^-}, \tag{5.14a}$$
$$d = t^+\frac{\partial}{\partial t^+} + t^-\frac{\partial}{\partial t^-}, \tag{5.14b}$$
$$k = (t^+)^2\frac{\partial}{\partial t^+} + (t^-)^2\frac{\partial}{\partial t^-}. \tag{5.14c}$$

Here h shifts the time coordinate, and d rescales all coordinates.

The $SL(2,\mathbb{R})$ symmetry of the near-horizon particle action reflects the $SL(2,\mathbb{R})$ isometry group of the near-horizon AdS_2 geometry. As pointed out in [11, 34], the trick of DFF to replace H by $H+K$ has a nice interpretation in AdS_2. To understand it, we must first review AdS_2 geometry.

INTERLUDE: ADS_2 GEOMETRY

On AdS_2, introduce global coordinates $u^{\pm}$ defined in terms of the coordinates of (5.13) by the relation

$$t^{\pm} = \tan u^{\pm}. \tag{5.15}$$

Then the AdS_2 metric takes the form

$$ds^2 = -\frac{Q}{4}\frac{du^+du^-}{\sin^2(u^+ - u^-)}. \tag{5.16}$$

In these coordinates, the global time generator is

$$h + k = \frac{\partial}{\partial u^+} + \frac{\partial}{\partial u^-}. \tag{5.17}$$

In figure 2 it is seen that the time coordinate conjugate to h is not a good global time coordinate on AdS_2, but the time coordinate conjugate to $h+k$ is. In fact, the generators h and d preserve the horizon, while $h+k$ preserves the boundary $u^+ = u^- + \pi$ (the right boundary in figure 2).

So in conclusion the DFF trick has a beautiful geometric interpretation in the black hole context. It is simply a coordinate transformation to "good" coordinates on AdS_2.

6. Quantum Mechanics on the Black Hole Moduli Space

6.1. THE BLACK HOLE MODULI SPACE METRIC

In this section we will consider five-dimensional $\mathcal{N} = 1$ supergravity with a single $U(1)$ charge coupled to the graviphoton and no vector multiplets.[13] We will use units with $M_p = L_p = 1$. The action is

$$S = \int d^5x\sqrt{g}[R - \frac{3}{4}F^2] + \frac{1}{2}\int A \wedge F \wedge F + \text{fermions}. \tag{6.1}$$

[13]Adding neutral hypermultiplets would not affect the discussion, since they decouple. Since these lectures were given, the case with additional vector multiplets was solved in [35], and the four-dimensional case was solved in [23]. The supersymmetry of cases with more than eight initial supersymmetries [36, 37] has not been worked out.

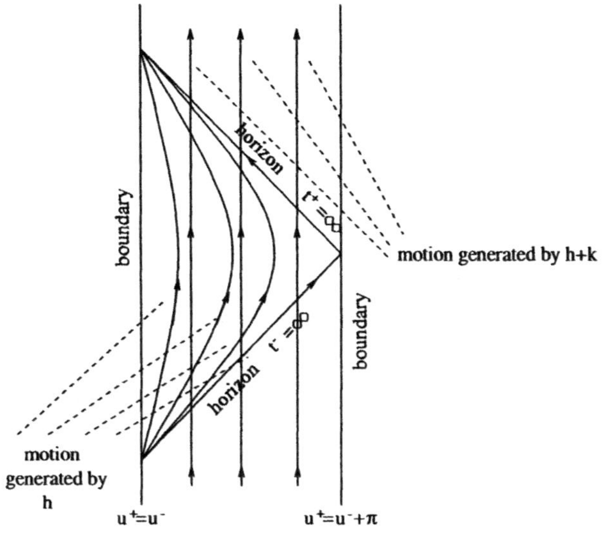

Figure 2. The geometry of AdS_2. The time conjugate to $h+k$ is a good global coordinate.

We can also get this from M-theory compactified on a Calabi-Yau with $b_2 = 1$ (the simplest example of such a threefold is the quintic). The black holes are then M2-branes wrapping Calabi-Yau two-cycles.

This system has a solution describing N static extremal black holes

$$ds^2 = -\psi^{-2}dt^2 + \psi d\vec{x}^2, \tag{6.2a}$$

$$A = \psi^{-1}dt, \tag{6.2b}$$

(*cf.* equation (5.1)) where ψ is the harmonic function on $\mathbb{R}^4$

$$\psi = 1 + \sum_{A=1}^{N} \frac{Q_A}{|\vec{x} - \vec{x}_A|^2}, \tag{6.2c}$$

and $\vec{x}_A$ is the $\mathbb{R}^4$ coordinate of the $A^{\underline{\text{th}}}$ black hole, whose charge is Q_A. Another picture of these holes is M2-branes wrapping Calabi-Yau cycles. The space of solutions is called the moduli space, which is parametrized by the $4N$ collective coordinates $\vec{x}_A$. The slow motion of such black holes is governed by the moduli space metric G_{AB}, so that the low energy effective action takes the form

$$S = \frac{1}{2}\int dt\dot{x}^A\dot{x}^B G_{AB}. \tag{6.3}$$

Note that due to the no-force condition there is no potential term in the action, and since $|\dot{\vec{x}}_A| \ll 1$, the higher order corrections can be neglected.

The first calculation of the moduli space metric of the four-dimensional Reissner-Nordström black holes was performed in [3, 4] and was generalized to dilaton black holes in [38]. The metric on the moduli space for the five-dimensional black holes (6.2) was derived in [14]. In order to find this metric, one starts with the following ansatz describing the linear order perturbation of the black hole solution (6.2)

$$ds^2 = -\psi^{-2}dt^2 + \psi d\vec{x}^2 + 2\psi^{-2}\vec{R}\cdot d\vec{x}dt, \tag{6.4a}$$

$$A = \psi^{-1}dt + (\vec{P} - \psi^{-1}\vec{R})\cdot d\vec{x}, \tag{6.4b}$$

where $\vec{P}$ and $\vec{R}$ are quantities that are first order in velocities. In equation (6.2c), $\vec{x}_A$ is replaced with $\vec{x}_A + \vec{v}_A t$. This is the most general Galilean-invariant ansatz to linear order. Then (roughly) one uses the equations of motion to solve for $\vec{P}$ and $\vec{R}$. Inserting this into the five-dimensional supergravity action gives the following result [14] for the action:

$$S = \frac{1}{2}\int dt \dot{x}^A \dot{x}^B G_{AB} = \frac{1}{4}\int dt \dot{x}^{Ak}\dot{x}^{Bl}(\delta^i_k\delta^j_l + I^{ri}_k I^{rj}_l)\partial_{Ai}\partial_{Bj}L, \tag{6.5}$$

where

$$L = -\int d^4x \psi^3, \tag{6.6}$$

with

$$\psi = 1 + \sum_{A=1}^{N}\frac{Q_A}{|\vec{x} - \vec{x}_A|^2}, \tag{6.7}$$

and I^r is a triplet of self-dual complex structures on $\mathbb{R}^4$ obeying equation (4.29). This Lagrangian has $\mathcal{N} = 4$ supersymmetry when Hermitian fermions $\lambda^{Ai} = \lambda^{Ai\dagger}$ are added.

6.2. THE NEAR-HORIZON LIMIT

6.2.1. *Spacetime geometry*

Taking the near-horizon limit of (6.2a) corresponds to neglecting the constant term in (6.2c). In figure 3 we have illustrated the resultant spatial geometry at a moment of fixed time for three black holes. Before the limit is taken (figure 3a), the geometry has an asymptotically flat region at large $|\vec{x}|$. Near the limit (figure 3b), as the origin is approached along a spatial trajectory, a single "throat" approximating that of a charge $\sum Q_A$ black hole is encountered. This throat region is an $AdS_2 \times S^3$ geometry with radii of

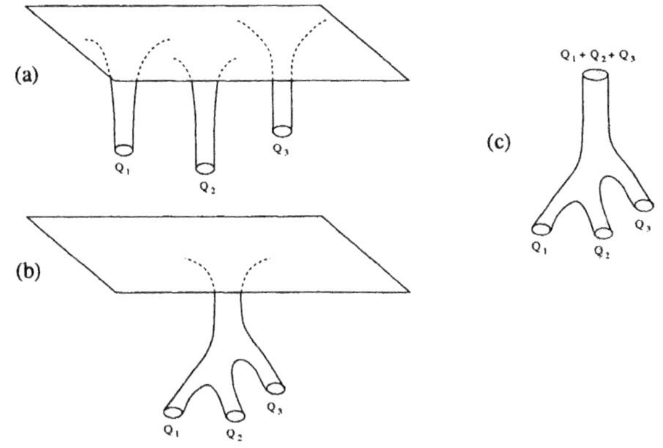

Figure 3. (a) Widely separated black holes. (b) Near-coincident black holes. (c) The near-horizon limit.

order $\sqrt{\sum Q_A}$. As one moves deeper inside the throat towards the horizon, the throat branches into smaller throats, each of which has smaller charge and correspondingly smaller radii. Eventually there are N branches with charge Q_A. At the end of each of these branches is an event horizon. When the limit is achieved (figure 3c), the asymptotically flat region moves off to infinity. Only the charge $\sum Q_A$ "trunk" and the many branches remain.

6.2.2. *Moduli space geometry*

It is also interesting to consider the near-horizon limit of the moduli space geometry. The metric is again given by (6.5), where one should neglect the constant term in the harmonic function (6.7). This is illustrated in figure 4 for the case of two black holes. Near the limit there is an asymptotically flat $\mathbb{R}^{4N}$ region corresponding to all N black holes being widely separated. This is connected to the near-horizon region where the black holes are strongly interacting, by tubelike regions which become longer and thinner as the limit is approached. When the limit is achieved, the near-horizon region is severed from the tubes and the asymptotically flat region.

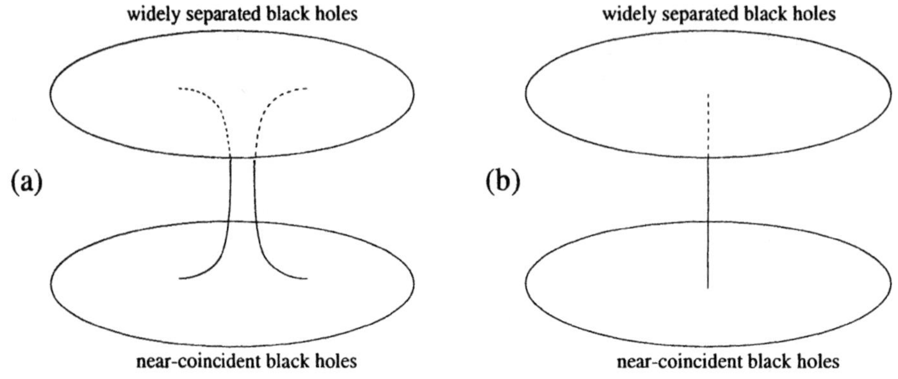

Figure 4. (a) Regions of the two-black hole moduli space. (b) The near-horizon limit.

6.3. CONFORMAL SYMMETRY

The near-horizon quantum mechanics has an $SL(2,\mathbb{R})$ conformal symmetry. The dilations D and special conformal transformations K are generated by

$$D = -\frac{1}{2}(x^{Ai}P_{Ai} + h.c.), \tag{6.8}$$

$$K = 6\pi^2 \sum_{A\neq B}^{N} \frac{Q_A^2 Q_B}{|\vec{x}_A - \vec{x}_B|^2}. \tag{6.9}$$

By splitting the potential L appearing in the metric (6.5) into pieces representing the 1-body, 2-body and 3-body interactions, one can show [14] that the conditions (4.34) and (4.36) are satisfied. Thus the $SL(2,\mathbb{R})$ symmetry can be extended to the full $D(2,1;0)$ superconformal symmetry as was described in section 4.4. This group is the special case of the $D(2,1;\alpha)$ superconformal groups for which there is an $SU(1,1|2)$ subgroup (in fact, $D(2,1;0) \cong SU(1,1|2) \rtimes SU(2)$), in agreement with [39].

So we have seen that there are noncompact regions of the near-horizon moduli space corresponding to coincident black holes. These regions are eliminated by the potential K in the modified Hamiltonian $L_0 = \frac{1}{2}(H+K)$, which is singular at the boundary of the noncompact regions. L_0 has a well defined spectrum with discrete eigenstates. A detailed description of the quantum states of this system remains to be found [40].

7. Discussion

Let us recapitulate. We have found that at low energies the quantum mechanics of N black holes divides into superselection sectors. One sector describes the dynamics of widely separated, non-interacting black holes. The other "near horizon" sector describes highly redshifted, near-coincident black holes and has an enhanced superconformal symmetry. Since they completely decouple from widely separated black holes, states of the near horizon theory are multi-black hole bound states.

It is instructive to compare this to an M-theoretic description of these black holes. In Calabi-Yau compactification of M theory to five dimensions, the black holes are described by M2-branes multiply wrapped around holomorphic cycles of the Calabi-Yau. In principle all the black hole microstates are described by quantum mechanics on the M2-brane moduli space, which at low energies should be the dual CFT_1 living on the boundary of AdS_2 [12]. In practice so far this problem has not been tractable. This moduli space has what could be called (in a slight abuse of terminology) a Higgs branch and a Coulomb branch. This Higgs branch is a sigma model whose target is the moduli space of a single multiply wrapped M2-brane worldvolume in the Calabi-Yau. In the Coulomb branch the M2-brane has fragmented into multiple pieces, and the branch is parametrized by the M2-brane locations. At finite energy the Coulomb branch connects to the Higgs branch at singular points where the M2-brane worldvolume degenerates.

At first one might think that the considerations of this paper correspond to the Coulomb branch, since the multi-black hole moduli space is parametrized by the black hole locations. However it is not so simple. The fact that the near horizon sector decouples from the sector describing non-interacting black holes strongly suggests that it is joined to the Higgs branch. Indeed in the D1/D5 black hole, there is a similar near-horizon region of the Coulomb branch which is not only joined to but is in fact a dual description of the singular regions of the Higgs branch [41, 42, 43, 44, 45]. We conjecture there is a similar story here: the near-horizon, multi-black hole quantum mechanics is dual to at least part of the Higgs branch of multiply wrapped M2-branes. Near-horizon microstates should therefore account for at least some of the internal black hole microstates. Exactly how much of the black hole microstructure is accounted for in this way remains to be understood.

Acknowledgments

We thank F. Cachazo, R. Jackiw, J. Maldacena, A. Maloney, G. Papadopoulos, L. Thorlacius, P. Townsend and especially M. Spradlin for enlightening discussions and communication. The research reviewed here was supported in part by NSERC and DOE grant DE-FGO2-91ER40654. We thank the organizers and especially Lárus Thorlacius for an excellent school, the participants for their stimulating questions, and NATO for financial support at the school.

A. Differential Geometry with Torsion

In this appendix, we give a brief summary of differential calculus with torsion, for the reader who is frustrated by the usual absence of such a discussion in most general relativity books.[14] Recall [47] that the covariant derivative of a tensor is given in terms of the (not necessarily symmetric) connection C^c_{ab}. The torsion $c^c{}_{ab}$ is just the antisymmetric part of the connection:

$$c^c{}_{ab} \equiv C^c_{[ab]} = \tfrac{1}{2}(C^c_{ab} - C^c_{ba}). \tag{A.1}$$

Either by direct computation, or by recalling that the difference between two connections is a tensor, one finds that the torsion is a true tensor. Of course, the torsion does contribute to the curvature tensor, and we remind the reader that many of the familiar symmetries of the curvature tensor are not obeyed in the presence of torsion. Also, if the symmetric part of the connection is given by the Levi-Civita connection, then the full connection annihilates the metric iff the fully covariant torsion tensor $c_{abc} = g_{ad}c^d{}_{bc}$ is completely antisymmetric.

Hopefully, the preceding paragraph was familiar. We now discuss the torsion in a tangent space formalism. As usual, the first step is to define the vielbein e^α_a, which is a basis of cotangent space vectors, labelled by $\alpha = 1, \ldots, n$, where n is the dimension of the manifold, obeying

$$\delta_{\alpha\beta} e^\alpha_a e^\beta_b = g_{ab}. \tag{A.2}$$

The vielbein e^α_a, and the inverse vielbein e^a_α which obeys

$$e^a_\alpha e^\beta_a = \delta^\beta_\alpha, \tag{A.3}$$

can then be used to map tensors into the tangent space; *e.g.* $V^\alpha \equiv V^a e^\alpha_a$.

The connection one-form $\Omega_a{}^\alpha{}_\beta$ is defined by demanding that the vielbein is covariantly constant:

$$\nabla_a e^\alpha_b \equiv \partial_a e^\alpha_b + \Omega_a{}^\alpha{}_\beta e^\beta_b - C^c_{ab} e^\alpha_c = 0. \tag{A.4}$$

Note that equation (A.4) is valid for any choice of connection, and does not imply that the metric is covariantly constant. The metric is covariantly constant iff $\delta_{\alpha\beta}$ is covariantly constant, which in turn holds iff the connection one-form $\Omega_{a\alpha\beta}$ is antisymmetric in the tangent space indices, where

[14] One excellent reference for physicists is [46].

we have lowered the middle index using the tangent space metric $\delta_{\alpha\beta}$. In other words, the familiar antisymmetry of the connection one-form [47] exists if and only if the metric is covariantly constant, whether or not there is torsion.

Equation (A.4) is easily solved for the connection one-form, giving

$$\Omega_a{}^\alpha{}_\beta = e^\alpha_b \partial_a e^b_\beta + C^c_{ab} e^\alpha_c e^b_\beta. \tag{A.5}$$

An immediate corollary of this, and the fact that the difference of two connections C^c_{ab} and C'^c_{ab} is a tensor, is that the difference between two connection one-forms is a tensor, and is, in fact, the same tensor as $C^c_{ab} - C'^c_{ab}$, but with the b and c indices lifted to the tangent bundle.[15]

The unique torsion-free connection one-form which annihilates the metric (*i.e.* that obtained from equation (A.4) using the Levi-Civita connection) is known as the spin connection, and is usually denoted $\omega_a{}^\alpha{}_\beta$. Given a completely antisymmetric torsion $c_{abc} = c_{[abc]}$, as in the first paragraph of this appendix, we define the connection one-form

$$\Omega^{+}_a{}^\alpha{}_\beta = \omega_a{}^\alpha{}_\beta + c^\alpha{}_{a\beta}, \tag{A.6}$$

where, of course, any required mapping between the tangent bundle and the spacetime is achieved by contracting with the vielbein.

As usual, spinors ψ are defined on the tangent bundle, and their covariant derivative is given by

$$\nabla_a \psi = \partial_a \psi - \frac{1}{4} \Omega_{a\alpha\beta} \gamma^{\alpha\beta} \psi, \tag{A.7}$$

where $\gamma^{\alpha\beta} \equiv \frac{1}{2}[\gamma^\alpha, \gamma^\beta]$ is a commutator of $SO(n)$ γ-matrices, which satisfy $\{\gamma^\alpha, \gamma^\beta\} = 2\delta^{\alpha\beta}$.

References

1. Hawking, S.W. (1975) Particle Creation by Black Holes, *Comm. Math. Phys.* **43**, 199–220.
2. Hawking, S.W. (1976) Black Holes and Thermodynamics, *Phys. Rev. D* **13**, 191–197.
3. Ferrell, F. and Eardley, D. (1987) Slow-Motion Scattering and Coalescence of Maximally Charged Black Holes, *Phys. Rev. Lett.* **59**, 1617–1620.
4. Gibbons, G.W. and Ruback, P.J. (1986) The Motion of Extreme Reissner-Nordstrom Black Holes in the Low Velocity Limit, *Phys. Rev. Lett.* **57**, 1492–1495.
5. Traschen, J. and Ferrell, R. (1992) Quantum Mechanical Scattering of Charged Black Holes, *Phys. Rev. D* **45**, 2628–2635.
6. Callan, C.G., Coleman, S. and Jackiw, R. (1970) A New Improved Energy-Momentum Tensor, *Ann. Phys. (NY)* **59**, 42–73.

[15] In this discussion, we are assuming that a vielbein has been chosen once and for all; we do not consider the effect of changing frames or coordinates.

7. Jackiw, R. (1972) Introducing Scale Symmetry, *Physics Today* **25**, 23–27.
8. Hagan, C.R. (1972) Scale and Conformal Transformations in Galilean-Covariant Field Theory, *Phys. Rev. D* **5**, 377–388.
9. Niederer, U. (1972) The Maximal Kinematical Invariance Group of the Free Schrödinger Equation, *Helv. Phys. Acta* **45**, 802–810.
10. de Alfaro, V., Fubini, S. and Furlan, G. (1976) Conformal Invariance in Quantum Mechanics, *Nuovo. Cim.* **34A**, 569–612.
11. Claus, P., Derix, M., Kallosh, R., Kumar, J., Townsend, P.K. and Van Proeyen, A. (1998) Black Holes and Superconformal Mechanics, *Phys. Rev. Lett.* **81**, 4553–4556.
12. Maldacena, J. (1998) The Large N Limit of Superconformal Field Theories and Supergravity, Adv. Theor. Math. Phys. **2**, 231–252.
13. Michelson, J. and Strominger, A. (1999) The Geometry of (Super) Conformal Quantum Mechanics, HUTP-99/A045, hep-th/9907191.
14. Michelson, J. and Strominger, A. (1999) Superconformal Multi-Black Hole Quantum Mechanics, *JHEP* **09**, 005.
15. Treiman, S.B., Jackiw, R. and Gross, D.J. (1972) *Lectures on current algebra and its applications*, Princeton University Press, Princeton.
16. Claus, P., Kallosh, R. and Van Proeyen, A. (1998) Conformal Symmetry on the World Volumes of Branes, KUL-TF-98/54, SU-ITP-98/67, hep-th/9812066.
17. Witten, E. (1981) Dynamical Breaking of Supersymmetry, *Nucl. Phys.* **B188**, 513–554.
18. Witten, E. (1982) Constraints on Supersymmetry Breaking, *Nucl. Phys.* **B202**, 253–316.
19. Witten, E. (1982) Supersymmetry and Morse Theory, *J. Diff. Geom.* **17**, 661–692.
20. Fubini, S. and Rabinovici, E. (1984) Superconformal Quantum Mechanics, *Nucl. Phys.* **B245**, 17–44.
21. Salomonson, P. and van Holten, J.W. (1982) Fermionic Coordinates and Supersymmetry in Quantum Mechanics, *Nucl. Phys.* **B196**, 509–531.
22. Gauntlett, J.P. (1993) Low-Energy Dynamics of Supersymmetric Solitons, *Nucl. Phys.* **B400**, 103–125.
23. Maloney, A., Spradlin, M. and Strominger, A. (1999) Superconformal Multi-Black Hole Moduli Spaces in Four Dimensions, HUTP-99/A055, hep-th/9911001.
24. Coles, R.A. and Papadopoulos, G. (1990) The Geometry of the One-Dimensional Supersymmetric Non-Linear Sigma Models, *Class. Quant. Grav.* **7**, 427–438.
25. Alvarez-Gaumé, L. (1983) Supersymmetry and the Atiyah-Singer Index Theorem, *Comm. Math. Phys.* **90**, 161–173.
26. Friedan, D. and Windey, P. (1984) Supersymmetric Derivation of The Atiyah-Singer Index and the Chiral Anomaly, *Nucl. Phys.* **B235 (FS11)**, 395–416.
27. Sevrin, A., Troost, W. and Van Proeyen, A. (1988) Superconformal Algebras in Two Dimensions with $N = 4$, *Phys. Lett.* **208B**, 447–450.
28. Gibbons, G.W., Papadopoulos, G. and Stelle, K.S. (1997) HKT and OKT Geometries on Soliton Black Hole Moduli Spaces, *Nucl. Phys.* **B508**, 623–658.
29. Gates, S.J. Jr., Hull C.M. and Roček, M. (1984) Twisted Multiplets and New Supersymmetric Non-linear σ-Models, *Nucl. Phys.* **B248**, 157–186.
30. Hull, C.M. (1999) The Geometry of Supersymmetric Quantum Mechanics, QMW-99-16, hep-th/9910028.
31. Grantcharov, G. and Poon, S.-Y. (1999) Geometry of Hyper-Kähler Connections with Torsion, math.DG/9908015.
32. Hellerman, S. and Polchinski, J. (1999) Supersymmetric Quantum Mechanics from Light Cone Quantization, NSF-ITP-99-101, hep-th/9908202.
33. Douglas, M., Polchinski, J. and Strominger, A. (1997) Probing Five-Dimensional Black Holes with D-Branes, *JHEP* **12**, 003.
34. Kallosh, R. (1999) Black Holes and Quantum Mechanics, hep-th/9902007.
35. Gutowski, J. and Papadopoulos, G. (1999) The Dynamics of Very Special Black Holes, hep-th/9910022.

36. Kaplan, D.M. and Michelson, J. (1997) Scattering of Several Multiply Charged Extremal $D = 5$ Black Holes, *Phys. Lett.* **B410** 125–130.
37. Michelson, J. (1998) Scattering of Four-Dimensional Black Holes, *Phys. Rev. D* **57** 1092–1097.
38. Shiraishi, K. (1993) Moduli Space Metric for Maximally-Charged Dilaton Black Holes, *Nucl. Phys.* **B402**, 399–410.
39. Gauntlett, J.P., Myers R.C. and Townsend, P.K. (1999) Black Holes of D=5 Supergravity, *Class. Quant. Grav.* **16**, 1–21.
40. Britto-Pacumio, R., Strominger A. and Volovich A., work in progress.
41. Maldacena, J. and Strominger, A. (1997) Semiclassical Decay of Near Extremal Fivebranes, *JHEP* **12**, 008.
42. Maldacena, J., Michelson, J. and Strominger, A. (1999) Anti-de Sitter Fragmentation, *JHEP* **03**, 011.
43. Seiberg, N. and Witten, E. (1999) The D1/D5 System and Singular CFT, *JHEP* **04**, 017.
44. Berkooz, M. and Verlinde, H. (1999) Matrix Theory, AdS/CFT and Higgs-Coulomb Equivalence, IASSNS-HEP-99/67, PUPT-1879, hep-th/9907100.
45. Aharony, O. and Berkooz, M. (1999) IR Dynamics of d=2, N=(4,4) Gauge Theories and DLCQ of "Little String Theories", PUPT-1886, RUNHETC-99-31, hep-th/9909101.
46. Nakahara, M. (1990) *Geometry, Topology and Physics*, Institute of Physics Publishing, Philadelphia.
47. Wald, R.M. (1984) *General Relativity*, The University of Chicago Press, Chicago.

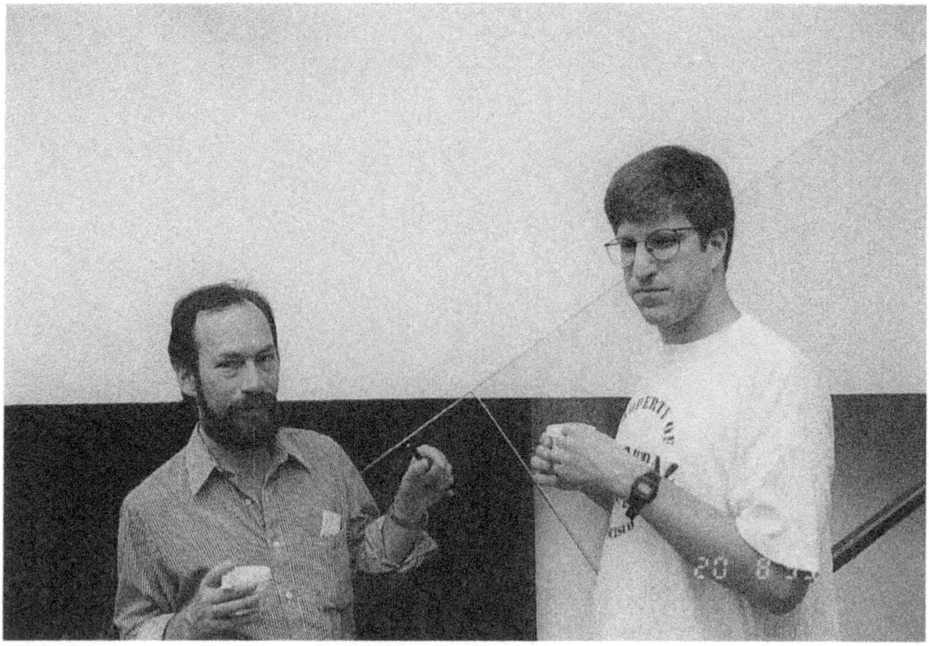

LARGE-N GAUGE THEORIES

YURI MAKEENKO
Institute of Theoretical and Experimental Physics,
B. Cheremushkinskaya 25,
117218 Moscow, Russia
email: makeenko@itep.ru

Abstract. Four pedagogical lectures on $O(N)$ vector models, large-N QCD, QCD in loop space, large-N reduction.

1. Introduction

The large-N limit of Gauge Theory was proposed in 1974 by 't Hooft [1] for quantum chromodynamics (QCD). The dimensionality of the gauge group $SU(N_c)$ was used as a parameter, considering N_c as a large number and performing an expansion in $1/N_c$.

The motivation was an expansion in the inverse number of field-components N in statistical mechanics where it is known as the $1/N$-expansion, and is a standard method for non-perturbative investigations.

The expansion of QCD in $1/N_c$ rearranges diagrams of perturbation theory in a way which is consistent with a string picture of strong interaction, whose phenomenological consequences agree with experiment. The accuracy of the leading-order term, which is often called large-N QCD or multicolor QCD, is expected to be of the order of the ratios of meson widths to their masses, *i.e.* about 10–15%.

L. Thorlacius and T. Jonsson (eds.), M-Theory and Quantum Geometry, 285-353.

While QCD is simplified in the large-N_c limit, it is still not yet solved. Generically, it is a problem of infinite matrices, rather than of infinite vectors as in the theory of second-order phase transitions in statistical mechanics.

We start these Lectures by showing how the $1/N$-expansion works for the O(N)-vector models, and describing some applications to the four-Fermi interaction and the nonlinear sigma model. Then we concentrate on the Yang–Mills theory at large N_c.

The methods described in these Lectures are developed mostly in the seventies and the beginning of the eighties. They are used over and over again when discussing a relation between Gauge Theory and String Theory.

The conformal invariance, described in Section 2 for the four-Fermi interaction, is used in the modern approaches based on the AdS/CFT correspondence [2] (discussed in the lectures by L. Thorlacius at this School). The notion of the large-N limit, described in Section 3, is widely used, in particular, in Matrix Theory [3] (discussed in the lectures by W. Taylor at this School). The loop equation, described in Section 4, has been recently applied [4] for studying the AdS/CFT correspondence. The large-N reduced models, described in Section 5, are applied to a nonperturbative formulation of superstring [5] and very recently to noncommutative gauge theory [6].

The content of my fifth lecture at this School, which was devoted to an application of the large-N methods to Matrix Theory at finite temperature, is not included in the text. It is mostly contained in Ref. [7].

2. O(N) Vector Models

The simplest models, which become solvable in the limit of a large number of field components, deal with a field which has N components forming an O(N) vector in an internal symmetry space. A model of this kind was first considered by Stanley [8] in statistical mechanics and is known as the spherical model. The extension to quantum field theory was done by Wilson [9] both for the four-Fermi and φ^4 theories.

In the framework of perturbation theory, the four-Fermi interaction is renormalizable only in $d = 2$ dimensions and is non-renormalizable for $d > 2$. The $1/N$-expansion resums perturbation-theory diagrams after which the four-Fermi interaction becomes renormalizable to each order in $1/N$ for $2 \leq d < 4$. An analogous expansion exists for the nonlinear O(N) sigma model.

The $1/N$ expansion of the vector models is associated with a resummation of Feynman diagrams. A very simple class of diagrams — the bubble graphs — survives to the leading order in $1/N$. This is why the large-N

limit of the vector models is solvable. Alternatively, the large-N solution is nothing but a saddle-point solution in the path-integral approach. The existence of the saddle point is due to the fact that N is large. This is to be distinguished from a perturbation-theory saddle point which is due to the fact that the coupling constant is small. Taking into account fluctuations around the saddle-point results in the $1/N$-expansion of the vector models.

We begin this Section with a description of the $1/N$-expansion of the N-component four-Fermi theory analyzing the bubble graphs. Then we introduce functional methods and construct the $1/N$-expansion of the O(N)-symmetric nonlinear sigma model. At the end we discuss the factorization in the O(N) vector models at large N.

2.1. FOUR-FERMI INTERACTION

The action of the O(N)-symmetric four-Fermi interaction in a d-dimensional Euclidean space is defined by

$$S\left[\bar{\psi},\psi\right] = \int d^d x \left(\bar{\psi}\,\hat{\partial}\,\psi + m\,\bar{\psi}\psi - \frac{G}{2}\left(\bar{\psi}\psi\right)^2\right). \tag{1}$$

Here $\hat{\partial} = \gamma_\mu \partial_\mu$ and $\psi = (\psi_1, \ldots, \psi_N)$ is a spinor field which forms an N-component vector in an internal symmetry space so that

$$\bar{\psi}\psi = \sum_{i=1}^{N} \bar{\psi}_i \psi_i \,. \tag{2}$$

In $d = 2$ this model was studied in the large-N limit in Ref. [10] and is often called the Gross–Neveu model.

The dimension of the four-Fermi coupling constant G is $\dim[G] = m^{2-d}$. For this reason, the perturbation theory for the four-Fermi interaction is renormalizable in $d = 2$ but is non-renormalizable for $d > 2$ (and, in particular, in $d = 4$). This is why the old Fermi theory of weak interactions was replaced by the modern electroweak theory, where the interaction is mediated by the $W^\pm$ and Z bosons.

The action (1) can be equivalently rewritten as

$$S\left[\bar{\psi},\psi,\chi\right] = \int d^d x \left(\bar{\psi}\,\hat{\partial}\,\psi + m\,\bar{\psi}\psi - \chi\,\bar{\psi}\psi + \frac{\chi^2}{2G}\right), \tag{3}$$

where χ is an auxiliary field. The two forms of the action, (1) and (3), are equivalent due to the equation of motion

$$\chi = G\bar{\psi}\psi \tag{4}$$

which can be derived by varying the action (3) with respect to χ.

In the path-integral quantization, where the partition function is defined by

$$Z = \int D\chi D\bar{\psi} D\psi \, \mathrm{e}^{-S[\bar{\psi},\psi,\chi]} \tag{5}$$

with $S[\bar{\psi},\psi,\chi]$ given by Eq. (3), the action (1) appears after performing the Gaussian integral over χ. Therefore, one alternatively gets

$$Z = \int D\bar{\psi} D\psi \, \mathrm{e}^{-S[\bar{\psi},\psi]} \tag{6}$$

with $S[\bar{\psi},\psi]$ given by Eq. (1).

The perturbative expansion of the O(N)-symmetric four-Fermi theory can be conveniently represented using the formulation (3) via the auxiliary field χ. Then the diagrams are of the type of those in Yukawa theory, and resemble the ones for QED with $\bar{\psi}$ and ψ being an analog of the electron-positron field and χ being an analog of the photon field. However, the auxiliary field $\chi(x)$ does not propagate, since it follows from the action (3) that

$$D_0\,(x-y) \equiv \langle\, \chi(x)\chi(y)\,\rangle_{\mathrm{Gauss}} = G\,\delta^{(d)}(x-y) \tag{7}$$

or

$$D_0\,(p) \equiv \langle\, \chi(-p)\chi(p)\,\rangle_{\mathrm{Gauss}} \;=\; G \tag{8}$$

in momentum space.

It is convenient to represent the four-Fermi vertex as the sum of two terms

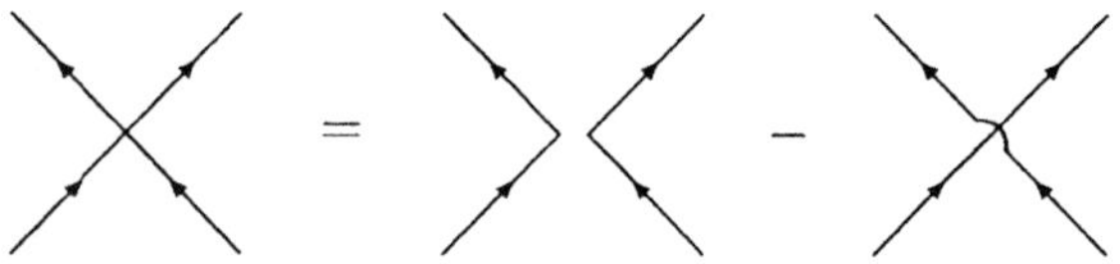

(9)

where the empty space inside the vertex is associated with the propagator (7) (or (8) in momentum space). The relative minus sign makes the vertex antisymmetric in both incoming and outgoing fermions as is prescribed by the Fermi statistics.

The diagrams that contribute to second order in G for the four-Fermi vertex are depicted, in these notations, in Figure 1. The O(N) indices propagate through the solid lines so that the closed line in the diagram in Figure 1b corresponds to the sum over the O(N) indices which results in a factor of N. Analogous one-loop diagrams for the propagator of the ψ-field are depicted in Figure 2.

Remark on the one-loop Gell-Mann–Low function of four-Fermi theory
Evaluating the diagrams in Figure 1 which are logarithmically divergent in $d = 2$, and noting that the diagrams in Figure 2 do not contribute to

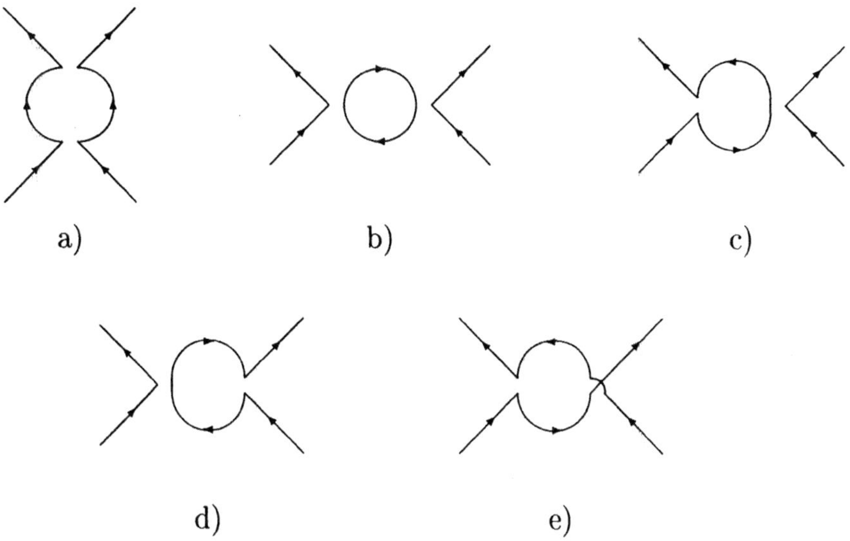

Figure 1. Diagrams of the second order of perturbation theory for the four-Fermi vertex. The diagram b) involves the sum over the O(N) indices.

Figure 2. One-loop diagrams for the propagator of the ψ-field. The diagram b) involves the sum over the O(N) indices.

the wave-function renormalization of the ψ-field, which emerges to the next order in G, one gets for the one-loop Gell-Mann–Low function

$$\mathcal{B}\left(G\right) = -\frac{\left(N-1\right)G^2}{2\pi}\,. \tag{10}$$

The four-Fermi theory in 2 dimensions is asymptotically free as was first noted by Anselm [11] and rediscovered in Ref. [10].

The vanishing of the one-loop Gell-Mann–Low function in the Gross–Neveu model for $N = 1$ is related to the same phenomenon in the Thirring model. The latter model is associated with the vector-like interaction $\left(\bar{\psi}\gamma_\mu\psi\right)^2$ of one species of fermions with γ_μ being the γ-matrices in 2 dimensions. Since a bispinor has in $d = 2$ only two components ψ_1 and ψ_2,

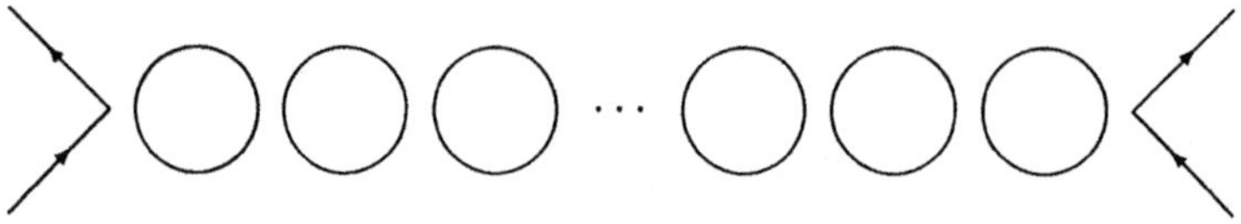

Figure 3. Bubble diagram which survives the large-N limit of the O(N) vector models.

both the vector-like and the scalar-like interaction (1) for $N = 1$ reduce to $\bar{\psi}_1\psi_1\bar{\psi}_2\psi_2$ since the square of a Grassmann variable vanishes. Therefore, these two models coincide. For the Thirring model, the vanishing of the Gell-Mann–Low function for any G was shown by Johnson [12] to all loops.

2.2. BUBBLE GRAPHS AS ZEROTH ORDER IN 1/N

The perturbation-theory expansion of the O(N)-symmetric four-Fermi theory contains, in particular, the diagrams of the type depicted in Figure 3 which are called *bubble graphs*. Since each bubble has a factor of N, the contribution of the n-bubble graph is $\propto G^{n+1}N^n$ which is of the order of

$$G^{n+1}N^n \sim G \tag{11}$$

as $N \to \infty$ since

$$G \sim \frac{1}{N}. \tag{12}$$

Therefore, all the bubble graphs are essential to the leading order in $1/N$.

Let us denote

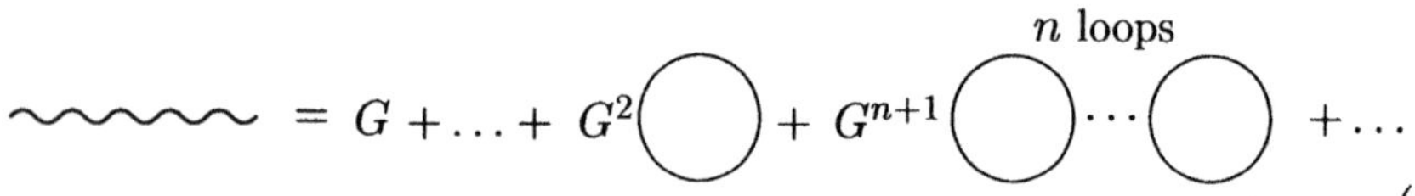

(13)

In fact the wavy line is nothing but the propagator D of the χ field with the bubble corrections included. The first term G on the RHS of Eq. (13) is nothing but the free propagator (8).

Summing the geometric series of the fermion-loop chains on the RHS of Eq. (13), one gets analytically[1]

$$D^{-1}(p) = \frac{1}{G} - N \int \frac{d^dk}{(2\pi)^d} \frac{\mathrm{sp}\left(\hat{k} + im\right)\left(\hat{k} + \hat{p} + im\right)}{(k^2 + m^2)\left((k+p)^2 + m^2\right)}. \tag{14}$$

[1]Recall that the free Euclidean fermionic propagator is given by $S_0(p) = (i\hat{p} + m)^{-1}$ due to Eqs. (3), (5) and the additional minus sign is associated with the fermion loop.

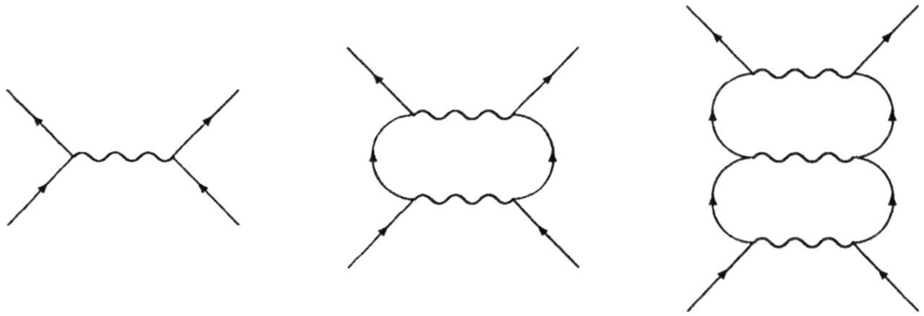

Figure 4. Some diagrams of the $1/N$-expansion for the O(N) four-Fermi theory. The wavy line represents the (infinite) sum of the bubble graphs (13).

This determines the exact propagator of the χ field at large N. It is $\mathcal{O}\left(N^{-1}\right)$ since the coupling G is included in the definition of the propagator.

The idea is now to change the order of summation of diagrams of perturbation theory using $1/N$ rather than G as the expansion parameter. Therefore, the zeroth-order propagator of the expansion in $1/N$ is defined as the sum over the bubble graphs (13) which is given by Eq. (14). Some of the diagrams of the new expansion for the four-Fermi vertex are depicted in Figure 4. The first diagram is proportional to G while the second and third ones are proportional to G^2 or G^3, respectively, and therefore are of order $\mathcal{O}(N^{-1})$ or $\mathcal{O}(N^{-2})$ with respect to the first diagram. The perturbation theory is thus rearranged as the $1/N$-expansion.

The general structure of the $1/N$-expansion is the same for all vector models, say, for the N-component nonlinear sigma model which is considered in Subsection 2.4.

The main advantage of the expansion in $1/N$ for the four-Fermi interaction, over the perturbation theory, is that it is renormalizable in $d < 4$ while the perturbation-theory expansion in G is renormalizable only in $d = 2$. Moreover, the $1/N$-expansion of the four-Fermi theory in $2 < d < 4$ demonstrates [9] an existence of an ultraviolet-stable fixed point, *i.e.* a nontrivial zero of the Gell-Mann–Low function.

In order to show that the $1/N$-expansion of the four-Fermi theory is renormalizable in $2 \leq d < 4$, let us analyze indices of the diagrams of the $1/N$-expansion. First of all, we shall get rid of an ultraviolet divergence of the integral over the d-momentum k in Eq. (14). The divergent part of the integral is proportional to Λ^{d-2} (logarithmically divergent in $d = 2$) with Λ being an ultraviolet cutoff. It can be canceled by choosing

$$G = \frac{g^2}{N} \Lambda^{2-d}, \qquad (15)$$

where g^2 is a proper dimensionless constant which is not necessarily positive since the four-Fermi theory is stable with either sign of G. The power of Λ in Eq. (15) is consistent with the dimension of G. This prescription works for $2 < d < 4$ where there is only one divergent term while another divergency $\propto p^2 \ln \Lambda$ emerges additionally in $d = 4$. This is why the consideration is not applicable in $d = 4$.

The propagator $D(p)$ is therefore finite, and behaves at large momenta $|p| \gg m$ as

$$D(p) \propto \frac{1}{|p|^{d-2}} . \tag{16}$$

The standard power-counting arguments then show that the only divergent diagrams appear in the propagators of the ψ and χ fields, and in the $\bar{\psi}$-χ-ψ three-vertex. These divergencies can be removed by a renormalization of the coupling g, mass, and wave functions of ψ and χ.

This completes a demonstration of renormalizability of the $1/N$-expansion for the four-Fermi interaction in $2 \leq d < 4$. For more detail, see Ref. [13].

2.3. SCALE AND CONFORMAL INVARIANCE OF FOUR-FERMI THEORY

The coefficient in Eq. (15) can easily be calculated in $d = 3$. To evaluate the divergent part of the integral in Eq. (14), we put $p = 0$ and $m = 0$. Remembering that the γ-matrices are 2×2 matrices in $d = 3$, we get

$$\int^{\Lambda} \frac{d^3k}{(2\pi)^3} \frac{\mathrm{sp}\,\hat{k}\hat{k}}{k^2 k^2} = 2 \int^{\Lambda} \frac{d^3k}{(2\pi)^3} \frac{1}{k^2} = \frac{1}{\pi^2} \int^{\Lambda} d|k| = \frac{\Lambda}{\pi^2} . \tag{17}$$

Note that the integral is linearly divergent in $d = 3$ and Λ is the cutoff for the integration over $|k|$. This divergence can be canceled by choosing G according to Eq. (15) with g equal to

$$g_* = \pi . \tag{18}$$

To calculate in $d = 3$ the coefficient of proportionality in Eq. (16), let us choose $G = \pi^2/N\Lambda$ as is prescribed by Eqs. (15), (18) and put in Eq. (14) $m = 0$ since we are interested in the asymptotics $|p| \gg m$. Then the RHS of Eq. (14) can be rearranged as

$$D^{-1}(p) = -2N \int \frac{d^3k}{(2\pi)^3} \left[\frac{k^2 + kp}{k^2(k+p)^2} - \frac{1}{k^2} \right] = \frac{N\,|p|}{8} . \tag{19}$$

The integral is obviously convergent.

Equation (19) (or (16) in d dimensions) is remarkable since it shows that the scale dimension of the field χ changes its value from $l_\chi = d/2$ in

a) b)

Figure 5. Diagrams for the $1/N$-correction to the ψ-field propagator (a) and the three-vertex (b).

perturbation theory to $l_\chi = 1$ in the zeroth order of the $1/N$ expansion (remember that the momentum-space propagator of a field with the scale dimension l is proportional to $|p|^{2l-d}$). This appearance of scale invariance in the $1/N$-expansion of the four-Fermi theory at $2 < d < 4$ was first pointed out by Wilson [9] and implies that the Gell-Mann–Low function $\mathcal{B}(g)$ has a zero at $g = g_*$ which is given in $d = 3$ by Eq. (18).

The (logarithmic) anomalous dimensions of the fields ψ, χ, and of the $\bar{\psi}$-χ-ψ three-vertex in $d = 3$ to order $1/N$ can be found as follows. The $1/N$-correction to the propagator of the ψ-field is given by the diagram depicted in Figure 5a). Since we are interested in an ultraviolet behavior, we can put again $m = 0$. Analytically, we have

$$S^{-1}(p) = i\hat{p} + \frac{8i}{N}\int^{\Lambda} \frac{d^3k}{(2\pi)^3} \frac{\hat{k}+\hat{p}}{|k|(k+p)^2}\,. \tag{20}$$

The (logarithmically) divergent contribution emerges from the domain of integration $|k| \gg |p|$ so we can expand the integrand in p. The p-independent term vanishes after integration over the directions of k so that we get

$$S^{-1}(p) = i\hat{p}\left[1 + \frac{8}{N}\int^{\Lambda} \frac{d^3k}{(2\pi)^3} \frac{1}{|k|^3}\right] = i\hat{p}\left[1 + \frac{2}{3\pi^2 N}\ln\frac{\Lambda^2}{p^2} + \text{finite}\right]. \tag{21}$$

The diagram, which gives a non-vanishing contribution to the three-vertex in order $1/N$, is depicted in Figure 5b. It reads analytically

$$\Gamma(p_1, p_2) = 1 + \frac{8}{N}\int^{\Lambda} \frac{d^3k}{(2\pi)^3} \frac{(\hat{k}+\hat{p}_1)(\hat{k}+\hat{p}_2)}{|k|(k+p_1)^2(k+p_2)^2}\,, \tag{22}$$

where p_1 and p_2 the incoming and outgoing fermion momenta, respectively. The logarithmic domain is $|k| \gg |p|_{\max}$ with $|p|_{\max}$ being the largest of $|p_1|$

and $|p_2|$. This gives

$$\Gamma(p_1, p_2) = 1 - \frac{2}{\pi^2 N} \ln \frac{\Lambda^2}{p_{\max}^2} + \text{finite}. \tag{23}$$

An analogous calculation of the $1/N$ correction for the field χ is a bit more complicated since it involves three two-loop diagrams (see Ref. [14]). The resulting expression for $D^{-1}(p)$ reads

$$\left(N\, D(p)\right)^{-1} = \frac{\Lambda}{g^2} + \left[-\frac{\Lambda}{\pi^2} + \frac{|p|}{8}\right] + \frac{1}{\pi^2 N}\left[2\Lambda - |p|\left(\frac{2}{3}\ln\frac{\Lambda^2}{p^2} + \text{finite}\right)\right]. \tag{24}$$

The linear divergence is canceled to order $1/N$ provided g is equal to

$$g_* = \pi\left(1 + \frac{1}{N}\right), \tag{25}$$

which determines g_* to order $1/N$. After this $D^{-1}(p)$ takes the form

$$D^{-1}(p) = \frac{N\,|p|}{8}\left(1 - \frac{16}{3\pi^2 N}\ln\frac{\Lambda^2}{p^2}\right). \tag{26}$$

To make all three expressions (21), (23), and (26) finite, we need logarithmic renormalizations of the wave functions of ψ- and χ-fields and of the vertex Γ. This can be achieved by multiplying them by the renormalization constants

$$Z_i(\Lambda) = 1 - \gamma_i \ln\frac{\Lambda^2}{\mu^2} \tag{27}$$

where μ stands for a reference mass scale and γ_i are anomalous dimensions. The index i stands for ψ, χ, or v for the ψ- and χ-propagators or the three-vertex Γ, respectively. We have, therefore, calculated

$$\begin{aligned} \gamma_\psi &= \frac{2}{3\pi^2 N}, \\ \gamma_{\mathrm{v}} &= -\frac{2}{\pi^2 N}, \\ \gamma_\chi &= -\frac{16}{3\pi^2 N} \end{aligned} \tag{28}$$

to order $1/N$. Due to Eq. (4) γ_χ coincides with the anomalous dimension of the composite fields $\bar{\psi}\psi$

$$\gamma_{\bar{\psi}\psi} = \gamma_\chi\,. \tag{29}$$

Note, that

$$Z_\psi^2 Z_{\mathrm{v}}^{-2} Z_\chi = 1\,. \tag{30}$$

This implies that the effective charge is not renormalized and is given by Eq. (25). Thus, the nontrivial zero of the Gell-Mann–Low function persists to order $1/N$ (and, in fact, to all orders of the $1/N$-expansion).

If g is chosen exactly at the critical point g_*, then the renormalization-group equations

$$\frac{\mu\, d \ln \Gamma_i}{d\mu} = \gamma_i\left(g^2\right), \tag{31}$$

where Γ_i stands generically either for vertices or for inverse propagators, possess the scale invariant solutions

$$\Gamma_i \propto \mu^{\gamma_i\left(g_*^2\right)}. \tag{32}$$

For the four-Fermi theory in $d = 3$, Eq. (32) yields

$$S(p) = \frac{1}{i\hat{p}}\left(\frac{p^2}{\mu^2}\right)^{\gamma_\psi}, \tag{33}$$

$$D(p) = \frac{8}{N\,|p|}\left(\frac{p^2}{\mu^2}\right)^{\gamma_\chi}, \tag{34}$$

$$\Gamma(p_1, p_2) = \left(\frac{\mu^2}{p_1^2}\right)^{\gamma_{\rm v}} f\left(\frac{p_2^2}{p_1^2}, \frac{p_1 p_2}{p_1^2}\right), \tag{35}$$

where f is an arbitrary function of the dimensionless ratios which is not determined by scale invariance. The indices here obey the relation

$$\gamma_{\rm v} = \gamma_\psi + \frac{1}{2}\gamma_\chi \tag{36}$$

which guarantees that Eq. (30), implied by scale invariance, is satisfied.

The indices γ_i are given to order $1/N$ by Eqs. (28). When expanded in $1/N$, Eqs. (33) and (34) obviously reproduce Eqs. (21) and (26). Therefore, one gets the exponentiation of the logarithms which emerge in the $1/N$-expansion. The calculation of the next terms of the $1/N$-expansion for the indices γ_i is contained in Ref. [15].

Scale invariance implies, in a renormalizable quantum field theory, more general conformal invariance as is first pointed out in Refs. [16, 17]. The conformal group in a d-dimensional space-time has $(d+1)(d+2)/2$ parameters as is illustrated by Table 1. More about the conformal group can be found in the lecture by Jackiw [18].

A heuristic proof [16] of the fact that scale invariance implies conformal invariance is based on the explicit form of the conformal current K^α_μ, which is associated with the special conformal transformation, via the energy-momentum tensor:

$$K^\alpha_\mu = \left(2x_\nu x^\alpha - x^2 \delta^\alpha_\nu\right)\theta_{\mu\nu}. \tag{37}$$

Group	Transformations		# Parameters
Lorentz	$\frac{d(d-1)}{2}$ rotations	$x'_\mu = \Omega_{\mu\nu} x_\nu$	$\frac{d(d-1)}{2}$
Poincaré	+ d translations	$x'_\mu = x_\mu + a_\mu$	$\frac{d(d+1)}{2}$
Weyl	+ 1 dilatation	$x'_\mu = \rho\, x_\mu$	$\frac{d^2+d+2}{2}$
Conformal	+ d special conformal	$\frac{x'_\mu}{(x')^2} = \frac{x_\mu}{x^2} + \alpha_\mu$	$\frac{(d+1)(d+2)}{2}$

TABLE 1. Contents and the number of parameters of groups of space-time symmetry.

Differentiating, we get

$$\partial_\mu K^\alpha_\mu = 2x^\alpha \theta_{\mu\mu} \,, \tag{38}$$

which is proportional to divergence of the dilatation current. Therefore, both the dilatation and conformal currents vanish simultaneously when $\theta_{\mu\nu}$ is traceless which is provided, in turn, by the vanishing of the Gell-Mann–Low function.

Conformal invariance completely fixes three-vertices as was first shown by Polyakov [19] for scalar theories. The proper formula for the four-Fermi theory reads [20]

$$\begin{aligned} \Gamma(p_1,p_2) &= \mu^{2\gamma_{\rm v}} \frac{\Gamma(\frac{d}{2})\Gamma(\frac{d}{2}-\gamma_{\rm v})}{\Gamma(\gamma_{\rm v})} \\ &\times \int \frac{d^d k}{\pi^{d/2}} \frac{\hat{k}+\hat{p}_1}{[(k+p_1)^2]^{1+\gamma_\chi/2}} \frac{\hat{k}+\hat{p}_2}{[(k+p_2)^2]^{1+\gamma_\chi/2}} \frac{1}{|k|^{d-2+2\gamma_\psi-\gamma_\chi/2}} \,, \end{aligned} \tag{39}$$

where the coefficient in the form of the ratio of the Γ-functions is prescribed by the normalization (33) and (34) and the indices are related by Eq. (36) but can be arbitrary otherwise[2].

[2]The only restriction $\gamma_\psi \geq 0$ is imposed by the Källén–Lehmann representation of the propagator while there is no such restriction on γ_χ since it is a composite field.

Equation (39), which results from conformal invariance, unambiguously fixes the function f in Eq. (35). In contrast to infinite-dimensional conformal symmetry in $d = 2$, the conformal group in $d > 2$ is less restrictive. It fixes only the tree-point vertex while, say, the four-point vertex remains an unknown function of two variables.

The integral on the RHS of Eq. (39) looks in $d = 3$ very much like that in Eq. (22) and can easily be calculated to the leading order in $1/N$ when only the region of integration over large momenta with $|k| \gtrsim |p|_{\max} \equiv \max\{|p_1|, |p_2|\}$ is essential to this accuracy.

Let us first note that the coefficient in front of the integral is $\propto \gamma_{\rm v} \sim 1/N$, so that one has to peak up the term $\sim 1/\gamma_{\rm v}$ in the integral for the vertex to be of order 1. This term comes from the region of integration with $|k| \gtrsim |p|_{\max}$. Recalling that $|p_1 - p_2| \lesssim |p|_{\max}$ in Euclidean space, one gets

$$\int \frac{d^3k}{2\pi} \frac{\hat{k}+\hat{p}_1}{[(k+p_1)^2]^{1+\gamma_\chi/2}} \frac{\hat{k}+\hat{p}_2}{[(k+p_2)^2]^{1+\gamma_\chi/2}} \frac{1}{|k|^{1+2\gamma_\psi-\gamma_\chi/2}} = \frac{1}{\gamma_{\rm v}\,(p^2_{\max})^{\gamma_{\rm v}}}, \tag{40}$$

where Eq. (36) has been used and

$$\Gamma(p_1,p_2) = \left(\frac{\mu^2}{p^2_{\max}}\right)^{\gamma_{\rm v}}. \tag{41}$$

While the integral in Eq. (40) is divergent in the ultraviolet for $\gamma_{\rm v} < 0$, this divergence disappears after the renormalization.

Equation (23) is reproduced by Eq. (40) when expanding in $1/N$. This dependence of the three-vertex solely on the largest momentum is typical for logarithmic theories in the ultraviolet region where one can put, say, $p_1 = 0$ without changing the integral with logarithmic accuracy. This is valid if the integral is fast convergent in infrared regions which is our case.

Remark on broken scale invariance

Scale (and conformal) invariance at a fixed point $g = g_*$ holds only for large momenta $|p| \gg m$. For smaller values of momenta, scale invariance is broken by masses. In fact, any dimensional parameter μ breaks scale invariance. If the bare coupling g is chosen in the vicinity of g_*, then scale invariance holds even in the massless case only for $|p| \gg \mu$ where $g(p)$ approaches g_* while it is broken if $|p| \lesssim \mu$.

2.4. NONLINEAR SIGMA MODEL

The nonlinear O(N) sigma model[3] in 2 Euclidean dimensions is defined by the partition function

$$Z = \int D\vec{n}\, \delta\left(\vec{n}^2 - \frac{1}{g^2}\right) \mathrm{e}^{-\frac{1}{2}\int d^2x (\partial_\mu \vec{n})^2} \tag{42}$$

where $\vec{n} = (n_1, \ldots, n_N)$ is an O(N) vector. While the action in Eq. (42) is pure Gaussian, the model is not free due to the constraint

$$\vec{n}^2(x) = \frac{1}{g^2}, \tag{43}$$

which is imposed on the $\vec{n}$ field via the (functional) delta-function.

The sigma model in $d = 2$ is sometimes considered as a toy model for QCD since it possesses:

1) asymptotic freedom [21];
2) instantons for $N = 3$ [22].

The action in Eq. (42) is $\sim N$ as $N \to \infty$ but the entropy, *i.e.* a contribution from the measure of integration, is also $\sim N$ so that a straightforward saddle point is not applicable.

To overcome this difficulty, we introduce an auxiliary field $u(x)$, which is ~ 1 as $N \to \infty$, and rewrite the partition function (42) as

$$Z \propto \int_\uparrow Du\,(x) \int D\vec{n}\,(x)\; \mathrm{e}^{-\frac{1}{2}\int d^2x \left[(\partial_\mu \vec{n})^2 - u\left(\vec{n}^2 - \frac{1}{g^2}\right)\right]}, \tag{44}$$

where the contour of integration over $u(x)$ is parallel to imaginary axis.

Doing the Gaussian integration over $\vec{n}$, we get

$$Z \propto \int_\uparrow Du\,(x)\; \mathrm{e}^{-\frac{N}{2}\mathrm{Sp}\ln\left(-\partial_\mu^2 + u(x)\right) + \frac{1}{2g^2}\int d^2x\, u(x)}. \tag{45}$$

The first term in the exponent is nothing but the sum of one-loop diagrams in 2 dimensions

$$\frac{N}{2}\mathrm{Sp}\ln\left(-\partial_\mu^2 + u\,(x)\right) = \sum_n \frac{1}{n} \tag{46}$$

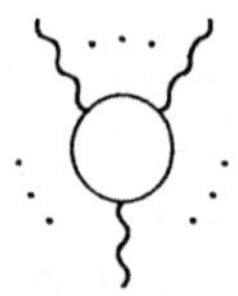

[3]The name comes from elementary particle physics where a nonlinear sigma model in 4 dimensions is used as an effective Lagrangian for describing low-energy scattering of the Goldstone π-mesons.

where the auxiliary field u is denoted again by the wavy line.

Now the path integral over $u(x)$ in Eq. (45) is a typical saddle-point one: the action $\sim N$ while the entropy ~ 1 since only one integration over u is left. The saddle-point equation for the nonlinear sigma model

$$\frac{1}{g^2} - NG\left(x, x; u_{\rm sp}\right) = 0 \tag{47}$$

while G is defined by

$$G\left(x, y; u\right) = \left\langle y \left| \frac{1}{-\partial_\mu^2 + u} \right| x \right\rangle . \tag{48}$$

The coupling g^2 in Eq. (47) is $\sim 1/N$ as is prescribed by the constraint (43) which involves a sum over N terms on the LHS. This guarantees that a solution to Eq. (47) exists. Next orders of the $1/N$-expansion for the 2-dimensional sigma model can be constructed analogously to the previous Subsections.

The $1/N$-expansion of the 2-dimensional nonlinear sigma model has many advantages over perturbation theory, which is usually constructed solving explicitly the constraint (43), say, choosing

$$n_N = \frac{1}{g}\sqrt{1 - g^2 \sum_{a=1}^{N-1} n_a^2} \tag{49}$$

and expanding the square root in g^2. Only $N-1$ dynamical degrees of freedom are left so that the O(N)-symmetry is broken in perturbation theory down to O($N-1$). The particles in perturbation theory are massless (like Goldstone bosons) and it suffers from infrared divergencies.

On the contrary, the solution to Eq. (47) has the form

$$u_{\rm sp} = m_{\rm R}^2 \equiv \Lambda^2 \, {\rm e}^{-\frac{4\pi}{Ng^2}} , \tag{50}$$

where Λ is an ultraviolet cutoff. Therefore, all N particles acquires the same mass $m_{\rm R}$ in the $1/N$-expansion so that the O(N) symmetry is restored. This appearance of mass is due to dimensional transmutation which says in this case that the parameter $m_{\rm R}$ rather than the renormalized coupling constant $g_{\rm R}^2$ is observable. The emergence of the mass cures the infrared problem.

To show that (50) is a solution to Eq. (47), let us look for a translationally invariant solution $u_{\rm sp}(x) \equiv m_{\rm R}^2$. Then Eq. (47) in the momentum space reads

$$\frac{1}{g^2} = N \int^{\Lambda} \frac{d^2k}{(2\pi)^2} \frac{1}{k^2 + m_{\rm R}^2} = \frac{N}{4\pi} \int_0^{\Lambda^2} \frac{dk^2}{k^2 + m_{\rm R}^2} = \frac{N}{4\pi} \ln \frac{\Lambda^2}{m_{\rm R}^2} . \tag{51}$$

The exponentiation results in Eq. (50).

Equation (51) relates the bare coupling g^2 and the cutoff Λ and allows us to calculate the Gell-Mann–Low function yielding

$$\mathcal{B}(g^2) \equiv \frac{\Lambda\, dg^2}{d\Lambda} = -\frac{Ng^4}{2\pi}\,. \tag{52}$$

The analogous one-loop perturbation-theory formula for any N reads [21]

$$\mathcal{B}(g^2) = -\frac{(N-2)g^4}{2\pi}\,. \tag{53}$$

Thus, the sigma-model is asymptotically free in 2-dimensions for $N > 2$ which is the origin of the dimensional transmutation. There is no asymptotic freedom for $N = 2$ since O(2) is Abelian.

2.5. LARGE-N FACTORIZATION IN VECTOR MODELS

The fact that a path integral has a saddle point at large N implies a very important feature of large-N theories — the factorization. It is a general property of the large-N limit and holds not only for the O(N) vector models. However, it is useful to illustrate it by these solvable examples.

The factorization at large N holds for averages of *singlet* operators, for example

$$\langle\, u\,(x_1)\ldots u\,(x_k)\,\rangle \equiv Z^{-1} \int_{\uparrow} Du\, \mathrm{e}^{\,-\frac{N}{2}\mathrm{Sp}\ln\left[-\partial_\mu^2+u\right]+\frac{1}{2g^2}\int d^2x\, u}\, u\,(x_1)\cdots u\,(x_k) \tag{54}$$

in the 2-dimensional sigma model.

Since the path integral has a saddle point at some $u\,(x) = u_{\rm sp}\,(x)$ which is, in fact, x-independent due to translational invariance, we get to the leading order in $1/N$:

$$\langle\, u\,(x_1)\ldots u\,(x_k)\,\rangle = u_{\rm sp}\,(x_1)\ldots u_{\rm sp}\,(x_k) + \mathcal{O}\left(\frac{1}{N}\right), \tag{55}$$

which can be written in the factorized form

$$\langle\, u\,(x_1)\ldots u\,(x_k)\,\rangle = \langle\, u\,(x_1)\,\rangle\ldots\langle\, u\,(x_k)\,\rangle + \mathcal{O}\left(\frac{1}{N}\right). \tag{56}$$

Therefore, u becomes "classical" as $N \to \infty$ in the sense of the $1/N$-expansion. This is an analog of the WKB-expansion in $\hbar = 1/N$. "Quantum" corrections are suppressed as $1/N$.

We shall return to discussing the large-N factorization in the next Section when considering the large-N limit of QCD.

3. Large-N QCD

The method of the $1/N$-expansion can be applied to QCD. This was done by 't Hooft [1] using the inverse number of colors for the gauge group $SU(N_c)$ as an expansion parameter.

For a $SU(N_c)$ gauge theory without virtual quark loops, the expansion goes in $1/N_c^2$ and rearranges diagrams of perturbation theory according to their topology. The leading order in $1/N_c^2$ is given by planar diagrams, which have a topology of a sphere, while the expansion in $1/N_c^2$ plays the role of a topological expansion. This reminds an expansion in the string coupling constant in string models of the strong interaction, which also has a topological character.

Virtual quark loops can be easily incorporated in the $1/N_c$-expansion. One distinguishes between the 't Hooft limit when the number of quark flavors N_f is fixed as $N_c \to \infty$ and the Veneziano limit [23] when the ratio N_f/N_c is fixed as $N_c \to \infty$. Virtual quark loops are suppressed in the 't Hooft limit as $1/N_c$ and lead in the Veneziano limit to the same topological expansion as dual-resonance models of strong interaction.

The simplification of QCD in the large-N_c limit is due to the fact that the number of planar graphs grows with the number of vertices only exponentially rather than factorially as do the total number of graphs. Correlators of gauge invariant operators factorize in the large-N_c limit which looks like the leading-order term of a "semiclassical" WKB-expansion in $1/N_c$.

We begin this Section with a description of the double-line representation of diagrams of QCD perturbation theory and rearrange it as the topological expansion in $1/N_c$. Then we discuss some properties of the $1/N_c$-expansion for a generic matrix-valued field.

3.1. INDEX OR RIBBON GRAPHS

In order to describe the $1/N_c$-expansion of the Yang–Mills theory, it is convenient to represent the gauge field by a Hermitean $N \times N$ matrix

$$A_\mu^{ij}(x) = g \sum_a A_\mu^a(x)[t^a]^{ij}. \tag{57}$$

Here $[t^a]^{ij}$ are the generators of the gauge group ($a = 1, \ldots, N_c^2 - 1$ for $SU(N_c)$) with the normalization

$$\operatorname{tr} t^a t^b = \frac{1}{2}\delta^{ab}. \tag{58}$$

The (Euclidean) action reads

$$S[A] = \int d^d x \frac{1}{2g^2} \operatorname{tr} F_{\mu\nu}^2, \tag{59}$$

where

$$F_{\mu\nu} = \partial_\mu A_\nu - \partial_\nu A_\mu - i[A_\mu, A_\nu] \tag{60}$$

is the non-Abelian field strength and g is the coupling constant.

The propagator of the matrix field $A^{ij}(x)$ has the form

$$\left\langle A_\mu^{ij}(x)\, A_\nu^{kl}(y) \right\rangle_{\text{Gauss}} = \frac{1}{2}\left(\delta^{il}\delta^{kj} - \frac{1}{N_c}\delta^{ij}\delta^{kl}\right) D_{\mu\nu}(x-y), \tag{61}$$

where we have assumed, as usual, a gauge fixing to define the propagator in perturbation theory. For instance, one has

$$D_{\mu\nu}(x-y) = \frac{g^2}{4\pi^2}\frac{\delta_{\mu\nu}}{(x-y)^2} \tag{62}$$

in the Feynman gauge.

Equation (61) can be immediately derived from the standard formula

$$\left\langle A_\mu^a(x)\, A_\nu^b(y) \right\rangle_{\text{Gauss}} = \delta^{ab} D_{\mu\nu}(x-y) \tag{63}$$

multiplying by the generators of the SU(N_c) gauge group according to the definition (57) and using the completeness condition

$$\sum_{a=1}^{N_c^2-1} (t^a)^{ij}(t^a)^{kl} = \frac{1}{2}\left(\delta^{il}\delta^{kj} - \frac{1}{N_c}\delta^{ij}\delta^{kl}\right) \quad \boxed{\text{for } \mathrm{SU}(N_c)}, \tag{64}$$

where the factor of 1/2 is due to the normalization (58) of the generators.

We concentrate in this Section only on the structure of diagrams in the index space, *i.e.* the space of the indices associated with the SU(N_c) group. We shall not consider, in most cases, space-time structures of diagrams which are prescribed by Feynman's rules.

Omitting at large N_c the second term in parentheses on the RHS of Eq. (61), we depict the propagator by the double line

$$\left\langle A_\mu^{ij}(x)\, A_\nu^{kl}(y) \right\rangle_{\text{Gauss}} \propto g^2\delta^{il}\delta^{kj} =$$

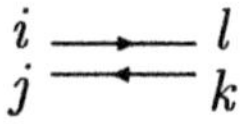

. (65)

Each line represents the Kronecker delta-symbol and has orientation which is indicated by arrows. This notation is obviously consistent with the space-time structure of the propagator which describes a propagation from x to y.

The arrows are due to the fact that the matrix A_μ^{ij} is Hermitean and its off-diagonal components are complex conjugate. The independent fields are, say, the complex fields A_μ^{ij} for $i > j$ and the diagonal real fields A_μ^{ii}. The arrow represents the direction of the propagation of the indices of the complex field A_μ^{ij} for $i > j$ while the complex-conjugate one, $A_\mu^{ji} = (A_\mu^{ij})^*$, propagates in the opposite direction. For the real fields A_μ^{ii}, the arrows are not essential.

The double-line notation appears generically in all models describing *matrix* fields in contrast to *vector* (in internal symmetry space) fields whose propagators are depicted by single lines as in the previous Section.

The three-gluon vertex, which is generated by the action (59), is depicted in the double-line notations as

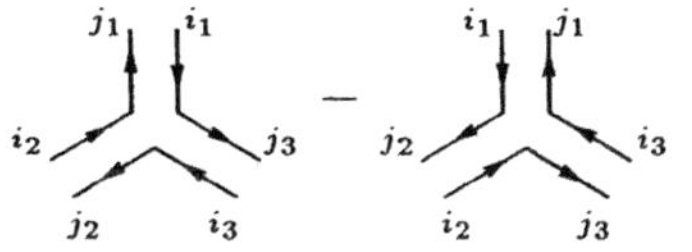

$$\propto g^{-2}\left(\delta^{i_1 j_3}\delta^{i_2 j_1}\delta^{i_3 j_2} - \delta^{i_1 j_2}\delta^{i_3 j_1}\delta^{i_2 j_3}\right) \tag{66}$$

where the subscripts 1, 2 or 3 refer to each of the three gluons. The relative minus sign is due to the commutator in the cubic in A term in the action (59). The color part of the three-vertex is antisymmetric under interchanging the gluons. The space-time structure, which reads in the momentum space as

$$\begin{aligned} &\gamma_{\mu_1\mu_2\mu_3}(p_1,p_2,p_3) \\ &= \delta_{\mu_1\mu_2}(p_1-p_2)_{\mu_3} + \delta_{\mu_2\mu_3}(p_2-p_3)_{\mu_1} + \delta_{\mu_1\mu_3}(p_3-p_1)_{\mu_2}\,, \end{aligned} \tag{67}$$

is antisymmetric as well. We consider all three gluons as incoming so that their momenta obey $p_1 + p_2 + p_3 = 0$. The full vertex is symmetric as is prescribed by Bose statistics.

The color structure in Eq. (66) can alternatively be obtained by multiplying the standard vertex

$$\Gamma^{a_1a_2a_3}_{\mu_1\mu_2\mu_3}(p_1,p_2,p_3) = f^{a_1a_2a_3}\gamma_{\mu_1\mu_2\mu_3}(p_1,p_2,p_3) \tag{68}$$

by $(t^{a_1})^{i_1j_1}(t^{a_2})^{i_2j_2}(t^{a_3})^{i_3j_3}$, with f^{abc} being the structure constants of the $\mathrm{SU}(N_c)$ group, and using the formula

$$f^{a_1a_2a_3}(t^{a_1})^{i_1j_1}(t^{a_2})^{i_2j_2}(t^{a_3})^{i_3j_3} = \frac{1}{2}\left(\delta^{i_1 j_3}\delta^{i_2 j_1}\delta^{i_3 j_2} - \delta^{i_1 j_2}\delta^{i_3 j_1}\delta^{i_2 j_3}\right), \tag{69}$$

which is a consequence of the completeness condition (64).

The four-gluon vertex involves six terms — each of them is depicted by a cross — which differ by interchanging of the color indices. We depict the color structure of the four-gluon vertex for simplicity in the case when $i_1 = j_2 = i$, $i_2 = j_3 = j$, $i_3 = j_4 = k$, $i_4 = j_1 = l$ but i, j, k, l take on

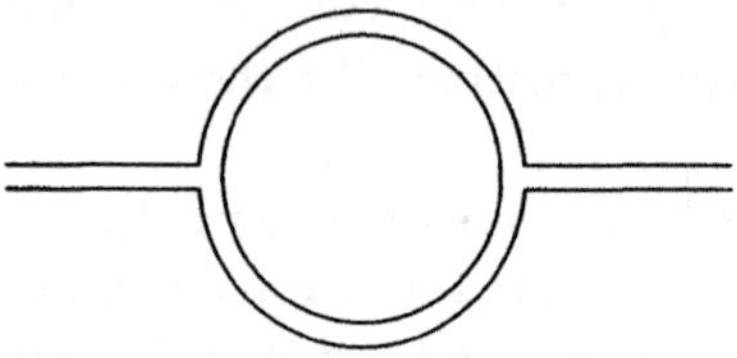

Figure 6. Double-line representation of a one-loop diagram for the gluon propagator. The sum over the N_c indices is associated with the closed index line. The relative contribution of this diagram is $\sim g^2 N_c \sim 1$.

different values. Then only the following term is left

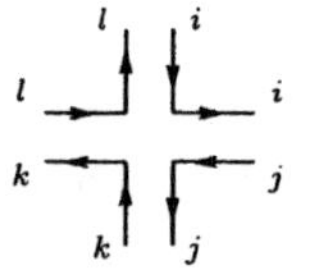

$$\propto g^{-2} \qquad (70)$$

and there are no deltas on the RHS since the color structure is fixed. In other words, we pick up only one color structure by equaling indices pairwise.

The diagrams of perturbation theory can now be completely rewritten in the double-line notation [1]. The simplest one which describes the one-loop correction to the gluon propagator is depicted in Figure 6.[4] This diagram involves two three-gluon vertices and a sum over the N_c indices which is associated with the closed index line. Therefore, the relative contribution of this diagram is $\sim g^2 N_c$.

In order for the large-N_c limit to be nontrivial, the bare coupling constant g^2 should satisfy

$$g^2 \sim \frac{1}{N_c}\,. \qquad (71)$$

This dependence on N_c is similar to Eqs. (12) and (47) for the vector models and is prescribed by the asymptotic-freedom formula

$$g^2 = \frac{24\pi^2}{11 N_c \ln\left(\Lambda/\Lambda_{QCD}\right)} \qquad (72)$$

of the pure SU(N_c) gauge theory.

Thus, the relative contribution of the diagram of Figure 6 is of order

$$\text{Figure 6} \;\sim\; g^2 N_c \;\sim\; 1 \qquad (73)$$

in the large-N_c limit.

[4]Here and in the most figures below the arrows of the index lines are omitted for simplicity.

The double lines of the diagram in Figure 6 can be viewed as bounding a piece of a plane. Therefore, these lines represent a two-dimensional object rather than a one-dimensional one as the single lines do in vector models. These double-line graphs are often called in mathematics the *ribbon* graphs or *fatgraphs*. We shall see below their connection with Riemann surfaces.

Remark on the U(N_c) gauge group

As is said above, the second term in the parentheses on the RHS of Eq. (64) can be omitted at large N_c. Such a completeness condition emerges for the U(N_c) group whose generators T^A $(A = 1, \ldots, N_c^2)$ are

$$T^A = \left(t^a, \frac{I}{\sqrt{2N}}\right), \qquad \operatorname{tr} T^A T^B = \frac{1}{2}\delta^{AB}. \tag{74}$$

They obey the completeness condition

$$\sum_{A=1}^{N_c^2} \left(T^A\right)^{ij} \left(T^A\right)^{kl} = \frac{1}{2}\delta^{il}\delta^{kj} \quad \boxed{\text{for } \mathrm{U}(N_c)}. \tag{75}$$

The point is that elements of both the SU(N_c) group and the U(N_c) group can be represented in the form $U = \exp iB$, where B is a general Hermitean matrix for U(N_c) and a traceless Hermitean matrix for SU(N_c).

Therefore, the double-line representation of perturbation-theory diagrams which is described in this Section holds, strictly speaking, only for the U(N_c) gauge group. However, the large-N_c limit of both the U(N_c) group and the SU(N_c) group is the same.

3.2. PLANAR AND NON-PLANAR GRAPHS

The double-line representation of perturbation theory diagrams in the index space is very convenient to estimate their orders in $1/N_c$. Each gluon propagator contributes a factor of g^2 Each three- or four-gluon vertex contributes a factor of g^{-2}. Each closed index line contributes a factor of N_c. The order of g in $1/N_c$ is given by Eq. (71).

Let us consider a typical diagram for the gluon propagator depicted in Figure 7. It has eight three-gluon vertices and four closed index lines which coincides with the number of loops. Therefore, the relative order of this diagram in $1/N_c$ is

$$\text{Figure 7} \sim \left(g^2 N_c\right)^4 \sim 1. \tag{76}$$

The diagrams of the type in Figure 7, which can be drawn on a sheet of a paper without crossing any lines, are called the *planar* diagrams. For

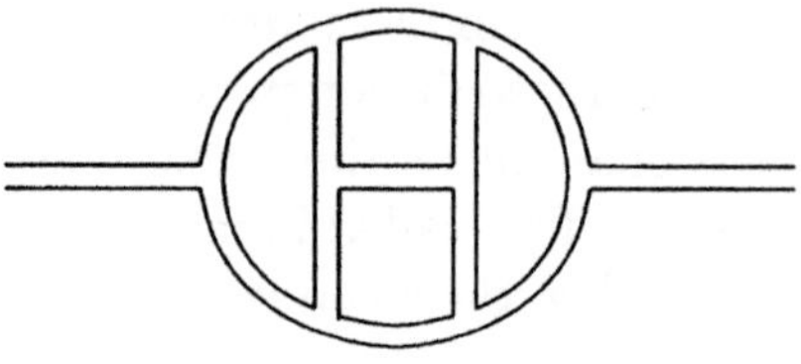

Figure 7. Double-line representation of a four-loop diagram for the gluon propagator. The sum over the N_c indices is associated with each of the four closed index lines whose number is equal to the number of loops. The contribution of this diagram is $\sim g^8 N_c^4 \sim 1$.

Figure 8. Double-line representation of a three-loop non-planar diagram for the gluon propagator. The diagram has six three-gluon vertices but only one closed index line (while three loops!). The order of this diagram is $\sim g^6 N_c \sim 1/N_c^2$.

such diagrams, an adding of a loop inevitably results in adding of two three-gluon (or one four-gluon) vertex. A planar diagram with n_2 loops has n_2 closed index lines. It is of order

$$n_2\text{-loop planar diagram} \sim \left(g^2 N_c\right)^{n_2} \sim 1\,, \tag{77}$$

so that all planar diagrams survive in the large-N_c limit.

Let us now consider a non-planar diagram of the type depicted in Figure 8. This diagram is a three-loop one and has six three-gluon vertices. The crossing of the two lines in the middle does not correspond to a four-gluon vertex and is merely due to the fact that the diagram cannot be drawn on a sheet of a paper without crossing the lines. The diagram has only one closed index line. The relative order of this diagram in $1/N_c$ is

$$\text{Figure 8} \sim g^6 N_c \sim \frac{1}{N_c^2}\,. \tag{78}$$

It is therefore suppressed at large N_c by $1/N_c^2$.

The non-planar diagram in Figure 8 can be drawn without line-crossing on a surface with one handle which is usually called in mathematics a torus or the surface of genus one. A plane is then equivalent to a sphere and has genus zero. Adding a handle to a surface produces a *hole* according to mathematical terminology. A general Riemann surface with h holes has genus h.

The above evaluations of the order of the diagrams in Figures 6–8 can now be described by the unique formula

$$\text{genus-}h \text{ diagram} \sim \left(\frac{1}{N_c^2}\right)^{\text{genus}} . \tag{79}$$

Thus, the expansion in $1/N_c$ rearranges perturbation-theory diagrams according to their topology [1]. For this reason, it is referred to as the *topological expansion* or the *genus expansion*. The general proof of Eq. (79) for an arbitrary diagram is given in Subsection 3.3.

Only planar diagrams, which are associated with genus zero, survive in the large-N_c limit. This class of diagrams is an analog of the bubble graphs in the vector models. However, the problem of summing the planar graphs is much more complicated than that of summing the bubble graphs. Nevertheless, it is simpler than the problem of summing all the graphs, since the number of the planar graphs with n_0 vertices grows at large n_0 exponentially [24, 25]

$$\#_{\mathrm{p}}\,(n_0) \equiv \# \text{ of planar graphs} \sim \text{const}^{n_0} , \tag{80}$$

while the number of all the graphs grows with n_0 factorially. There is no dependence in Eq. (80) on the number of external lines of a planar graph which is assumed to be much less than n_0.

It is instructive to see the difference between the planar diagrams and, for instance, the ladder diagrams which describe e^+e^- elastic scattering in QED. Let the ladder has n rungs. Then there are $n!$ ladder diagrams, but only one of them is planar. This simple example shows why the number of planar graphs is much smaller than the number of all graphs, most of which are non-planar.

We shall discuss in these Lectures what is known about solving the problem of summing the planar graphs.

Equation (79) holds, strictly speaking, only for the relative order while the contribution of tree diagrams to a connected n-point Green's function is $\sim (g^2)^{n-1}$ which is its natural order in $1/N_c$. In order to make contributions of all planar diagrams to be of the same order ~ 1 in the large-N_c limit, independently of the number of external lines, it is convenient to contract the Kronecker deltas associated with external lines.

Let us do this in a cyclic order as is depicted in Figure 9 for a generic connected diagram with three external gluon lines. The extra deltas which are added to contract the color indices are depicted by the single lines. They can be viewed as a *boundary* of the given diagram. The actual size of the boundary is not essential — it can be shrunk to a point. Then a bounded piece of a plane will be topologically equivalent to a sphere with a puncture.

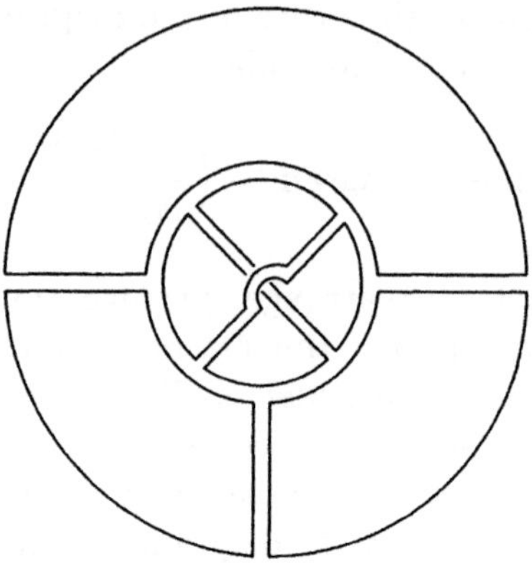

Figure 9. Generic index diagram with $n_0 = 10$ vertices, $n_1 = 10$ gluon propagators, $n_2 = 4$ closed index lines, and $B = 1$ boundary. The color indices of the external lines are contracted by the Kronecker deltas (represented by the single lines) in a cyclic order. The extra factor of $1/N_c$ is due to the normalization (81). Its order in $1/N_c$ is $\sim 1/N_c^2$ in accord with Eq. (79).

I shall prefer to draw planar diagrams in a plane with an extended boundary (boundaries) rather than in a sphere with a puncture (punctures).

It is clear from the graphic representation that the diagram in Figure 9 is associated with the trace over the color indices of the three-point Green's function

$$G^{(3)}_{\mu_1\mu_2\mu_3}(x_1, x_2, x_3) \equiv \frac{1}{N_c}\left\langle \operatorname{tr}\left[A_{\mu_1}(x_1)\, A_{\mu_2}(x_2)\, A_{\mu_3}(x_3)\right]\right\rangle . \tag{81}$$

We have introduced here the factor $1/N_c$ to make G_3 of $\mathcal{O}(1)$ in the large-N_c limit. Therefore, the contribution of the diagram in Figure 9 having one boundary should be divided by N_c.

The extension of Eq. (81) to multi-point Green's functions is obvious:

$$G^{(n)}_{\mu_1\cdots\mu_n}(x_1, \ldots, x_n) \equiv \frac{1}{N_c}\left\langle \operatorname{tr}\left[A_{\mu_1}(x_1)\cdots A_{\mu_n}(x_n)\right]\right\rangle . \tag{82}$$

The factor $1/N_c$, which normalizes the trace, provides the natural normalization

$$G^{(0)} = 1 \tag{83}$$

of the averages.

Though the two terms in the index-space representation (66) of the three-gluon vertex look very similar, their fate in the topological expansion is quite different. When the color indices are contracted anti-clockwise, the first term leads to the planar contributions to $G^{(3)}$, the simplest of which is

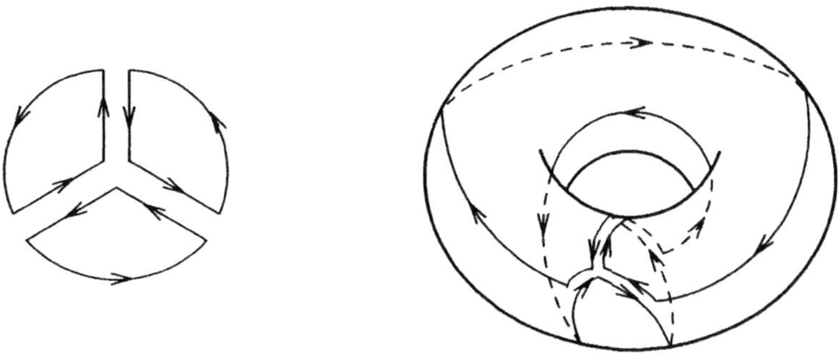

a) b)

Figure 10. Planar a) and non-planar b) contributions of the two color structures in Eq. (66) for three-gluon vertex to $G^{(3)}$ in the lowest order of perturbation theory.

depicted in Figure 10a. The anti-clockwise contraction of the color indices in the second term leads to a non-planar graph in Figure 10b which can be drawn without crossing of lines only on a torus. Therefore, the two color structures of the three-gluon vertex contribute to different orders of the topological expansion. The same is true for the four-gluon vertex.

Remark on oriented Riemann surfaces

Each line of an index graph of the type depicted in Figure 9 is oriented. This orientation continues along a closed index line while the pairs of index lines of each double line have opposite orientations. The overall orientation of the lines is prescribed by the orientation of the external boundary which we choose to be, say, anti-clockwise like in Figure 10a. Since the lines are oriented, the faces of the Riemann surface associated with a given graph are oriented too — all in the same way — anti-clockwise. Vice versa, such an orientation of the Riemann surfaces unambiguously fixes the orientation of all the index lines. This is the reason why we omit the arrows associated with the orientation of the index lines: their directions are obvious.

Remark on cyclic-ordered Green's functions

The cyclic-ordered Green's functions (82) naturally arise in the expansion of the trace of the path-ordered non-Abelian phase factor for a closed contour.

Figure 11. Example of a) connected and b) disconnected planar graphs.

One gets

$$\left\langle \frac{1}{N_c} \operatorname{tr} \mathbf{P} \, \mathrm{e}^{\, i \oint_\Gamma dx^\mu A_\mu(x)} \right\rangle$$

$$= \sum_{n=0}^{\infty} i^n \oint_\Gamma dx_1^{\mu_1} \int_{x_1}^{x_1} dx_2^{\mu_2} \ldots \int_{x_1}^{x_{n-1}} dx_n^{\mu_n} \, G^{(n)}_{\mu_1 \cdots \mu_n}(x_1, \ldots, x_n) \,. \qquad (84)$$

The reason is because the ordering along a closed path implies the cyclic-ordering in the index space.

Remark on generating functionals for planar graphs

By connected or disconnected planar graphs we mean, respectively, the graphs which were connected or disconnected before the contraction of the color indices as is illustrated by Figure 11. The graph in Figure 11a is connected planar while the graph in Figure 11b is disconnected planar.

The usual exponential relation between the generating functionals $W[J]$ and $Z[J]$ for connected graphs and all graphs, does not hold for the planar graphs. The reason is that an exponentiation of such a connected planar diagram for the cyclic-ordered Green's functions (82) can give disconnected non-planar ones.

The generating functionals for all and connected planar graphs can be constructed [26] by means of introducing non-commutative sources $\boldsymbol{j}_\mu(x)$. "Non-commutative" means that there is no way to transform $\boldsymbol{j}_{\mu_1}(x_1)\boldsymbol{j}_{\mu_2}(x_2)$ into $\boldsymbol{j}_{\mu_2}(x_2)\boldsymbol{j}_{\mu_1}(x_1)$. This non-commutativity of the sources reflects the cyclic-ordered structure of the Green's functions (82) which possess only cyclic symmetry.

Using the short-hand notations

$$\boldsymbol{j} \circ A \equiv \sum_\mu \int d^d x \, \boldsymbol{j}_\mu(x) A_\mu(x) \,, \qquad (85)$$

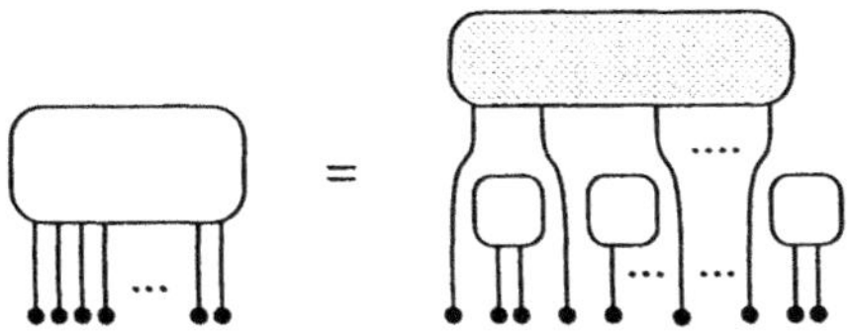

Figure 12. Graphic derivation of Eq. (89): $Z[j]$ is denoted by an empty box, $W[j]$ is denoted by a shadow box, j is denoted by a filled circle. By picking a leftmost external line of a planar graph, we end up with a connected planar graph, whose remaining external lines are somewhere to the right interspersed by disconnected planar graphs. It is evident that $jZ[j]$ plays the role of a new source for the connected planar graph. If we instead pick up the rightmost external line, we get the inverse order $Z[j]j$, which results in Eq. (90).

where the symbol $\circ$ includes the sum over the d-vector (or whatever available) indices except for the color ones, we write down the definitions of the generating functionals for all planar and connected planar graphs, respectively, as

$$Z[j] \equiv \sum_{n=0}^{\infty} \left\langle \frac{1}{N_c} \operatorname{tr} (j \circ A)^n \right\rangle \tag{86}$$

and

$$W[j] \equiv \sum_{n=0}^{\infty} \left\langle \frac{1}{N_c} \operatorname{tr} (j \circ A)^n \right\rangle_{\text{conn}} . \tag{87}$$

The planar contribution to the Green's functions (82) and their connected counterparts can be obtained, respectively, from the generating functionals $Z[j]$ and $W[j]$ applying the non-commutative derivative which is defined by

$$\frac{\delta}{\delta j_\mu(x)} j_\nu(y) f(j) = \delta_{\mu\nu} \delta^{(d)}(x-y) f(j) , \tag{88}$$

where f is an arbitrary function of j's. In other words the derivative picks up only the leftmost variable.

The relation which replaces the usual one for planar graphs is

$$Z[j] = W[jZ[j]], \tag{89}$$

while the cyclic symmetry says

$$W[jZ[j]] = W[Z[j]j] . \tag{90}$$

A graphic derivation of Eqs. (89) and (90) is given in Figure 12. In other words, given $W[j]$, one should construct an inverse function as the solution to the equation

$$j_\mu(x) = J_\mu(x) W[j], \tag{91}$$

after which Eq. (89) says

$$Z[j] = W[J]. \tag{92}$$

More about this approach to the generating functionals for planar graphs can be found in Ref. [27].

An iterative solution to Eq. (91) for the Gaussian case can be easily found. In the Gaussian case, only $G^{(2)}$ is nonvanishing which yields

$$W[j] = 1 - g^2 j \circ D \circ j, \tag{93}$$

where the propagator D is given by Eq. (62). Using Eq. (89), we get explicitly

$$Z[j] = 1 - g^2 \int d^d x\, d^d y\, D_{\mu\nu}(x-y)\, j_\mu(x)\, Z[j]\, j_\nu(y)\, Z[j]. \tag{94}$$

While this equation for $Z[j]$ is quadratic, its solution can be written only as a continued fraction due to the non-commutative nature of the variables. In order to find it, we rewrite Eq. (94) as

$$Z[j] = \frac{1}{1 + g^2 \int d^d x\, d^d y\, D_{\mu\nu}(x-y)\, j_\mu(x)\, Z[j]\, j_\nu(y)}, \tag{95}$$

whose iterative solution reads [26]

$$Z[j] = \cfrac{1}{1 + g^2 j \cfrac{\circ D \circ}{1 + g^2 j \cfrac{\circ D \circ}{1 + g^2 j \cfrac{\circ D \circ}{\vdots} j} j} j}. \tag{96}$$

3.3. TOPOLOGICAL EXPANSION AND QUARK LOOPS

It is easy to incorporate quarks in the topological expansion. A quark field belongs to the fundamental representation of the gauge group SU(N_c) and its propagator is represented by a single line

$$\langle \psi_i \bar{\psi}_j \rangle \propto \delta_{ij} \quad = \quad i \longrightarrow j \quad . \tag{97}$$

The arrow indicates, as usual, the direction of propagation of a (complex) field ψ. We shall omit these arrows for simplicity.

The diagram for the gluon propagator which involves one quark loop is depicted in Figure 13a. It has two three gluon vertices and no closed index

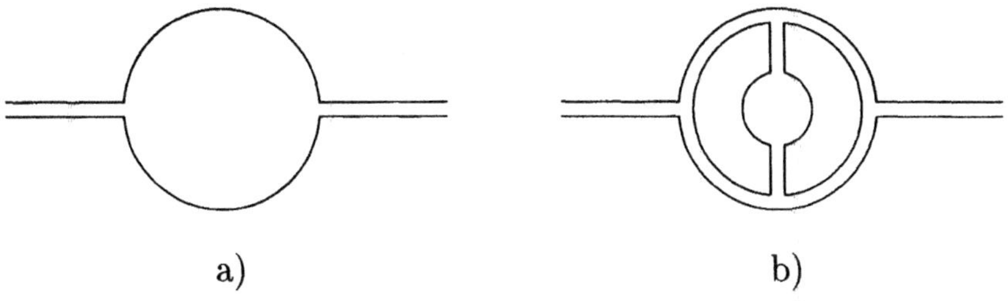

Figure 13. Diagrams for the gluon propagator with a quark loop which is represented by the single lines. The diagram a) involves one quark loop and has no closed index lines so that its order is $\sim g^2 \sim 1/N_c$. The diagram b) involves three loops one of which is a quark loop. Its relative order is $\sim g^6 N_c^2 \sim 1/N_c$.

lines so that its order in $1/N_c$ is

$$\text{Figure 13a} \sim g^2 \sim \frac{1}{N_c}. \tag{98}$$

Analogously, the relative order of a more complicated tree-loop diagram in Figure 13b, which involves one quark loop and two closed index lines, is

$$\text{Figure 13b} \sim g^6 N_c^2 \sim \frac{1}{N_c}. \tag{99}$$

It is evident from this consideration that quark loops are not accompanied by closed index lines. One should add a closed index line for each quark loop in order for a given diagram with L quark loops to have the same double-line representation as for pure gluon diagrams. Therefore, given Eq. (79), diagrams with L quark loops are suppressed at large N_c by

$$L \text{ quark loops} \sim \left(\frac{1}{N_c}\right)^{L+2\cdot\text{genus}}. \tag{100}$$

The single-line representation of the quark loops is similar to the one of the external boundary in Figure 9. Moreover, such a diagram emerges when one calculates perturbative gluon corrections to the vacuum expectation value of the quark operator

$$O_\Gamma = \frac{1}{N_c}\bar{\psi}\Gamma\psi, \tag{101}$$

where Γ stands for one of the combinations of the gamma-matrices:

$$\Gamma = \mathrm{I},\ \gamma_5,\ \gamma_\mu,\ i\gamma_\mu\gamma_5,\ \Sigma_{\mu\nu} = \frac{1}{2i}[\gamma_\mu, \gamma_\nu], \ldots. \tag{102}$$

The factor of $1/N_c$ is introduced in (101) to make it $\mathcal{O}(1)$ in the large-N_c limit. Therefore, the external boundary can be viewed as a single line associated with valence quarks. The difference between virtual quark loops and external boundaries is that each of the latter comes along with the factor of $1/N_c$ due to the definitions (82) and (101).

In order to prove Eqs. (79) and its quark counterpart (100), let us consider a generic diagram in the index space which has $n_0^{(3)}$ three-point vertices (either three-gluon or quark-gluon ones), $n_0^{(4)}$ four-gluon vertices, n_1 propagators (either gluon or quark ones), n_2 closed index lines, L virtual quark loops and B external boundaries. Its order in $1/N_c$ is

$$\frac{1}{N_c^B}(g^2)^{n_1-n_0^{(3)}-n_0^{(4)}}N_c^{n_2} \sim N_c^{n_2-n_1+n_0-B} \tag{103}$$

where the total number of vertices $n_0 = n_0^{(3)} + n_0^{(4)}$ is introduced. The extra factor of $1/N_c^B$ is due to the extra normalization factor of $1/N_c$ in operators associated with external boundaries.

The exponent on the RHS of Eq. (103) can be expressed via the Euler characteristic χ of a given graph of genus h. Let us first mention that a proper Riemann surface, which is associated with a given graph, is open and has $B+L$ boundaries (represented by single lines). This surface can be closed by attaching a cap to each boundary. The single lines then become double lines together with the lines of the boundary of each cap. We have already considered this procedure when deducing Eq. (100) from Eq. (79).

The number of faces for a closed Riemann surface constructed in such a manner is $n_2 + L + B$, while the number of edges and vertices are n_1 and n_0, respectively. Euler's theorem says that

$$\chi \equiv 2 - 2h = n_2 + L + B - n_1 + n_0 \,. \tag{104}$$

Therefore the RHS of Eq. (103) can be rewritten as

$$N_c^{n_2-n_1+n_0-B} = N_c^{2-2h-L-2B} \,. \tag{105}$$

We have thus proven that the order in $1/N_c$ of a generic graph does not depend on its order in the coupling constant and is completely expressed via the genus h and the number of virtual quark loops L and external boundaries B by

$$\text{generic graph} \sim \left(\frac{1}{N_c}\right)^{2h+L+2(B-1)} \,. \tag{106}$$

For $B = 1$, we recover Eqs. (79) and (100).

Remark on the order of gauge action

We see from Eq. (82) that the natural variables for the large-N_c limit are the matrices A_μ which include the factor of g (see Eq. (57)). In these variables, the action (59) is $\mathcal{O}(N_c^2)$ at large N_c, since g^2 is $\sim 1/N_c$ and the trace is $\sim N_c$.

This result can be anticipated from the free theory because the kinetic part of the action involves the sum over $N_c^2 - 1$ free gluons. Therefore, the non-Abelian field strength (60) is ~ 1 for $g^2 \sim 1/N_c$.

The fact that the action is $\mathcal{O}(N_c^2)$ in the large-N_c limit is a generic property of the models describing matrix fields. It will be crucial for developing saddle-point approaches at large N_c which are considered below.

Remark on phenomenology of multicolor QCD

While $N_c = 3$ in the real world, there are phenomenological indications that $1/N_c$ may be considered as a small parameter. We have already mentioned some of them in the text — the simplest one is that the ratio of the ρ-meson width to its mass, which is $\sim 1/N_c$, is small. Considering $1/N_c$ as a small parameter immediately leads to qualitative phenomenological consequences which are preserved by the planar diagrams associated with multicolor QCD, but are violated by the non-planar diagrams.

The most important consequence is the relation of the $1/N_c$-expansion to the topological expansion in the dual-resonance model of hadrons. Vast properties of hadrons are explained by the dual-resonance model. A very clear physical picture behind this model is that hadrons are excitations of a string with quarks at the ends.

I shall briefly list some consequences of multicolor QCD:

1) The "naive" quark model of hadrons emerges at $N_c = \infty$. Hadrons are built out of (valence or constituent) quark and antiquark $q\bar{q}$, while exotic states like $qq\bar{q}\bar{q}$ do not appear.
2) The partial width of decay of the ϕ-meson, which is built out of $s\bar{s}$ (the strange quark and antiquark), into K^+K^- is $\sim 1/N_c$, while that into $\pi^+\pi^-\pi^0$ is $\sim 1/N_c^2$. This explains Zweig's rule. The masses of the ρ- and ω-mesons are degenerate at $N_c = \infty$.
3) The coupling constant of meson-meson interaction is small at large N_c.
4) The widths of glueballs are $\sim 1/N_c^2$, *i.e.* they should be even narrower than mesons built out of quarks. The glueballs do not interact or mix with mesons at $N_c = \infty$.

All these hadron properties (except the last one) approximately agree with experiment, and were well-known even before 1974 when multicolor QCD was introduced. Glueballs are not yet detected experimentally (maybe because of their property listed in the item 4).

3.4. LARGE-N_C FACTORIZATION

The vacuum expectation values of several colorless or white operators, which are singlets with respect to the gauge group, factorize in the large-N_c limit of QCD (or other matrix models). This property is similar to that already discussed in Subsection 2.5 for the vector models.

The simplest gauge-invariant operator in a pure SU(N_c) gauge theory is the square of the non-Abelian field strength:

$$O\,(x) = \frac{1}{N_c}\,\mathrm{tr}\,F^2_{\mu\nu}\,(x)\,. \tag{107}$$

The normalizing factor is the same as in Eqs. (81), (82), which provides the natural normalization

$$\left\langle \frac{1}{N_c}\,\mathrm{tr}\,F^2_{\mu\nu}\,(x) \right\rangle = \left\langle \frac{g^2}{2N_c}\,F^a_{\mu\nu}\,(x)\,F^a_{\mu\nu}\,(x) \right\rangle \sim 1\,. \tag{108}$$

The contribution of all planar graphs to the average on the LHS of Eq. (108) is of order 1 in accord with the general formula (106) for $B=1$.

In order to verify the factorization in the large-N_c limit, let us consider the index space diagrams for the average of the product of two colorless operators $O\,(x_1)$ and $O\,(x_2)$ given by (107). It involves a factorized part when gluons are emitted and absorbed by the same operators. The contribution of the factorized part is or order 1 as above.

Alternatively, the connected correlator of the two operators is associated with the general formula (106) for two boundaries $B=2$. Its contribution is suppressed by $1/N_c^2$ in the large-N_c limit. For this correlator, at least one gluon line is emitted and absorbed by different operators $O\,(x_1)$ and $O\,(x_2)$. Notice, that these graphs themselves are planar, while the suppression comes from the number of boundaries.

This example illustrates the general property that only (planar) diagrams with gluon lines emitted and absorbed by the same operators survive as $N_c \to \infty$. Since correlations between the colorless operators $O\,(x_1)$ and $O\,(x_2)$ are of order $1/N_c^2$, the *factorization* property holds as $N_c \to \infty$:

$$\begin{aligned}&\left\langle \frac{1}{N_c}\,\mathrm{tr}\,F^2\,(x_1)\,\frac{1}{N_c}\,\mathrm{tr}\,F^2\,(x_2) \right\rangle \\ &\quad = \left\langle \frac{1}{N_c}\,\mathrm{tr}\,F^2\,(x_1) \right\rangle \left\langle \frac{1}{N_c}\,\mathrm{tr}\,F^2\,(x_2) \right\rangle + \mathcal{O}\left(\frac{1}{N_c^2}\right).\end{aligned} \tag{109}$$

For a general set of gauge-invariant operators $O_1, \ldots, O_n$, the factorization property can be represented by

$$\langle\, O_1 \cdots O_n \,\rangle = \langle\, O_1 \,\rangle \cdots \langle\, O_n \,\rangle + \mathcal{O}\left(\frac{1}{N_c^2}\right). \tag{110}$$

This is analogous to Eq. (56) for the vector models.

The factorization in large-N_c QCD was first discovered by A.A. Migdal in the late seventies. An important observation that the factorization implies a semiclassical nature of the large-N_c limit of QCD was done by Witten [28]. We shall discuss this in the next two Subsections.

The factorization property also holds for gauge-invariant operators constructed from quarks like in Eq. (101) as a consequence of Eq. (106).

Remark on factorization beyond perturbation theory

The large-N_c factorization (110) has been shown above to all orders of perturbation theory. It can be also verified at all orders of the strong coupling expansion in the SU(N_c) lattice gauge theory. A non-perturbative proof of the factorization will be given in the next Section by using quantum equations of motion (the loop equations).

3.5. THE MASTER FIELD

The large-N_c factorization in QCD assumes that gauge-invariant objects behave as c-numbers, rather than as operators. Likewise the vector models, this suggests that the path integral is dominated by a saddle point.

We already saw in Subsection 2.5 that the factorization in the vector models does not mean that the fundamental field itself, for instance $\vec{n}$ in the sigma-model, becomes "classical". It is the case, instead, for a singlet composite field.

We are now going to apply a similar idea to the Yang–Mills theory whose partition function reads

$$Z = \int DA_\mu^a \, \mathrm{e}^{-S}\,. \tag{111}$$

The action, $\sim N_c^2$, is large as $N_c \to \infty$, but the "entropy" is also $\sim N_c^2$ due to the $N_c^2 - 1$ integrations over A_μ^a:

$$DA_\mu^a \sim \mathrm{e}^{N_c^2}\,. \tag{112}$$

Consequently, the saddle-point equation of the large-N_c Yang–Mills theory is *not* the classical one.

The idea is to rewrite the path integral over A_μ for the Yang–Mills theory as that over a colorless composite field $\Phi\,[A]$, likewise it was done in Subsection 2.4 for the sigma-model. The expected new path-integral representation of the partition function (111) would be something like

$$Z \propto \int D\Phi \, \frac{1}{\frac{\partial \Phi[A]}{\partial A_\mu^a}} \, \mathrm{e}^{-N_c^2 S[\Phi]}\,. \tag{113}$$

The Jacobian

$$\frac{\partial \Phi\,[A]}{\partial A_\mu^a} \equiv \mathrm{e}^{\,-N_c^2 J[\Phi]} \tag{114}$$

in Eq. (113) is related to the old entropy factor, so that $J\,[\Phi] \sim 1$ in the large-N_c limit.

The original partition function (111) can be then rewritten as

$$Z \propto \int D\Phi\, \mathrm{e}^{\,N_c^2 J[\Phi]-N_c^2 S[\Phi]}, \tag{115}$$

where $S\,[\Phi]$ represents the Yang–Mills action in the new variables. The new "entropy" factor $D\Phi$ is $\mathcal{O}(1)$ because the variable $\Phi\,[A]$ is a color singlet. The large parameter N_c enters Eq. (115) only in the exponent. Therefore, the saddle-point equation can be immediately written:

$$\frac{\delta S}{\delta \Phi} = \frac{\delta J}{\delta \Phi}\,. \tag{116}$$

Remembering that Φ is a functional of A_μ: $\Phi \equiv \Phi\,[A]$, we rewrite the saddle-point equation (116) as

$$\frac{\delta S}{\delta A_\nu^a} = \left(\nabla_\mu F_{\mu\nu}\right)^a = \frac{\delta J}{\delta A_\nu^a}\,. \tag{117}$$

It differs from the classical Yang–Mills equation by the term on the RHS coming from the Jacobian (114).

Given $J\,[\Phi]$ which depends on the precise from of the variable $\Phi\,[A]$, Eq. (117) has a solution

$$A_\mu(x) = A_\mu^{\mathrm{cl}}(x)\,. \tag{118}$$

Let us first assume that there exists only one solution to Eq. (117). Then the path integral is saturated by a single configuration (118), so that the vacuum expectation values of gauge-invariant operators are given by their values at this configuration:

$$\langle\, O\,\rangle = O\left(A_\mu^{\mathrm{cl}}(x)\right). \tag{119}$$

The factorization property (110) will obviously be satisfied.

An existence of such a classical field configuration in multicolor QCD was conjectured by Witten [28]. It was discussed in the lectures by Coleman [29] who called it the *master field.* Equation (117) which determines the master field is often referred to as the master-field equation.

A subtle point with the master field is that a solution to Eq. (117) is determined only up to a gauge transformation. To preserve gauge invariance, it is more reasonable to speak about the whole gauge orbit as a solution of

Eq. (117). However, this will not change Eq. (119) since the operator O is gauge invariant.

The conjecture about an existence of the master field has surprisingly rich consequences. Since vacuum expectation values are Poincaré invariant, the RHS of Eq. (119) does. This implies that $A_\mu^{\rm cl}(x)$ must itself be Poincaré invariant up to a gauge transformation: a change of $A_\mu^{\rm cl}(x)$ under translations or rotations can be compensated by a gauge transformation. Moreover, there must exist a gauge in which $A_\mu^{\rm cl}(x)$ is space-time independent: $A_\mu^{\rm cl}(x) = A_\mu^{\rm cl}(0)$. In this gauge, rotations must be equivalent to a global gauge transformation, so that $A_\mu^{\rm cl}(0)$ transforms as a Lorentz vector.

In fact, the idea about such a master field in multicolor QCD may be incorrect as was pointed out by Haan [30]. The conjecture about an existence of only one solution to the master-field equation (117) seems to too strong. If several solutions exists, one needs an additional averaging over these solutions. This is a very delicate matter, since this additional averaging must still preserve the factorization property. One might better think about this situation as if $A_\mu^{\rm cl}(0)$ would be an operator in some Hilbert space rather than a c-valued function. Such an operator-valued master field is sometimes called the master field in the *weak* sense, while the above conjecture about a single classical configuration of the gauge field, which saturates the path integral, is called the master field in the *strong* sense.

The concept of the master field is rather vague until a precise form of the composite field $\Phi[A]$, and consequently the Jacobian $\Phi[A]$ that enters Eq. (117), is not defined. However, what is important is that the master field (in the weak sense) is space-time independent. This looks like a simplification of the problem of solving large-N_c QCD. A Hilbert space, in which the operator $A_\mu^{\rm cl}(0)$ acts, should be specified by $\Phi[A]$. We shall consider in the next Subsection a realization of these ideas for the case of $\Phi[A]$ given by the trace of the non-Abelian phase factor for closed contours.

Remark on non-commutative probability theory

An adequate mathematical language for describing the master field in multicolor QCD (and, generically, in matrix models at large N_c) was found by I. Singer in 1994. It is based on the concept of free random variables of non-commutative probability theory, introduced by Voiculescu [31]. How to describe the master field in this language and some other applications of non-commutative free random variables to the problems of planar quantum field theory are discussed in Refs. [32, 33].

3.6. $1/N_C$ AS SEMICLASSICAL EXPANSION

A natural candidate for the composite operator $\Phi[A]$ from the previous Subsection is given by the trace of the non-Abelian phase factor for closed contours — the Wilson loop. It is labeled by the loop C in the same sense as the field $A_\mu(x)$ is labeled by the point x, so we shall use the notation

$$\Phi(C) \equiv \Phi[A] = \frac{1}{N_c} \operatorname{tr} \mathbf{P}\, e^{i \oint_C dx^\mu A_\mu(x)} . \tag{120}$$

Nobody up to now managed to reformulate QCD at finite N_c in terms of $\Phi(C)$ in the language of path integral. This is due to the fact that self-intersecting loops are not independent (they are related by the so-called Mandelstam relations [34]), and the Jacobian is huge. The reformulation was done [35] in the language of Schwinger–Dyson or loop equations which will be described in the next Section.

Schwinger–Dyson equations are a convenient way of performing the semiclassical expansion, which is an alternative to the path integral. Let us illustrate an idea how to do this by an example of the φ^3 theory. The RHS of the Schwinger–Dyson equations is proportional to the Planck's constant $\hbar$. In the semiclassical limit $\hbar \to 0$, we get

$$\left(-\partial_1^2 + m^2\right) \langle \varphi(x_1) \ldots \varphi(x_n) \rangle + \frac{\lambda}{2} \left\langle \varphi^2(x_1) \ldots \varphi(x_n) \right\rangle = 0 , \tag{121}$$

whose solution is of the factorized form

$$\langle \varphi(x_1) \ldots \varphi(x_n) \rangle = \langle \varphi(x_1) \rangle \ldots \langle \varphi(x_n) \rangle + \mathcal{O}(\hbar) \tag{122}$$

provided that

$$\langle \varphi(x) \rangle \equiv \varphi_{\rm cl}(x) \tag{123}$$

obeys

$$\left(-\partial^2 + m^2\right) \varphi_{\rm cl}(x) + \frac{\lambda}{2} \varphi_{\rm cl}^2(x) = 0 . \tag{124}$$

Equation (124) is nothing but the classical equation of motion for the φ^3 theory, which specifies extrema of the action entering the path integral. Thus, we have reproduced, using the Schwinger–Dyson equations, the well-known fact that the path integral is dominated by a classical solution as $\hbar \to 0$. It is also clear how to perform the semiclassical expansion in $\hbar$ in the language of the Schwinger–Dyson equations: one should solve them by iterations.

The reformulation of multicolor QCD in terms of the loop functionals $\Phi(C)$ is, in a sense, a realization of the idea of the master field in the weak sense, when the master field acts as an operator in the space of loops.

Remark on the large-N_c limit as statistical averaging

There is yet another, pure statistical, explanation why the large-N_c limit is a "semiclassical" limit for the collective variables $\Phi(C)$. The matrix $U^{ij}[C_{xx}]$, that describes the parallel transport along a closed contour C_{xx}, can be reduced by the gauge transformation to

$$U[C_{xx}] = \Omega[C_{xx}] \operatorname{diag}\left(e^{ig\alpha_1(C)}, \ldots, e^{ig\alpha_{N_c}(C)}\right) \Omega^\dagger[C_{xx}]. \tag{125}$$

Then $\Phi(C)$ reads

$$\Phi(C) = \frac{1}{N_c} \sum_{j=1}^{N_c} e^{ig\alpha_j(C)}. \tag{126}$$

The phases $\alpha_j(C)$ are gauge invariant and normalized so that $\alpha_j(C) \sim 1$ as $N_c \to \infty$. For simplicity we omit below all the indices (including space ones) except color.

The commutator of Φ's can be estimated using the representation (126). Since $[\alpha_i, \alpha_j] \propto \delta_{ij}$, one gets

$$[\Phi(C), \Phi(C')] \sim g^2 \frac{1}{N_c} \sim \frac{1}{N_c^2} \tag{127}$$

in the limit (71), *i.e.* the commutator can be neglected as $N_c \to \infty$, and the field $\Phi(C)$ becomes classical.

Note that the commutator (127) is of order $1/N_c^2$. One factor $1/N_c$ is because of g in the definition (126) of $\Phi(C)$, while the other has a deep reason. Let us image the summation over j in Eq. (126) as some statistical averaging. It is well-known in statistics that such averages fluctuate weakly as $N_c \to \infty$, so that the dispersion is of order $1/N_c$. It is the factor which emerges in the commutator (127).

We see that the factorization is valid only for the gauge-invariant quantities which involve the averaging over the color indices, like that in Eq. (126). There is no reason to expect factorization for gauge invariants which do not involve this averaging, for instance for the phases $\alpha_j(C)$. Moreover, their commutator is ~ 1, so that $\alpha_j(C)$'s strongly fluctuate even at $N_c = \infty$. An explicit example of such strongly fluctuating gauge-invariant quantities was first constructed in Ref. [30].

The résumé to this Remark is that the factorization is due to the additional statistical averaging in the large-N_c limit. There is no reason to assume an existence of the master field in the strong sense in order to explain the factorization.

4. QCD in Loop Space

QCD can be entirely reformulated in terms of the colorless composite field $\Phi(C)$ — the trace of the Wilson loop for closed contours. This fact involves two main steps:

i) All the observables are expressed via $\Phi(C)$.
ii) Dynamics is entirely reformulated in terms of $\Phi(C)$.

This approach is especially useful in the large-N_c limit where everything is expressed via the vacuum expectation value of $\Phi(C)$ — the Wilson loop average. Observables are given by summing the Wilson loop average over paths with the same weight as in free theory. The Wilson loop average obeys itself a close functional equation — the loop equation.

We begin this Section with presenting the formulas which relate observables to the Wilson loops. Then we translate quantum equation of motion of Yang–Mills theory into loop space. We derive the closed equation for the Wilson loop average as $N_c \to \infty$ and discuss its various properties, including a non-perturbative regularization. Finally, we briefly comment on what is known about solutions of the loop equation.

4.1. OBSERVABLES IN TERMS OF WILSON LOOPS

All observables in QCD can be expressed via the Wilson loops $\Phi(C)$ defined by Eq. (120). This property was first advocated by Wilson [36] on a lattice. Calculation of QCD observables can be divided in two steps:

1) Calculation of the Wilson loop averages for arbitrary contours.
2) Summation of the Wilson loop averages over the contours with some weight depending on a given observable.

At finite N_c, observables are expressed via the n-loop averages

$$W_n(C_1, \ldots, C_n) = \langle \Phi(C_1) \cdots \Phi(C_n) \rangle \,, \tag{128}$$

which are analogous to the n-point Green functions for φ^3 theory. The appropriate formulas for the continuum theory can be found in Ref. [37].

Great simplifications occur in these formulas at $N_c = \infty$, when all observables are expressed only via the one-loop average

$$W(C) = \langle \Phi(C) \rangle \equiv \left\langle \frac{1}{N_c} \operatorname{tr} \mathbf{P}\, \mathrm{e}^{i \oint_C dx^\mu A_\mu} \right\rangle . \tag{129}$$

This is associated with the quenched approximation.

For example, the average of the product of two colorless quark vector currents (101) is given at large N_c by

$$\langle \bar\psi \gamma_\mu \psi(x_1)\, \bar\psi \gamma_\nu \psi(x_2) \rangle = \sum_{C \ni x_1, x_2} J_{\mu\nu}(C) \langle \Phi(C) \rangle \,, \tag{130}$$

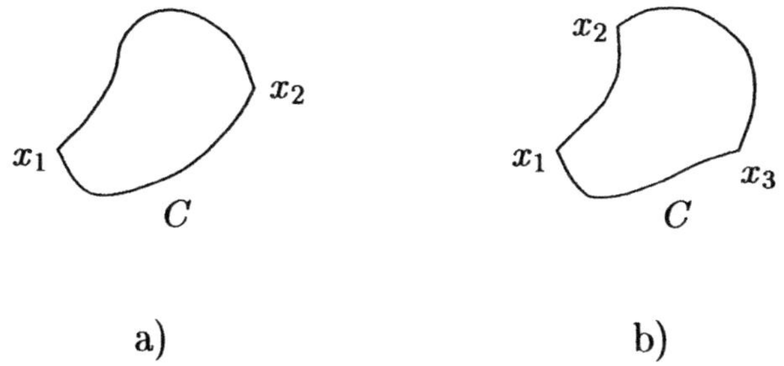

Figure 14. Contours in the sum over paths representing observables: a) in Eq. (130) and b) in Eq. (131). The contour a) passes two nailed points x_1 and x_2. The contour b) passes three nailed points x_1, x_2, and x_3.

where the sum runs over contours C passing through the points x_1 and x_2 as is depicted in Figure 14a. An analogous formula for the (connected) correlators of three quark scalar currents reads

$$\langle \bar{\psi}\psi\,(x_1)\,\bar{\psi}\psi\,(x_2)\,\bar{\psi}\psi\,(x_3)\rangle_{\text{conn}} = \sum_{C\ni x_1,x_2,x_3} J\,(C)\,\langle\,\Phi\,(C)\,\rangle\,, \tag{131}$$

where the sum runs over contours C passing through the three points x_1, x_2, and x_3 as is depicted in Figure 14b. A general (connected) correlator of n quark currents is given by a similar formula with C passing through n points $x_1, \ldots, x_n$ (some of them may coincide).

The weights $J_{\mu\nu}\,(C)$ in Eq. (130) and $J\,(C)$ in Eq. (131) are completely determined by free theory. If quarks were scalars rather than spinors, then we would get

$$J\,(C) = \mathrm{e}^{-\frac{1}{2}m^2\tau - \frac{1}{2}\int_0^\tau dt\,\dot{z}_\mu^2(t)} = \mathrm{e}^{-mL(C)} \qquad \boxed{\text{scalar quarks}}\,, \tag{132}$$

where $L(C)$ is the length of the (closed) contour C. For spinor quarks, an additional disentangling of the gamma-matrices is needed (ee, *e.g.*, Ref. [38]).

Remark on renormalization of Wilson loops

Perturbation theory for $W(C)$ can be obtained by expanding the path-ordered exponential in the definition (129) in g (see Eq. (84)) and averaging over the gluon field A_μ. Because of ultraviolet divergencies, we need a (gauge invariant) regularization. After such a regularization has been introduced, the Wilson loop average for a smooth contour C of the type in Figure 15a reads

$$W\,(C) = \mathrm{e}^{-g^2\frac{(N_c^2-1)}{8\pi N_c}\frac{L(C)}{a}}\,W_{\text{ren}}\,(C)\,, \tag{133}$$

where a is the cutoff, $L(C)$ is the length of C, and $W_{\text{ren}}\,(C)$ is finite when expressed via the renormalized charge g_{R}. The exponential factor is due to

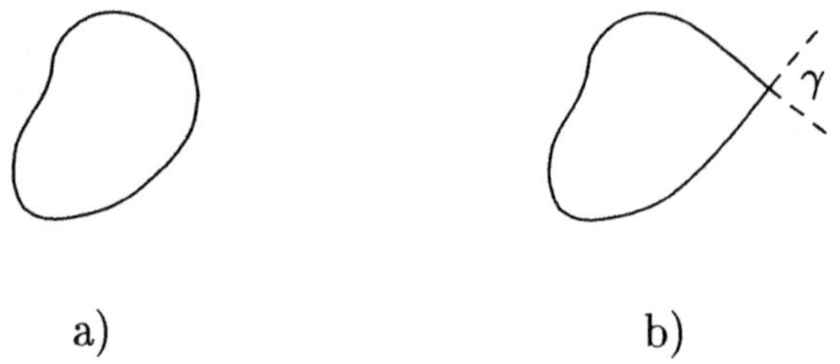

Figure 15. Examples of a) smooth contour and b) contour with a cusp. The tangent vector to the contour jumps through angle γ at the cusp.

the renormalization of the mass of a heavy test quark. This factor does not emerge in the dimensional regularization where $d = 4-\varepsilon$. The multiplicative renormalization of the smooth Wilson loop was shown in Refs. [39, 40, 41].

If the contour C has a cusp (or cusps) but no self-intersections as is illustrated by Figure 15b, then $W(C)$ is still multiplicatively renormalizable [42]:

$$W(C) = Z(\gamma)\, W_{\text{ren}}(C)\,, \tag{134}$$

while the (divergent) factor $Z(\gamma)$ depends on the cusp angle (or angles) γ (or γ's) and $W_{\text{ren}}(C)$ is finite when expressed via the renormalized charge g_{R}.

4.2. SCHWINGER–DYSON EQUATIONS FOR WILSON LOOP

Dynamics of (quantum) Yang–Mills theory is described by the quantum equation of motion

$$\nabla_{\mu}^{ab} F_{\mu\nu}^{b}(x) \overset{\text{w.s.}}{=} \hbar \frac{\delta}{\delta A_{\nu}^{a}(x)} \tag{135}$$

which is understood in the weak sense, *i.e.* for the averages

$$\left\langle \nabla_{\mu}^{ab} F_{\mu\nu}^{b}(x)\, Q[A] \right\rangle = \hbar \left\langle \frac{\delta}{\delta A_{\nu}^{a}(x)} Q[A] \right\rangle . \tag{136}$$

The standard set of Schwinger–Dyson equations of Yang–Mills theory emerges when the functional $Q[A]$ is chosen in the form of the product of A's as in Eq. (82).

Strictly speaking, the last statement is incorrect, since we have not added, in Eqs. (135) and (136), contributions coming from the variation of gauge-fixing and ghost terms in the Yang–Mills action. However, these two contributions are mutually cancelled for gauge-invariant functionals $Q[A]$. We shall deal below only with such gauge-invariant functionals (the Wilson loops). This is why we have not considered the contribution of the gauge-fixing and ghost terms.

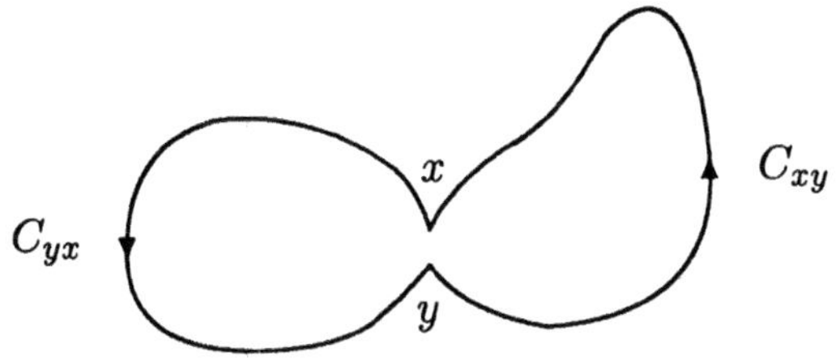

Figure 16. Contours C_{yx} and C_{xy} which enter the RHS's of Eqs. (137) and (141).

It is also convenient to use the matrix notation (57), when Eq. (135) for the Wilson loop takes on the form

$$\left\langle \frac{1}{N_c} \operatorname{tr} \mathbf{P}\, \nabla_\mu F_{\mu\nu}(x)\, \mathrm{e}^{\,i \oint_C d\xi^\mu A_\mu} \right\rangle = \left\langle \frac{g^2}{2N_c} \operatorname{tr} \frac{\delta}{\delta A_\nu(x)} \mathbf{P}\, \mathrm{e}^{\,i \oint_C d\xi^\mu A_\mu} \right\rangle, \tag{137}$$

where we have restored the units with $\hbar = 1$.

The variational derivative on the RHS can be calculated by virtue of the formula

$$\frac{\delta A_\mu^{ij}(y)}{\delta A_\nu^{kl}(x)} = \delta_{\mu\nu}\, \delta^{(d)}(x-y) \left(\delta^{il}\delta^{kj} - \frac{1}{N_c} \delta^{ij}\delta^{kl} \right) \tag{138}$$

which is a consequence of

$$\frac{\delta A_\mu^{a}(y)}{\delta A_\nu^{b}(x)} = \delta_{\mu\nu}\, \delta^{(d)}(x-y)\, \delta^{ab}\,. \tag{139}$$

The second term in the parentheses in Eq. (138) — same as in Eq. (64) — is because A_μ is a matrix from the adjoint representation of $\mathrm{SU}(N_c)$.

By using Eq. (138), we get for the variational derivative on RHS of Eq. (137):

$$\begin{aligned} \operatorname{tr} \frac{\delta}{\delta A_\nu(x)} \mathbf{P}\, \mathrm{e}^{\,i \oint_C d\xi^\mu A_\mu} &= i \oint_C dy_\nu\, \delta^{(d)}(x-y) \times \\ \Bigg[\frac{1}{N_c} \operatorname{tr} \mathbf{P}\, \mathrm{e}^{\,i \int_{C_{yx}} d\xi^\mu A_\mu} \frac{1}{N_c} \operatorname{tr} \mathbf{P}\, \mathrm{e}^{\,i \int_{C_{xy}} d\xi^\mu A_\mu} &- \frac{1}{N_c^3} \operatorname{tr} \mathbf{P}\, \mathrm{e}^{\,i \int_{C} d\xi^\mu A_\mu} \Bigg]. \end{aligned} \tag{140}$$

The contours C_{yx} and C_{xy}, which are depicted in Figure 16, are the parts of the loop C: from x to y and from y to x, respectively. They are always closed due to the presence of the delta-function. It implies that x and y should be the same points of *space* but not necessarily of the *contour* (*i.e.* they may be associated with different values of the parameter σ).

We finally rewrite Eq. (137) as

$$\left\langle \frac{1}{N_c} \operatorname{tr} \mathbf{P} \, \nabla_\mu F_{\mu\nu}(x) \, e^{i \oint_C d\xi^\mu A_\mu} \right\rangle$$
$$= i\lambda \oint_C dy_\nu \, \delta^{(d)}(x-y) \left[\langle \Phi(C_{yx}) \, \Phi(C_{xy}) \rangle - \frac{1}{N^2} \langle \Phi(C) \rangle \right] \qquad (141)$$

where we have introduced

$$\lambda = \frac{g^2 N_c}{2}. \qquad (142)$$

Notice that the RHS of Eq. (141) is completely represented via the (closed) Wilson loops.

4.3. PATH AND AREA DERIVATIVES

As we already mentioned, the RHS of Eq. (141) is completely represented via the (closed) Wilson loops. It is crucial for the loop-space formulation of QCD that the LHS of Eq. (141) can also be represented in loop space as some operator applied to the Wilson loop. To do this we need to develop a differential calculus in loop space.

Loop space consists of arbitrary continuous closed loops, C. They can be described in a parametric form by the functions $x_\mu(\sigma) \in L_2$,[5] where $\sigma_0 \leq \sigma \leq \sigma_f$ and $\mu = 1, \ldots, d$, which take on values in a d-dimensional Euclidean space. The functions $x_\mu(\sigma)$ can be discontinuous, generally speaking, for an arbitrary choice of the parameter σ. The continuity of the loop C implies a continuous dependence on parameters of the type of proper length.

The functions $x_\mu(\sigma) \in L_2$ which are associated with the elements of loop space obey the following restrictions:

i) The points $\sigma = \sigma_0$ and $\sigma = \sigma_f$ are identified: $x_\mu(\sigma_0) = x_\mu(\sigma_f)$ — the loops are closed.

ii) The functions $x_\mu(\sigma)$ and $\Lambda_{\mu\nu} x_\nu(\sigma) + \alpha_\mu$, with $\Lambda_{\mu\nu}$ and α_μ independent of σ, represent the same element of the loop space — rotational and translational invariance.

iii) The functions $x_\mu(\sigma)$ and $x_\mu(\sigma')$ with $\sigma' = f(\sigma)$, $f'(\sigma) \geq 0$ describe the same loop — reparametrization invariance.

An example of functionals which are defined on the elements of loop space is the Wilson loop average (129) or, more generally, the n-loop average (128).

The differential calculus in loop space is built out of the path and area derivatives.

[5]Let us remind that L_2 stands for the Hilbert space of functions $x_\mu(\sigma)$ whose square is integrable over the Lebesgue measure: $\int_{\sigma_0}^{\sigma_f} d\sigma x_\mu^2(\sigma) < \infty$.

The *area derivative* of a functional $\mathcal{F}(C)$ is defined by the difference

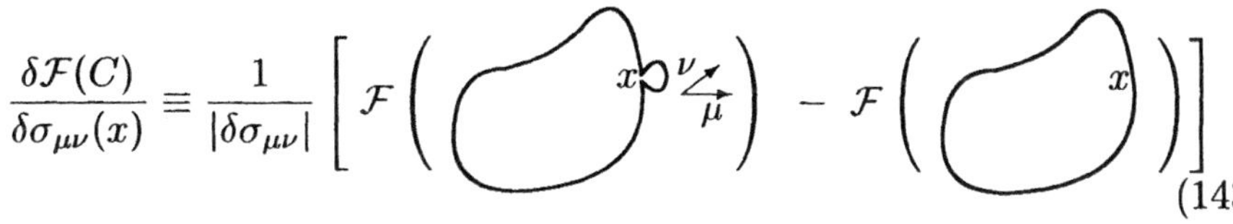

(143)

where an infinitesimal loop $\delta C_{\mu\nu}(x)$ is attached to a given loop at the point x in the $\mu\nu$-plane and $|\delta\sigma_{\mu\nu}|$ stands for the area enclosed by the $\delta C_{\mu\nu}(x)$. For a rectangular loop $\delta C_{\mu\nu}(x)$, one gets $\delta\sigma_{\mu\nu} = dx_\mu \wedge dx_\nu$, where the symbol $\wedge$ implies antisymmetrization.

Analogously, the *path derivative* is defined by

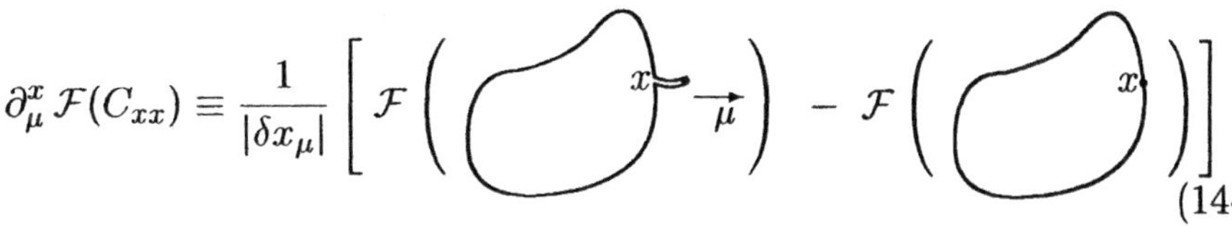

(144)

where δx_μ is an infinitesimal path along which the point x is shifted from the loop and $|\delta x_\mu|$ stands for the length of the δx_μ.

These two differential operations are well-defined for so-called functionals of the Stokes type which satisfy the backtracking condition — they do not change when an appendix passing back and forth is added to the loop at some point x:

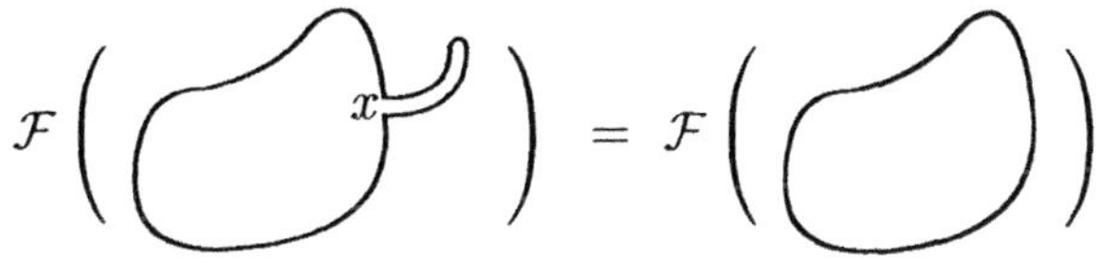

(145)

This condition is equivalent to the Bianchi identity of Yang–Mills theory and is obviously satisfied by the Wilson loop (129) due to the properties of the non-Abelian phase factor. Such functionals are known in mathematics as Chen integrals.

A simple example of the Stokes functional is the area of the minimal surface, $A_{\min}(C)$. It obviously satisfies Eq. (145). Otherwise, the length $L(C)$ of the loop C is not a Stokes functional, since the lengths of contours on the LHS and RHS of Eq. (145) are different.

For the Stokes functionals, the variation on the RHS of Eq. (143) is proportional to the area enclosed by the infinitesimally small loop $\delta C_{\mu\nu}(x)$ and does not depend on its shape. Analogously, the variation on the RHS of Eq. (144) is proportional to the length of the infinitesimal path δx_μ and does not depend on its shape.

If x is a regular point (like any point of the contour for the functional (129)), the RHS of Eq. (144) vanishes due to the backtracking condition (145). In order for the result to be nonvanishing, the point x should be a *marked* (or irregular) point. A simple example of the functional with a marked point x is

$$\Phi^a[C_{xx}] \equiv \frac{1}{N_c} \operatorname{tr} \left(t^a \mathbf{P} \, \mathrm{e}^{i \int_{C_{xx}} d\xi^\mu A_\mu(\xi)} \right) \tag{146}$$

with the SU(N_c) generator t^a inserted in the path-ordered product at the point x.

The area derivative of the Wilson loop is given by the Mandelstam formula

$$\frac{\delta}{\delta\sigma_{\mu\nu}(x)} \frac{1}{N_c} \operatorname{tr} \mathbf{P} \, \mathrm{e}^{i \oint_C d\xi^\mu A_\mu} = \frac{i}{N_c} \operatorname{tr} \mathbf{P} \, F_{\mu\nu}(x) \, \mathrm{e}^{i \oint_C d\xi^\mu A_\mu} . \tag{147}$$

In order to prove it, it is convenient to choose $\delta C_{\mu\nu}(x)$ to be a rectangle in the $\mu\nu$-plane and straightforwardly use the definition (143). The sense of Eq. (147) is very simple: $F_{\mu\nu}$ is a curvature associated with the connection A_μ.

The functional on the RHS of Eq. (147) has a marked point x, and is of the type in Eq. (146). When the path derivative acts on such a functional according to the definition (144), the result reads

$$\partial_\mu^x \frac{1}{N_c} \operatorname{tr} \mathbf{P} \, B(x) \, \mathrm{e}^{i \oint_C d\xi^\mu A_\mu} = \frac{1}{N_c} \operatorname{tr} \mathbf{P} \, \nabla_\mu B(x) \, \mathrm{e}^{i \oint_C d\xi^\mu A_\mu} , \tag{148}$$

where

$$\nabla_\mu B = \partial_\mu B - i \, [A_\mu, B] \tag{149}$$

is the covariant derivative in the adjoint representation.

Combining Eqs. (147) and (148), we finally represent the expression on the LHS of Eq. (137) (or Eq. (141)) as

$$\frac{1}{N_c} \operatorname{tr} \mathbf{P} \, \nabla_\mu F_{\mu\nu}(x) \, \mathrm{e}^{i \oint_C d\xi^\mu A_\mu} = \partial_\mu^x \frac{\delta}{\delta\sigma_{\mu\nu}(x)} \frac{i}{N_c} \operatorname{tr} \mathbf{P} \, \mathrm{e}^{i \oint_C d\xi^\mu A_\mu} , \tag{150}$$

i.e. via the action of the path and area derivatives on the Wilson loop. It is therefore rewritten in loop space.

A résumé of the results of this subsection is presented in Table 2 as a vocabulary for translation of Yang–Mills theory from the language of ordinary space in the language of loop space.

Ordinary space		Loop space	
$\Phi\,[A]$	phase factor	$\Phi\,(C)$	loop functional
$F_{\mu\nu}\,(x)$	field strength	$\frac{\delta}{\delta\sigma_{\mu\nu}(x)}$	area derivative
∇^x_μ	covariant derivative	∂^x_μ	path derivative
$\nabla\wedge F=0$	Bianchi identity		Stokes functionals
$\nabla_\mu F_{\mu\nu}$ $=\delta/\delta A_\nu$	Schwinger-Dyson equations		Loop equations

TABLE 2. Vocabulary for translation of Yang–Mills theory from ordinary space in loop space.

Remark on Bianchi identity for Stokes functionals

The backtracking relation (145) can be equivalently represented as

$$\epsilon_{\mu\nu\lambda\rho}\,\partial^x_\mu\,\frac{\delta}{\delta\sigma_{\nu\lambda}\,(x)}\Phi\,(C)\;=\;0\,, \tag{151}$$

by choosing the appendix in Eq. (145) to be an infinitesimal straight line in the ρ-direction and geometrically applying the Stokes theorem. Using Eqs. (147) and (148), Eq. (151) can in turn be rewritten as

$$\epsilon_{\mu\nu\lambda\rho}\frac{1}{N_c}\,\mathrm{tr}\,\mathbf{P}\,\nabla_\mu F_{\nu\lambda}\,(x)\;\mathrm{e}^{\,i\oint_C d\xi^\mu\,A_\mu}\;=\;0\,. \tag{152}$$

Therefore, Eq. (151) represents the Bianchi identity in loop space.

Remark on relation to variational derivative

The standard variational derivative, $\delta/\delta x_\mu(\sigma)$, can be expressed via the path and area derivatives by the formula

$$\frac{\delta}{\delta x_\mu(\sigma)}=\dot{x}_\nu(\sigma)\frac{\delta}{\delta\sigma_{\mu\nu}(x(\sigma))}+\sum_{i=1}^{m}\partial^{x_i}_\mu\delta(\sigma-\sigma_i)\,, \tag{153}$$

where the sum on the RHS is present for the case of a functional having m marked (irregular) points $x_i \equiv x(\sigma_i)$. A simplest example of the functional with m marked points is just a function of m variables $x_1, \ldots, x_m$.

By using Eq. (153), the path derivative can be calculated as the limiting procedure

$$\partial_\mu^{x(\sigma)} = \int\limits_{\sigma-0}^{\sigma+0} d\sigma' \frac{\delta}{\delta x_\mu(\sigma')} \,. \tag{154}$$

The result is obviously nonvanishing only when ∂_μ^x is applied to a functional with $x(\sigma)$ being a marked point.

It is nontrivial that the area derivative can also be expressed via the variational derivative [40]:

$$\frac{\delta}{\delta\sigma_{\mu\nu}(x(\sigma))} = \int\limits_{\sigma-0}^{\sigma+0} d\sigma'(\sigma'-\sigma)\frac{\delta}{\delta x_\mu(\sigma')}\frac{\delta}{\delta x_\nu(\sigma)} \,. \tag{155}$$

The point is that the six-component quantity, $\delta/\delta\sigma_{\mu\nu}(x(\sigma))$, is expressed via the four-component one, $\delta/\delta x_\mu(\sigma)$, which is possible because the components of $\delta/\delta\sigma_{\mu\nu}(x(\sigma))$ are dependent due to the loop-space Bianchi identity (151).

4.4. LOOP EQUATIONS

By virtue of Eq. (150), Eq. (141) can be represented completely in loop space:

$$\begin{aligned} &\partial_\mu^x \frac{\delta}{\delta\sigma_{\mu\nu}(x)} \left\langle \Phi(C) \right\rangle \\ &= \lambda \oint\limits_C dy_\nu \, \delta^{(d)}(x-y) \left\langle \left[\Phi(C_{yx}) \, \Phi(C_{xy}) - \frac{1}{N_c^2} \Phi(C) \right] \right\rangle , \end{aligned} \tag{156}$$

or, using the definitions (128) and (129) of the loop averages, as

$$\partial_\mu^x \frac{\delta}{\delta\sigma_{\mu\nu}(x)} W(C) = \lambda \oint\limits_C dy_\nu \, \delta^{(d)}(x-y) \left[W_2(C_{yx}, C_{xy}) - \frac{1}{N_c^2} W(C) \right] . \tag{157}$$

This equation is not closed. Having started from $W(C)$, we obtain another quantity, $W_2(C_1, C_2)$, so that Eq. (157) connects the one-loop average with a two-loop one. This is similar to the case of the (quantum) φ^3-theory, whose Schwinger–Dyson equations connect the n-point Green functions with different n. We shall derive this complete set of equations for the n-loop averages in this Subsection later on.

However, the two-loop average factorizes in the large-N_c limit:

$$W_2(C_1, C_2) = W(C_1) W(C_2) + \mathcal{O}\left(\frac{1}{N_c^2}\right), \tag{158}$$

as was discussed in Subsection 3.4. Keeping the constant λ (defined by Eq. (142)) fixed in the large-N_c limit as is prescribed by Eq. (71), we get [35]

$$\partial_\mu^x \frac{\delta}{\delta \sigma_{\mu\nu}(x)} W(C) = \lambda \oint_C dy_\nu \, \delta^{(d)}(x-y) \, W(C_{yx}) \, W(C_{xy}) \tag{159}$$

as $N_c \to \infty$.

Equation (159) is a closed equation for the Wilson loop average in the large-N_c limit. It is referred to as the *loop equation.*

To find $W(C)$, Eq. (159) should be solved in the class of Stokes functionals with the initial condition

$$W(0) = 1 \tag{160}$$

for loops which are shrunk to points.

The factorization (158) can itself be derived from the chain of loop equations. Proceeding as before, we get

$$\begin{aligned}
&\frac{1}{\lambda} \partial_\mu^x \frac{\delta}{\delta \sigma_{\mu\nu}(x)} W_n(C_1, \ldots, C_n) \\
&= \oint_{C_1} dy_\nu \, \delta^{(d)}(x-y) \left[W_{n+1}(C_{xy}, C_{yx}, \ldots, C_n) - \frac{1}{N_c^2} W_n(C_1, \ldots, C_n) \right] \\
&\quad + \sum_{j \geq 2} \frac{1}{N_c^2} \oint_{C_j} dy_\nu \, \delta^{(d)}(x-y) \left[W_{n-1}\left(C_1 C_j, \ldots, \underline{C_j}, \ldots, C_n\right) \right. \\
&\qquad \left. - W_n(C_1, \ldots, C_n) \right].
\end{aligned} \tag{161}$$

Here x belongs to C_1; $C_1 C_j$ stands for the joining of C_1 and C_j; $\underline{C_j}$ means that C_j is omitted.

Equation (161) looks like the Schwinger–Dyson equation for the φ^3-theory. Moreover, the number of colors N_c enters Eq. (161) simply as a scalar factor N_c^{-2}, likewise Plank's constant $\hbar$ enters in the φ^3-theory. It is the major advantage of the use of loop space. What is said in Subsection 3.6 about the "semiclassical" nature of the $1/N_c$-expansion of QCD is explicitly realized in Eq. (161). Its expansion in $1/N_c$ is straightforward.

At $N_c = \infty$, Eq. (161) is simplified to

$$\partial_\mu^x \frac{\delta}{\delta \sigma_{\mu\nu}(x)} W_n(C_1, \ldots, C_n) = \lambda \oint_{C_1} dy_\nu \, \delta^{(d)}(x-y) W_{n+1}(C_{yx}, C_{xy}, \ldots, C_n). \tag{162}$$

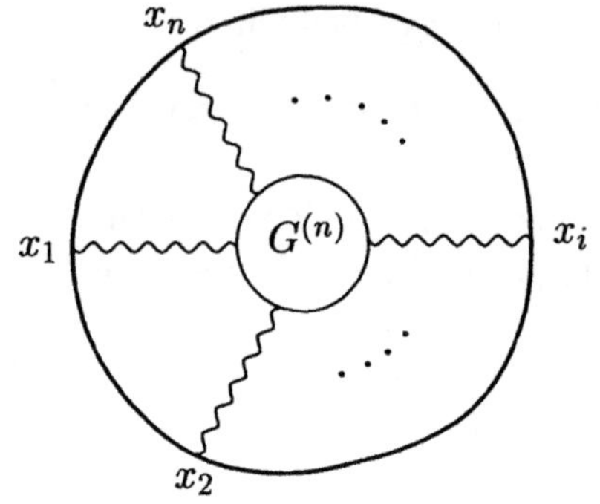

Figure 17. Graphic representation of the terms on the RHS of Eq. (164).

This equation possesses a factorized solution

$$W_n(C_1,\ldots,C_n) = W(C_1)\cdots W(C_n) + \mathcal{O}\left(\frac{1}{N_c^2}\right) \tag{163}$$

provided $W(C)$ obeys Eq. (159) which plays the role of a "classical" equation in the large-N_c limit. Thus, we have given a non-perturbative proof of the large-N_c factorization of the Wilson loops.

4.5. RELATION TO PLANAR DIAGRAMS

The perturbation-theory expansion of the Wilson loop average can be calculated from Eq. (84) which we represent in the form

$$W(C) = 1 + \sum_{n=2}^{\infty} i^n \oint\limits_C dx_1^{\mu_1} \oint\limits_C dx_2^{\mu_2} \ldots \oint\limits_C dx_n^{\mu_n}$$
$$\times\theta_c(1,2,\ldots,n)\, G^{(n)}_{\mu_1\mu_2\cdots\mu_n}(x_1,x_2,\ldots,x_n)\,, \tag{164}$$

where $\theta_c(1,2,\ldots,n)$ orders the points $x_1, \ldots, x_n$ along contour in the cyclic order and $G^{(n)}_{\mu_1\cdots\mu_n}$ is given by Eq. (82). This θ-function has the meaning of the propagator of a heavy test particle which lives in the contour C.

We assume, for definitiveness, the dimensional regularization throughout this Subsection to make all the integrals well-defined.

Each term on the RHS of Eq. (164) can be conveniently represented by the diagram in Figure 17, where the integration over the contour C is associated with each point x_i lying in the contour C.

These diagrams are analogous to those discussed in Subsection 3.2 with one external boundary — the Wilson loop in the given case. In the large-N_c

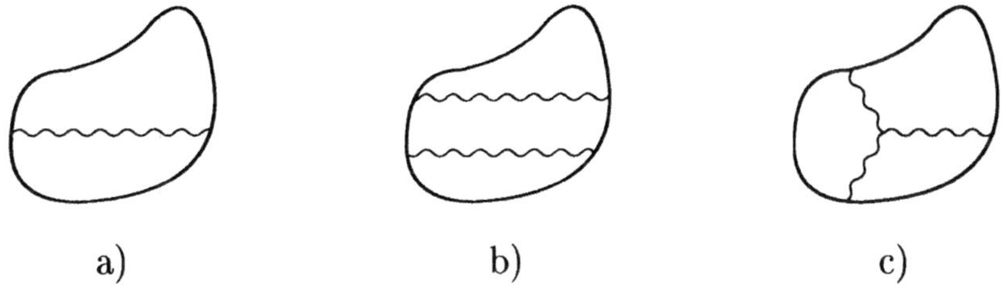

Figure 18. Planar diagrams for $W(C)$: a) of order λ with gluon propagator, and of order λ^2 b) with two noninteracting gluons and c) with the three-gluon vertex. Diagrams of order λ^2 with one-loop insertions to gluon propagator are not drawn.

limit, only planar diagrams survive. Some of them, which are of the lowest order in λ, are depicted in Figure 18.

The large-N_c loop equation (159) describes the sum of the planar diagrams. Its iterative solution in λ reproduces the set of planar diagrams for $W(C)$ provided the initial condition (160) and some boundary conditions for asymptotically large contours are imposed.

Equation (164) can be viewed as an ansatz for $W(C)$ with some unknown functions $G^{(n)}_{\mu_1\cdots\mu_n}(x_1,\ldots,x_n)$ to be determined by the substitution into the loop equation. To preserve symmetry properties of $W(C)$, the functions $G^{(n)}$ must be symmetric under a cyclic permutation of the points 1, …, n and depend only on $x_i - x_j$ (translational invariance). A main advantage of this ansatz is that it automatically corresponds to a Stokes functional, due to the properties of vector integrals, and the initial condition (160) is satisfied.

The action of the area and path derivatives on the ansatz (164) is easily calculable. For instance, the area derivative reads

$$\begin{aligned}\frac{\delta W(C)}{\delta\sigma_{\mu\nu}(z)} = {} & \sum_{n=1}^{\infty} i^n \oint_C dx_1^{\mu_1}\ldots\oint_C dx_n^{\mu_n}\,\theta_c(1,2,\ldots,n) \\ & \times\Big[\Big(\partial^z_\mu\delta_{\nu\alpha}-\partial^z_\nu\delta_{\mu\alpha}\Big)G^{(n+1)}_{\alpha\mu_1\cdots\mu_n}(z,x_1,\ldots,x_n) \\ & \quad+(\delta_{\mu\beta}\delta_{\nu\alpha}-\delta_{\mu\alpha}\delta_{\nu\beta})\,G^{(n+2)}_{\alpha\beta\mu_1\cdots\mu_n}(z,z,x_1,\ldots,x_n)\Big]\,. \end{aligned} \tag{165}$$

The analogy with the Mandelstam formula (147) is obvious.

More about solving the loop equation by the ansatz (164) can be found in Refs. [37, 43, 44].

4.6. LOOP-SPACE LAPLACIAN AND REGULARIZATION

The loop equation (159) is *not* yet entirely formulated in loop space. It is a d-vector equation whose both sides depend explicitly on the point x which does not belong to loop space. The fact that we have a d-vector equation for a scalar quantity means, in particular, that Eq. (159) is overspecified.

A practical difficulty in solving Eq. (159) is that the area and path derivatives, $\delta/\delta\sigma_{\mu\nu}(x)$ and ∂_μ^x, which enter the LHS are complicated, generally speaking, non-commutative operators. They are intimately related to the Yang–Mills perturbation theory where they correspond to the non-Abelian field strength $F_{\mu\nu}$ and the covariant derivative ∇_μ. However, it is not easy to apply these operators to a generic functional $W(C)$ which is defined on elements of loop space.

A much more convenient form of the loop equation can be obtained by integrating both sides of Eq. (159) over dx_ν along the same contour C, which yields

$$\oint\limits_C dx_\nu\, \partial_\mu^x \frac{\delta}{\delta\sigma_{\mu\nu}(x)} W(C) = \lambda \oint\limits_C dx_\mu \oint\limits_C dy_\mu\, \delta^{(d)}(x-y)\, W(C_{yx})W(C_{xy})\,. \tag{166}$$

Now both the operator on the LHS and the functional on the RHS are scalars without labeled points and are well-defined in loop space. The operator on the LHS of Eq. (166) can be interpreted as an infinitesimal variation of elements of loop space.

Equations (159) and (166) are completely equivalent. A proof of equivalence of scalar Eq. (166) and original d-vector Eq. (159) is based on the important property of Eq. (159) whose both sides are identically annihilated by the operator ∂_ν^x. It is a consequence of the identity

$$\nabla_\mu \nabla_\nu F_{\mu\nu} = -\frac{i}{2}\left[F_{\mu\nu}, F_{\mu\nu}\right] = 0 \tag{167}$$

in the ordinary space. Due to this property, the vanishing of the contour integral of some vector is equivalent to vanishing of the vector itself, so that Eq. (159) can in turn be deduced from Eq. (166).

Equation (166) is associated with the so-called second-order Schwinger–Dyson equation

$$\int d^d x\, \nabla_\mu F_{\mu\nu}^a(x) \frac{\delta}{\delta A_\nu^a(x)} \stackrel{\text{w.s.}}{=} \hbar \int d^d x\, d^d y\, \delta^{(d)}(x-y) \frac{\delta}{\delta A_\nu^a(y)} \frac{\delta}{\delta A_\nu^a(x)} \tag{168}$$

in the same sense as Eq. (159) is associated with Eq. (135). It is called "second order" since the RHS involves two variational derivatives with respect to A_ν.

The operator on the LHS of Eq. (166) is a well-defined object in loop space. When applied to regular functionals which do not have marked points, it can be represented, using Eqs. (154) and (155), in an equivalent form

$$\Delta \equiv \oint_C dx_\nu \, \partial_\mu^x \frac{\delta}{\delta \sigma_{\mu\nu}(x)} = \int_{\sigma_0}^{\sigma_f} d\sigma \int_{\sigma-0}^{\sigma+0} d\sigma' \frac{\delta}{\delta x_\mu(\sigma')} \frac{\delta}{\delta x_\mu(\sigma)} \,. \tag{169}$$

As was first pointed out by Gervais and Neveu [45], this operator is nothing but a functional extension of the Laplace operator, which is known in mathematics as the Levy operator.[6] Equation (166) can be represented in turn as an (inhomogeneous) functional Laplace equation

$$\Delta W(C) = \lambda \oint_C dx_\mu \oint_C dy_\mu \, \delta^{(d)}(x-y) \, W(C_{yx}) W(C_{xy}) \,. \tag{170}$$

We shall refer to this equation as the loop-space Laplace equation.

The form (170) of the loop equation is convenient for a non-perturbative ultraviolet regularization.

The idea is to start from the regularized version of Eq. (168), replacing the delta-function on the RHS by the kernel of the regularizing operator:

$$\delta^{ab} \delta^{(d)}(x-y) \stackrel{\text{Reg.}}{\Longrightarrow} \left\langle y \left| \boldsymbol{R}^{ab} \right| x \right\rangle = \boldsymbol{R}^{ab} \, \delta^{(d)}(x-y) \tag{171}$$

with

$$\boldsymbol{R}^{ab} = \left(\mathrm{e}^{\, a^2 \nabla^2 / 2} \right)^{ab} , \tag{172}$$

where ∇_μ is the covariant derivative in the adjoint representation. The regularized version of Eq. (168) is

$$\int d^d x \, \nabla_\mu F^a_{\mu\nu}(x) \frac{\delta}{\delta A^a_\nu(x)} \stackrel{\text{w.s.}}{=} \hbar \int d^d x \, d^d y \left\langle y \left| \boldsymbol{R}^{ab} \right| x \right\rangle \frac{\delta}{\delta A^b_\nu(y)} \frac{\delta}{\delta A^a_\nu(x)} \,. \tag{173}$$

To translate Eq. (173) in loop space, we use the path-integral representation

$$\left\langle y \left| \boldsymbol{R}^{ab} \right| x \right\rangle = \int\limits_{\substack{r(0)=x \\ r(a^2)=y}} Dr(t) \, \mathrm{e}^{-\frac{1}{2} \int_0^{a^2} dt \, \dot{r}^2(t)} \, 2 \, \mathrm{tr} \left[t^a U(r_{yx}) t^b U(r_{xy}) \right] \tag{174}$$

with

$$U(r_{yx}) = \mathbf{P} \, \mathrm{e}^{\, i \int_x^y dr_\mu A_\mu(r)} , \tag{175}$$

[6]See the book by Levy [46] and a review [47].

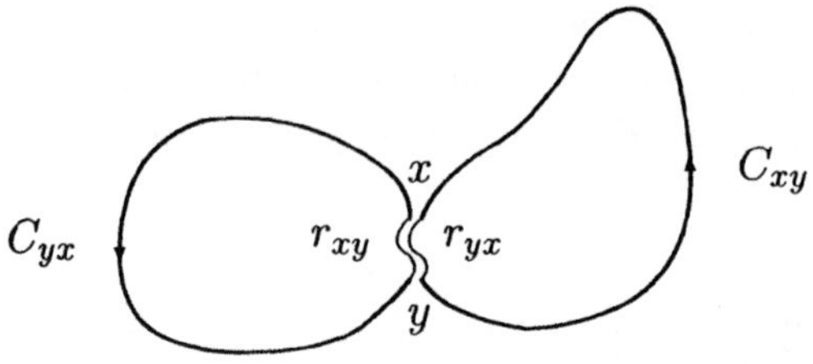

Figure 19. Contours $C_{yx}r_{xy}$ and $C_{xy}r_{yx}$ which enter the RHS's of Eqs. (176) and (177).

where the integration is over regulator paths $r_\mu(t)$ from x to y whose typical length is $\sim a$.

Calculating the variational derivatives on the RHS of Eq. (173), using Eq. (174) and the completeness condition (64), we get as $N \to \infty$:

$$\int d^d x\, d^d y \left\langle y \left| \boldsymbol{R}^{ab} \right| x \right\rangle \frac{\delta}{\delta A_\nu^b(y)} \frac{\delta}{\delta A_\nu^a(x)} \Phi(C) = \lambda \oint_C dx_\mu \oint_C dy_\mu$$

$$\times \int_{\substack{r(0)=x \\ r(a^2)=y}} \mathcal{D}r(t)\, \mathrm{e}^{-\frac{1}{2}\int_0^{a^2} dt \dot{r}^2(t)}\, \Phi(C_{yx}r_{xy})\, \Phi(C_{xy}r_{yx}) \,, \tag{176}$$

where the contours $C_{yx}r_{xy}$ and $C_{xy}r_{yx}$ are depicted in Figure 19. Averaging over the gauge field and using the large-N_c factorization, we arrive at the regularized loop-space Laplace equation [48]

$$\Delta W(C) = \lambda \oint_C dx_\mu \oint_C dy_\mu \int_{\substack{r(0)=x \\ r(a^2)=y}} \mathcal{D}r(t)\, \mathrm{e}^{-\frac{1}{2}\int_0^{a^2} dt\, \dot{r}^2(t)}\, W(C_{yx}r_{xy}) W(C_{xy}r_{yx}) \tag{177}$$

which manifestly recovers Eq. (170) when $a \to 0$.

The constructed regularization is non-perturbative while perturbatively reproduces regularized Feynman diagrams. An advantage of this regularization of the loop equation is that the contours $C_{yx}r_{xy}$ and $C_{xy}r_{yx}$ on the RHS of Eq. (177) both are closed and do not have marked points if C does not have. Therefore, Eq. (177) is written entirely in loop space.

Remark on functional Laplacian

It is worth noting that the representation of the functional Laplacian on the RHS of Eq. (169), which involves the standard variational derivatives, is defined for a wider class of functionals than Stokes functionals. It is easier to deal with the whole operator Δ, rather than separately with the area and path derivatives.

The functional Laplacian is parametric invariant and possesses a number of remarkable properties. While a finite-dimensional Laplacian is an operator of the second order, the functional Laplacian is that of the first order and satisfies the Leibnitz rule

$$\Delta(UV) = \Delta(U)V + U\Delta(V) . \tag{178}$$

The functional Laplacian can be approximated [49] in loop space by a (second-order) partial differential operator in such a way to preserve these properties in the continuum limit. This loop-space Laplacian can be inverted to determine a Green function $G(C, C')$ in the form of a sum over surfaces $S_{C,C'}$ connecting two loops which is analogous to the sum-over-path representation of the Green function of the ordinary Laplacian. The standard perturbation theory can then be recovered by iterating Eq. (170) (or its regularized version (177)) in λ with the Green function of the loop-space Laplacian.

4.7. SURVEY OF NON-PERTURBATIVE SOLUTIONS

While the loop equations were proposed long ago, not much is known about their non-perturbative solutions. We briefly list some of the results.

It was shown in Ref. [50] that area law

$$W(C) \equiv \langle \Phi(C) \rangle \propto \mathrm{e}^{-K \cdot A_{\min}(C)} \tag{179}$$

satisfies the large-N_c loop equation for asymptotically large C. However, a self-consistency equation for K, which should relate it to the bare charge and the cutoff, was not investigated. In order to do this, one needs more detailed information about the behavior of $W(C)$ for intermediate loops.

The *free* bosonic Nambu–Goto string which is defined as a sum over surfaces spanned by C

$$W(C) = \sum_{S:\partial S = C} \mathrm{e}^{-K \cdot A(S)}, \tag{180}$$

with the action being the area $A(S)$ of the surface S, is *not* a solution for intermediate loops. Consequently, QCD does not reduce to this kind of string, as was originally expected in Refs. [51, 52, 53, 54, 55]. Roughly speaking, the ansatz (180) is not consistent with the factorized structure on the RHS of Eq. (159).

Nevertheless, it was shown that if a free string satisfies Eq. (159), then the same interacting string satisfies the loop equations for finite N_c. Here "free string" means, as usual in string theory, that only surfaces of genus zero are present in the sum over surfaces, while surfaces or higher genera are associated with a string interaction. The coupling constant of this interaction is $\mathcal{O}\left(N_c^{-2}\right)$.

A formal solution of Eq. (159) for all loops was found by Migdal [56] in the form of a fermionic string

$$W(C) = \sum_{S:\partial S=C} \int D\psi \, \mathrm{e}^{-\int d^2\xi \left[\bar{\psi}\sigma_k \partial_k \psi + \bar{\psi}\psi m^4 \sqrt{g}\right]}, \tag{181}$$

where the world sheet of the string is parametrized by the coordinates ξ_1 and ξ_2 for which the 2-dimensional metric is conformal, *i.e.* diagonal. The field $\psi(\xi)$ describes 2-dimensional elementary fermions (elves) living in the surface S, and m stands for their mass. Elves were introduced to provide factorization which now holds due to some remarkable properties of 2-dimensional fermions. For large loops, the internal fermionic structure becomes frozen, so that the empty string behavior (179) is recovered. For small loops, the elves are necessary for asymptotic freedom. However, it is unclear whether or not the string solution (181) is practically useful for a study of multicolor QCD, since the methods of dealing with the string theory in four dimensions are not yet developed.

A very interesting solution of the large-N_c loop equation on a lattice was found by Eguchi and Kawai [57]. They showed that the SU(N_c) gauge theory on an infinite lattice reduces at $N_c = \infty$ to the model on a hypercube. The equivalence is possible only at $N_c = \infty$, when the space-time dependence is absorbed by the internal symmetry group. More about this large-N_c reduction will be said in the next Section.

4.8. WILSON LOOPS IN QCD$_2$

Two-dimensional QCD is popular since the paper by 't Hooft [58] as a simplified model of QCD$_4$.

One can always choose the axial gauge $A_1 = 0$, so that the commutator in the non-Abelian field strength (60) vanishes in two dimensions. Therefore, there is no gluon self-interaction in this gauge and the theory looks, at the first glance, like the Abelian one.

The Wilson loop average in QCD$_2$ can be straightforwardly calculated via the expansion (164) where only disconnected (free) parts of the correlators $G^{(n)}$ for even n should be left, since there is no interaction. Only the planar structure of color indices contributes at $N_c = \infty$. Diagrammatically, the diagrams of the type depicted in Figure 18a and Figure 18b are relevant for contours without self-intersections, while that in Figure 18c should be omitted in two dimensions.

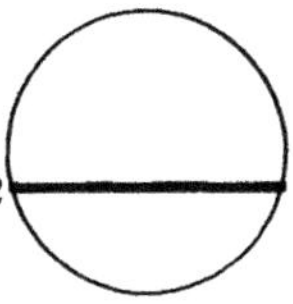

Figure 20. Graphic representation of the contour integral on the LHS of Eq. (186) in the axial gauge. The bold line represents the gluon propagator (184) with $x_2 = y_2$ due to the delta-function.

The color structure of the relevant planar diagrams can be reduced by using the completeness condition (64) at large N_c. We have

$$\begin{aligned} W(C) &= 1 + \sum_k^{\infty} (-\lambda)^k \oint_C dx_1^{\mu_1} \oint_C dx_2^{\nu_1} \cdots \oint_C dx_{2k-1}^{\mu_k} \oint_C dx_{2k}^{\nu_k} \\ &\times \theta_c(1, 2, \ldots, 2k)\, D_{\mu_1\nu_1}(x_1 - x_2) \cdots D_{\mu_k\nu_k}(x_{2k-1} - x_{2k}), \end{aligned} \tag{182}$$

where the points $x_1, \ldots, x_{2k}$ are still cyclic ordered along the contour. We can exponentiate the RHS of Eq. (182) to get finally

$$W(C) = \mathrm{e}^{-\frac{\lambda}{2} \oint_C dx^\mu \oint_C dy^\nu D_{\mu\nu}(x-y)}. \tag{183}$$

This is the same formula as in the Abelian case if λ stands for e^2.

The propagator $D_{\mu\nu}(x, y)$ is, strictly speaking, the one in the gauge $A_1 = 0$ which reads

$$D_{\mu\nu}(x - y) = \frac{1}{2} \delta_{\mu 2} \delta_{\nu 2} |x_1 - y_1| \, \delta^{(1)}(x_2 - y_2) \,. \tag{184}$$

However, the contour integral on the RHS of Eq. (183) is gauge invariant, and we can simply choose instead

$$D_{\mu\nu}(x - y) = \delta_{\mu\nu} \frac{1}{4\pi} \ln \frac{\ell^2}{(x - y)^2}, \tag{185}$$

where ℓ is an arbitrary parameter of the dimension of length. Nothing depends on it because the contour integral of a constant vanishes.

The contour integral in the exponent on the RHS of Eq. (183) can be graphically represented as is depicted in Figure 20, where $x_2 = y_2$ due to the delta-function in Eq. (184) and the bold line represents $|x_1 - y_1|$. This gives

$$\oint_C dx^\mu \oint_C dy^\nu D_{\mu\nu}(x - y) = A(C) \tag{186}$$

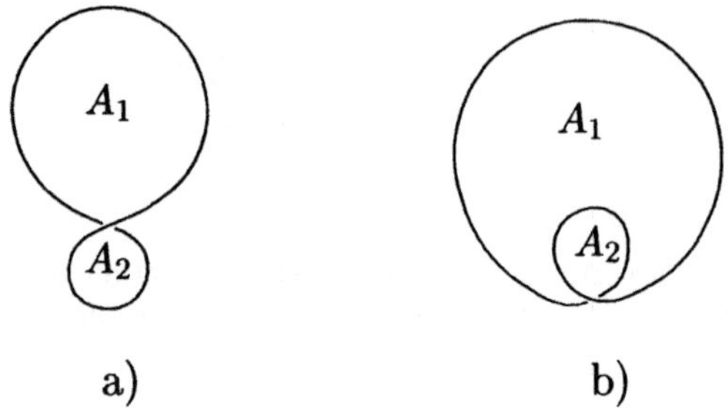

Figure 21. Contours with one self-intersection: A_1 and A_2 stand for the areas of the proper windows. The total area enclosed by the contour in Figure a) is A_1+A_2. The areas enclosed by the exterior and interior loops in Figure b) are A_1+A_2 and A_2, respectively, while the total area of the surface with the folding is A_1+2A_2.

where $A\,(C)$ is the area enclosed by the contour C. We get finally

$$W\,(C) = \mathrm{e}^{-\frac{\lambda}{2}A(C)} \tag{187}$$

for the contours without self-intersections.

Therefore, area law holds in two dimensions both in the non-Abelian and Abelian cases. This is, roughly speaking, because of the form of the two-dimensional propagator (185) which falls down with the distance only logarithmically in the Feynman gauge.

The difference between the Abelian and non-Abelian cases shows up for the contours with self-intersections.

We first note that the simple formula (186) does *not* hold for contours with arbitrary self-intersections.

The simplest contours with one self-intersection are depicted in Figure 21. There is nothing special about the contour in Figure 21a. Equation (186) still holds in this case with $A\,(C)$ being the total area $A\,(C) = A_1 + A_2$.

The Wilson loop average for the contour in Figure 21a coincides both for the Abelian and non-Abelian cases and equals

$$W\,(C) \;=\; \mathrm{e}^{-\frac{\lambda}{2}(A_1+A_2)}\,. \tag{188}$$

This is nothing but the exponential of the total area.

For the contour in Figure 21b, we get

$$\oint_C dx^\mu \oint_C dy^\nu D_{\mu\nu}\,(x-y) = A_1 + 4A_2\,. \tag{189}$$

This is easy to understand in the axial gauge where the ends of the propagator line can lie both on the exterior and interior loops, or one end at the

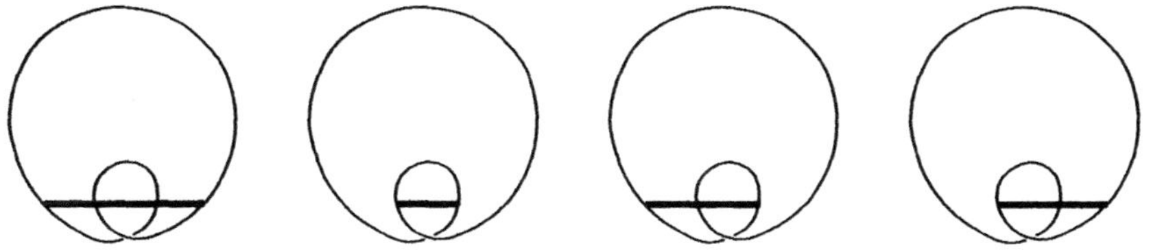

Figure 22. Three type of contribution in Eq. (189) The ends of the propagator line lie both on a) exterior and b) interior loops, or c), d) one end on the exterior loop and another end on the interior loop.

exterior loop and the other end on the interior loop. These cases are illustrated by Figure 22. The contributions of the diagrams in Figure 22a,b,c,d are $A_1 + A_2$, A_2, A_2, and A_2, respectively. The result given by Eq. (189) is obtained by summing over all four diagrams.

For the contour in Figure 21b, the Wilson loop average is

$$W(C) = \mathrm{e}^{-\frac{\lambda}{2}(A_1+4A_2)} \tag{190}$$

in the Abelian case and

$$W(C) = (1 - \lambda A_2)\, \mathrm{e}^{-\frac{\lambda}{2}(A_1+2A_2)} \tag{191}$$

in the non-Abelian case at $N_c = \infty$. They coincide only to the order λ as they should. The difference to the next orders is because only the diagrams with one propagator line connecting the interior and exterior loops are planar and, therefore, contribute in the non-Abelian case. Otherwise, the diagram is non-planar and vanishes as $N_c \to \infty$. Notice, that the exponential of the total area $A(C) = A_1 + 2A_2$ of the surface with the folding, which is enclosed by the contour C, appears in the exponent for the non-Abelian case. The additional pre-exponential factor could be associated with an entropy of foldings of the surface.

The Wilson loop averages (188) and (191) in QCD_2 at large N_c as well as the ones for contours with arbitrary self-intersections, which have a generic form

$$W(C) = P(A_1, \ldots, A_n)\, \mathrm{e}^{-\frac{\lambda}{2} Area} \tag{192}$$

where P is a polynomial of the areas of individual windows and *Area* is the total area of the surface with foldings, were first calculated in Ref. [59] by solving the two-dimensional loop equation and in Ref. [60] by applying the non-Abelian Stokes' theorem.

Remark on the string representation

A nice property of QCD_2 at large N_c is that the exponential of the area enclosed by the contour C emerges[7] for the Wilson loop average $W(C)$. This is as it should for the Nambu–Goto string (180). However, the additional pre-exponential factors (like that in Eq. (191)) are very difficult to interpret in the stringy language. They may become negative for large loops which is impossible for a bosonic string. This explicitly demonstrates in $d = 2$ the statement of the previous subsection that the Nambu–Goto string is not a solution of the large-N_c loop equation.

5. Large-N Reduction

The large-N_c reduction was first discovered by Eguchi and Kawai [57] who showed that the Wilson lattice gauge theory on a d-dimensional hypercubic lattice is equivalent at $N_c = \infty$ to the one on a hypercube with periodic boundary conditions. This construction is based on an extra $(Z_{N_c})^d$-symmetry which the reduced theory possesses to each order of the strong coupling expansion.

Soon after it was recognized that a phase transition occurs in the reduced model with decreasing the coupling constant, so that this symmetry is broken in the weak coupling regime. To cure the construction at weak coupling, the quenching prescription was proposed by Bhanot, Heller and Neuberger [61] and elaborated by many authors. An elegant alternative reduction procedure based on twisting prescription was advocated by Gonzalez-Arroyo and Okawa [62]. Each of these prescriptions results in the reduced model which is fully equivalent to multicolor QCD, both on the lattice and in the continuum.

While the reduced models look as a great simplification, since the space-time is reduced to a point, they still involve an integration over d infinite matrices which is in fact a continual path integral. It is not clear at the moment whether or not this is a real simplification of the original theory which can make it solvable. Nevertheless, the reduced models are useful and elegant representations of the original theory at large N_c.

We shall start this Section by a simplest example of a pure matrix scalar theory. The quenched reduced model for this case was proposed by Parisi [63] on the lattice end elaborated by Gross and Kitazawa [64] in the continuum, while the twisted reduced model was advocated by Eguchi and Nakayama [65]. Then we concentrate on the Eguchi–Kawai reduction of Yang–Mills theory.

[7]This is not true, as is already discussed, in the Abelian case for contours with self-intersections.

5.1. REDUCTION OF SCALAR FIELD

Let us begin with a simplest example of a pure matrix scalar theory on a lattice whose partition function is defined by the path integral

$$Z = \int \prod_{x} \prod_{i \geq j} d\varphi_x^{ij} \, \mathrm{e}^{\sum_x N_c \operatorname{tr}\left(-V[\varphi_x] + \sum_\mu \varphi_x \varphi_{x+a\hat{\mu}}\right)} . \tag{193}$$

Here φ_x is a $N_c \times N_c$ Hermitean matrix field with x running over sites of a hypercubic lattice and $V[\varphi]$ is some interaction potential, say

$$V[\varphi] = \frac{M}{2}\varphi^2 + \frac{\lambda_3}{3}\varphi^3 + \frac{\lambda_4}{4}\varphi^4 . \tag{194}$$

The prescription of the large-N_c reduction is formulated as follows. We substitute

$$\varphi_x \rightarrow S_x \Phi S_x^\dagger , \tag{195}$$

where

$$[S_x]^{kj} = \mathrm{e}^{\,ip_k^\mu x_\mu} \delta^{kj} = \operatorname{diag}\left(\mathrm{e}^{\,ip_1^\mu x_\mu}, \ldots, \mathrm{e}^{\,ip_{N_c}^\mu x_\mu}\right) \tag{196}$$

is a diagonal unitary matrix which eats the coordinate dependence, so that Φ does *not* depend on x.

The averaging of a functional $F[\varphi_x]$ which is defined with the same weight as in Eq. (193),

$$\left\langle F[\varphi_x] \right\rangle \equiv \frac{1}{Z} \int \prod_x d\varphi_x \, \mathrm{e}^{\sum_x N_c \operatorname{tr}\left(-V[\varphi_x] + \sum_\mu \varphi(x)\varphi(x+a\hat{\mu})\right)} F[\varphi_x] , \tag{197}$$

can be calculated at $N_c = \infty$ by

$$\left\langle F[\varphi_x] \right\rangle \rightarrow a^{N_c d} \int_{-\frac{\pi}{a}}^{\frac{\pi}{a}} \prod_{\mu=1}^{d} \prod_{i=1}^{N_c} \frac{dp_i^\mu}{2\pi} \left\langle F[S_x \Phi S_x^\dagger] \right\rangle_{\text{Reduced}} \tag{198}$$

where the average on the RHS is calculated [63] for the *quenched reduced model* whose averages are defined by

$$\left\langle F[\Phi] \right\rangle_{\text{Reduced}} \equiv \frac{1}{Z_{\text{Reduced}}} \times \int \prod_{i \geq j} d\Phi_{ij} \, \mathrm{e}^{\,-N_c \operatorname{tr} V[\Phi] + N_c \sum_{ij} |\Phi_{ij}|^2 \sum_\mu \cos\left((p_i^\mu - p_j^\mu)a\right)} F[\Phi] . \tag{199}$$

The partition function of the reduced model reads

$$Z_{\text{Reduced}} = \int \prod_{i \geq j} d\Phi_{ij} \, \mathrm{e}^{\,-N_c \operatorname{tr} V[\Phi] + N_c \sum_{ij} |\Phi_{ij}|^2 \sum_\mu \cos\left((p_i^\mu - p_j^\mu)a\right)} \tag{200}$$

which can be deduced, modulo the volume factor, from the partition function (193) by the substitution (195).

Notice that the integration over the momenta p_i^μ on the RHS of Eq. (198) is taken *after* the calculation of averages in the reduced model. Such variables are usually called *quenched* in statistical mechanics which clarifies the terminology.

Since $N_c \to \infty$ it is not necessary to integrate over the quenched momenta in Eq. (198). The integral should be recovered if p_i^μ's would be uniformly distributed in a d-dimensional hypercube. Moreover, a similar property holds for the matrix integral over Φ as well, which can be substituted by its value at the saddle point configuration Φ_s:

$$\left\langle F[\varphi_x] \right\rangle \;\to\; F[S_x \Phi_s S_x^\dagger]\,, \tag{201}$$

where the momenta p_i^μ are uniformly distributed in the hypercube. Therefore, this saddle point configuration plays the role of a master field in the sense of Subsection 3.5.

In order to show how Eq. (198) works, let us demonstrate how the planar diagrams of perturbation theory for the scalar matrix theory (193) are recovered in the quenched reduced model.

The quenched reduced model (200) is of the general type discussed in Section 3. The propagator is given by

$$\left\langle \Phi_{ij}\Phi_{kl} \right\rangle_{\text{Gauss}} = \frac{1}{N_c} G\left(p_i - p_j\right) \delta_{il}\delta_{kj} \tag{202}$$

with

$$G\left(p_i - p_j\right) = \frac{1}{M - \sum_\mu \cos\left((p_i^\mu - p_j^\mu)a\right)}\,. \tag{203}$$

It is convenient to associate the momenta p_i and p_j in Eq. (203) with each of the two index lines representing the propagator and carrying, respectively, indices i and j. Remember, that these lines are oriented for a Hermitean matrix Φ and their orientation can be naturally associated with the direction of the flow of the momentum. The total momentum carried by the double line is $p_i - p_j$.

The simplest diagram which represents the correction of the second order in λ_3 to the propagator is depicted Figure 23. The momenta p_i and p_j flows along the index lines i and j while the momentum p_k circulates along the index line k. The contribution of the diagram in Figure 23 reads

$$\frac{\lambda_3^2}{N_c^2} G\left(p_i - p_j\right)^2 \sum_k G\left(p_i - p_k\right) G\left(p_k - p_j\right), \tag{204}$$

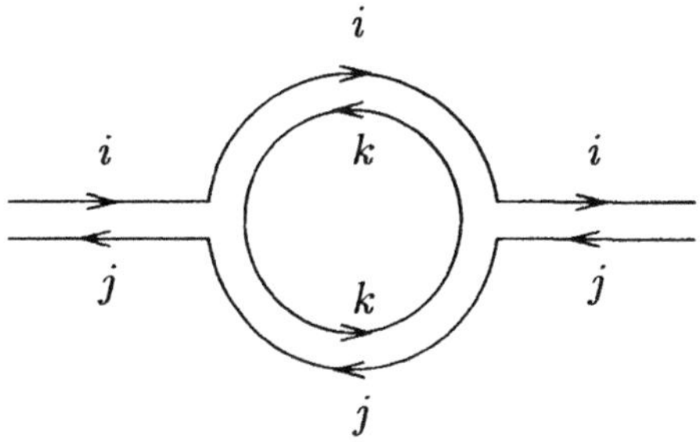

Figure 23. Simplest planar diagram of the second order in λ_3 for the propagator in the quenched reduced model (200). The momentum p_i flows along the index line i. The momentum $p_i - p_j$ is associated with the double line ij.

where the summation over the index k is just a standard one over indices forming a closed loop.

In order to show that the quenched-model result (204) reproduces the correction to the propagator in the original theory on an infinite lattice, we pass to the variables of the total momenta flowing along the double lines:

$$p_i - p_j = p; \quad p_k - p_j = q; \quad p_i - p_k = p - q\,, \tag{205}$$

which is obviously consistent with the momentum conservation at each of the two vertices of the diagram in Figure 23. Since p_k's are uniformly distributed in the hypercube, the summation over k can be substituted as $N_c \to \infty$ by the integral

$$\frac{1}{N_c}\sum_k f\left(p_k\right) \Rightarrow a^d \int_{-\frac{\pi}{a}}^{\frac{\pi}{a}} \frac{d^d q}{(2\pi)^d} f\left(q\right). \tag{206}$$

The prescription (198) then gives the correct expression

$$a^d \frac{\lambda_3^2}{N_c} G\left(p\right)^2 \int_{-\frac{\pi}{a}}^{\frac{\pi}{a}} \frac{d^d q}{(2\pi)^d} G\left(q\right) G\left(p-q\right) \tag{207}$$

for the second-order contribution of the perturbation theory for the propagator on the lattice.

It is now clear how a generic planar diagram is recovered by the reduced model. We first represent the diagram by the double lines and associate the momentum p_i^μ with an index line carrying the index i. Then we write down the expression for the diagram in the reduced model with the propagator (203). Passing to the momenta flowing along the double lines, similar to Eq. (205), we get an expression which coincides with the integrand of the Feynman diagram for the theory on the whole lattice. It is crucial that such a change of variables can always be done for a planar diagram consistently with the momentum conservation at each vertex. The last step is that the summation over indices of closed index lines reproduces the integration over momenta associated with each of the loops according to Eq. (206). It is assumed that the number of loops is much less than N_c which is always true for a given diagram since N_c is infinite.

We thus have shown how planar diagrams of the lattice theory defined by the partition function (193) are recovered by the reduced model (200). The lattice was needed only as a regularization to make all integrals well-defined and was not crucial in the consideration. This construction can be formulated directly for the continuum theory [64, 66] where the propagator turns into

$$G\left(p_i - p_j\right) \;=\; \frac{1}{\left(p_i - p_j\right)^2 + m^2} \tag{208}$$

and a Lorentz-invariant regularization can be achieved by choosing $p^2 < \Lambda^2$.

Remark on the twisted reduced model

An alternative reduction procedure is based on the twisting prescription [62]. We again perform the unitary transformation (195) with the matrices S_x being expressed via a set of d (unitary) $N_c \times N_c$ matrices Γ_μ by

$$S_x = \Gamma_1^{x_1/a}\Gamma_2^{x_2/a}\Gamma_3^{x_3/a}\Gamma_4^{x_4/a} \tag{209}$$

where the coordinates of the (lattice) vector x_μ are measured in the lattice units. The matrices Γ_μ are explicitly constructed in Ref. [62] and commute by

$$\Gamma_\mu \Gamma_\nu = Z_{\mu\nu}\Gamma_\nu\Gamma_\mu \tag{210}$$

with $Z_{\mu\nu} = Z^\dagger_{\nu\mu}$ being elements of Z_{N_c}.

For the twisting reduction prescription, Eq. (198) is valid providing the average on the RHS is calculated for the *twisted reduced model* which is defined by the partition function [65]

$$Z_{\mathrm{TRM}} = \int d\Phi \, \mathrm{e}^{-N_c \operatorname{tr} V[\Phi] + N_c \sum_\mu \operatorname{tr} \Gamma_\mu \Phi \Gamma_\mu^\dagger \Phi} \,. \tag{211}$$

We can change the order of Γ's in Eq. (209) defining a more general path-dependent factor

$$S_x = \mathbf{P} \prod_{l \in C_{x\infty}} \Gamma_\mu \,. \tag{212}$$

The path-ordered product in this formula runs over all links $l = (z, \mu)$ forming a path $C_{x\infty}$ from infinity to the point x.

Due to Eq. (210), changing the form of the path multiplies S_x by the Abelian factor

$$Z(C) = \prod_{\square\in S:\partial S=C} Z_{\mu\nu}(\square) \tag{213}$$

where (μ, ν) is the orientation of the plaquette $\square$. The product runs over any surface spanned by the closed loop C which is obtained by passing the original path forward and the new path backward. Due to the Bianchi identity

$$\prod_{\square\in cube} Z_{\mu\nu}(\square) = 1 \tag{214}$$

where the product goes over six plaquettes forming a 3-dimensional cube on the lattice, the product on the RHS of Eq. (213) does not depend on the form of the surface S and is a functional of the loop C.

It is now easy to see that under this change of the path we get

$$[S_x]_{ij}[S_x^\dagger]_{kl} \to |Z(C)|^2 \, [S_x]_{ij}[S_x^\dagger]_{kl} \tag{215}$$

and the path-dependence is canceled because $|Z(C)|^2 = 1$. This is a general property which holds for the twisting reduction prescription of any even (*i.e.* invariant under the center Z_{N_c}) representation of $SU(N_c)$.

5.2. REDUCTION OF YANG–MILLS FIELD

The statement of the Eguchi–Kawai reduction of the Yang–Mills field says that the theory on a d-dimensional space-time is equivalent at $N_c = \infty$ to the reduced model which is nothing but its reduction to a point. The action of the reduced model is given by

$$S_{\rm EK} = \frac{1}{2g^2\Lambda^d} \operatorname{tr}\, [A_\mu, A_\nu]^2 \,, \tag{216}$$

where A_μ are d space-independent matrices and Λ is a dimensionful parameter.

A naive statement of the Eguchi–Kawai reduction is that the averages coincide in both theories, for example,

$$\left\langle \frac{1}{N_c} \operatorname{tr} \mathbf{P}\, \mathrm{e}^{i\oint d\xi^\mu A_\mu(\xi)} \right\rangle_{\rm d-dim} = \left\langle \frac{1}{N_c} \operatorname{tr} \mathbf{P}\, \mathrm{e}^{i\oint d\xi^\mu A_\mu} \right\rangle_{\rm EK} \tag{217}$$

where the LHS is calculated with the action (59) and the RHS is calculated with the reduced action (216). Strictly speaking, this naive statement is valid only in $d = 2$ or supersymmetric case for the reason which will be explained in a moment.

The precise equivalence is valid only if the average of open Wilson loops vanish in the reduced model:

$$\left\langle \frac{1}{N_c} \operatorname{tr} \mathbf{P} \, \mathrm{e}^{\, i \int_{C_{yx}} d\xi^\mu A_\mu} \right\rangle_{\mathrm{EK}} = 0 \,, \tag{218}$$

as it does in the d-dimensional theory due to the local gauge invariance under which

$$\left(\mathbf{P} \, \mathrm{e}^{\, i \int_{C_{yx}} d\xi^\mu A_\mu(\xi)} \right)_{ij} \to \left(\Omega^\dagger(y) \mathbf{P} \, \mathrm{e}^{\, i \int_{C_{yx}} d\xi^\mu A_\mu(\xi)} \Omega(x) \right)_{ij} . \tag{219}$$

The point is that this gauge invariance transforms in the reduced model into (global) rotation of the reduced field by constant matrices Ω:

$$A_\mu \to \Omega^\dagger A_\mu \Omega \,. \tag{220}$$

which does *not* guarantee such vanishing in the reduced model.

There exists, however, a symmetry of the reduced action (216) under the shift of A_μ by a unit matrix[8]:

$$A_\mu^{ij} \to A_\mu^{ij} + a_\mu \delta^{ij} \,, \tag{221}$$

which is often called the R^d-symmetry. Under the transformation (221), we get

$$\left(\mathbf{P} \, \mathrm{e}^{\, i \int_{C_{yx}} d\xi^\mu A_\mu} \right)_{ij} \to \mathrm{e}^{\, i (y^\mu - x^\mu) a_\mu} \left(\mathbf{P} \, \mathrm{e}^{\, i \int_{C_{yx}} d\xi^\mu A_\mu} \right)_{ij} \tag{222}$$

which guarantees, if the symmetry is not broken, the vanishing of the open Wilson loops

$$W_{\mathrm{EK}} \left(C_{yx} \right) \equiv \left\langle \frac{1}{N_c} \operatorname{tr} \mathbf{P} \, \mathrm{e}^{\, i \int_{C_{yx}} d\xi^\mu A_\mu} \right\rangle_{\mathrm{EK}} = 0 \tag{223}$$

in the reduced model.

The equivalence of the two theories can then be shown using the loop equation which reads for the reduced model

$$\begin{aligned} \partial_\mu^x \frac{\delta}{\delta \sigma_{\mu\nu}(x)} W_{\mathrm{EK}}(C) &= \left\langle \frac{1}{N_c} \operatorname{tr} \mathbf{P} [A_\mu, [A_\mu, A_\nu]] \, \mathrm{e}^{\, i \oint_{C_{xx}} d\xi^\mu A_\mu} \right\rangle_{\mathrm{EK}} \\ &= \lambda \Lambda^d \left\langle \frac{1}{N_c} \operatorname{tr} \mathbf{P} \frac{\partial}{\partial A_\nu} \, \mathrm{e}^{\, i \oint_{C_{xx}} d\xi^\mu A_\mu} \right\rangle_{\mathrm{EK}} \\ &= \lambda \Lambda^d \oint_C dy_\nu \, W_{\mathrm{EK}}(C_{yx}) \, W_{\mathrm{EK}}(C_{xy}) \,. \end{aligned} \tag{224}$$

[8]This symmetry is rigorously defined on a lattice where it is associated with a direction-dependent Z_{N_c} transformation.

The RHS is pretty much similar to the one in Eq. (159) while $\delta^{(d)}(x-y)$ is missing.

This delta function can be recovered if the R^d symmetry is not broken since

$$W_{\rm EK}\left(C_{yx}\right) \sim \frac{\delta^{(d)}(x-y)}{\delta^{(d)}(0)} W_{\rm EK}\left(C_{yx}\right) \tag{225}$$

due to Eq. (223) for the open loops.

This is not a rigorous argument since a regularization is needed. What actually happens is the following. If we smear the delta function introducing

$$\delta_{\Lambda}^{(d)}(x) = \left(\frac{\Lambda}{\sqrt{2\pi}}\right)^{d} \mathrm{e}^{-x^2\Lambda^2/2}, \tag{226}$$

then

$$\frac{1}{\delta_{\Lambda}^{(d)}(0)} \left(\delta_{\Lambda}^{(d)}(0)\right)^2 \propto \Lambda^d \,\mathrm{e}^{-x^2\Lambda^2} \to \delta^{(d)}(x), \tag{227}$$

reproducing the delta function.

5.3. R^D-SYMMETRY IN PERTURBATION THEORY

Since N_c is infinite, the R^d-symmetry can be broken spontaneously. The point is that the large-N_c limit plays the role of a statistical averaging, as is mentioned already in Subsection 3.6, and phase transitions are possible for infinite number of degrees of freedom. This phenomenon occurs in perturbation theory of the reduced model for $d \geq 3$.

The perturbation theory can be constructed expanding the fields around solutions of the classical equation

$$[A_\mu, [A_\mu, A_\nu]] = 0\,. \tag{228}$$

Any diagonal matrix

$$A_\mu^{\rm cl} \equiv p_\mu = \mathrm{diag}\left\{p_\mu^{(1)}, \ldots, p_\mu^{(N_c)}\right\} \tag{229}$$

is a solution to Eq. (228).

The perturbation theory of the reduced model can be constructed expanding around the classical solution (229):

$$A_\mu = A_\mu^{\rm cl} + g A_\mu^{\rm q}\,, \tag{230}$$

where $A_\mu^{\rm q}$ is off-diagonal.

Substituting (230) into the action (216), we get

$$S_{\rm EK} = \mathrm{tr}\left\{\frac{1}{2}[p_\mu, A_\nu^{\rm q}]^2 - \frac{1}{2}[p_\mu, A_\mu^{\rm q}]^2\right\} + \text{higher orders}\,. \tag{231}$$

To fix the gauge symmetry (220), it is convenient to add

$$S_{\text{g.f.}} = \operatorname{tr}\left\{\frac{1}{2}[p_\mu, A^{\text{q}}_\mu]^2 + [p_\mu, b][p_\mu, c]\right\}, \tag{232}$$

where b and c are ghosts.

The sum of (231) and (232) gives

$$S_2 = \operatorname{tr}\left\{\frac{1}{2}[p_\mu, A^{\text{q}}_\nu]^2 + [p_\mu, b][p_\mu, c]\right\} \tag{233}$$

up to quadratic order in A^{q}_μ.

Doing the Gaussian integral over A^{q}_ν, we get at the one-loop level:

$$\int dp_\mu dA^{\text{q}}_\mu \,\mathrm{e}^{-S_2} \ldots = \int \prod_{k=1}^{N} dp^{(k)}_\mu \prod_{i<j} \left[(p^{(i)}_\mu - p^{(j)}_\mu)^2\right]^{1-d/2} \ldots, \tag{234}$$

where the integration over p_μ accounts for equivalent classical solutions.

For $d = 1$ the product on the RHS of Eq. (234) reproduces the Vandermonde determinant. For $d = 2$ it vanishes and does not affect dynamics. For $d \geq 3$ the measure is singular and the eigenvalues collapse. This leads us to a spontaneous breakdown of the R^d in perturbation theory.

The equivalence between the $N_c = \infty$ Yang–Mills theory on a whole space and the reduced model can be provided [61] introducing a quenching prescription similar to the one described in Subsection 5.1. Then no collapse of eigenvalues happens and d-dimensional planar graphs are reproduced by the reduced model. More about the quenching prescription in Yang–Mills theory can be found in the reviews [44, 67] and cited there original papers.

Remark on supersymmetric case

In a supersymmetric gauge theory, there is an extra contribution from fermions to the exponent on the RHS of Eq. (234). Since the integration over fermions results in the extra factor $[(p^{(i)}_\mu - p^{(j)}_\mu)^2]^{\operatorname{tr} I/2}$, this yields finally the exponent $1-d/2+ \operatorname{tr} I/2$. It vanishes in $d = 4$ for either Majorana or Weyl fermions and in $d = 10$ for the Majorana–Weyl fermions. Therefore, the R^d-symmetry is not broken and no quenching is needed in the supersymmetric case [68, 5].

5.4. TWISTED REDUCED MODEL

The continuum version of the twisted reduced model can be constructed [69] by substituting $A_\mu \to A_\mu - \gamma_\mu$ into the action (216), where the matrices γ_μ obey the commutation relation

$$[\gamma_\mu, \gamma_\nu] = B_{\mu\nu} I, \tag{235}$$

where $B_{\mu\nu}$ is an antisymmetric tensor and d is even. This is possible only for infinite Hermitean matrices (operators). An example of such matrices is x and p operators in quantum mechanics. Eq. (235) is a continuum version of Eq. (210).

The Wilson loop averages in the twisted reduced model are defined by

$$W_{\mathrm{TEK}}\left(C_{yx}\right) = \left\langle \frac{1}{N_c} \operatorname{tr} \mathbf{P}\, \mathrm{e}^{-i\int_{C_{yx}} d\xi^\mu \gamma_\mu} \frac{1}{N_c} \operatorname{tr} \mathbf{P}\, \mathrm{e}^{i\int_{C_{yx}} d\xi^\mu A_\mu} \right\rangle_{\mathrm{TEK}} . \tag{236}$$

They vanish for open loops which is provided by the vanishing of the trace of the path-ordered exponential of γ_μ in this definition. For closed loops this factor does not vanish and is needed to provide the equivalence with d-dimensional Yang–Mills perturbation theory, since the classical extrema of the twisted reduced model are $A_\mu^{\mathrm{cl}} = \gamma_\mu$ and the perturbation theory is constructed expanding around this classical solution.

The proof of the equivalence can be done using the loop equation quite similarly to that of Subsection 5.2 for the Eguchi–Kawai model with an unbroken R^d symmetry.

Acknowledgements

I would like to thank the organizers for the warm hospitality in Iceland. This work is supported in part by the grant RFFI 97–02–17927.

References

1. G. 't Hooft, "A planar diagram theory for strong interactions," *Nucl. Phys.* **B72** (1974) 461.
2. J. Maldacena, "The large N limit of superconformal field theories and supergravity," *Adv. Theor. Math. Phys.* **2** (1998) 231 hep-th/9711200.
3. T. Banks, W. Fischler, S. H. Shenker and L. Susskind, "M theory as a matrix model: a conjecture," *Phys. Rev.* **D55** (1997) 5112, hep-th/9610043.
4. N. Drukker, D. J. Gross and H. Ooguri, "Wilson loops and minimal surfaces," *Phys. Rev.* **D60** (1999) 125006, hep-th/9904191.
5. N. Ishibashi, H. Kawai, Y. Kitazawa and A. Tsuchiya, "A large-N reduced model as superstring," *Nucl. Phys.* **B498** (1997) 467, hep-th/9612115.
6. H. Aoki, N. Ishibashi, S. Iso, H. Kawai, Y. Kitazawa and T. Tada, "Noncommutative Yang-Mills in IIB matrix model," hep-th/9908141.
7. Y. Makeenko, "Formulation of matrix theory at finite temperature," *Forschr. Phys* **48** (2000) 171, hep-th/9903030.
8. H.E. Stanley, "Spherical model as the limit of infinite spin dimensionality," *Phys. Rev.* **176** (1968) 718.
9. K.G. Wilson, "Quantum field-theory models in less than 4 dimensions," *Phys. Rev.* **D7** (1973) 2911.
10. D.J. Gross and A. Neveu, "Dynamical symmetry breaking in asymptotically free field theories," *Phys. Rev.* **D10** (1974) 3235.
11. A.A. Anselm, "A model of a field theory with nonvanishing renormalized charge," *Sov. Phys. JETP* **9** (1959) 608.

12. K. Johnson, "Solution of the equations for the Green's functions of a two dimensional relativistic field theory", *Nuovo Cim.* **20** (1961) 773.
13. G. Parisi, "The theory of non-renormalizable interaction," *Nucl. Phys.* **B100** (1975) 368.
14. W. Chen, Y. Makeenko, and G.W. Semenoff, "Four-fermion theory and the conformal bootstrap," *Ann. Phys.* **228** (1993) 341.
15. J.A. Gracey, "Calculation of exponent η to $O(1/N^2)$ in the $O(N)$ Gross Neveu model," *Int. J. Mod. Phys.* **A6** (1991) 395, 2755(E).
16. G. Mack and A. Salam, "Finite-component field representation of the conformal group," *Ann. Phys.* **53** (1969) 144.
17. D. Gross and J. Wess, "Scale invariance, conformal invariance, and the high-energy behavior of scattering amplitudes," *Phys. Rev.* **D2** (1970) 753.
18. R. Jackiw, "Field theoretic investigations in current algebra," in *Lectures on Current Algebra and its Applications* by S.B. Treiman, R. Jackiw, and D.J. Gross, Princeton Univ. Press, 1972, p. 97; also in *Current Algebra and Anomalies* by S.B. Treiman, R. Jackiw, B. Zumino, and E. Witten, World Sci., 1985, p. 81.
19. A.M. Polyakov, "Conformal symmetry of critical fluctuations," *JETP Lett.* **12** (1970) 381.
20. A.A. Migdal, "On hadronic interactions at small distances," *Phys. Lett.* **37B** (1971) 98.
21. A.M. Polyakov, "Interaction of Goldstone particles in two dimensions. Applications to ferromagnets and massive Yang–Mills fields," *Phys. Lett.* **59B** (1975) 79.
22. A.A. Belavin and A.M. Polyakov, "Metastable states of two-dimensional isotropic ferromagnet," *JETP Lett.* **22** (1975) 245.
23. G. Veneziano, "Some aspects of a unified approach to gauge, dual and Gribov theories," *Nucl. Phys.* **B117** (1976) 519.
24. W.T. Tuttle, "A census of planar triangulations," *Can. J. Math.* **14** (1962) 21.
25. J. Koplik, A. Neveu and S. Nussinov, "Some aspects of the planar perturbation series," *Nucl. Phys.* **B123** (1977) 109.
26. P. Cvitanović, "Planar perturbation expansion," *Phys. Lett.* **99B** (1981) 49.
27. P. Cvitanović, P.G. Lauwers, and P.N. Scharbach, "The planar sector of field theories," Nucl. Phys. **B203** (1982) 385.
28. E. Witten, "The 1/N expansion in atomic and particle physics, in *Recent developments in gauge theories*," eds. G. 't Hooft et al., Plenum, 1980, p.403.
29. S. Coleman, "1/N," in Proc. of Erice Int. School of Subnuclear Physics 1979, Plenum, N.Y., 1982, p. 805 (Reprinted in S. Coleman, *Aspects of symmetry*, Cambridge Univ. Press, 1985, pp.351–402).
30. O. Haan, "Large N as a thermodynamic limit," *Phys. Lett.* **106B** (1981) 207.
31. D.V. Voiculescu, K.J. Dykema and A. Nica, "Free Random Variables," AMS, Providence 1992.
32. M.R. Douglas, "Stochastic master fields," *Phys. Lett.* **B344** (1995) 117.
33. R. Gopakumar and G.J. Gross, "Mastering the master field," Nucl. Phys. **B451** (1995) 379.
34. S. Mandelstam, "Charge-monopole duality and the phases of non-Abelian gauge theories," *Phys. Rev.* **D19** (1979) 2391.
35. Yu.M. Makeenko and A.A. Migdal, "Exact equation for the loop average in multicolor QCD," *Phys. Lett.* **88B** (1979) 135.
36. K.G. Wilson, "Confinement of quarks," *Phys. Rev.* **D10** (1974) 2445.
37. Yu.M. Makeenko and A.A. Migdal, "Quantum chromodynamics as dynamics of loops," *Nucl. Phys.* **B188** (1981) 269.
38. R.A. Brandt, F. Neri and D. Zwanziger, "Lorentz invariance from classical particle paths in quantum field theory of electric and magnetic charge," *Phys. Rev.* **D19** (1979) 1153.
39. J.L. Gervais and A. Neveu, "The slope of the leading Regge trajectory in quantum chromodynamics," *Nucl. Phys.* **B163** (1980) 189.

40. A.M. Polyakov, "Gauge fields as rings of glue," *Nucl. Phys.* **B164** (1980) 171.
41. V.S. Dotsenko and S.N. Vergeles, "Renormalizability of phase factors in non-Abelian gauge theory," *Nucl. Phys.* **B169** (1980) 527.
42. R.A. Brandt, F. Neri and M. Sato, "Renormalization of loop functions for all loops," *Phys. Rev.* **D24** (1981) 879.
43. R.A. Brandt, A. Gocksch, M. Sato, and F. Neri, "Loop space," *Phys. Rev.* **D26** (1982) 3611.
44. A.A. Migdal, "Loop equations and 1/N expansion," *Phys. Rep.* **102** (1983) 199.
45. J.L. Gervais and A. Neveu, "Local harmonicity of the Wilson loop integral in classical Yang–Mills theory," *Nucl. Phys.* **B153** (1979) 445.
46. P. Lévy, "Probleèms concrets d'analyse fonctionnelle," Paris, 1951.
47. M.N. Feller, "Infinite-dimensional elliptic equations and operators of the type by P. Lèvy," *Sov. J. Usp. Mat. Nauk.* **41** (1986) 97.
48. M.B. Halpern and Yu.M. Makeenko, "Continuum-regularized loop-space equation," *Phys. Lett.* **218B** (1989) 230.
49. Yu.M. Makeenko, "Polygon discretization of the loop-space equation," *Phys. Lett.* **212B** (1988) 221.
50. Yu.M. Makeenko and A.A. Migdal, "Self-consistent area law in QCD," *Phys. Lett.* **97B** (1980) 235.
51. J.L. Gervais and A. Neveu, "The quantum dual string wave functional in Yang–Mills theories," *Phys. Lett.* **80B** (1979) 255.
52. Y. Nambu, "QCD and the string model," Phys. Lett. **80B** (1979) 372.
53. A.M. Polyakov, "String representations and hidden symmetries for gauge fields," *Phys. Lett.* **82B** (1979) 247.
54. D. Foerster, "Yang–Mills theory — a string theory in disguise," *Phys. Lett.* **87B** (1979) 87.
55. T. Eguchi, "Strings in U(N) lattice gauge theory," *Phys. Lett.* **87B** (1979) 91.
56. A.A. Migdal, "QCD = Fermi string theory," *Nucl. Phys.* **189** (1981) 253.
57. T. Eguchi and H. Kawai, "Reduction of dynamical degrees of freedom in the large-N gauge theory," *Phys. Rev. Lett.* **48** (1982) 1063.
58. G. 't Hooft, "A two-dimensional model for mesons," *Nucl. Phys.* **B75** (1974) 461.
59. V.A. Kazakov and I.K. Kostov, "Non-linear strings in two-dimensional $U(\infty)$ theory," *Nucl. Phys.* **B176** (1980) 199.
60. N. Bralić, "Exact computation of loop averages in two-dimensional Yang–Mills theory," *Phys. Rev.* **D22** (1980) 3090.
61. G. Bhanot, U. Heller and H. Neuberger, "The quenched Eguchi–Kawai model," *Phys. Lett.* **113B** (1982) 47.
62. A. Gonzalez-Arroyo and M. Okawa, "The twisted Eguchi-Kawai model: A reduced model for large N lattice gauge theory," *Phys. Rev.* **D27** (1983) 2397.
63. G. Parisi, "A simple expression for planar field theories," *Phys. Lett.* **112B** (1982) 463.
64. D.J. Gross and Y. Kitazawa, "A quenched momentum prescription for large-N theories," *Nucl. Phys.* **B206** (1982) 440.
65. T. Eguchi and R. Nakayama, "Simplification of quenching procedure for large N spin models," *Phys. Lett.* **122B** (1983) 59.
66. S. Das and S. Wadia, "Translation invariance and a reduced model for summing planar diagrams in QCD," *Phys. Lett.* **B117** (1982) 228.
67. S.R. Das, "Some aspects of large-N theories," *Rev. Mod. Phys.* **59** (1987) 235.
68. R.L. Mkrtchyan and S.B. Khokhlachev, "Reduction of the $U(\infty)$ theory to a model of random matrices", *JETP Lett.* **37** (1983) 160.
69. A. Gonzalez-Arroyo and C. P. Korthals Altes, "Reduced model for large N continuum field theories," *Phys. Lett.* **B131** (1983) 396.

17 8'99

INTRODUCTION TO RANDOM SURFACES

THORDUR JONSSON
Raunvisindastofnun Haskolans,
University of Iceland,
Dunhaga 3, 107 Reykjavik,
Iceland
email: thjons@raunvis.hi.is

Abstract. We begin by discussing random walks and branched polymers and show how the continuum properties of such objects are obtained from discrete approximations. We then review the theory of dynamically triangulated surfaces and prove the nonscaling of the string tension. We construct the continuum limit of lattice surfaces explicitly and show that the theory is dominated by branched polymer configurations unless the action depends on the extrinsic curvature or higher powers of the intrinsic curvature.

1. Introduction

Statistical models of discretized random surfaces have been studied since the early 1980s. The original motivation came from two-dimensional quantum gravity and the high temperature expansion of lattice gauge theories [1, 2, 3, 4]. It was soon realized that random surfaces provide a natural mathematical framework for quantizing strings [5, 6]. Random surfaces also have applications to various problems in statistical physics.

In these lectures we give an introduction to the theory of random surfaces, concentrating on general theoretical results. However, much of our knowledge about random surface models comes from numerical simlations which are discussed in the lectures of J. Jurkiewicz. Higher dimensional random manifolds are also of interest, mainly in connection with quantum gravity in more than two dimensions. A general reference for these lectures is the monograph [7] which also contains an extensive list of references.

In the next section we discuss simple theories of random geometrical objects, random paths and branched polymers where one can illustrate the

L. Thorlacius and T. Jonsson (eds.), M-Theory and Quantum Geometry, 355-380.

general techniques in a straightforward fashion. We then describe the theory of randomly triangulated surfaces which is a discretization of Polyakov's string theory. We show that the string tension of the discretized theory is bounded from below by a positive constant which suggests that the model is dominated by tree-like surfaces. We then treat lattice surfaces and prove that the simplest model with area action has he same critical behaviour as a model of branched polymers. The philosophy is to discretize the geometrical objects directly rather than discretizing a particular parametrization of the objects.

The ultimate goal of the theory of random surfaces is to give a mathematical meaning to functional integrals over surfaces and higher dimensional geometrical objects. This goal has turned out to be elusive but in recent years considerable progress has been made, in particular in the theory of two-dimensional quantum gravity where one is interested in integrating over the Riemannian metrics on the sphere and higher genus surfaces. This problem is discussed in the lectures of J. Ambjørn. In the next section we review how one can integrate over paths in a rigorous fashion.

2. Random paths

The Euclidean propagator $G(x, y)$ for a boson of mass m has the path integral representation

$$G(x,y) = \int_{\omega:x\to y} e^{-S(\omega)}\,\mathcal{D}\omega \tag{1}$$

where the action $S(\omega)$, given by

$$S(\omega) = m\int_{\omega} ds, \tag{2}$$

is a multiple of the length of the path ω. Clearly the length of the path is independent of parametrization so here we have the first and simplest example of a geometric theory. In Eq. (1) we integrate over *geometric paths* rather than parametrized paths. By a geometric path we mean an equivalence class of parametrized paths related by reparametrizations.

The theory of the Wiener integral [8] makes it rather easy to give a meaning to the functional integral above. We begin by reviewing the Wiener integral and then discretize the functional integral and show how this leads to the well known continuum result

$$G(x,y) = (-\Delta + m^2)^{-1}(x,y) \tag{3}$$

where Δ is the Laplacian in $\mathbf{R}^d$. We do this in two different ways which yield the same result. The reason for using two different methods is that one wishes to use analogous methods for random surfaces where it is by no means evident that they are equivalent. Indeed, using discretization is often the only way of constructing functional integrals over more than one dimensional objects.

Let us denote by $\Omega_t(x,y)$ the set of all continuous paths from x to y parametrized by the interval $[0,t]$. Let $\Omega(x,y)$ denote the union of the sets $\Omega_t(x,y)$ over all positive t and denote by $\tilde{\Omega}(x,y)$ the equivalence classes of paths in $\Omega(x,y)$ under smooth reparametrizations. If $\omega \in \Omega(x,y)$ we denote its equivalence class by $[\omega]$. Our functional integral is over the set $\tilde{\Omega}(x,y)$ and the goal is to define suitable discrete or finite dimensional subsets of $\tilde{\Omega}(x,y)$ which become dense in $\tilde{\Omega}(x,y)$ as a cutoff is removed.

Let us denote the Wiener measure on $\Omega_t(x,y)$ by $D^t\omega$, normalized so that

$$\begin{aligned} \int D^t\omega &= \frac{1}{(2\pi t)^{d/2}} e^{-(x-y)^2/2t} \\ &\equiv K_t(x,y), \end{aligned} \tag{4}$$

which is the usual heat kernel. This measure is uniquely characterized by the integrals over cylinder functions, i.e. functions which only depend on the location of the path at a finite number of times. These integrals are given by

$$\begin{aligned} \int f(\omega(t_1),\ldots,\omega(t_n))\, D^t\omega &= \int K_{t_1}(x,x_1)K_{t_2-t_1}(x_1,x_2)\ldots K_{t-t_n}(x_n,y) \\ &\quad \times f(x_1,\ldots,x_n)\, dx_1\ldots dx_n. \end{aligned} \tag{5}$$

In order to give a meaning to Eq. (1) we are interested in the related measure on $\Omega(x,y)$ defined as

$$D\omega = e^{-tm^2}\, D^t\omega\, dt, \tag{6}$$

where dt is a Lebesgue measure on $\mathbf{R}_+$. The total mass of this measure is the propagator $G(x,y)$. We then obtain a measure $D[\omega]$ on $\tilde{\Omega}(x,y)$ by integrating over equivalence classes. If F is a function on $\tilde{\Omega}(x,y)$ we define

$$\int F([\omega])\, D[\omega] = \int F([\omega])\, D\omega \tag{7}$$

since F is also a well defined function on $\Omega(x,y)$.

2.1. LATTICE PATHS

We take $a > 0$ and let $\tilde{\Omega}^a(x,y)$ denote all paths from x to y on the hypercubic lattice $a\mathbf{Z}^d$ with lattice spacing a, i.e.

$$\tilde{\Omega}^a(x,y) = \{(x_0,x_1,\ldots,x_{n-1},x_n) : x_i \in a\mathbf{Z}^d,\ |x_i - x_{i-1}| = a\}, \tag{8}$$

$x_0 = x$ and $x_n = y$. It is clear that $\tilde{\Omega}^a(x, y)$ is contained in $\tilde{\Omega}(x, y)$ in a natural fashion. As $a \to 0$ we can approximate any path in $\tilde{\Omega}(x, y)$ arbitrarily well in a natural metric on $\tilde{\Omega}(x, y)$ by a path from $\tilde{\Omega}^a(x, y)$.

We would like to construct a suitable weight factor $W_a(\tilde{\omega})$ for lattice paths such that

$$\lim_{a \to 0} \sum_{\tilde{\omega} \in \tilde{\Omega}^a} F(\tilde{\omega}) W_a(\tilde{\omega}) = \int F([\omega])\, D[\omega] \tag{9}$$

for any bounded integrable function F on the space of geometric paths (here we have dropped the endpoints x and y from our notation). Eq. (9) makes it precise how the lattice paths become uniformly dense in $\tilde{\Omega}$ as $a \to 0$.

The construction of the weight factor for lattice paths is a standard calculation so we shall be brief. The first thing to consider is the entropy of paths, i.e. how many paths there are of a given length from x to y. The number of paths with n steps originating at x is clearly $(2d)^n$. The number of n step paths with both endpoints fixed has the same growth up to a power of n. The natural choice for the lattice action is therefore

$$S_1(\tilde{\omega}) = \beta(a) n \tag{10}$$

where $\beta(a)$ is a parameter to be chosen and n is the number of steps in $\tilde{\omega} \in \tilde{\Omega}^a$. Since the dimension of the propagator is $2 - d$ it is natural to expect the weight factor to have the form

$$W_a(\tilde{\omega}) = a^{2-d} e^{-S_1(\tilde{\omega})}. \tag{11}$$

If the propagator has an anomalous dimension η the power of a in Eq. (11) is $2 - d - \eta$. Defining a lattice propagator with this weight factor,

$$G^a(x, y) = \sum_{\tilde{\omega}: x \to y} W_a(\tilde{\omega}), \tag{12}$$

we find that its Fourier transform, defined by

$$\widehat{G^a}(k) = a^d \sum_{x \in a\mathbf{Z}^d} G^a(x, 0) e^{-ik \cdot x}, \tag{13}$$

is equal to

$$\left(m^2(a) + 2a^{-2} \sum_{j=1}^{d} (1 - \cos(ak_j)) \right)^{-1}, \tag{14}$$

where k lies in the first Brillouin zone and

$$m^2(a) = \frac{e^{a\beta(a)} - 2d}{a^2}. \tag{15}$$

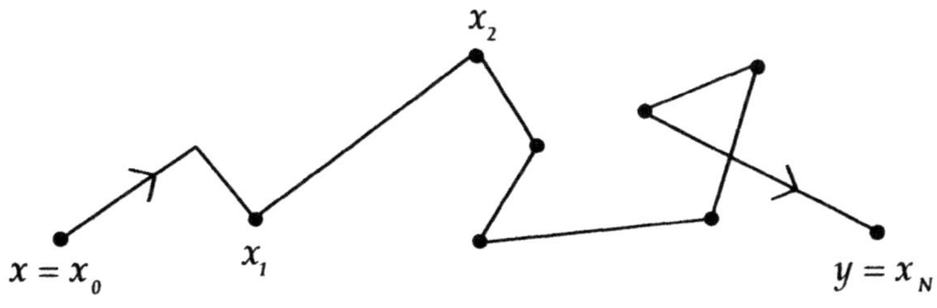

Figure 1. A piecewise linear path from x to y.

The convergence of the Fourier transform of the lattice propagator to the Fourier transform of the continuum propagator with mass m_* is now obtained by choosing

$$\beta(a) = a^{-1} \log(2d + m_*^2 a^2). \tag{16}$$

It follows by a straightforward argument that the propagator converges pointwise in position space at noncoinciding points. One can prove a stronger result, namely that the suitably normalized counting measure on the lattice paths $\tilde{\Omega}^a$ converges (weakly) to the Wiener measure on all geometric paths as $a \to 0$ [9]. This result establishes Eq. (9).

2.2. DYNAMICALLY TRIANGULATED PATHS

Regarding a path as a manifold we can triangulate it in a trivial way and approximate it by n straight line segments connecting points on the path. A one dimensional triangulation is uniquely characterized by its number of 1-simplexes n. Here we regard the triangulation as the fundamental object as we now explain. Denote the vertices $0, 1, \ldots, n$ and map them into $\mathbf{R}^d$, $i \mapsto x_i$. In view of the fact that the Wiener measure on paths parametrized by $[0, 1]$ can formally be written

$$D\omega = e^{-\int_0^1 \dot{\omega}^2 \, dt} \prod_{t \in [0,1]} d\omega(t) \tag{17}$$

it is natural to consider the action

$$S_2(\omega) = a^{-2} \sum_{i=1}^{n} (x_i - x_{i-1})^2 \tag{18}$$

as an action functional for the piecewise linear paths. Here a is a cutoff parameter which is taken to 0 in the continuum limit.

We can regard the piecewise linear paths as geometric paths where we have fixed the parametrization such that the line segment from x_i to x_{i+1} is linearly parametrized by the interval $[i/n, (i+1)/n]$. If we denote these picewise linear paths by $\Omega_{1,n}$ we can define a measure D_n on $\Omega_{1,n}$ which is an approximation to the Wiener measure on all geometric paths by the formula

$$D_n\omega = Ce^{-S_2(\omega)} \prod_{i=1}^{n-1} dx_i, \tag{19}$$

where C is a normalization factor. As before we are considering paths with both endpoints fixed so we only integrate over the internal vertices but all our considerations generalize to the case of paths with arbitrary endpoints.

It is natural to extend the above considerations and study paths with an independent intrinsic metric $g = e^2$ which is simply a positive function on the path. We then obtain a one-dimensional analogue of the Polyakov functional integral for random surfaces

$$Z(x,y) = \int_{\omega:x\to y} e^{-\tilde{S}(\omega,e)}\,\mathcal{D}\omega\mathcal{D}e \tag{20}$$

where the action is given by

$$\tilde{S}(\omega,e) = \int_0^1 \left(\frac{\dot{\omega}^2}{2e} + m^2 e\right) dt. \tag{21}$$

We will see that $Z(x,y)$ is equal to the free propagator $G(x,y)$ considered above. The action $\tilde{S}(\omega,e)$ is invariant under the group of orientation preserving diffeomorphisms of the unit interval

$$(\omega,e) \mapsto (\omega\circ\xi, \dot{\xi}\, e\circ\xi). \tag{22}$$

Using the diffeomorphism invariance we can always gauge transform to a constant intrinsic metric e and this constant labels the orbits of the pairs (ω,e) under the diffeomorphism group. We can therefore identify the diffeomorphism classes with the set $\Omega_1 \times \mathbf{R}_+$. In order to discretize the functional integral we now discretize this path space by $\Omega_{1,n} \times \{a^2 n\}$ where a is a cutoff. We will see that the correct continuum limit is obtained when $n \to \infty$ such that the product $a^2 n$ is constant. This scaling reflects the fact that a random walker travels on average a distance $\sqrt{n}$ if he takes n steps of unit length. The length of paths therefore scales differently from the distance in embedding space and this discrepancy in scaling gives the Hausdorff dimension of random paths the value 2. For a detailed discussion of the notion of Hausdorff dimension we refer to the lectures of J. Ambjørn and J. Jurkiewicz. The intuitive notion of a Hausdorff dimension for a random geometric structure is the following. If the structure defines a lengthscale ℓ and has volume (area, length) V then the Hausdorff dimension is d_H if

$$V \sim \ell^{d_H} \tag{23}$$

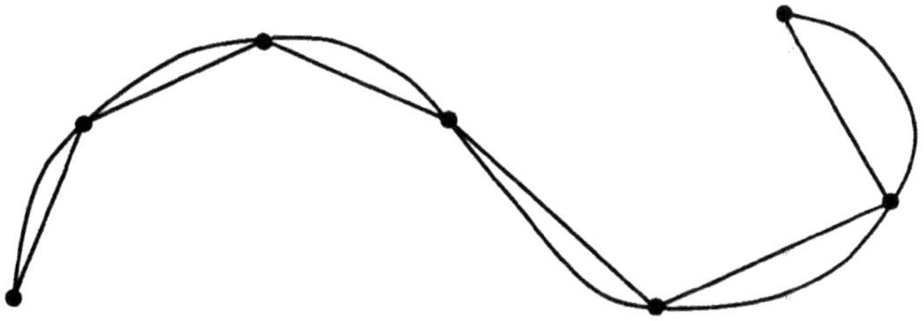

Figure 2. A path ω in $\Omega_1(x,y)$ and its approximation ω_N in $\Omega_{1,n}(x,y)$.

as $\ell \to \infty$.

At this point we adopt the same strategy as with the lattice paths and look for a weight factor $W(n,a)$ such that

$$\begin{aligned} \langle F \rangle^a \equiv & \sum_{n=1}^{\infty} W(n,a) \int F(\omega, na^2) \\ & \times \exp\left(-\frac{1}{2a^2} \sum_{n=1}^{n} (x_i - x_{-1})^2 - \frac{1}{2} m^2 n a^2 \right) \prod_{i=1}^{n-1} \frac{dx_i}{a^d} \end{aligned} \tag{24}$$

converges to $\int F\, D[\omega]$ as $a \to 0$. We have divided the Lebesgue measures dx_i by the factors a^d in order to make the measure dimensionless. Otherwise this factor would show up in $W(n,a)$. The gaussian form of the action makes it particularly easy to determine the weight factor. In the case $F = 1$ we can carry out all the gaussian integrations explicitly and obtain

$$\langle 1 \rangle^a = \sum_{n=1}^{\infty} W(n,a) \frac{(2\pi)^{nd/2}}{(2\pi n)^{d/2}} e^{-(x-y)^2/2a^2 n - m^2 n a^2/2}. \tag{25}$$

If we take

$$W(n,a) = \frac{1}{2} a^{2-d} (2\pi)^{-nd/2}, \tag{26}$$

then we notice that the sum (25) converges to the integral

$$\frac{1}{2} \int_0^{\infty} e^{-(x-y)^2/2t - \frac{1}{2} m^2 t} (2\pi t)^{-d/2}\, dt = G(x,y), \tag{27}$$

as $a \to 0$ with $a^2 n = t$. The same convergence can be established with an arbitrary bounded measurable function F on the space of geometric paths [9].

There is nothing special about the gaussian action (18). It just makes the calculation easy. The same convergence is obtained with an action of the form

$$S_\varphi(\omega) = \sum_{i=1}^{n} \varphi((x_i - x_{i-1})^2/a^2) - \frac{1}{2} \sigma^2 m^2 a^2 n \tag{28}$$

for any continuous function φ for which

$$\int x^2 e^{-\varphi(x)}\, dx < \infty. \tag{29}$$

In Eq. (28) σ^2 is a constant which only depends on φ. This result is a consequence of the central limit theorem [10].

The two methods we have illustrated here for discretizing the functional integral over geometric paths can both be extended to calculate integrals over random paths with a curvature dependent action [11] as well as to fermionic paths [12]. Self-avoiding paths do not have an action which is local on the paths and functional integrals over such paths cannot be treated by these methods.

3. Branched polymers

Branched polymers are the simplest generalization of random paths. There is no known continuum theory of branched polymers comparable to the Wiener integral for paths. Here we will apply the methods of the previous section to the problem of calculating the continuum limit of statistical sums over branched polymers. These results have a value in their own right and they are particularly relevant for the theory of random surfaces because some random surface theories are dominated by branched polymer configurations.

3.1. EXTRINSIC PROPERTIES

Let $\mathcal{B}$ be the collection of all finite, rooted connected tree graphs where the links meeting at each vertex are cyclically ordered. By 'rooted' we mean that one vertex is singled out and called the root. This is for technical convenience only. If $b \in \mathcal{B}$ we let $V(b)$ denote the collection of all vertices in b. For $i \in V(b)$ we let σ_i denote the order of i, i.e. the number of links meeting at i. We assume, again for technical convenience, that the order of the root is 1.

In this section we study the statistical properties of tree graphs embedded in a Euclidean space. A branched polymer ω is a tree graph b as defined above together with a mapping

$$\omega : V(b) \mapsto \mathbf{R}^d. \tag{30}$$

We denote the image of i under ω by x_i.

We will discuss directly a general action for branched polymers of the form

$$S(\omega) = \sum_{(ij)} \varphi(|x_i - x_j|), \tag{31}$$

where the sum is over pairs of nearest neighbour vertices and the function φ satisfies

$$c \equiv \int e^{-\varphi(|x|)}\, dx \ < \infty. \tag{32}$$

To each vertex i we also associate a weight factor $w(\sigma_i)$ which only depends on the order of i. It simplifies the notation not to include an explicit cutoff parameter (so effectively $a = 1$) but we reintroduce it when we take the continuum limit.

The simplest object to study is the one point function $G(\mu)$ for branched polymers which is defined by mapping the root to a fixed point in $\mathbf{R}^d$ and integrating over the position of the remaining vertices:

$$G(\mu) = \sum_{b\in\mathcal{B}} \prod_{i\in V(b)} w(\sigma_i) e^{-\mu|b|} \int e^{-S(\omega)} \prod_{i\in V'(b)} dx_i. \tag{33}$$

Here $|b|$ denotes the number of links in b and the prime on $V'(b)$ signifies the omission of the root. One can think of the parameter μ as the cosmological constant for branched polymers.

We can integrate successively over the vertices and find

$$\int e^{-S(\omega)} \prod_{i\in V'(b)} dx_i = c^{|b|}. \tag{34}$$

It follows that

$$G(\mu) = \sum_{n=1}^{\infty} \beta_n e^{-\mu n} \tag{35}$$

where

$$\beta_n = c^n \sum_{b\in\mathcal{B},\, |b|=n} \ \prod_{i\in V(b)} w(\sigma_i). \tag{36}$$

In order for $G(\mu)$ to be finite we must take $\mu > \mu_0$ where μ_0 is the radius of convergence of $\sum_n \beta_n z^n$. Since

$$\#\{b \in \mathcal{B} : |b| = n\} = \frac{(2n)!}{n!(n+1)!} \sim n^{-3/2} 4^n, \tag{37}$$

it follows that $\mu_0 < \infty$ provided the series $\sum_n w(n) z^n$ has a positive convergence radius which we assume to exist and denote by r_0.

We note that $G(\mu)$ is analytic for $\mu > \mu_0$ but singular at μ_0. The same is true for the n point function which is defined by fixing n of the vertices in the imbedding space $\mathbf{R}^d$. We can now determine the generic critical

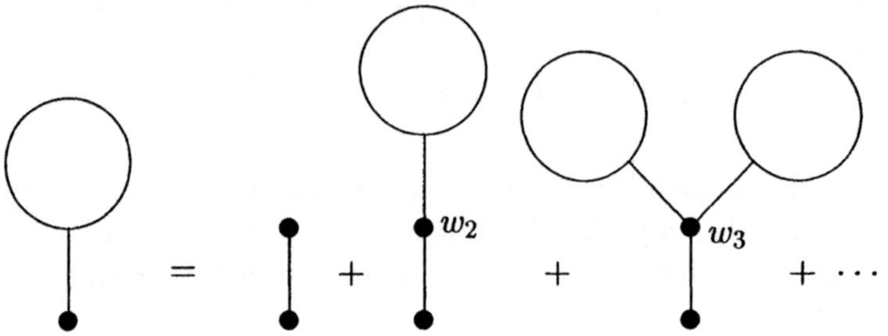

Figure 3. Graphical illustration of the identity (38).

behaviour of $G(\mu)$ at $\mu = \mu_0$. Let k denote the order of the vertex next to the root. Then we can write

$$G(\mu) = e^{-\mu} \sum_{k=1}^{\infty} w(k) G(\mu)^k. \tag{38}$$

We can assume that $w(1) = 1$ by multiplying all weight factors with a constant if necessary. The one point function $G(\mu)$ tends to 0 as $\mu \to \infty$. It follows that, for $\mu > \mu_0$, $G(\mu)$ is the smallest positive solution of the equation

$$e^{\mu} = \frac{1 + f(G)}{G} \equiv F(G), \tag{39}$$

where

$$f(G) = \sum_n w(n+1) G^n \tag{40}$$

has convergence radius $r_0 > 0$. There are now two cases to consider:

(i) $F(G)$ has a minimum at some $G_0 < r_0$,
(ii) $F(G)$ is decreasing in the interval $[0, r_0]$ and and $e^{\mu_0} = r_0$.

The case (i) is generic and occurs e.g. if f is a polynomial or if the weight factors $w(n)$ decrease sufficiently fast with n. In this case the minimum of $F(G)$ at G_0 is quadratic since all the weights are positive. It follows that $G_0 = G(\mu_0)$ and

$$G(\mu_0) - G(\mu) \sim (\mu - \mu_0)^{1/2} \tag{41}$$

as $\mu \to \mu_0$.

In the nongeneric case one can get a continuous range of critical exponents for G [13]. Allowing the weight factors to have both signs in the generic case leads to so called multicritical branched polymers where the minimum of G can have any order [14]. In the remainder of this section we consider the generic case with positive weight factors.

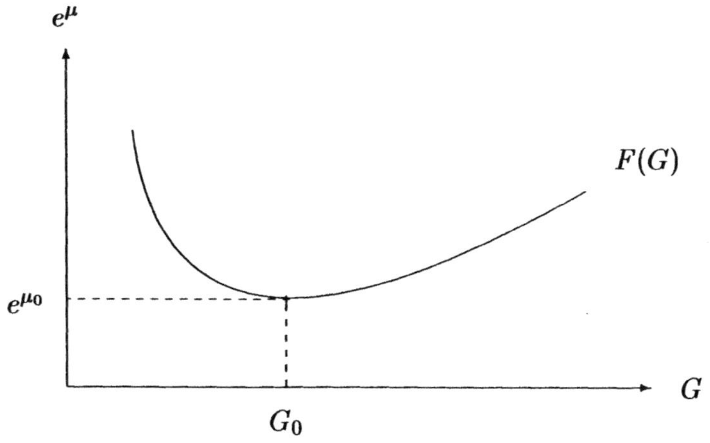

Figure 4. Graphical illustration of the identity (39) in the generic case (i).

The two point function $G^2_\mu(x,y)$ is defined in a similar way as the one point function by summing over tree graphs with two marked vertices and keeping two points corresponding to the marked vertices fixed in the integration. We note that there is a unique path on any tree graph connecting two marked vertices so we have an effective random path connecting the two vertices with outgrowths sitting at each vertex. Evaluating the sum gives

$$G^2_\mu(x,y) = \frac{(1+f(G(\mu)))^2}{f'(G(\mu))} G^{RP}_{\bar{\mu}}(x,y) \tag{42}$$

where $G^{\rm RP}_{\bar{\mu}}(x,y)$ is the propagator for piecewise linear random paths discussed in the previous section and $\bar{\mu}$ is a coupling constant renormalized by summing over the outgrowths,

$$\bar{\mu} = -\log\left(e^{-\mu} f'(G(\mu))\right). \tag{43}$$

The prefactor in (42) is associated with outgrowths at the endpoints. It is straightforward to check that

$$\bar{\mu} \sim (\mu - \mu_0)^{\frac{1}{2}}. \tag{44}$$

Furthermore, it is straightforward to show, using the properties of $G^{\rm RP}$, that the two point function decays exponentially

$$G^2_\mu(x,y) \sim e^{-m_b(\mu)|x-y|} \tag{45}$$

for large separation $|x-y|$ where the mass tends to zero at the critical point according to

$$m_b(\mu) \sim \bar{\mu}^{1/2} \sim (\mu-\mu_0)^{1/4}. \tag{46}$$

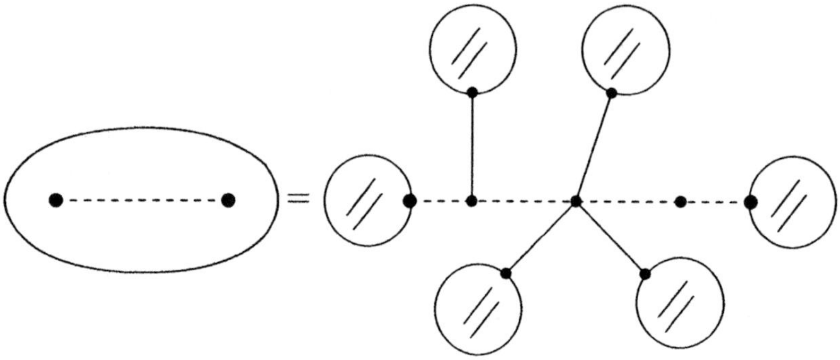

Figure 5. Graphical representation of the two-point function for branched polymers. The dashed line represents the unique shortest path between the marked vertices. The shaded blobs represent contributions from polymers rooted in vertices on this path. Each of the blobs equals the one point function.

One can construct the continuum limit of the two point function in exactly the same way as for random walks by introducing a cutoff parameter a and scaling appropriately. We find that there is no anaomalous scaling, i.e. $\eta = 0$ and the critical exponent of the mass $\nu = 1/4$ according to (46).

It is customary in random geometric theories to define the susceptibility χ as the integral of the two point function. This terminology comes from the analogy with spin systems. In the present case the susceptibility is

$$\chi(\mu) = \int G_\mu^2(x, y)\, dy. \tag{47}$$

On a rooted polymer b we can choose a second vertex in exactly $N_v(b) - 1$ ways where $N_v(b)$ is the the number of vertices in b. If we differentiate the one point function with respect to μ we bring down a factor of $|b|$ so

$$\chi(\mu) \sim -\frac{d}{d\mu} G(\mu) \sim (\mu - \mu_0)^{-1/2} \tag{48}$$

since $N_v(b) = |b| + 1$. It follows that the critical exponent γ of the susceptibilty is $1/2$ and Fisher's scaling relation

$$\gamma = (2 - \eta)\nu \tag{49}$$

holds for the critical exponents.

3.2. INTRINSIC PROPERTIES

All random walks consisting of n steps are identical as far as their intrinsic properties are concerned. However, this is not the case for surfaces and higher dimensional objects which have a rich intrinsic structure. The intrinsic structure is what we are interested in for applications to quantum gravity, i.e. what do the random geometric structures look like 'seen from inside'? In this subsection we study briefly the intrinsic properties of branched polymers.

Let $\mathcal{B}_2(n)$ denote the collection of all branched polymers with two marked vertices separated by n links. We define the intrinsic two point function as

$$G^I_\mu(n) = \sum_{b\in\mathcal{B}_2(n)} \prod_{i\in V(b)} w(\sigma_i) \int e^{-S(b)-\mu|b|} \prod_{i\in V'(b)} dx_i. \tag{50}$$

It is straightforward to show that

$$G^I_\mu(n) \sim e^{-m_I(\mu)n} \tag{51}$$

where the intrinsic mass $m_I(\mu)$ scales to 0 at the critical point as

$$m_I(\mu) \sim (\mu - \mu_0)^{1/2}, \tag{52}$$

i.e. the intrinsic critical exponent for mass is $\nu_I = 1/2$ which is different from the value found for the critical exponent associated with the mass of the extrinsic two point function in the previous subsection. The continuum limit of the two point function can be constructed in the same way as before:

$$G^I_{\text{cont}}(r) = \lim_{a\to 0} a^\lambda G^I_{\mu(a)}(r/a), \tag{53}$$

where r is an arbitrary real number, a is a scaling parameter going to zero along a sequence so that r/a runs through integers and $\mu(a) \to \mu_0$ as $a \to 0$. The exponent λ has the value 0 so we have an anomalous dimension $\eta_I = 1$ since the underlying dimension is $d = 1$.

One can check that the scaling relations $d_H\nu = 1$ and $\gamma = \nu(2-\eta)$ hold for the intrinsic exponents. In particular we see that the intrinsic Hausdorff dimension d^I_H of branched polymers is 2. This means that the average number of vertices at distance n from the root on a rooted polymer is proportional to n for large n provided there is at least one vertex at distance n from the root. This result was first established by Kolmogorov in the 1930s, see e.g. [15], but rediscovered by physicists recently [14]. We note that the intrinsic Hausdorff dimension $d^I_H = 2$ is smaller than the extrinsic Hausdorff dimension $d_H = 4$. This is a special case of an inequality

$$d^I_H \le d_H \tag{54}$$

between intrinsic and extrinsic Hausdorff dimensions which is expected to be valid generically [16]. The intuitive reason is that the distance between two points on a random manifold measured in imbedding space is always smaller than the distance between the same points measured along the imbedded manifold.

4. Dynamicaly triangulated surfaces

In this section we turn to random surfaces and begin by discussing a discretization of the functional integral

$$Z = \int e^{-S_P(g,X)} \mathcal{D}g\mathcal{D}X \tag{55}$$

where the Polyakov action is defined by

$$S_P(g,X) = \int_\sigma \left(\Lambda + \frac{1}{2} g^{ab} \frac{\partial X^\kappa}{\partial \xi^a} \frac{\partial X^\kappa}{\partial \xi^b} \right) \sqrt{g}\, d^2\xi. \tag{56}$$

Here g is an intrinsic metric on the two-dimensional surface σ with coordinates ξ^a and κ is a space-time index which runs from 1 to d. The constant Λ is the cosmological constant.

4.1. DEFINITIONS

The integral over (equivalence classes of) metrics will be replaced by a sum over discrete triangulations. A triangulation is a finite collection of 2-simplexes some of whose sides are pairwise identified in such a way that the resulting complex is a connected surface, possibly with a boundary which corresponds to those links that are not identified with any other link. We think of the triangles as being equilateral in the intrinsic metric of the corresponding piecewise linear surface. The triangulations form a discrete subset of the set of all metrics and the hypothesis is that they become dense in the space of all metrics as the sidelength of the triangles goes to zero in the same way as the lattice paths become uniformly dense in the space of all paths as the lattice spacing tends to zero.

If T is a triangulation we let ∂T denote its boundary and $|T|$ the number of triangles in T. We let $V(T)$ denote the vertices in T. We denote by $\mathcal{T}(n_1, \ldots, n_b)$ the collection of all triangulations with b boundary components of length $n_1, \ldots, n_b$. The length of a boundary component is the number of links it contains. Different boundary components may share vertices.

We now define what we mean by a surface. A surface based on a triangulation $T \in \mathcal{T}(n_1, \ldots, n_b)$ is a mapping

$$X : V(T) \mapsto \mathbf{R}^d \tag{57}$$

where the vertex i is mapped to the point x_i. We think of each triangle Δ in T as being mapped to a triangle in $\mathbf{R}^d$ defined by the images of the vertices of Δ. The boundary ∂T is mapped to a union of polygonal loops $\gamma_1, \ldots, \gamma_b$. The discretization of the Polyakov action is

$$S(X,T) = \frac{1}{2}\sum_{(ij)}(x_i - x_j)^2 + \mu|T|, \tag{58}$$

where the sum runs over nearest neighbour vertices. We will denote the first term on the right hand side, the gaussian term, by $S_T(X)$.

The basic quantities of interest are the loop functions. The b-loop function is defined as

$$G_\mu(\gamma_1, \ldots, \gamma_b) = \sum_{T \in \mathcal{T}(n_1,\ldots,n_b)} e^{-\mu|T|} G^{(T)}(\gamma_1, \ldots, \gamma_b), \tag{59}$$

where

$$G^{(T)}(\gamma_1, \ldots, \gamma_b) = \int e^{-S_T(X)} \prod_i dx_i \tag{60}$$

and in Eq. (60) the integration is only over the internal vertices. The images of the boundary vertices are fixed at the corners of the loops γ_j. Note that a coupling constant in front of the gaussian action can be scaled out and just renormalizes the cosmological constant μ.

4.2. BASIC PROPERTIES

Unfortunately the loop functions as defined above are not convergent for any value of μ. The reason is that the number of triangulations with a fixed number A of triangles grows faster than exponentially with A. No sensible action for surfaces can suppress this divergence. We must therefore restrict the summation in Eq. (59) to surfaces of a fixed genus. If $N_{h,b}(A)$ is the number of surfaces of genus h with b boundary components made up of A triangles, then

$$N_{h,b}(A) \sim A^{\gamma_h + b - 3} e^{\mu_0} \tag{61}$$

where $\gamma_h = (5h-1)/2$ and the constant μ_0 is independent of h and b. In the sequel we restrict our attention to planar surfaces. Giving a meaning to the sum over all genera is a nontrivial problem which is not completely solved [7, 17].

Let us consider a surface with one degenerate boundary component which consists of a single vertex. Then it is easy to carry out the gaussian integrations in Eq. (60) and we obtain

$$\int e^{-\frac{1}{2}\sum_{ij}(x_i - x_j)^2} \prod_i dx_i = (\det C_T)^{-d/2} (2\pi)^{d(N_v(T)-1)/2} \tag{62}$$

where $N_v(T)$ is the number of vertices in T and C_T is the modified adjacency matrix of T. This is a matrix indexed by the internal vertices, i.e. the vertices which we integrate over, and

$$(C_T)_{ij} = \begin{cases} \sigma_i, & \text{if } i = j, \\ -\#\text{ links from } i \text{ to } j, & \text{if } i \neq j. \end{cases} \tag{63}$$

Here σ_i denotes the order of the vertex i. The integration formula (62) generalizes to surfaces with an arbitrary boundary and also allows us to extend the model to any real value of the imbedding dimension d. This result shows that for any d there is a constant μ_d such that the loop functions are convergent for $\mu > \mu_d$ and analytic in μ but divergent for $\mu < \mu_d$. The loop functions depend continuously on the coordinates of the boundary loops. This result holds for a large class of action functionals [18].

The two point (or two loop) function decays exponentially in the separation of the points in imbedding space and thereby defines a mass. The integrated two point function defines the susceptibility $\chi(\mu)$ as we discused for branched polymers. The value of the critical exponent γ for random surface models with a local action is always bounded from above by 1/2. This is a consequence of the inequality

$$-\frac{d\chi}{d\mu} \geq \chi^3 \tag{64}$$

which follows from two observations. The derivative of χ can be represented as the double integral over the three point function and the three point function is smaller than a product of three two point functions with a common vertex which is integrated over. We refer to [7] for details.

4.3. THE STRING TENSION

The mass in a random surface theory can be regarded as a measure of the response of a surface to being stretched in some direction. The string tension is in a similar way a measure of the response of a surface to being stretched simultaneously in two orthogonal directions. Now we make this more precise.

Let $\gamma(L_1, L_2)$ be a square loop in $\mathbf{R}^d$, lying in a two-dimensional plane, with sides of length L_1 and L_2. We assume that the sides have n_1 and n_2 vertices and we are interested in studying the behaviour of the one-loop function $G_\mu(\gamma(L_1, L_2))$ as L_1 and L_2 grow with $L_i/n_i = c$ fixed. We define the string tension as

$$\tau(\mu) = \lim_{L_1, L_2 \to \infty} -\frac{\log G_\mu(\gamma(L_1, L_2))}{L_1 L_2}. \tag{65}$$

The above limit exists and is independent of the constant c. If we do not fix the L_i/n_i ratio we obtain a boundary dependent contribution to τ. We can regard the string tension as the free energy per unit area of a surface spanning a large loop of area A. In a gauge theory the analogous quantity is a measure of the linear growth of the static potential between heavy quarks.

The existence of the limit defining the string tension follows from the subadditivity of $\log G_\mu(\gamma(L_1, L_2))$ in L_1 and L_2. The subadditivity is a consequence of the inequality

$$G_\mu(\gamma(L_1 + L_1', L_2)) \geq G_\mu(\gamma(L_1, L_2)) G_\mu(\gamma(L_1', L_2)). \tag{66}$$

The inequality (66) does in fact not hold unless we impose a condition on the surfaces at the boundary loops but the surfaces which satisfy the boundary condition yield the same string tension as the full ensemble. The details can be found in [7].

The most important fact about the string tension is the inequality [19]

$$\tau(\mu) \geq 1 \tag{67}$$

for any $\mu \geq \mu_d$ and $d \geq 2$. This inequality shows that the physical string tension which one would like to construct in a continuum limit

$$\tau_{\text{phys}} = \lim_{\mu \to \mu_0} \frac{\tau(\mu)}{a^2(\mu)} \tag{68}$$

is infinite. Here $a(\mu)$ is a cutoff parameter which tends to 0 as $\mu \to \mu_d$. This result shows that the surfaces shrink to esentially one-dimensional tubes as the critical point is approached. Branching of the tubes is not suppressed by the positive string tension so the inequality (67) can be taken as an indication that the loop functions of the model are dominated by branched polymer-like surfaces. In view of the importance of this result we will outline the proof.

For simplicity let us take $L_1 = L_2 = L$ and $n_1 = n_2 = n$. Let $T \in \mathcal{T}(4n)$ and let $X : T \mapsto \mathbf{R}^d$ be a surface contributing to $G_\mu(\gamma(L, L))$, i.e.

$$X(\partial T) = \gamma(L, L). \tag{69}$$

For each such T let $X(\gamma, T)$ be the surface that minimizes the gaussian action S_T with the boundary condition (69). For an arbitrary surface X based on T which satisfies the same boundary condition we can write $X = X(\gamma, T) + X'$, where $X'(\partial T) = 0$. Since the action is quadratic in the coordinates of the vertices and $X(\gamma, T)$ minimizes the action we find

$$S_T(X) = S_T(X(\gamma, T)) + S_T(X'). \tag{70}$$

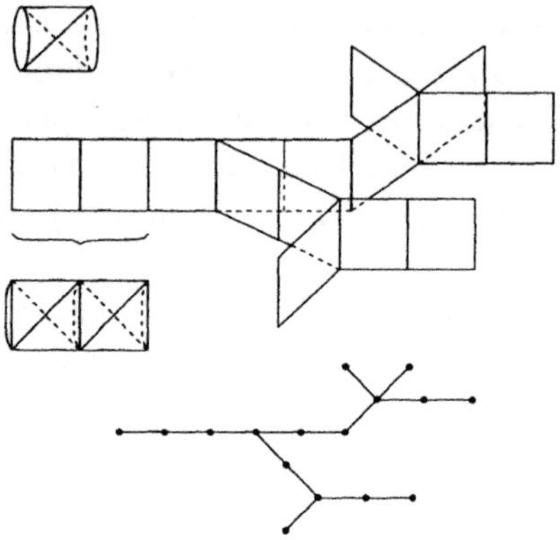

Figure 6. Branched polymer-like triangulations that are expected to dominate the loop functions in $d > 1$. The triangulation is constructed from elementary building blocks each of which consists of two pairs of triangles as indicated in the figure.

We denote the first term on the right hand side above by $S_{T,\min}$. Then we can write

$$G_\mu(\gamma(L,L)) = \sum_{T \in \mathcal{T}(4n)} e^{-\mu|T|} e^{-S_{T,\min}} G_\mu^{(T)}(O_n), \tag{71}$$

where O_n is a degenerate loop in $\mathbf{R}^d$ consisting of $4n$ coiniciding points.

The desired lower bound on the string tension follows now from a very simple lower bound on the minimal action. If we have a triangle with sides of length ℓ_1, ℓ_2, ℓ_3 then the area A of the tringle always satisfies the inequality

$$\ell_1^2 + \ell_2^2 \geq 4A. \tag{72}$$

The minimal action is simply the sum over the lengths of the links squared in the minimal surface $X(\gamma, T)$. Hence,

$$S_{T,\min} = \frac{1}{2} \sum_i \ell_i^2 \geq A_{\min} \geq L^2 \tag{73}$$

where $A_{\min}$ is the area of the minimal surface. Since the minimal surface spans a loop of area L^2 its area is larger than L^2. We conclude that

$$G_\mu(\gamma(L,L)) \leq e^{-L^2} G_\mu(O_n). \tag{74}$$

The desired result follows now from the bound

$$G_\mu(O_n) \leq e^{c(\mu)n} \tag{75}$$

where $c(\mu)$ is a constant. The proof of this bound is slightly technical but only uses the continuity of the loop functions in the coupling constant μ and the coordinates of the boundary loops [7].

4.4. FURTHER RESULTS

The results discussed above for the randomly triangulated surface model are valid in any imbedding dimension d. Of course we need $d \geq 2$ for the definition of the string tension to make sense. Here we list some of the most important results obtained for specific dimensions.

In the case $d = 0$ which corresponds to pure Euclidean gravity in two space-time dimensions the model can be solved exactly in a very strong sense and the continuum limit of the loop functions constructed explicitly [20]. For $d = 1/2$ which corresponds to an Ising model on a random surface all the critical exponents can be calculated [21] and the same holds true for $d = 1$.

The model becomes rather simple for $d = -2$ since the determinant of the adjacency matrix appears to the power 1, and again all critical exponents can be calculated. The model can be coupled to a variety of two-dimensional models (multiple Ising models, Potts models etc.). It appears that the model is dominated by branched polymers when coupled to matter fields with central charge $c > 1$. We note that for surfaces imbedded in Euclidean space each imbedding dimension corresponds to a free gaussian field on the surface and therefore contributes 1 to the central charge of fields on the surface. There is considerable numerical evidence that the critical behaviour only depends on the total central charge of matter fields on the sufaces. This is also predicted by the KPZ-theory [22, 23, 24].

For $d \to -\infty$ the weight of smooth regular surfaces in the loop functions increases but there is no evidence for a sharp transition to smooth surfaces at any fixed negative value of d. One can see intuitively that smooth surfaces should have a large weight for large negative d because the determinant of the adjacency matrix equals the number of spanning trees on the surface and regular surfaces have the largest number of spanning trees. This also agrees with the result that branched polymers dominate for $d > 1$ because tree-like surfaces have the smallest number of spanning trees.

5. Lattice surfaces

In this section we study the simplest lattice random surface model. This model was originaly introduced in order to study lattice gravity and the strong coupling expansion of lattice gauge theories. The main objective here is to show that the model is equivalent to a branched polymer model.

5.1. DEFINITIONS

A plaquette p is an elementary square in the lattice $\mathbf{Z}^d$. The boundary of a plaquette consists of four links each of which connects two lattice points which are nearest neighbours. A lattice surface S is constructed by gluing together a finite number of plaquettes from $\mathbf{Z}^d$ along links (one or more) which they share, in such a way that the resulting complex is an orientable simplicial surface. We allow plaquettes to be used arbitrarily often and the surfaces can intersect and overlap themselves without any restriction. We require surfaces to be connected and the links of plaquettes that are not identified with any other link form the boundary ∂S of the surface S. We let $|S|$ denote the area of S, i.e. the number of plaquettes in S.

On the lattice we have the same problem as for triangulations namely that the number of surfaces with a given area A (and one fixed vertex to remove translational degeneracy) grows superexponentially with A. We are therefore forced to consider sufaces with fixed topology so in the sequel we restrict our attention to surfaces with the topology of a sphere with a number of holes which correspond to the boundary components. We will refer to such surfaces as planar surfaces.

We let $\mathcal{S}(\gamma_1, \ldots, \gamma_n)$ denote the collection of all planar surfaces with boundary components $\gamma_1, \ldots, \gamma_n$ and define the loop functions as

$$G_\beta(\gamma_1, \ldots, \gamma_n) = \sum_{S \in \mathcal{S}(\gamma_1, \ldots, \gamma_n)} e^{-\beta|S|}. \tag{76}$$

One can show that there is a positive constant β_d which only depends on the dimension d such that all the loop functions are convergent for $\beta > \beta_d$ and divergent for $\beta < \beta_d$. The statistical model defined in this way has been called the planar lattice random surface (PLRS) model.

Let γ be a lattice loop and γ_ℓ the translate of γ by ℓ lattice spacings in a coordinate direction. Then one can prove that there is a positive mass $m(\beta)$ such that

$$G_\beta(\gamma, \gamma_\ell) \sim e^{-m(\beta)\ell} \tag{77}$$

as $\ell \to \infty$. Similarly if $\gamma_{L \times L}$ is a square $L \times L$ loop in a coordinate plane we define the string tension by

$$G_\beta(\gamma_{L\times L}) \sim e^{-\tau(\beta)L^2}. \tag{78}$$

The existence of the mass and the string tension is established by subadditivity arguments in very much the same fashion as we explained for dynamically tringulated surfaces. The susceptibility $\chi(\beta)$ is defined in the same way as before

$$\begin{aligned} \chi(\beta) &= \sum_{x \in \mathbf{Z}^d} G_\beta(\gamma, \gamma_x) \\ &\sim -\frac{d}{d\beta} G_\beta(\gamma), \end{aligned} \tag{79}$$

where γ_x is the translate of γ by the lattice vector x. We will now establish the critical behaviour of these quantities as $\beta \to \beta_d$.

5.2. CRITICAL BEHAVIOUR

We begin by introducing an auxiliary class of lattice surfaces. The set

$$\mathcal{S}^{(2)}(\gamma_1, \ldots, \gamma_n) \subset \mathcal{S}(\gamma_1, \ldots, \gamma_n) \tag{80}$$

is defined to consist of those lattice surfaces that cannot be made disconnected by cutting the surfaces along two links which form a closed loop. We will call such closed loops *bottlenecks*. The surfaces that satisfy this condition give rise to another lattice surface theory where we can define all the same quantities as in our original theory, mass, string tension etc. We denote these quantities in the new theory by the same symbol as before but add a superscript (2) and therefore write $\beta_d^{(2)}, m^{(2)}(\beta), \tau^{(2)}(\beta)$ etc. We expect the new surface theory to have the same critical behaviour as the original one since the constraint is a local one. This universality has been checked in numerical simulations.

With the definitions we made above one can prove the following result: *If the susceptibility of the PLRS-model $\chi(\beta)$ diverges as $\beta \to \beta_d$ and $\chi^{(2)}(\beta) \to \infty$ as $\beta \to \beta_d^{(2)}$ then all the critical exponents of the PLRS-model are the same as for branched polymers, $\tau(\beta_d) > 0$ and the continuum limit of the two loop correlation function equals the free propagator $(-\Delta + m^2)^{-1}$.*

We will now outline the proof of the above result. Let γ_ℓ be a closed lattice loop of length 2 made up of two copies of a link ℓ. Let us introduce the notation

$$G_\beta(\gamma_\ell) = G(\beta), \quad G_\beta^{(2)}(\gamma_\ell) = G^{(2)}(\beta). \tag{81}$$

These functions are independent of the link ℓ by the lattice symmetry. Take a surface $S \in \mathcal{S}^{(2)}(\gamma_{L\times L})$. If we cut this surface open along some links and thereby create holes with boundaries of length 2 we can close the surface by gluing to the holes surfaces from $\mathcal{S}(\gamma_\ell)$ (suitably translated and/or rotated). By this process we obtain a surface with the same boundary as the original surface and a number of bottlenecks.

Conversely, any surface in $\mathcal{S}(\gamma_{L\times L})$ can be decomposed into a collection of maximal outgrowths with a boundary of length 2 and a base surface with no bottlenecks. This allows us to express the one loop function in the PLRS-model as the one loop function in the auxiliary model with no bottlenecks but with a renormalized coupling constant:

$$G_\beta(\gamma_{L\times L}) = \sum_{S \in \mathcal{S}^{(2)}(\gamma_{L\times L})} e^{-\beta|S|}(1 + G(\beta))^{N_\ell(S)}, \tag{82}$$

where $N_\ell(S)$ is the number of links in S. Noting that $N_\ell(S) = 2L + 2|S|$ we obtain

$$G_\beta(\gamma_{L\times L}) = (1 + G(\beta))^{2L} G^{(2)}_{\bar{\beta}}(\gamma_{L\times L}), \tag{83}$$

where the renormalized coupling is given by

$$\bar{\beta} = \beta - 2\log(1 + G(\beta)). \tag{84}$$

Since $\bar{\beta} \geq \beta_d^{(2)}$ it follows that $G(\beta_d) < \infty$ and Eq. (83) implies that for all values of β we have the identity

$$\tau(\beta) = \tau^{(2)}(\bar{\beta}). \tag{85}$$

By a similar argument one can show that

$$G(\beta) = G^{(2)}(\bar{\beta}), \tag{86}$$

a result we will use below.

We are now ready to relate the susceptibilities in the two models. We know that the susceptibilities diverge at the same rate as the first derivative of the one loop function so we adopt the definitions

$$\chi(\beta) = -\frac{d}{d\beta}G(\beta), \quad \chi^{(2)}(\bar{\beta}) = -\frac{d}{d\bar{\beta}}G^{(2)}(\bar{\beta}). \tag{87}$$

It follows by the chain rule that

$$\chi(\beta) = \left(1 + \frac{2\chi(\beta)}{1 + G(\beta)}\right)\chi^{(2)}(\bar{\beta}), \tag{88}$$

and solving for $\chi(\beta)$ yields

$$\chi(\beta) = \frac{(1 + G(\beta))\chi^{(2)}(\bar{\beta})}{1 + G(\beta) - 2\chi^{(2)}(\bar{\beta})}. \tag{89}$$

Since we are assuming that the susceptibilities in both models diverge at their respective critical points we conclude immediately that

$$\bar{\beta}(\beta_d) > \beta_d^{(2)} \tag{90}$$

and the denominator in Eq. (89) vanishes for $\beta = \beta_d$, i.e.

$$1 + G(\beta_d) = 2\chi^{(2)}(\bar{\beta}(\beta_d)). \tag{91}$$

Differentiating Eq. (89) with respect to β and isolating the most divergent term on the right hand side we find that

$$-\chi'(\beta) \sim \chi^3(\beta), \tag{92}$$

which implies that the critical exponent of the susceptibility in the PLRS-model $\gamma = 1/2$. We see that the general upper bound on the exponent is saturated in this case and γ has the same value as for branched polymers.

We now proceed to analyse the two-loop function along similar lines. Let us consider a surface $S \in \mathcal{S}(\gamma_1, \gamma_2)$ with two boundary loops γ_1 and γ_2 both consisting of two links. This surface may contain two different kinds of bottlenecks. If we travel along the surface from γ_1 to γ_2 we necessarily cross some of the bottlenecks. We will refer to these as intermediate bottlenecks. The bottlenecks which are not intermediate just renormalize the coupling as in our decomposition of the surfaces contributing to the one loop function. If we let γ be the first intermediate bottleneck as we travel along the surface from γ_1 to γ_2 we can write

$$G_\beta(\gamma_1, \gamma_2) = (1 + G(\beta))^2 G^{(2)}_{\bar{\beta}}(\gamma_1, \gamma_2) + \sum_{\gamma} G^{(2)}_{\bar{\beta}}(\gamma_1, \gamma) G_\beta(\gamma, \gamma_2), \qquad (93)$$

where the first term on the right hand side above corresponds to surfaces that do not have any intermediate bottleneck.

Now note that there is a one to one corespondence between loops of length 2 and pairs (x, e_j) where x is a lattice point and e_j is a unit vector in the direction of one of the coordinate axes. The vertices of the loop are simply x and $x + e_j$. It follows (by translational invariance) that we can write

$$G_\beta(\gamma_1, \gamma_2) = G_\beta(x_1 - x_2; j_1, j_2) \qquad (94)$$

where γ_k corresponds to (x_k, j_k), $k = 1, 2$. This form of the two-loop function makes it natural to write it as a matrix

$$G_\beta(\gamma_1, \gamma_2) = \mathbf{G}_\beta(x_1 - x_2)_{j_1 j_2}. \qquad (95)$$

Introducing the Fourier transform of the matrix, defined by

$$\widehat{\mathbf{G}}_\beta(p) = \sum_{x \in \mathbf{Z}^d} e^{-ip\cdot x} \mathbf{G}_\beta(x) \qquad (96)$$

(and similarly for the two-loop function without bottlenecks), we can rewrite Eq. (93) as

$$\widehat{\mathbf{G}}_\beta(p) = \widehat{\mathbf{G}}^{(2)}_{\bar{\beta}}(p) + \widehat{\mathbf{G}}^{(2)}_{\bar{\beta}}(p)\widehat{\mathbf{G}}_\beta(p). \qquad (97)$$

Here we have absorbed the factor $(1 + G(\beta))^2$ into the two-loop function and matrix multiplication is understood in Eq. (97). We can now invert the matrix $1 - \widehat{\mathbf{G}}^{(2)}_{\bar{\beta}}(p)$ and solve for $\widehat{\mathbf{G}}_\beta(p)$. Since the two-loop function $\widehat{\mathbf{G}}^{(2)}_{\bar{\beta}}(p)$ is analytic at $\bar{\beta} = \bar{\beta}(\beta_d)$ we can deduce the critical behaviour of the two-loop function in the PLRS-model.

Due to the matrix form of the two-loop function the calculation is slightly tedious, see [7], but the result is that the Fourier transform $\widehat{\mathbf{G}}_\beta(p)$ is analytic in each component of p in a disc of radius $r \sim (\beta - \beta_d)^{1/4}$. We conclude immediately that he critical exponent of the mass is $1/4$ as we found for branched polymers.

With this knowledge it is straightforward to construct the continuum limit of the two loop function and we find

$$\lim_{a \to 0} a^2 \widehat{\mathbf{G}}_\beta(a)(ap) = \frac{1}{m^2 + p^2}\mathbf{I}, \tag{98}$$

where $\mathbf{I}$ is the matrix all of whose matrix elements are equal to 1. This completes the outline of the proof of the equivalence of the PLRS-model with a model of branched polymers.

6. Conclusion

In these lectures we have studied the two simplest models of discretized random surfaces, namely the dynamically triangulated surfaces with gaussian action and lattice surfaces with area action. In both cases we have seen that the continuum limit is dominated by narrow tree-like surface configurations which makes the models equivalent to branched polymers. This is not rigorously proven in the case of randomly triangulated surfaces but the evidence is overwhelming. Much of this evidence comes from numerical simulations which we have not discussed here. In fact there are excellent reasons to believe that the two models are equivalent. The lesson is that in order to construct a genuine random surface theory one should consider theories which suppress narrow outgrowths on surfaces.

The entropy of surfaces has two origins. First of all there is the intrinsic entropy asociated with the internal arrangement of plaquettes and triangles and secondly there is the imbedding entropy which has to do with how the surface lies in the ambient space. It seems like in the simplest cases it is the imbedding entropy that dominates and drives the collapse of the surfaces to branched polymers since the polymer-like surfaces have the largest imbedding entropy.

Just as the study of discrete random walks suggested how to discretize random surfaces we can look to random paths for guidance as to how we should modify the random surface models in order to avoid the dominance of branched polymers. It is known that an extrinsic curvature term in the action of random paths can shift the Hausdorff dimension from 2 to 1 [11] so that smooth paths dominate and the same end can be achieved by introducing fermionic degrees of freedom for the paths [12]. One would like to do the same for random surfaces and there are several promising ways to do this. Here we will only discuss briefly extrinsic curvature actions in thc context of randomly triangulated surfaces. Some of the same effects can be obtained by adding higher powers of the intrinsic curvature to the action. Fermionic degrees of freedom for random surfaces are not very well-understood but see [25] for some recent results.

The simplest extra term one can add to the Polyakov action in order to enhance the contribution of smooth surfaces is the extrinsic curvature action [26]

$$S_E(X,g) = \int (D_a n_i^\kappa)^2 \sqrt{g}\, d^2\xi, \tag{99}$$

where D_a is the covariant derivative in the normal bundle of the surface X and n_i^κ are the components of $d-2$ orthonormal vector fields on the surface. For surfaces imbedded into $\mathbf{R}^3$ there is a simple discretization of this action which can be written

$$S_{DE}(T,X) = \sum_{(\Delta,\Delta')} (1 - n_\Delta \cdot n_\Delta) \tag{100}$$

where the sum runs over pairs of neighbouring triangles in the triangulation T and n_Δ is a normal vector to Δ. There is some numerical evidence that an action of this form will cause a transition to a phase of smooth surfaces but analytic understanding is lacking [27, 28].

Another possibility is the so-called gonihedric action [29]

$$S_{\rm Gon}(T,X) = \sum_{(ij)} |x_i - x_j|\, \theta_{ij}^\zeta, \tag{101}$$

where the sum runs over nearest neighbour pairs of vertices and θ_{ij} is the dihedral angle between the two triangles meeting along the link from i to j. It is known that the exponent ζ must be taken smaller than $(d-2)/2$ in order for this model to be well-behaved [30].

References

1. D. Weingarten, Pathological lattice field theory for interacting strings, *Phys. Lett. B* **90** (1980) 280; Euclidean quantum gravity on the lattice, *Nucl. Phys. B* **210** (1982) 229.
2. T. Eguchi and H. Kawai, The number of random surfaces on the lattice and the large-N limit, *Phys. Lett. B* **110** (1980) 143; Planar random surfaces on the lattice, *Phys. Lett. B* **114** (1982) 247.
3. J. Fröhlich, C. E. Pfister and T. Spencer, On the statistical mechanics of surfaces, In *Stochastic processes in quantum theory and statistical physics*, Lecture notes in physics **173**, Springer-Verlag, Berlin-Heidelberg-New York (1982).
4. B. Durhuus, J. Fröhlich and T. Jonsson, Self-avoiding and planar random surfaces on the lattice, *Nucl. Phys. B* **225** (1983) 18¡5.
5. J. Ambjørn, B. Durhuus and J. Fröhlich, Diseases of triangulated random surface models and possible cures, *Nucl. Phys. B* **257** (1985) 433.

6. F. David, A model of random surfaces with non-trivial critical behaviour, *Nucl. Phys. B* **257** (1985) 543; Planar diagrams, two-dimensional lattice gravity and surface models, *Nucl. Phys. B* **257** (1985) 45. and
7. J. Ambjørn, B. Durhuus and T. Jonsson, *Quantum geometry - a statistical field theory approach*, Cambridge University Press, Cambridge (1997).
8. B. Simon, *Functional integration and quantum physics,* Academic Press, New York-San Fransisco-London (1979).
9. B. Durhuus and T. Jonsson, *Discrete approximations to integrals over unparametrized paths*, to appear.
10. P. Billingsley, *Convergence of probabilty measures*, John Wiley & Sons, New York (1968).
11. J. Ambjørn, B. Durhuus and T. Jonsson, Statistical mechanics of paths with curvature dependent action, *J. Phys. A Math. Gen.* **21** (1988) 891.
12. J. Ambjørn, B. Durhuus and T. Jonsson, A random walk representation of the Dirac propagator, *Nucl. Phys. B* **330** (1990) 509.
13. P. Bialas and Z. Burda, Phase transitions in fluctuating branched geometry, *Phys. Lett. B* **384** (1996) 75.
14. J. Ambjørn, B. Durhuus and T. Jonsson, Summing over all genera for $d > 1$: a toy model, *Phys. Lett. B* **244** (1990) 403.
15. K. B. Athreya and P. E. Ney, *Branching processes*, Springer-Verlag, Berlin-Heidelberg-New York (1972).
16. T. Jonsson, Intrinsic and extrinsic geometry of random surfaces, *Phys. Lett. B* **278** (1992) 89.
17. F. David, Phases of the large-N matrix models and non-perturbative effects in $2D$ gravity, *Nucl. Phys. B* **348** (1991) 507.
18. J. Ambjørn, B. Durhuus, J. Fröhlich and P. Orland, The appearance of critical dimensions in string theories, *Nucl. Phys. B* **270** (1986) 457.
19. J. Ambjørn and B. Durhuus, Regularized bosonic strings need extrinsic curvature, *Phys. Lett. B* **118** (1987) 253.
20. J. Ambjørn, J. Jurkiewicz and Y. M. Makeenko, Multiloop correlators for two-dimensional quantum gravity, *Phys. Lett. B* **251** (1990) 517.
21. V. A. Kazakov, Ising model on a dynamical planar random lattice: Exact solution, *Phys. Lett. A* **119** (1986) 140.
22. V. G. Knizhnik, A. M. Polyakov and A. B. Zamolodchikov, Fractal structute of 2d-quantum gravity, *Mod. Phys. Lett. A* **3** (1988) 819.
23. F. David, Conformal field theories coupled to 2-d gravity in the conformal gauge, *Mod. Phys. Lett. A* **3** (1988) 1651.
24. J. Distler and H. Kawai, Conformal field theory and 2-D quantum gravity, *Nucl. Phys. B* **321** (1989) 509.
25. Z. Burda, J. Jurkiewicz and A. Krzywicki, Wilson fermions on a randomly triangulated manifold, *Phys. Rev. D* **60** (1999) 105029.
26. A. M. Polyakov, Fine structure of strings, *Nucl. Phys. B* **268** (1986) 406.
27. J. Ambjørn, A. Irbäck, J. Jurkiewicz and B. Petersson, The theory of dynamical random surfaces with extrinsic curvature, *Nucl. Phys. B* **393** (1993) 571.
28. J. F. Wheater, From polymer membranes to strings, *J. Phys. A Math. Gen.* **27** (1994) 3223.
29. R. V. Ambartzumian, G. K. Savvidy, K. G. Savvidy and G. S. Sukiasian, Alternative model of random surfaces, *Phys. Lett. B* **275** (1992) 99.
30. B. Durhuus and T. Jonsson, On subdivision invariant actions for random surfaces, *Phys. Lett. B* **297** (1992) 271.

LORENTZIAN AND EUCLIDEAN QUANTUM GRAVITY – ANALYTICAL AND NUMERICAL RESULTS

J. AMBJØRN
The Niels Bohr Institute,
Blegdamsvej 17, DK-2100 Copenhagen Ø, Denmark
email: ambjorn@nbi.dk

J. JURKIEWICZ
Institute of Physics, Jagellonian University,
ul. Reymonta 4, PL-30 059, Kraków 16, Poland
email: jurkiewi@thrisc.if.uj.edu.pl

AND

R. LOLL
Albert Einstein Institute
(Max-Planck-Institute for Gravitational Physics),
Am Mühlenberg 1, D-14476 Golm, Germany
email: loll@aei-potsdam.mpg.de

Abstract. We review some recent attempts to extract information about the nature of quantum gravity, with and without matter, by quantum field theoretical methods. More specifically, we work within a covariant lattice approach where the individual space-time geometries are constructed from fundamental simplicial building blocks, and the path integral over geometries is approximated by summing over a class of piece-wise linear geometries. This method of "dynamical triangulations" is very powerful in 2d, where the regularized theory can be solved explicitly, and gives us more insights into the quantum nature of 2d space-time than continuum methods are presently able to provide. It also allows us to establish an explicit relation between the Lorentzian- and Euclidean-signature quantum theories. Analogous regularized gravitational models can be set up in higher dimensions. Some analytic tools exist to study their state sums, but, unlike in 2d, no complete analytic solutions have yet been constructed. However, a great advantage of our approach is the fact that it is well-suited for numerical simulations. In the second part of this review we describe the relevant Monte Carlo techniques, as well as some of the physical results that have

L. Thorlacius and T. Jonsson (eds.), M-Theory and Quantum Geometry, 381-449.

been obtained from the simulations of Euclidean gravity. We also explain why the Lorentzian version of dynamical triangulations is a promising candidate for a non-perturbative theory of quantum gravity.

1. Introduction

There is at present no satisfactory theory of four-dimensional quantum gravity. This is partly related to the conceptual questions that arise when dealing with fluctuating geometries without reference to any background metric. Addressing these seems to call for a genuinely non-perturbative formulation of quantum gravity. The need for a non-perturbative approach persists in formulations where gravity appears embedded into a larger theory, such as string or M-theory. Although there have been attempts to identify non-perturbative structures within these unified theories (see [1, 2] and many successive papers), they so far seem to have raised as many questions as they have been able to answer. In these lectures we will review an alternative non-perturbative approach to quantum gravity. It is more conservative in spirit, in that it does not conjecture the existence of any radically new physical principles or symmetries. This so-called "dynamical triangulations" approach[1] can be formulated entirely within the framework of ordinary quantum field theory, while taking into account that the governing symmetry principle is reparametrization- (diffeomorphism-) invariance.

Our ansatz can be made in any dimension, and has been solved analytically in two space-time dimensions. In higher dimensions one so far has to rely largely on numerical simulations, and up to now only the case of Euclidean signature has been studied in detail. Although gravity in two dimensions has no genuine dynamical degrees of freedom, it does provide a testing ground for addressing some of the conceptual problems of quantum gravity explicitly, for example, how to define a notion of reparametrization-invariant length in a fluctuating geometry or how to define correlation functions without reference to a background geometry. In 2d Euclidean gravity, a number of questions can be answered by continuum functional methods (within Liouville quantum field theory), but some of the most interesting problems involving geodesic distances can only be addressed in the non-

[1]The method of dynamical triangulations was introduced in the context of string theory and 2d quantum gravity in [3, 4, 5], and subsequently extended to higher-dimensional Euclidean quantum gravity [6, 7]. An extensive review covering the developments up to 1996 can be found in the book [8]. A more recent summary is contained in [9], while the review [10] deals with a variety of lattice approaches to four-dimensional quantum gravity, including dynamical triangulations. The use of dynamical-triangulations methods in Lorentzian gravity was pioneered in [11, 12, 13].

perturbative setting of dynamical triangulations. The fact that in this latter approach quantum gravity is obtained as the scaling limit of a lattice theory shows that, contrary to common belief, discrete methods can be used also for reparametrization-invariant theories, as long as the discretization takes place directly on the space of geometries (that is, on the space of metrics modulo diffeomorphisms).

Another point that has been analyzed in the two-dimensional model is the analytic continuation between geometries of Lorentzian and Euclidean signature. Canonical quantization attempts of (Lorentzian) gravity usually take as their starting point globally hyperbolic manifolds equipped with a non-degenerate Lorentzian metric (which gives rise to a causal structure). However, it is *a priori* unclear to what extent these essentially classical structures should be preserved in the quantum theory. For example, in a covariant path-integral approach it is not obvious that a causality constraint should be imposed on each individual space-time configuration contributing to the amplitude. This idea was first advocated by Teitelboim [14], and more recently in [15, 16]. One argument in its favour is that it seems hard to imagine how any notion of causality could emerge in the full quantum theory unless it had been imposed in some form on the individual histories in the first place.

Two-dimensional quantum gravity is an ideal testing ground for such ideas, since it can be solved explicitly. We will show that by restricting the state sum to discrete geometries with a causal structure one obtains a theory of quantum gravity with the following features [11, 12].

(1) The expectation values of reparametrization-invariant space-time distances have canonical dimensions. In other words, in spite of strong fluctuations in the geometry, there is still a sense in which the quantum space-time is two-dimensional.

(2) When matter with conformal charge $c \leq 1$ is coupled to this Lorentzian quantum gravity theory, both property (1) and the properties of the conformal field theory (critical exponents etc.) remain unchanged.

There is an alternative way of arriving at the same theory, namely, by starting from the geometric configurations of the 2d *Euclidean* gravity theory, and removing its “baby universes” in a systematic manner. (The proliferation of these branching structures is responsible for most of the fractal and geometric properties of the Euclidean theory.) In this way one creates a many-to-1 correspondence between Euclidean and Lorentzian geometries. Lorentzian 2d quantum gravity appears in this construction as a “renormalized” version of the Euclidean quantum theory [17].

Loosening the requirement of causality for the individual space-time histories, one may allow for changes in the topology of the *spatial* (constant-time) slices as a function of time, without changing the overall topology of *space-time*. This is of course tantamount to reintroducing geometries with baby universes. We will show below that once this process is permitted, it will totally dominate the structure of space-time. The resulting theory (which is equivalent to Euclidean Liouville quantum gravity) has the following features.

(1) The expectation values of *geodesic* space-time distances have anomalous dimensions, and the intrinsic Hausdorff dimension of quantum space-time is four, and not two [18, 19].

(2) When matter with conformal charge $c \leq 1$ is coupled to this Euclidean quantum gravity theory, both the fractal properties of the gravitational sector and the critical exponents of the conformal field theory are changed.

We therefore may say that the interaction between gravity and matter is weak in the Lorentzian 2d gravity model, at least as long as $c \leq 1$. Allowing for baby-universe creation leads to a strong coupling between matter and gravity: the fractal properties of space-time become a function of the matter content and in turn the back-reaction of the fluctuating geometry changes the critical properties of the matter. The coupling of Ising models to 2d gravity provides a particularly clear geometric illustration of the role played by the baby universes [20].

If one transcends the so-called $c = 1$ barrier, even Lorentzian gravity exhibits a strong gravity-matter interaction, leading to a change in the fractal properties of space-time [13]. In the Euclidean case (i.e. including baby universes), it has been known for a long time that beyond $c = 1$ the space-time disintegrates completely into fractals. These so-called branched polymers can be viewed as trees of baby universes of cut-off size. However, this is *not* what happens in the Lorentzian case. One *does* observe a phase transition for $c > 1$ (to be precise, the transition must take place somewhere in the interval $1 < c < 4$ [13, 21, 22]), but the new geometry is less pathological than the branched polymers. In particular, the critical matter exponents retain their Onsager values.

Many of the results quoted above for $c > 1$ have been obtained by Monte Carlo simulations. Numerical methods have been very successful in the study of discretized quantum gravity. Moreover, in two-dimensional quantum gravity there has been a fruitful interaction between theory and "experiment". Non-rigorous calculations have been verified or falsified by numerical "experiments", and "observations" have inspired theoretical progress. In the Lorentzian case, visualizations of the computer-generated space-time geometries have been useful in understanding the influence of matter on geometry [23].

In higher dimensions, only partial results have been obtained by analytical methods. These include asymptotic estimates for the partition functions of dynamically triangulated Euclidean gravity in 3d and 4d [24, 25], and qualitative descriptions of the extreme branched-polymer and crumpled phases [25, 26, 27]. However, as already mentioned, our regularized quantum gravity models can readily be studied by means of computer simulations. In a first step, one determines the phase diagram of the discretized theory, in order to locate potential critical points where a continuum limit can be obtained. One can then use standard finite-size tools to study the scaling of various observables. This method has been very successful and has enabled us to study both the fractal properties of space-time, and the critical exponents of matter coupled to fluctuating geometries.

The rest of this paper is organized as follows. In Section 2.1 we describe and solve the simplest discretized Lorentzian model of 2d gravity, using a "Lorentzian" version of the method of dynamical triangulations. In Section 2.2, the corresponding continuum limit is obtained, while in Section 3 we demonstrate how the inclusion of baby universes changes the structure of quantum space-time. A non-perturbative definition of Euclidean quantum gravity for arbitrary dimension d by means of dynamical triangulations is given in Section 4, where we also discuss the inclusion of matter fields. Section 5 outlines the principles for numerical simulations of the model, with special emphasis on 2d gravity. In Subsection 5.2, we describe various possibilities of defining notions of (fractal) "dimensionality" in the framework of Euclidean quantum gravity. We discuss why these "observables" are particularly well-suited for use in numerical calculations. Subsection 5.3 provides the interpretation of the numerical results obtained in 2d Euclidean quantum gravity. Section 6 describes the numerical approach to higher-dimensional Euclidean quantum gravity. Finally, in Section 7, we outline the future perspectives for both Euclidean and Lorentzian lattice gravity.

2. Lorentzian gravity in 2d

2.1. THE DISCRETE MODEL

In order to solve two-dimensional quantum gravity in a path-integral formulation, one has to "count" geometries. Since for a fixed space-time topology the gravitational action consists only of the cosmological term, all geometries of a fixed space-time volume contribute with the same weight[2]. In the simplest model of two-dimensional gravity, no *spatial* topology changes are allowed either. For simplicity, we choose the spatial slices (at constant "time") to be circles, so that the overall topology of space-time is of the

[2]Maybe surprisingly, in the framework of dynamical triangulations also higher-dimensional gravity can be reduced to a pure counting problem, c.f. Section 4.3.

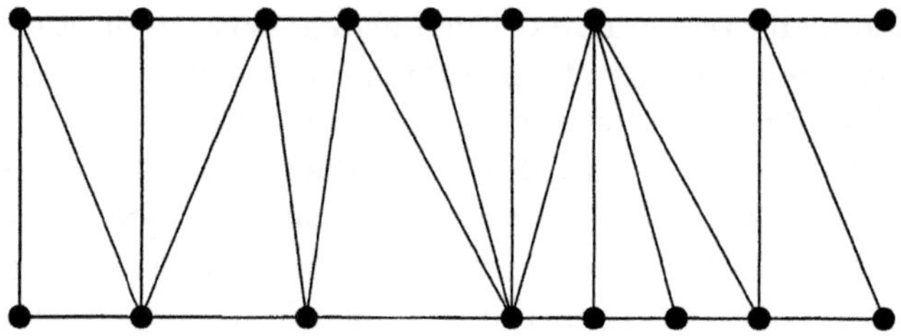

Figure 1. The propagation of a spatial slice from step t to step $t+1$. The ends of the strip should be joined to form a band with topology $S^1 \times [0,1]$.

form $S^1 \times [0,1]$ (for other spatial boundary conditions, see [28]). This mimicks the situation in classical gravity, where one usually works with globally hyperbolic space-times [29]. We will use a natural class of triangulations of the cylinder $S^1 \times [0,1]$, to which we will assign edge lengths (i.e. a discretized metric) in such a way that each two-dimensional geometry carries a discrete causal structure. Our simplicial space-times will have a preferred foliation into a discrete set of circular slices, consisting only of space-like edges. The foliation parameter t can be interpreted as a discrete version of "proper time". The main task of our dynamical-triangulations approach is to count the geometries in this class, that is, to perform the state sum in the regularized context, and then attempt to take a continuum limit.

The geometry of each spatial slice is uniquely characterized by the length assigned to it. In the discretized version, the length L will be "quantized" in units of a lattice spacing a, i.e. $L = l \cdot a$ where l is an integer. A slice will thus be defined by l vertices and l links connecting them. To obtain a 2d geometry, we will evolve this spatial loop in discrete steps. This leads to a preferred notion of (discrete) "time" t, where each loop represents a slice of constant t. The propagation from time-slice t to time-slice $t+1$ is governed by the following rule: each vertex i at time t is connected to k_i vertices at time $t+1$, $k_i \geq 1$, by links which are assigned squared edge lengths $-a^2$. The k_i vertices, $k_i > 1$, at time-slice $t+1$ will be connected by $k_i - 1$ consecutive space-like links, thus forming $k_i - 1$ triangles. Finally the right boundary vertex in the set of k_i vertices will be identified with the left boundary vertex of the set of k_{i+1} vertices. In this way we get a total of $\sum_{i=1}^{l}(k_i - 1)$ vertices (and also links) at time-slice $t+1$ and the two spatial slices are connected by $\sum_{i=1}^{l} k_i \equiv l_t + l_{t+1}$ triangles (see Fig. 1).

The elementary building blocks of a geometry are therefore triangles with one space- and two time-like edges. We define them to be flat in the interior. A consistent way of assigning interior angles to such Minkowskian triangles is described in [30]. The angle between two time-like edges is $\gamma_{tt} = -\arccos\frac{3}{2}$, and between a space- and a time-like edge $\gamma_{st} = \frac{\pi}{2} + \frac{1}{2}\arccos\frac{3}{2}$, summing up to $\gamma_{tt} + 2\gamma_{st} = \pi$. The sum over all angles around a vertex with j incoming and k outgoing time-like edges (by definition $j, k \geq 1$) is given by $2\pi + (4 - j - k) \arccos \frac{3}{2}$. The regular triangulation of flat Minkowski space corresponds to $j = k = 2$ at all vertices. The volume of a single triangle is given by $\frac{\sqrt{5}}{4} a^2$.

One may view these geometries as a subclass of all possible triangulations that allow for the introduction of a causal structure. Namely, if we think of all time-like links as being future-directed, a vertex v' lies in the future of a vertex v iff there is an oriented sequence of time-like links leading from v to v'. Two arbitrary vertices may or may not be causally related in this way.

In quantum gravity we are instructed to sum over all geometries connecting, say, two spatial boundaries of length L_1 and L_2, with the weight of each geometry g given by

$$e^{iS[g]}, \qquad S[g] = \Lambda \int \sqrt{-\det g} \quad \text{(in 2d)}, \tag{1}$$

where Λ is the cosmological constant. In our discretized model the boundaries will be characterized by integers l_1 and l_2, the number of vertices or links at the two boundaries. The path-integral amplitude for the propagation from geometry l_1 to l_2 will be the sum over all interpolating surfaces of the kind described above, with a weight given by the discretized version of (1). Let us call the corresponding amplitude $G_\lambda^{(1)}(l_1, l_2)$. We thus have

$$G_\lambda^{(1)}(l_1, l_2) = \sum_{t=1}^{\infty} G_\lambda^{(1)}(l_1, l_2; t), \tag{2}$$

$$G_\lambda^{(1)}(l_1, l_2; t) = \sum_{l=1}^{\infty} G_\lambda^{(1)}(l_1, l; 1)\, l\, G_\lambda^{(1)}(l, l_2, t-1), \tag{3}$$

$$G_\lambda^{(1)}(l_1, l_2; 1) = \frac{1}{l_1} \sum_{\{k_1, \ldots, k_{l_1}\}} e^{i\lambda a^2 \sum_{i=1}^{l_1} k_i}, \tag{4}$$

where λ denotes the *bare* cosmological constant[3] (we have absorbed the finite triangle volume factor), and where t denotes the total number of time-slices connecting l_1 and l_2.

From a combinatorial point of view it is convenient to mark a vertex on the entrance loop in order to get rid of the factors l and $1/l$ in (3) and (4), that is,

$$G_\lambda(l_1, l_2; t) \equiv l_1 G_\lambda^{(1)}(l_1, l_2; t) \tag{5}$$

[3]One obtains the renormalized (continuum) cosmological constant Λ in (1) by an additive renormalization, see below.

(the unmarking of a point may be thought of as the factoring out by (discrete) spatial diffeomorphisms). Note that $G_\lambda(l_1, l_2; 1)$ plays the role of a transfer matrix, satisfying

$$G_\lambda(l_1, l_2; t_1 + t_2) = \sum_l G_\lambda(l_1, l; t_1)\, G_\lambda(l, l_2; t_2) \tag{6}$$

$$G_\lambda(l_1, l_2; t + 1) = \sum_l G_\lambda(l_1, l; 1)\, G_\lambda(l, l_2; t). \tag{7}$$

Knowing $G_\lambda(l_1, l_2; 1)$ allows us to find $G_\lambda(l_1, l_2; t)$ by iterating (7) t times. This program is conveniently carried out by introducing the generating function for the numbers $G_\lambda(l_1, l_2; t)$,

$$G_\lambda(x, y; t) \equiv \sum_{k,l} x^k\, y^l\, G_\lambda(k, l; t), \tag{8}$$

which we can use to rewrite (6) as

$$G_\lambda(x, y; t_1 + t_2) = \oint \frac{dz}{2\pi i\, z}\, G_\lambda(x, z^{-1}; t_1) G_\lambda(z, y; t_2), \tag{9}$$

where the contour should be chosen to include the singularities in the complex z-plane of $G_\lambda(x, z^{-1}; t_1)$ but not those of $G_\lambda(z, y; t_2)$.

One can either view the introduction of $G_\lambda(x, y; t)$ as a purely technical device or take x and y as boundary cosmological constants,

$$x = \mathrm{e}^{i\lambda_i a}, \qquad y = \mathrm{e}^{i\lambda_o a}, \tag{10}$$

such that $x^k = \mathrm{e}^{i\lambda_i a\, k}$ becomes a boundary cosmological term, and similarly for $y^l = \mathrm{e}^{i\lambda_o a\, l}$. Let us for notational convenience define

$$g = \mathrm{e}^{i\lambda a^2} \tag{11}$$

(not to be confused with the symbol for the continuum metric). For the technical purpose of counting we view x, y and g as variables in the complex plane. In general the function

$$G(x, y; g; t) \equiv G_\lambda(x, y; t) \tag{12}$$

will be analytic in a neighbourhood of $(x, y, g) = (0, 0, 0)$.

From the definitions (4) and (5) it follows by standard techniques of generating functions that we may associate a factor g with each triangle, a factor x with each vertex on the entrance loop and a factor y with each vertex on the exit loop, leading to

$$G(x, y; g; 1) = \sum_{k=0}^{\infty} \left(gx \sum_{l=0}^{\infty} (gy)^l \right)^k - \sum_{k=0}^{\infty} (gx)^k = \frac{g^2 xy}{(1 - gx)(1 - gx - gy)}. \tag{13}$$

Formula (13) is simply a book-keeping device for all possible ways of evolving from an entrance loop of any length in one step to an exit loop of any length. The subtraction of the term $1/(1-gx)$ has been performed to exclude the degenerate cases where either the entrance or the exit loop is of length zero.

From (13) and eq. (9), with $t_1 = 1$, we obtain

$$G(x,y;g;t) = \frac{gx}{1-gx}\, G(\frac{g}{1-gx}, y; g; t-1). \tag{14}$$

This equation can be iterated and the solution written as

$$G(x,y;g;t) = F_1^2(x)F_2^2(x)\cdots F_{t-1}^2(x)\frac{g^2xy}{[1-gF_{t-1}(x)][1-gF_{t-1}(x)-gy]}, \tag{15}$$

where $F_t(x)$ is defined iteratively by

$$F_t(x) = \frac{g}{1-gF_{t-1}(x)}, \quad F_0(x) = x. \tag{16}$$

Let F denote the fixed point of this iterative equation. By standard techniques one readily obtains

$$F_t(x) = F\,\frac{1-xF+F^{2t-1}(x-F)}{1-xF+F^{2t+1}(x-F)}, \quad F = \frac{1-\sqrt{1-4g^2}}{2g}. \tag{17}$$

Inserting (17) in eq. (15), we can write

$$G(x,y;g,t) = \frac{F^{2t}(1-F^2)^2\, xy}{(A_t - B_t x)(A_t - B_t(x+y) + C_t xy)} \tag{18}$$

$$= \frac{F^{2t}(1-F^2)^2\, xy}{\Big[(1-xF)-F^{2t+1}(F-x)\Big]\Big[(1-xF)(1-yF)-F^{2t}(F-x)(F-y)\Big]}, \tag{19}$$

where the time-dependent coefficients are given by

$$A_t = 1-F^{2t+2}, \quad B_t = F(1-F^{2t}), \quad C_t = F^2(1-F^{2t-2}). \tag{20}$$

The combined region of convergence to the expansion in powers $g^k x^l y^m$, valid for all t is

$$|g| < \frac{1}{2}, \quad |x| < 1, \quad |y| < 1. \tag{21}$$

2.2. THE CONTINUUM LIMIT

The path integral formalism we are using here is very similar to the one used to represent the free particle as a sum over paths. Also there one performs a summation over geometric objects (the paths), and the path integral itself serves as the propagator. From the particle case it is known that the bare mass undergoes an additive renormalization (even for the free particle), and that the bare propagator is subject to a wave-function renormalization (see [8] for a review). The same is true in two-dimensional Euclidean gravity, treated in the formalism of dynamical triangulations [8]. The coupling constants with positive mass dimension, i.e. the cosmological constant and the boundary cosmological constants, undergo an additive renormalization, while the partition function itself (i.e. the Hartle-Hawking-like wave function) undergoes a multiplicative wave-function renormalization. We therefore expect the bare coupling constants λ, λ_i and λ_0 to behave as

$$\lambda = \frac{C_\lambda}{a^2} + \tilde{\Lambda}, \qquad \lambda_i = \frac{C_{\lambda_i}}{a} + \tilde{X}, \qquad \lambda_o = \frac{C_{\lambda_o}}{a} + \tilde{Y}, \tag{22}$$

where $\tilde{\Lambda}, \tilde{X}, \tilde{Y}$ denote the renormalized cosmological and boundary cosmological constants. If we introduce the notation

$$g_c = \mathrm{e}^{iC_\lambda}, \qquad x_c = \mathrm{e}^{iC_{\lambda_i}}, \qquad y_c = \mathrm{e}^{iC_{\lambda_o}}, \tag{23}$$

for critical values of the coupling constants, it follows from (10) and (11) that

$$g = g_c\, \mathrm{e}^{ia^2\tilde{\Lambda}}, \qquad x = x_c\, \mathrm{e}^{ia\tilde{X}}, \qquad y = y_c\, \mathrm{e}^{ia\tilde{Y}}. \tag{24}$$

The wave-function renormalization will appear as a multiplicative cut-off dependent factor in front of the bare "Green's function" $G(x, y; g; t)$,

$$G_{\tilde{\Lambda}}(\tilde{X}, \tilde{Y}; T) = \lim_{a \to 0} a^\eta G(x, y; g; t), \tag{25}$$

where $T = a\,t$, and where the critical exponent η should be chosen so that the right-hand side of eq. (25) exists. In general this will only be possible for particular choices of g_c, x_c and y_c in (25).

The basic relation (6) can survive the limit (25) only if $\eta = 1$, since we have assumed that the boundary lengths L_1 and L_2 have canonical dimensions and satisfy $L_i = a\,l_i$.

A closer analysis reveals that only at $g_c = 1/2$ one can obtain a sensible continuum limit. It corresponds to a purely imaginary bare cosmological constant $\lambda_c := C_\lambda / a^2 = -i \ln 2 / a^2$. If we want to approach this point from the region in the complex g-plane where $G(x, y; g; t)$ converges it is natural to choose the renormalized coupling $\tilde{\Lambda}$ imaginary as well, $\tilde{\Lambda} = i\Lambda$, i.e.

$$\lambda = i\, \frac{\ln \frac{1}{2}}{a^2} + i\Lambda. \tag{26}$$

One obtains a well-defined scaling limit (corresponding to $\Lambda \in \mathbb{R}$) by letting $\lambda \to \lambda_c$ along the imaginary axis. The Lorentzian form for the continuum propagator is obtained by an analytic continuation $\Lambda \to -i\Lambda$ in the *renormalized* coupling of the resulting Euclidean expressions.

At this stage it may seem that we are surreptitiously reverting to a fully Euclidean model. We could of course equivalently have conducted the entire discussion up to this point in the "Euclidean sector", by omitting the factor of $-i$ in the exponential (1) of the action, choosing λ positive real and taking all edge lengths equal to 1. However, from a purely Euclidean point of view there would not have been any reason for restricting the state sum to a subclass of geometries admitting a causal structure. The associated preferred notion of a discrete "time" allows us to define an "analytic continuation in time". Because of the simple form of the action in two dimensions, the rotation

$$\int dx\, dt \sqrt{-\det g_{lor}} \to i \int dx\, dt_{eu} \sqrt{\det g_{eu}} \tag{27}$$

to Euclidean metrics in our model is equivalent to the analytic continuation of the cosmological constant Λ.

From (18) or (19) it follows that we can only get macroscopic loops in the limit $a \to 0$ if we simultaneously take $x, y \to 1$. (For $g_c = -1/2$, one needs to take $x, y \to -1$. The continuum expressions one obtains are identical to those for $g_c = 1/2$.) Again the critical points correspond to purely imaginary bare boundary cosmological coupling constants. We will allow for such imaginary couplings and thus approach the critical point $\lambda_i = \lambda_o = 0$ from the region of convergence of $G(x, y; g; t)$, i.e. via real, positive X, Y where

$$\lambda_i = iXa, \qquad \lambda_o = iYa. \tag{28}$$

Again X and Y have an obvious interpretation as positive boundary cosmological constants in a Euclidean theory, which may be analytically continued to imaginary values to reach the Lorentzian sector.

Summarizing, we have

$$g = \frac{1}{2} \mathrm{e}^{-\Lambda a^2} \to \frac{1}{2}(1 - \Lambda a^2), \quad \text{(i.e. } F = 1 - a\sqrt{\Lambda}) \tag{29}$$

as well as

$$x = \mathrm{e}^{-Xa} \to 1 - aX, \qquad y = \mathrm{e}^{-aY} \to 1 - aY, \tag{30}$$

where the arrows $\to$ in (29) and (30) should be viewed as analytic coupling constant redefinitions of Λ, X and Y, which we have performed to get rid of factors of $1/2$ etc. in the formulas below. With the definitions (29) and (30) it is straightforward to perform the continuum limit of $G(x, y; g; t)$ as

$(x,y,g) \to (x_c,y_c,g_c) = (1,1,1/2)$, yielding

$$G_\Lambda(X,Y;T) = \frac{4\Lambda\, e^{-2\sqrt{\Lambda}T}}{(\sqrt{\Lambda}+X) + e^{-2\sqrt{\Lambda}T}(\sqrt{\Lambda}-X)} \times \frac{1}{(\sqrt{\Lambda}+X)(\sqrt{\Lambda}+Y) - e^{-2\sqrt{\Lambda}T}(\sqrt{\Lambda}-X)(\sqrt{\Lambda}-Y)}. \quad (31)$$

For $T \to \infty$ one finds

$$G_\Lambda(X,Y;T) \stackrel{T\to\infty}{\longrightarrow} \frac{4\Lambda\, e^{-2\sqrt{\Lambda}T}}{(X+\sqrt{\Lambda})^2(Y+\sqrt{\Lambda})}, \quad (32)$$

while the limit for $T \to 0$ is

$$G_\Lambda(X,Y;T\!=\!0) = \frac{1}{X+Y}, \quad (33)$$

in accordance with the expectation that

$$G_\Lambda(L_1,L_2;T\!=\!0) = \delta(L_1 - L_2). \quad (34)$$

The general expression for $G_\Lambda(L_1,L_2;T)$ can be computed as the inverse Laplace transform of formula (31), yielding

$$G_\Lambda(L_1,L_2;T) = \frac{e^{-[\coth\sqrt{\Lambda}T]\sqrt{\Lambda}(L_1+L_2)}}{\sinh\sqrt{\Lambda}T}\, \frac{\sqrt{\Lambda L_1 L_2}}{L_2}\, I_1\!\left(\frac{2\sqrt{\Lambda L_1 L_2}}{\sinh\sqrt{\Lambda}T}\right), \quad (35)$$

where $I_1(x)$ is a modified Bessel function of the first kind. The asymmetry between L_1 and L_2 arises because the entrance loop has a marked point, whereas the exit loop has not. The amplitude with both loops marked is obtained by multiplying with L_2, while the amplitude with no marked loops is obtained after dividing (35) by L_1. Quite remarkably, our highly non-trivial expression (35) agrees with the loop propagator obtained from a bona-fide continuum calculation in proper-time gauge of pure 2d gravity by Nakayama [31].

The basic result (31) for $G_\Lambda(X,Y;T)$ can be derived by taking the continuum limit of the recursion relation (14). By inserting (29) and (30) in eq. (14) and expanding to first order in the lattice spacing a we obtain

$$\frac{\partial}{\partial T} G_\Lambda(X,Y;T) + \frac{\partial}{\partial X}\Big[(X^2-\Lambda) G_\Lambda(X,Y;T)\Big] = 0. \quad (36)$$

This is a standard first-order partial differential equation which should be solved with the boundary condition (33) at $T=0$, since this expresses the natural condition (34) on $G_\Lambda(L_1,L_2)$. The solution is thus

$$G_\Lambda(X,Y;T) = \frac{\bar{X}^2(T;X)-\Lambda}{X^2-\Lambda}\, \frac{1}{\bar{X}(T;X)+Y}, \quad (37)$$

where $\bar{X}(T;X)$ is the solution to the characteristic equation

$$\frac{d\bar{X}}{dT} = -(\bar{X}^2 - \Lambda), \quad \bar{X}(T=0) = X. \tag{38}$$

It is readily seen that the solution is indeed given by (31) since we obtain

$$\bar{X}(T) = \sqrt{\Lambda}\, \frac{(\sqrt{\Lambda} + X) - \mathrm{e}^{-2\sqrt{\Lambda}T}(\sqrt{\Lambda} - X)}{(\sqrt{\Lambda} + X) + \mathrm{e}^{-2\sqrt{\Lambda}T}(\sqrt{\Lambda} - X)}. \tag{39}$$

If we interpret the propagator $G_\Lambda(L_1, L_2; T)$ as the matrix element between two boundary states of a Hamiltonian evolution in "time" T,

$$G_\Lambda(L_1, L_2; T) = \langle L_1 | \mathrm{e}^{-\hat{H}T} | L_2 \rangle \tag{40}$$

we can, after an inverse Laplace transformation, read off the functional form of the Hamiltonian operator $\hat{H}$ from (36),

$$\hat{H}(L, \frac{\partial}{\partial L}) = -L\frac{\partial^2}{\partial L^2} + L\Lambda. \tag{41}$$

The corresponding Hamiltonian for the propagator of unmarked loops is given by

$$\hat{H}_u(L, \frac{\partial}{\partial L}) = -L\frac{\partial^2}{\partial L^2} - 2\frac{\partial}{\partial L} + \Lambda L. \tag{42}$$

The Hamiltonian (42) is Hermitian with respect to the natural measure $L\,\mathrm{d}L$, which has its origin in the basic completeness relation (3) for the transfer matrix with unmarked entrance and exit loops. If one wants to construct a unitary evolution with respect to the "time"-parameter T appearing in the transfer-matrix approach, one can simply exponentiate $(i\hat{H}T)$.

However, we should point out that there is an alternative to the analytic continuation $T \to -iT$ if one wants to relate the Euclidean and Lorentzian sectors of the theory. The combination $\sqrt{\Lambda}T$ appearing as an argument in (35) arises in taking the continuum limit of powers of the form F^t in expressions like (18), (19), where F is defined in (17). There are two aspects to a possible analytic continuation of F^t. The power t in F^t should clearly not be continued, since it is simply an integer counting the number of iterations of the transfer matrix. However, the function F itself does refer to the action, because the dimensionless coupling constant $g = \mathrm{e}^{i\lambda a_t a_l}$ is the action for a single Lorentzian triangle. (For added clarity we have distinguished between the lattice spacings in time- and space-directions, and called them a_t and a_l.) From the expression for F in terms of g in (17), we have $F = 1 - \sqrt{a_t a_l \Lambda}$.

The analytic continuation of F in time, from Euclidean to Lorentzian time, corresponds to the substitution $a_t \to -i\, a_t$ under the square-root sign, and thus becomes equivalent to the continuation $\Lambda \to -i\Lambda$ in the cosmological constant, as already remarked below eq. (27). The subtleties associated with the analytic continuation in the "time"-parameter T appearing in a transfer-matrix formulation of quantum gravity were first discussed in [32, 33] in the context of a square-root action formulation. Similar difficulties will also be present in higher-dimensional gravity, where the analytic continuation from Euclidean metrics to Lorentzian metrics cannot be absorbed by a continuation in Λ alone. To conclude, it is not obvious how to choose the correct analytic continuation back to Lorentzian signature, once the continuum limit has been taken. The continuation $T \to -iT$ leads to a unitary theory. For the continuation $\Lambda \to -i\Lambda$, we have not yet found a scalar product which makes the corresponding evolution operator unitary.

3. Topology changes and Euclidean quantum gravity

3.1. BABY UNIVERSE CREATION

In our non-perturbative regularization of 2d quantum gravity we have so far not included the possibility of topology changes of space. We will now show that *if* one allows for spatial topology changes, one is led in an essentially unambiguous manner to a *different* theory of two-dimensional quantum gravity, where space-time is much more fractal, and which agrees with Euclidean 2d quantum gravity as defined by Euclidean dynamical triangulations or Liouville theory.

By a topology change of *space* in our Lorentzian setting we have in mind the following: a baby universe may branch off at some time T and develop in the future, where it will eventually disappear in the vacuum, but it is not allowed to rejoin the "parent" universe and thus change the overall topology of the two-dimensional manifold. This is a restriction we impose to be able to compare with the analogous calculation in usual 2d Euclidean quantum gravity.

It is well known that such a branching leads to additional complications, compared with the Euclidean situation, in the sense that, in general, no continuum Lorentzian metrics which are smooth and non-degenerate everywhere can be defined on such space-times (see, for example, [34] and references therein). These considerations do not affect the cosmological term in the action, but lead potentially to contributions from the Einstein-Hilbert term at the singular points where a branching or pinching occurs.

We have so far ignored the curvature term in the action since it gives merely a constant contribution in the absence of topology change. We will continue to do so in the slightly generalized setting just introduced. The continuum results of [34] suggest that the contributions from the two singular points associated with each branching of a baby universe (one at the branching point and one at the tip of the baby universe where it contracts to a point) cancel in the action. The physical geometry of these configurations may seem slightly contrived, but they may well be important in the quantum theory of gravity and deserve further study. However, for the moment our main motivation for introducing them is to make contact with the usual non-perturbative Euclidean path-integral results.

We will use the rest of this section to demonstrate the following: once we allow for spatial topology changes,

(1) this process completely dominates and changes the critical behaviour of the discretized theory, and
(2) the disc amplitude $W_\Lambda(L)$ (the Hartle-Hawking wave function) is uniquely determined, almost without any calculations.

Our starting point will be the discretized model introduced in Sec. 2.1. In this model we do not directly have a disc amplitude like the Hartle-Hawking wave functional. However, as discussed at the beginning of this section, the degeneracy of the metric at a point in the interior of the disc is always compensated (in the sense of complex contributions to the action) by the degeneracy of the metric at the point where the baby universe branched off. We can thus define the disc amplitude in Lorentzian gravity as

$$w^{(b)}(x,g) := \sum_t G^{(b)}(x, l_2 = 1; g; t) = G^{(b)}(x, l_2 = 1; g), \tag{43}$$

and the continuum version as

$$W_\Lambda^{(b)}(X) = \int \mathrm{d}T \, G_\Lambda^{(b)}(X, L_2 = 0; T) = G_\Lambda^{(b)}(X, L_2 = 0) \tag{44}$$

where the superscript $^{(b)}$ indicates the "bare" Lorentzian model without spatial topology changes. It follows that

$$w^{(b)}(x,g) \to a^{-1} W_\Lambda^{(b)}(X). \tag{45}$$

There are a number of ways to implement the creation of baby universes, some more natural than others, but they all agree in the continuum limit, as will be clear from the general arguments provided below. Here we discuss only the simplest way of implementing such a change. This is shown in Fig. 2: stepping forward from t to $t+1$ from a loop of length l_1 we create a baby universe of length $l < l_1$ by pinching it off non-locally from the main branch.

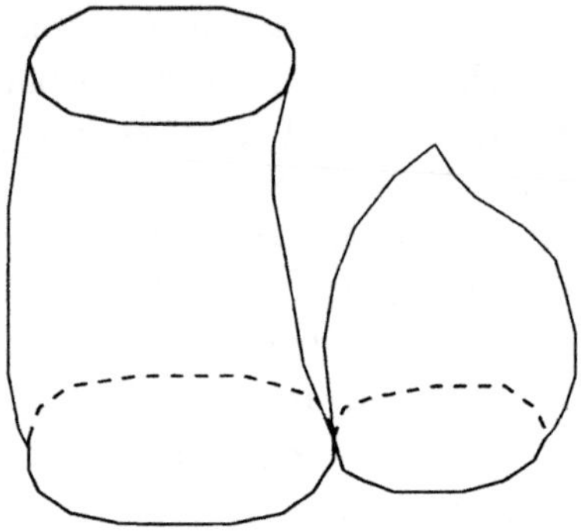

Figure 2. A "baby universe" created by a global pinching.

Accounting for the new possibilities of evolution in each step according to fig. 2, the new and old transfer matrices are related by

$$G_\lambda(l_1,l_2;1)=G_\lambda^{(b)}(l_1,l_2;1)+\sum_{l=1}^{l_1-1} l_1 w(l_1-l,g)\,G_\lambda^{(b)}(l,l_2;1). \tag{46}$$

The factor l_1 in the sum comes from the fact that the "pinching" shown in fig. 2 can take place at any of the l_1 vertices. As before, the new transfer matrix leads to new amplitudes $G_\lambda(l_1,l_2;t)$, satisfying

$$G_\lambda(l_1,l_2;t_1+t_2)=\sum_l G_\lambda(l_1,l;t_1)G_\lambda(l,l_2;t_2), \tag{47}$$

and in particular

$$G_\lambda(l_1,l_2;t)=\sum_l G_\lambda(l_1,l;1)\;G_\lambda(l,l_2;t-1). \tag{48}$$

Performing a (discrete) Laplace transformation of eq. (48) leads to

$$G(x,y;g;t)=\oint\frac{dz}{2\pi i\,z}\left[G_\lambda^{(b)}(x,z^{-1};1)+x\frac{\partial}{\partial x}\Big(w(x;g)G_\lambda^{(b)}(x,z^{-1};1)\Big)\right]G(z,y;g;t-1), \tag{49}$$

or, using the explicit form of the transfer matrix $G_\lambda^{(b)}(x,z;1)$, formula (13),

$$G(x,y;g;t)=\Big[1+x\frac{\partial w(x,g)}{\partial x}+xw(x,g)\frac{\partial}{\partial x}\Big]\frac{gx}{1-gx}\,G\Big(\frac{g}{1-gx},y;g;t-1\Big). \tag{50}$$

At this point neither the disc amplitude $w(x, g)$ nor $G(x, y; g; t)$ are known. We will now show that they are uniquely determined if we assume that the boundary length scales canonically with the lattice spacing, $L = a\,l$, implying a renormalized boundary cosmological constant X with the dimension of mass, $x = x_c(1 - aX)$. In addition we assume that the dimension of the renormalized cosmological constant Λ is canonical too, $g = g_c(1 - \frac{1}{2}\Lambda a^2)$. Somewhat related arguments have been presented in different settings in [35, 36].

It follows from relation (47) that we need

$$G_\lambda(l_1, l_2, t) \xrightarrow{a \to 0} a\, G_\Lambda(L_1, L_2; T). \tag{51}$$

It is important for the following discussion that $G_\lambda(l_1, l_2; t)$ cannot contain a non-scaling part since from first principles (sub-additivity) it has to decay exponentially in t. By a Laplace transformation, using $x = x_c(1 - aX)$ in the scaling limit, we thus conclude that

$$G_\lambda(x, l_2, t) \xrightarrow{a \to 0} G_\Lambda(X, L_2, T), \tag{52}$$

and further, by a Laplace transformation in L_2,

$$G_\lambda(x, y; t) \xrightarrow{a \to 0} a^{-1} G_\Lambda(X, Y; T). \tag{53}$$

We will now show that the scaling of $w(x, g)$ is quite restricted too. The starting point is a combinatorial identity which the disc amplitude has to satisfy. The arguments are valid both for the disc amplitude in Euclidean quantum gravity and the disc amplitude we have introduced for our model in (43). We will assume the general scaling

$$w(x, g) = w_{ns}(x, g) + a^\eta W_\Lambda(X) + \text{less singular terms} \tag{54}$$

for the disc amplitude. In the case $\eta < 0$ the first term is considered absent (or irrelevant). However, if $\eta > 0$ a term like w_{ns} will generically be present, since any slight redefinition of the coupling constants of the model will produce such a term if it was not there from the beginning.

We will introduce an explicit mark in the bulk of $w(x, g)$ by differentiating with respect to g. This leads to the combinatorial identity

$$g\, \frac{\partial w(x, g)}{\partial g} = \sum_t \sum_l G(x, l; g; t)\, l\, w(l, g), \tag{55}$$

or, after the usual Laplace transform,

$$g\, \frac{\partial w(x, g)}{\partial g} = \sum_t \oint \frac{dz}{2\pi i\, z}\, G(x, z^{-1}; g; t)\, \frac{\partial w(z, g)}{\partial z}. \tag{56}$$

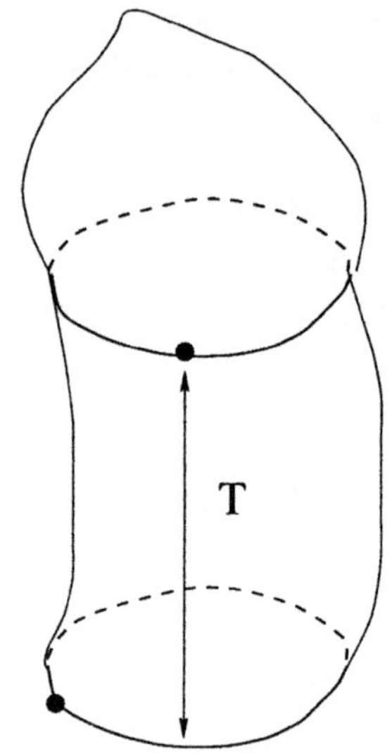

Figure 3. Marking a vertex in the bulk of $W_\Lambda(X)$. The mark has a distance T from the boundary loop, which itself has one marked vertex.

The situation is illustrated in Fig. 3. A given mark has a distance t (T in the continuum) to the entrance loop. In the figure we have drawn all points which have the same distance to the entrance loop and which form a connected loop. In the bare model these are all the points at distance t. In the case where baby universes are allowed (which we have not included in the figure), there can be many disconnected loops at the same distance. Let us assume a general scaling

$$T = a^{\varepsilon} t, \qquad \varepsilon > 0, \tag{57}$$

for the time variable T in the continuum limit. Above we saw that the bare model without baby universe creation corresponded to $\varepsilon = 1$. With the generalization (57) we account for the fact that by allowing for baby universes we have introduced an explicit asymmetry between the time- and space-directions.

Inserting (54) and (57) into eq. (56) we obtain

$$\frac{\partial w_{ns}}{\partial g} - 2a^{\eta-2}\frac{\partial W_\Lambda(X)}{\partial \Lambda} = \\ \frac{1}{a^{\varepsilon}}\int \mathrm{d}T \int dZ\, G_\Lambda(X,-Z;T)\left[\frac{\partial w_{ns}}{\partial z} - a^{\eta-1}\frac{1}{z_c}\frac{\partial W_\Lambda(Z)}{\partial Z}\right], \tag{58}$$

where $(x,g) = (x_c, g_c)$ in the non-singular part.

From eq. (58) and the requirement $\epsilon > 0$ it follows that the only consistent choices for η are

1: $\eta < 0$, i.e.

$$a^{\eta-2}\frac{\partial W_\Lambda(X)}{\partial\Lambda} = \frac{a^{\eta-1}}{2a^{\varepsilon}}\int \mathrm{d}T \int dZ\, G_\Lambda(X,-Z;T)\,\frac{1}{z_c}\frac{\partial W_\Lambda(Z)}{\partial Z}, \quad (59)$$

in which case we get $\varepsilon = 1$; and

2: $1 < \eta < 2$. Here formula (58) splits into the two equations

$$-a^{\eta-2}\frac{\partial W_\Lambda(X)}{\partial\Lambda} = \frac{1}{2a^{\varepsilon}}\frac{\partial w_{ns}}{\partial z}\bigg|_{z=x_c}\int \mathrm{d}T \int dZ\, G_\Lambda(X,-Z;T), \quad (60)$$

and

$$\frac{\partial w_{ns}}{\partial g}\bigg|_{g=g_c} = -\frac{a^{\eta-1}}{a^{\varepsilon}}\int \mathrm{d}T \int dZ\, G_\Lambda(X,-Z;T)\,\frac{1}{z_c}\frac{\partial W_\Lambda(Z)}{\partial Z}. \quad (61)$$

We are led to the conclusion that $\varepsilon = 1/2$ and $\eta = 3/2$, which are precisely the values found in Euclidean 2d gravity. Let us further remark that eq. (60) in this case becomes

$$-\frac{\partial W_\Lambda(X)}{\partial\Lambda} = \text{const.}\; G_\Lambda(X, L_2 = 0), \quad (62)$$

which differs from (44) by a derivative with respect to the cosmological constant. We will explain the reason for this difference below. Finally, eq. (61) becomes

$$\int \mathrm{d}T \int dZ\, G_\Lambda(X,-Z;T)\,\frac{\partial W_\Lambda(Z)}{\partial Z} = \text{const.}, \quad (63)$$

which will be satisfied automatically if $\eta = 3/2$ and $\varepsilon = 1/2$, as we will show below.

We will now analyze a possible scaling limit of (50), assuming the canonical scaling $x = x_c(1-aX)$ and $g = g_c(1-\frac{1}{2}\Lambda a^2)$. In order that the equation have a scaling limit at all, x_c, g_c and $w_{ns}(x_c, g_c)$ must satisfy two relations which can be determined straightforwardly from (50). The remaining continuum equation reads

$$\begin{aligned} a^{\varepsilon}\frac{\partial}{\partial T}G_\Lambda(X,Y;T) &= -a\frac{\partial}{\partial X}\Big[(X^2-\Lambda)G_\Lambda(X,Y;T)\Big] \\ &\quad -a^{\eta-1}\frac{\partial}{\partial X}\Big[W_\Lambda(X)G_\Lambda(X,Y;T)\Big]. \end{aligned} \quad (64)$$

The first term on the right-hand side of eq. (64) is precisely the one we have already encountered in our original model, while the second term is due to the creation of baby universes. Clearly the case $\eta < 0$ (in fact $\eta \leq 1$) is inconsistent with the presence of the second term, i.e. the creation of baby universes. However, since $\eta < 2$, the last term on the right-hand side of (64) will always dominate over the first term. *Once we allow for the creation of baby universes, this process will completely dominate the continuum limit.* In addition we get $\varepsilon = \eta - 1$, in agreement with (61). It follows that $\eta > 1$ and we conclude that $\varepsilon = 1/2, \eta = 3/2$ are the only possible scaling exponents if we allow for the creation of baby universes. These are precisely the scaling exponents obtained from two-dimensional Euclidean gravity in terms of dynamical triangulations, as we have already remarked. The topology changes of space have induced an anomalous dimension for T. If the second term on the right-hand side of (64) had been absent, this would have led to $\varepsilon = 1$, and the time T scaling in the same way as the spatial length L.

In summary, in the case $(\eta, \varepsilon) = (3/2, 1/2)$ eq. (64) leads to the continuum equation

$$\frac{\partial}{\partial T} G_\Lambda(X, Y; T) = -\frac{\partial}{\partial X}\Big[W_\Lambda(X) G_\Lambda(X, Y; T)\Big], \tag{65}$$

which, combined with eq. (62), determines the continuum disc amplitude $W_\Lambda(X)$. Integrating (65) with respect to T and using that $G_\Lambda(L_1, L_2; T{=}0) = \delta(L_1{-}L_2)$, i.e.

$$G_\Lambda(X, L_2{=}0; T{=}0) = 1, \tag{66}$$

we obtain

$$-1 = \frac{\partial}{\partial X}\Big[W_\Lambda(X) \frac{\partial}{\partial \Lambda} W_\Lambda(X)\Big]. \tag{67}$$

Since $W_\Lambda(X)$ has length dimension $-3/2$, i.e. $W_\Lambda^2(X) = X^3 F(\sqrt{\Lambda}/X)$, the general solution must be of the form

$$W_\Lambda(X) = \sqrt{-2\Lambda X + b^2 X^3 + c^2 \Lambda^{3/2}}. \tag{68}$$

From the very origin of $W_\Lambda(X)$ as the Laplace transform of a disc amplitude $W_\Lambda(L)$ which is bounded, it follows that $W_\Lambda(X)$ has no singularities or cuts for $\mathrm{Re}\, X > 0$. This requirement fixes the constants b, c in (68) such that

$$W_\Lambda(X) = b\Big(X - \frac{\sqrt{2}}{b\sqrt{3}}\sqrt{\Lambda}\Big)\sqrt{X + \frac{2\sqrt{2}}{b\sqrt{3}}\sqrt{\Lambda}}, \tag{69}$$

where the constant b is determined by the model-dependent constant in (62). This expression for the disc amplitude agrees after a rescaling of the cosmological constant with the amplitude $W_\Lambda^{(eu)}(X)$ from 2d Euclidean quantum gravity,

$$W_\Lambda(X) = (X - \frac{1}{2}\sqrt{\Lambda})\sqrt{X + \sqrt{\Lambda}}. \tag{70}$$

With $W_\Lambda(X)$ substituted into (65), the resulting equation is familiar from the usual theory of 2d Euclidean quantum gravity, where it has been derived in various ways [18, 35, 36], with T playing the role of *geodesic distance* between the initial and final loop.

Before showing that the anomalous scaling of the proper time – once baby universes are allowed – leads to an intrinsic fractal space-time dimension of four (rather than two), let us comment on the difference between the equations for the amplitudes (59)-(61) for $(\eta, \varepsilon) = (-1, 1)$ and $(\eta, \varepsilon) = (3/2, 1/2)$ respectively. In the first case there are no baby universes and eq. (59) entails that only *macroscopic loops* at a distance T from the entrance loop are important (as illustrated in Fig. 3). On the other hand, the term $\partial W_\Lambda(Z)/\partial Z$ which describes the presence of these macroscopic loops is absent in eq. (60). This is consistent with eq. (62), which shows explicitly that the length of the upper loop in Fig. 3 remains at the cut-off scale, and therefore can never become macroscopic. It is also consistent with the abundance of baby universes, since at any point in space-time the probability for creating a little "tip" of cut-off size will dominate. At the same time, the right-hand side of eq. (59), that is, eq. (63), will play no role when $1 < \eta < 2$, being simply equal to a constant. This latter property is satisfied automatically, as can be seen by using an equation analogous to (65) for the exit instead of the entrance loop. Thus eq. (63) becomes proportional to

$$\int_0^\infty \mathrm{d}T \frac{\partial}{\partial T}\, G(X, L_2\!=\!0; T) = \text{const.}, \tag{71}$$

proving our previous assertion.

3.2. THE FRACTAL DIMENSION OF EUCLIDEAN 2D GRAVITY

If we allow for baby universe creation, the fractal structure of space-time is determined by (65) and (70), where T is the time separating the entrance and exit loops. As already mentioned, these are exactly the equations governing Euclidean quantum gravity, if one replaces T by the geodesic distance between the two loops.

One can solve eq. (65) in the same way as its Lorentzian analogue (36) was solved by (37)-(39). We have

$$G_\Lambda(X, Y; T) = \frac{W_\Lambda(\bar{X}(T; X))}{W_\Lambda(X)} \frac{1}{\bar{X}(T; X) + Y}, \tag{72}$$

where $\bar{X}(T; X)$ is the solution to the characteristic equation

$$\frac{d\bar{X}}{dT} = -W_\Lambda(\bar{X}), \qquad \bar{X}(T=0) = X. \tag{73}$$

This equation can be solved in terms of elementary functions. In particular, one finds

$$\bar{X}(T; X=\infty) \propto \sqrt{\Lambda}\coth^2\left(\alpha\Lambda^{1/4}T\right), \quad \alpha = \frac{1}{2}\sqrt{3/2}. \tag{74}$$

We may now define a *two-point function* by contracting the entrance loop to a point and closing the exit loop by the disc amplitude. This is shown in Fig. 3, except that now the entrance loop must be contracted[4]. The two-point function $G_\Lambda(T)$ in Euclidean gravity can be interpreted as the (unnormalized) average over all geometries with two marked points which are separated by a geodesic distance T. At the discretized level it is defined by an equation analogous to (55), but without summing over t,

$$G(t;g) = \sum_l G(l_1, l; g; t)\; l\; w(l,g). \tag{75}$$

As in eq. (58), the continuum limit will be dominated by the non-scaling part of $w(l;g)$, i.e. by small l, and the continuum two-point function is simply given by

$$G_\Lambda(T) \sim G_\Lambda(L_1{=}0, L_2{=}0; T). \tag{76}$$

Using (72) and the Laplace transforms of G_Λ and W_Λ we obtain

$$G_\Lambda(T) = \oint \frac{\mathrm{d}X}{2\pi i} \oint \frac{\mathrm{d}Y}{2\pi i} \frac{W_\Lambda(\bar{X}(T;X))}{W_\Lambda(X)} \frac{1}{\bar{X}(T;X)+Y}. \tag{77}$$

With the help of the characteristic equation (73) this leads to

$$G_\Lambda(T) \sim \frac{\mathrm{d}}{\mathrm{d}T} \oint \frac{\mathrm{d}X}{2\pi i} \frac{\bar{X}(T;X)}{W_\Lambda(X)}. \tag{78}$$

The contour of integration can be deformed to infinity and we obtain from (74) that

$$G_\Lambda(T) \sim \Lambda^{3/4} \frac{\cosh \alpha\Lambda^{1/4}T}{\sinh^3 \alpha\Lambda^{1/4}T}. \tag{79}$$

This two-point function may be viewed as the partition function for universes with two marked points separated by a geodesic distance T. Since

[4]One cannot *a priori* contract both loops, since the two points in the two-point function are separated by a geodesic distance T, and there may be many points at distance T from the entrance point, as shown in Fig. 3. They will in general form several connected loops. However, after solving the model, it turns out that one *can* just contract the exit loop and obtain the two-point function. The reason is that $w(x,g)$ contains a non-universal part, which again implies that a typical disc amplitude in Fig. 3 will be of cut-off size. This is precisely the contents of eq. (60), whose non-scaling part is given by $\partial w_{ns}/\partial z|_{z=x_c}$.

we wanted to solve *Euclidean* gravity, this is our final result. Given (79), we can calculate the average space-time volume of such a universe,

$$\langle V \rangle_T = T^4 F\Big(\Lambda^{1/4} T\Big), \qquad F(0) > 0. \tag{80}$$

The function F is again expressible in terms of elementary functions and one finds

$$\langle V \rangle_T \approx T^4 \quad \text{for} \quad T < 1/\Lambda^{1/4}. \tag{81}$$

This formula shows that the intrinsic fractal dimension of 2d Euclidean quantum space-time is four, whereas a similar derivation in the case of Lorentzian gravity yields [11]

$$\langle V \rangle_T \approx T^2 \quad \text{for} \quad T < 1/\Lambda^{1/2}. \tag{82}$$

The discrepancy in dimension can be explained purely in terms of the baby universe structure: *since at each point of the two-dimensional Lorentzian surface a baby universe can branch off, the resulting fractal Euclidean space-time has twice the intrinsic dimension.*

4. Euclidean quantum gravity

4.1. SOME GENERALITIES

In the last section we arrived at the 2d Euclidean gravity theory through an "extension" of the Lorentzian model. Euclidean quantum gravity can of course be defined independently, as the quantization of classical gravity on the space of all Riemannian metrics (of positive definite signature) instead of the space of (indefinite-signature) Lorentzian metrics. In two dimensions, Euclidean gravity has a well-defined continuum path-integral formulation. Choosing a conformal gauge-fixing leads to the so-called Liouville gravity. Certain aspects of this theory can be solved by a bootstrap approach. In higher dimensions the path-integral approach to Euclidean quantum gravity is problematic, since the Einstein-Hilbert action is unbounded from below. There are various ways of tackling this problem: analytically continuing the unstable modes, adding stabilizing higher-derivative terms to the action, or defining the theory non-perturbatively via a lattice regularization, such that the action is bounded for any finite lattice volume.

One example of the latter is the dynamical-triangulations approach, which has the added bonus of being exactly soluble by combinatorial methods in two dimensions. Its continuum limit agrees with continuum Liouville quantum gravity, wherever the two formulations can be compared. We are in fact in the unusual situation that the lattice approach can address and answer more questions than the continuum methods. We will not review the combinatorial approach here since the results (for pure gravity) were already obtained in Section 3, starting from Lorentzian gravity. For a detailed description we refer to [8], chapter 4. The generalization of this Euclidean lattice path integral to higher dimensions is straightforward and shares two virtues with the 2d case: calculating the partition function for gravity is again turned into a combinatorial problem, and the model is well-suited for numerical simulations. We have by now a good qualitative understanding of the phase structure of Euclidean dynamically-triangulated gravity in $d = 3, 4$, although complete analytical solutions of the discretized models are still missing. However, the combinatorial nature of the partition function gives us some hope that progress can be made also in these cases.

4.2. DYNAMICAL TRIANGULATIONS

This lattice approach shares many elements with lattice regularizations of ordinary quantum field theory. The main difference lies in the fact that the geometric degrees of freedom become dynamical and the lattices are therefore no longer part of the inert background structure. The geometric quantum fluctuations must be taken properly into account when building discretized models of matter interacting with quantum gravity. The field-theoretical, non-perturbative Euclidean path integral of such a theory takes the general form

$$Z[G, \Lambda, \{\beta_i\}] = \int \mathcal{D}[g_{\mu\nu}]\mathcal{D}\phi_i \; e^{-S_{matter}[g_{\mu\nu},\phi_i;\{\beta_i\}]-S_{EH}[g_{\mu\nu};G,\Lambda]}, \tag{83}$$

where the integration is over equivalence classes of metrics $[g_{\mu\nu}]$ and matter fields ϕ_i. The action contains a matter part S_{matter}, depending on a set of matter couplings $\{\beta_i\}$, and a purely geometric part, given by the Einstein-Hilbert action with a cosmological term,

$$S_{EH}[g_{\mu\nu}; G, \Lambda] = \frac{1}{16\pi G} \int d^d x \, \sqrt{\det g}\,(2\Lambda - R). \tag{84}$$

In (83) we have omitted possible boundary terms. Except in $d = 2$, expressions of the kind (83) have remained formal, due to the absence of a suitable diffeomorphism-invariant integration measure. The lattice formulation is an attempt to remedy this situation, by using an intermediate regularization of the non-perturbative path integral (83) (see [37, 8] for reviews). Defining a discrete regularization consists of several steps:

- a discretization of the individual metric manifolds, together with a definition of discretized geometric "observables", such as lengths, volumes and (scalar) curvature. These are necessary for obtaining a discretized version of the action, and for analyzing the physical properties of the theory in terms of scaling relations and correlation functions.

- a suitable choice of an integration measure on the space of discretized geometries (i.e. equivalence classes of metrics), such that the discrete path integral converges.
- a discretization of the matter sector, which will be closely related to standard lattice-regularizations in field theory.

Let us now describe the dynamical-triangulations regularization of Euclidean quantum gravity. It consists in replacing the d-dimensional Riemannian metric continuum manifold by a simplicial manifold constructed from equilateral d-dimensional simplices of (geodesic) edge length a. (A simplex is a point in $d = 0$, an edge in $d = 1$, a triangle in $d = 2$, a tetrahedron in $d = 3$, etc.) Using Regge's prescription [38], all quantities can be expressed as functions of the squared edge lengths. For example, the curvature depends on local deficit angles, which in turn are expressible in terms of edge lengths. A simplicial complex is obtained by gluing together d-simplices pairwise along $(d-1)$-dimensional faces (which are themselves $(d-1)$-simplices). Since, unlike in Regge calculus, our edge lengths are not variable, all d-simplices have the same size, and the total volume of the simplicial complex is simply proportional to the number N_d of such cells. Each d-simplex is built from simplices of lower dimensionality. It contains $d+1$ 0-simplices (vertices), $(d+1)d/2$ 1-simplices (links) etc. A lower-dimensional subsimplex is in general shared by a number of d-simplices, called the *order* of the subsimplex. A simplicial complex is a simplicial manifold if the neighbourhood of any p-simplex ($p < d$) has the topology of a $(d-p-1)$-dimensional sphere. Physically the manifold requirement may be viewed as a regularity condition at the cut-off scale, which will be convenient to use in our construction. The numbers N_k of (sub-)simplices of dimension $k \leq d$ are not independent, but (due to the regularity requirement) must satisfy a set of so-called Dehn-Sommerville relations, namely,

$$N_i = \sum_{k=1}^{d} (-1)^{k+d} \begin{pmatrix} k+1 \\ i+1 \end{pmatrix} N_k, \tag{85}$$

together with the Euler constraint

$$\sum_{k=0}^{d} (-1)^k N_k = \chi. \tag{86}$$

For fixed Euler number χ and $d = 2$, all N_k can be expressed as functions of the single variable N_2, say. For $d = 3, 4$, two of the N_k are independent.

A triangulation T together with an assignment of geodesic edge lengths and flat simplex interiors may be viewed as a piecewise linear manifold, and provides an explicit coordinate-independent representation of a metric manifold. In this way each triangulation corresponds to a unique equivalence class of metrics (albeit of piecewise-linear, and not of differentiable type).

When it comes to numerical simulations, it is often convenient to assign a label to each vertex. From the list of vertex labels for all d-simplices the whole manifold can be reconstructed. Since the labels themselves have no physical meaning, the labelling introduces a redundancy. Invariance under permutations of the labels may loosely be regarded as a discrete analogue of the diffeomorphism invariance of a differentiable manifold. Also the lower-dimensional subsimplices are characterized by their vertex labels. Moreover, the regularity requirement implies that we cannot have two different (sub-)simplices with the same set of vertex labels. For each triangulation T with N_0 vertices the number of different labellings equals $N_0!/C(T)$, where $C(T)$ is the order of the automorphism group of T. We can therefore distinguish between labelled triangulations $\tilde{T}$ and abstract unlabelled triangulations T. As mentioned above, different T's (with fixed topology) correspond to different equivalence classes of piecewise linear metrics assigned to the underlying manifold, and allow us to work directly with a reparameterization-invariant set of geometries.

Since the simplices are flat on the inside, curvature is located (distributionally) at simplices of lower dimension. Circulating around a $(d-2)$-dimensional simplex, the contributing angles will in general not add up to 2π. The resulting deficit angle depends on the number of simplices meeting at the subsimplex. We conclude that the curvature is concentrated at the $(d-2)$-dimensional subsimplices of the triangulation. The Einstein-Hilbert action for a dynamically triangulated manifold in dimension d assumes the simple form

$$S_{EH} = \kappa_d N_d - \kappa_{d-2} N_{d-2}, \tag{87}$$

with the two dimensionless coupling constants κ_d and κ_{d-2}. As usual in lattice regularizations, the lattice spacing a has disappeared from the formulation and will have to be reintroduced in the scaling limit. We are using the Einstein-Hilbert action because of its simplicity; one could in principle consider also the inclusion of higher-order curvature terms.

With each simplicial lattice described above one can associate a dual lattice, whose vertices are located at the centres of the simplices of the original lattice. In a similar way we can associate to each p-simplex a dual object of dimension $d-p$. For example, the dual links connect the centres of neighbouring simplices, and the dual of a $(d-2)$-simplex is a closed loop (a two-dimensional object) whose length is equal to the number of d-simplices of the original lattice which share the $(d-2)$-simplex. Since in d dimensions each simplex has $d+1$ neighbours, the dual lattices have the form of graphs of a scalar ϕ^{d+1}-theory (that is, all their vertices are $(d+1)$-valent), but with a local d-dimensional topological structure.

The simplicial structures described above possess a natural notion of length for any path connecting two points, since the equivalence class of metrics is uniquely determined. To simplify matters, we will only consider certain sets of discretized paths on the simplicial manifold. The first set is given by paths which run along the links of the original simplicial lattice, and the other by paths running along the links of the dual lattice. In either case we may define a distance between points on the lattice or its dual as the number of edges of the shortest path connecting the two. At first glance these definitions seem different from the standard notion of a geodesic distance, but they all coincide in the scaling limit, up to trivial numerical factors[5]. Numerical tests of this assumption will be discussed below.

4.3. THE FUNCTIONAL INTEGRAL

The association of triangulations with equivalence classes of metrics motivates the use of the discrete sum over d-dimensional triangulations as a discretized analogue of the diffeomorphism-invariant integration measure in (83),

$$\int \mathcal{D}[g_{\mu\nu}] \to \sum_T \frac{1}{C(T)}, \tag{88}$$

where the sum is taken over unlabelled simplicial manifolds. The need for including the symmetry factor $C(T)$ becomes apparent when one rewrites the right-hand side of (88) as a sum over labelled triangulations $\tilde{T}$,

$$\sum_T \frac{1}{C(T)} \to \sum_{\tilde{T}} \frac{1}{N_0(\tilde{T})!}. \tag{89}$$

In order that a discretized path integral with this choice of measure lead to a theory with a well-defined thermodynamic limit, the number of triangulations with fixed volume N_d should grow at most exponentially with N_d as $N_d \to \infty$. This is not the case unless we fix the space-time topology (usually to that of a sphere S^d); otherwise the growth is factorial. This property has been proven for $d = 2$ and arbitrary topology [39] and for simply connected manifolds in $d = 3, 4$ [24, 25].

[5]The situation is the same as for a regular 2d quadratic lattice in flat space: if we are only allowed to connect vertices along the lattice links, the lattice distance between different lattice points can differ by as much as a factor $\sqrt{2}$ from the Euclidean distance in flat space. However, in the scaling limit the rotational symmetry of the original field theory will be restored on the lattice and the two notions of distance will only differ by an overall factor.

It is worthwhile pointing out that the simple choices (87) for the action and (88) for the measure lead to a partition function of the form

$$Z(g_d, g_{d-2}) = \sum_T \frac{1}{C(T)}\, g_d^{N_d(T)}\, g_{d-2}^{N_{d-2}(T)}, \tag{90}$$

where $g_d = -\log \kappa_d$ and $g_{d-2} = \log \kappa_{d-2}$. Eq. (90) shows that the partition function is the generating function for the number of triangulations (of fixed topology) with given numbers N_d and N_{d-2} of d- and $(d-2)$-simplices. We thus reach the surprising conclusion that *even in dimension* $d > 2$, quantum gravity can be formulated as a (relatively simple) counting problem.

4.4. INCLUSION OF MATTER FIELDS

The discretization of matter fields coupled to dynamical triangulations is achieved by standard lattice field-theoretical methods. The simplest types of matter fields one can study are either scalar fields or (Potts) spin fields, carrying a discrete space-time label. One may also combine several fields of this type. The matter fields can be located at the vertices of the triangulation or at the centres of the d-simplices (that is, at the dual vertices). The interactions are typically of the form of nearest-neighbour interactions, where the "nearest neighbours" are the vertices that are one link length (or one dual link length) away from the original vertex. We expect these two formulations to become equivalent in the scaling limit. Some typical examples of matter actions are

$$S_I = \frac{\beta}{2} \sum_{\{ij\}} (\delta_{\sigma_i \sigma_j} - 1), \tag{91}$$

where the σ_i are a set of Potts spins, or the Gaussian action

$$S_g = \sum_{\{ij\}} (\phi_i - \phi_j)^2 \tag{92}$$

for a massless scalar field ϕ. Note that we did not need to include a coupling constant in front of the action (92). The massless scalar field can always be rescaled by a factor, which can then be absorbed by a redefinition of the coupling constants of the geometric sector. The coupling of Ising spins (Potts spins σ_i with $i = 1, 2$) and Gaussian fields to the 2d Lorentzian gravity model proceeds in a manner identical to (91) and (92).

In higher dimensions we will also consider the coupling to gauge fields. As usual these are associated with the one-dimensional edges of the triangulation, and there are again two alternative formulations, depending on whether the links or the dual links are used. The gauge field action is more complicated and we postpone its discussion to Section 6. The inclusion of fermionic degrees of freedom on a random manifold remains an open problem, particularly in higher dimensions. It requires the definition of a spin connection on a simplicial manifold (c.f. [40]). The problem was solved in $d = 2$ [41] . In this case one can prove that a system of Wilson fermions on a triangulated manifold can be "bosonized" and represented as a system of Ising spins on the manifold [42, 41].

The discretized path-integral measure contains also a sum over matter fields. In the case of spin variables, the sum is simply taken over all possible spin configurations. For continuous fields like the scalar fields above, one may consider non-trivial integration measures which introduce an additional coupling to geometry. This possibility does not exist in field theories on fixed, regular lattices, where such a dependence is always trivial. It leads to some subtleties in the case of gauge fields, as we will discuss later.

The simplest and most extensively studied example of a dynamically triangulated theory is that of Euclidean gravity on a two-sphere. The fundamental building blocks in this case are equilateral triangles (2-simplices). Triangles are glued together pairwise along edges (1-simplices), and each triangle has exactly three neighbours. The dual lattice is thus equivalent to a planar ϕ^3-diagram. The curvature is localized at the vertices (0-simplices) which in general are shared by many triangles, each contributing $\pi/3$ to the total angle around the vertex. The regularity requirement introduced above prohibits configurations where a vertex is its own neighbour or where two vertices are connected by more than one link (in other words, closed loops of link length one or two are forbidden).

The Lorentzian gravity model introduced in Section 2.1 may be viewed as a restricted version of the dynamically triangulated Euclidean model, since the triangulations contributing to the Lorentzian state sum constitute a subset of those appearing in the Euclidean system defined above. Recall also our construction of Euclidean from Lorentzian gravity in Section 3.1, by allowing for additional baby-universe branching. Again the set of all such geometries is a subset of all 2d simplicial manifolds, but both continuum theories coincide. What is at work in this latter case is "universality", which ensures that the continuum limit is to a large extent independent of the short-distance details characterizing the class of triangulations we sum over. The universality properties of Euclidean 2d quantum gravity are well studied. For example, one may relax the manifold regularity condition to obtain a much larger class of simplicial complexes, whose continuum limit is still Euclidean quantum gravity. Only drastic modifications, like the suppression of baby universes, can move the system to a different universality class with a different critical behaviour and therefore a different continuum limit. Also around the fixed point leading to Lorentzian 2d gravity one finds an independence of short-distance details. Universality with respect to a change of fundamental building blocks and the inclusion of higher curvature has been demonstrated in [28]. It is also encountered in a recently developped procedure for obtaining Lorentzian from Euclidean quantum gravity by removing baby universes [17]. There one ends up with a generalized class of triangulations (compared to the original Lorentzian model), but the continuum limit is still the same.

5. Numerical setup

5.1. MONTE CARLO METHOD AND ERGODIC MOVES

Even in two dimensions, there are a number of issues of the matter-coupled theory that presently can only be addressed by numerical methods. Lorentzian gravity coupled to Ising spins has not yet been solved analytically. In matter-coupled Euclidean 2d quantum gravity, analytical considerations have not yet led to a determination of the fractal dimension of space-time (there are various suggestions leading to different answers), nor do we know what happens beyond the infamous $c = 1$ barrier, where analytical calculations break down. In higher dimensions, we do not even have analytical solutions of the pure-gravity models. In all of these situations, numerical simulations of the systems come in handy. They can answer specific questions and lead to unexpected results which in turn can inspire further analytical work.

Numerical simulations of simplicial gravity have been the subject of a number of reviews (see [9, 43] for annual updates and [44] for more information on the computer codes used in the programs). Here we will only sketch the methods and use the simplest case of 2d Euclidean gravity as an illustration. Most of what we will have to say carries over to 2d Lorentzian gravity with only minor modifications.

As explained in the previous section, the discretized theory can be described by the partition function

$$Z = \sum_{\tilde{T}} \frac{1}{N_0(\tilde{T})!} e^{-S_{EH}(\tilde{T})} \sum_{\phi_i} e^{-S_{matter}(\phi_i)}, \tag{93}$$

where S_{EH} is the discretized Einstein-Hilbert action (87), and the first sum is taken over all labelled triangulations of fixed spherical topology. For $d = 2$, the geometric part of the action simplifies and takes the form

$$S_{EH} = \mu N_2, \tag{94}$$

up to an additive constant proportional to the Euler number χ, with $\mu = (\kappa_2 - \kappa_0)/2$. In numerical simulations, it is simpler to use the labelled instead of the unlabelled triangulations. The partition function (93) is the analogue of the grand canonical partition function in statistical mechanics. The cosmological constant κ_d plays the role of a thermodynamic potential for the number of simplices. For general d, we may rewrite (93) as

$$Z = \sum_{N_d} e^{-\kappa_d N_d} Z_{N_d}, \tag{95}$$

where Z_{N_d} can be interpreted as the partition function at fixed volume. A gravity-matter configuration C is uniquely specified by a geometry in the form of a labelled triangulation and by the values of all matter field variables. Each configuration enters the statistical sum with a probability proportional to $e^{-S(C)}$, where $S(C) = S_{EH} + S_{matter}$. As usual in statistical mechanics, physical information can be obtained by measuring the averages of various operators in this ensemble. The fact that each configuration has a real positive weight makes it possible to study the system (93) by Monte Carlo methods. The goal is to construct a numerical "generator" which produces configurations C with a probability $P(C) \sim e^{-S(C)}$.

Except for very few cases, where a direct generation of all configurations is possible (e.g. for a conformal charge $c = -2$ in 2d), the standard way of obtaining such a generator is by means of a stochastic process (a Markov chain), which can be regarded as a random walk in configuration space. Since the random walk takes place on a computer, each step corresponds to the real time needed to perform such an operation. The stochastic process is defined by a function $W(C \to C') \geq 0$, giving the probability for a transition from a configuration C to C' in one step. Assuming for simplicity that the configuration space is discrete, we have a normalization condition

$$\sum_{C'} W(C \to C') = 1. \tag{96}$$

In most cases $W(C \to C')$ is chosen to vanish outside some "neighbourhood" of C. Starting from an initial state $p_0(C)$, the system after n steps is characterized by a probability distribution $p_n(C)$, where

$$p_{n+1}(C) = \sum_{C'} p_n(C') W(C' \to C). \tag{97}$$

The transition probability W must satisfy two basic requirements,

1. *ergodicity:* any two configurations can be joined by a finite number of steps, and
2. *detailed balance:* the condition

$$P(C)W(C \to C') = P(C')W(C' \to C), \tag{98}$$

relating $W(C \to C')$ to $W(C' \to C)$.

It follows from (98) that $W(C \to C')$ and $W(C' \to C)$ are either both zero or both non-zero in which case they satisfy

$$\frac{W(C \to C')}{W(C' \to C)} = \frac{P(C')}{P(C)}. \tag{99}$$

These two requirements guarantee that the stochastic process has a unique asymptotic probability distribution $p_\infty(C) \sim P(C)$, which is the only eigenstate of the transition matrix W with eigenvalue 1,

$$\sum_{C'} p_\infty(C')W(C' \to C) = p_\infty(C). \tag{100}$$

All other eigenvalues are strictly smaller than 1. This implies that – independent of the initial configuration – the system will reach the asymptotic distribution after infinitely many steps. Note that the asymptotic distribution has the desired probability distribution. The rate at which the system approaches this limiting distribution depends on the other eigenvalues of W. The contributions from other eigenstates decay exponentially with the number of steps. The second-largest eigenvalue λ_1 provides us with an estimate of the autocorrelation time $T = -1/\log \lambda_1$. When the number n of steps is $\gg T$, we can assume that the distribution is asymptotic. Typically the autocorrelation time behaves like $T \sim f^\delta$, where f counts the number of degrees of freedom of the system and δ is a dynamical exponent, depending on the details of the algorithm.

In a practical implementation the system starts from some configuration C_0. During the first step, it changes to C_1 with probability $W(C_0 \to C_1)$, or remains at C_0 if the change is not accepted. After $n \gg T$ steps it reaches a configuration $C_n = C^{(1)}$ with a probability proportional to $P(C^{(1)})$. This configuration is the starting point of a new process, during which another sufficiently large number of steps is performed and a new configuration $C^{(2)}$ generated. Repeating this process we create a (finite) set of configurations $\{C^{(1)}, \ldots, C^{(N)}\}$, where N depends on the computer time we spend on the project. The average of any operator in the ensemble of configurations defined by the probability distribution $P(C)$ is approximated by an average over this finite sample of "typical" configurations.

The requirements listed above by no means define the stochastic process uniquely. We are interested in efficient algorithms which minimize δ, and can produce a large number of independent configurations in the shortest possible time. There is no simple way to guess at the outset whether an algorithm is efficient or not. Each problem must be treated individually and the autocorrelation time measured numerically. There are some general guidelines which one usually follows when creating a new algorithm. As discussed above, the efficiency depends on the choice of the transition probabilities $W(C \to C')$. We would like the algorithm to have a high "mobility", that is, a high probability that a configuration of the system will change at each step. This means that the set $\{C'\}$ of configurations which can be reached from a given C should be limited. If it is too large, each transition probability will be small, implying that the configuration will most likely not change. On the other hand, the set $\{C'\}$ must be large enough to ensure ergodicity.

The detailed-balance condition (99) implies that if both transitions are to be reasonably probable, we must choose the set $\{C'\}$ such that its elements have similar probabilities to that of C (equivalently, have a small action difference $|S(C) - S(C')|$). On a fixed lattice, small differences in the action are usually realized by considering at each step only local changes in the field variables, for instance, changing only one variable, while keeping all the others fixed. When a more complicated change is attempted, the action difference is in general large and proportional to the volume of the system. Local changes are not very efficient when the typical fluctuations are long-ranged, as happens close to a continuous phase transition. Creating a Monte Carlo algorithm which is at the same time ergodic and has a reasonably small autocorrelation time even in the critical region is an art.

The first step in constructing an algorithm for simplicial gravity is to define a method of coding the configurations. From a numerical point of view it is natural to work with labelled rather than unlabelled triangulations, because otherwise it is almost impossible to keep track of the (dynamical) connectivity. To code the geometric structure of such a configuration it is in principle sufficient to have a list of the vertex labels of all simplices (of all dimensions) of the triangulation. Two vertices are neighbours if they belong to the same simplex. From this list we can reconstruct the complete geometry of the system. The fact that we have a list of simplices means in practice that we have also assigned labels to the simplices. Two simplices are neighbours if they share a $(d-1)$-simplex or, equivalently, d vertices. In addition, we can make a list of all subsimplices and count their order. The process of reconstruction may still be complicated and it is often useful to keep at each step even more information, in the extreme case the lists of all subsimplices. The more information we keep, the easier it is to reconstruct local properties of the geometry. However, it also means that during geometry updates more data will have to be changed.

The next step consists in constructing the stochastic process described above. In our case this amounts to defining a set of "moves", which connect configurations with different geometry and matter content. Usually these two changes are performed separately: changes of the matter fields are performed for a fixed geometric configuration, by techniques very similar to those used for fixed lattices. For spin systems there exist highly effective cluster algorithms [45] with very short autocorrelation times. Cluster algorithms are special in that they permit global changes of the spin configurations. For Gaussian fields, no such algorithms exist and the field variables are usually updated sequentially. (There are some methods to shorten their autocorrelation time.) Generally speaking, the autocorrelation times for Gaussian fields are much longer than those for spin systems. It is therefore necessary to repeat the updating of all fields on the manifold several times in order to obtain independent field configurations.

In addition, we must define a set of geometric moves which generate local changes of the geometry, while preserving its underlying topology. Such moves are necessary even in the absence of matter. "Local changes" means that only a small region of the manifold is affected at each step. Again, it is important that the set of moves be ergodic in the space of all possible triangulations. For simplicial manifolds, one possible choice is given by the so-called Alexander moves [46]. In practical applications most algorithms for a d-dimensional geometry are based on a finite subset of moves containing $d+1$ operations. We will describe this set of operations in the case of 2d and postpone its generalization to higher d to Section 6.

The first operation is called a flip and involves two neighbouring triangles. The triangles are denoted by their vertex labels, $\{123\}$ and $\{124\}$, and share the link $\{12\}$. It is important that the four labels are all distinct (excluding $\{3\} = \{4\}$, say) and that the vertices $\{3\}$ and $\{4\}$ are not connected by a link. The flip consists in replacing this configuration by the two triangles $\{341\}$ and $\{342\}$. In other words, the link $\{12\}$ is "flipped" to the link $\{34\}$. The restrictions imposed above guarantee that this move does not produce a pathological triangulation. The inverse of the flip operation is again a flip. Note that the flip move takes place entirely inside the link loop $\{1324\}$, whose boundary it leaves unchanged. Since two simplices are replaced by two others, the flip is sometimes called a $(2,2)$-move. A flip can be performed almost everywhere in the manifold, giving rise to a different labelled manifold whose connectivity is changed locally. If matter fields are present at the vertices, one usually assumes that their values are unchanged. The transition weights $W(C \to C')$ and $W(C' \to C)$ for a flip and its inverse can easily be computed from the detailed-balance condition. We will not be more specific here, since this depends on the details of the geometric coding (in particular, on how links are selected). A flip move leaves the numbers of triangles and vertices unchanged.

The second move adds a new vertex $\{4\}$ inside a triangle $\{123\}$. This move replaces the old triangle by the three triangles $\{124\}$, $\{234\}$ and $\{314\}$, and leaves the manifold outside the "boundary" of $\{123\}$ unaffected. It is also known as a $(1,3)$-move. The new vertex is of order three, since it is shared by three triangles. The third move is the inverse of the $(1,3)$-

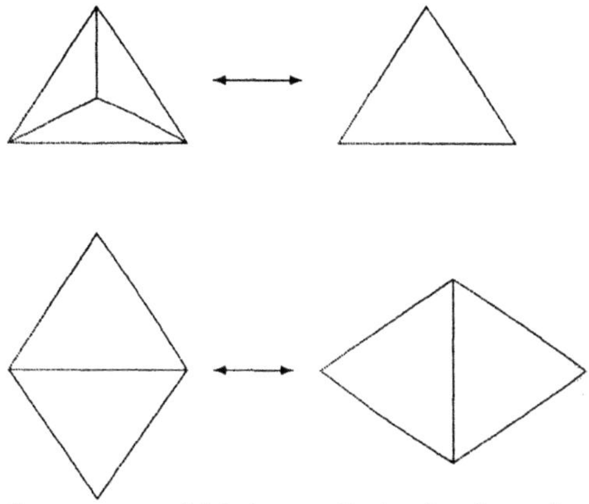

Figure 4. A set of three moves which is ergodic in the class of two-dimensional triangulations of fixed topology. The first diagram shows the $(3,1)$-move and its inverse. The second diagram shows the $(2,2)$- or flip move. By itself, this move is ergodic in the class of triangulations of fixed volume N_2 and topology.

move. It is a $(3,1)$-move where a vertex of order three is removed, and the three triangles which share it are replaced by a single triangle. The $(1,3)$-move can be performed on each triangle of a manifold, but its inverse needs a vertex of coordination number three. It is obvious that in both cases the configurations before and after the move are two different labelled triangulations. If matter fields are present, a new field must be created at the new vertex generated during a $(1,3)$-move. This must be taken into account when the transition weights are calculated. The three types of moves are depicted in Fig. 4.

The three operations just described leave the manifold topology unchanged. They form an ergodic set, which means that any two simplicial manifolds of the same topology can be related by a finite sequence of moves. In 2d one can construct such a sequence explicitly. The set of ergodic moves is not unique, and alternative sets of local moves have been used in applications. For example, a point-splitting algorithm is described in [8]. The point-splitting move and its inverse (illustrated in Fig. 5) change the volume N_2 by ± 2. The $(1,3)$- and $(3,1)$-moves are special cases and the $(2,2)$-move can be realized as a sequence of two point-splitting moves.

Our discussion so far suggests that we may set up a numerical simulation by generating configurations according to the probability distribution (93), and use the resulting sample of configurations to measure the quantities of interest. However, this is not really feasible, since the system described by (93) is an open system in the sense that arbitrarily large configurations may be produced by using one of the sets of geodesic moves described above. In practice, we must limit this size because of the obvious memory restrictions of a computer. A simple solution is to generate a set of configurations

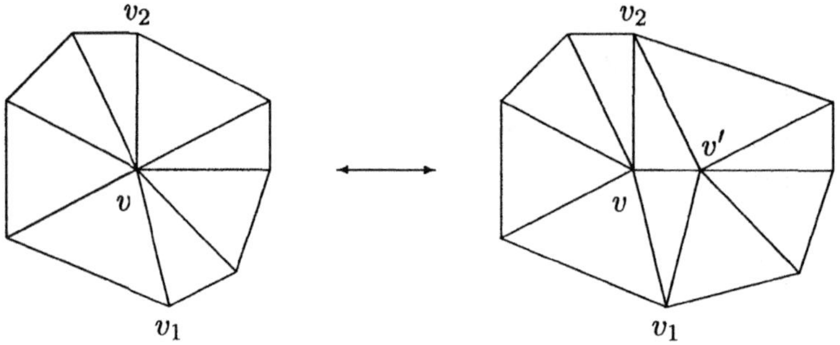

Figure 5. The point-splitting moves constitute an alternative set of ergodic moves.

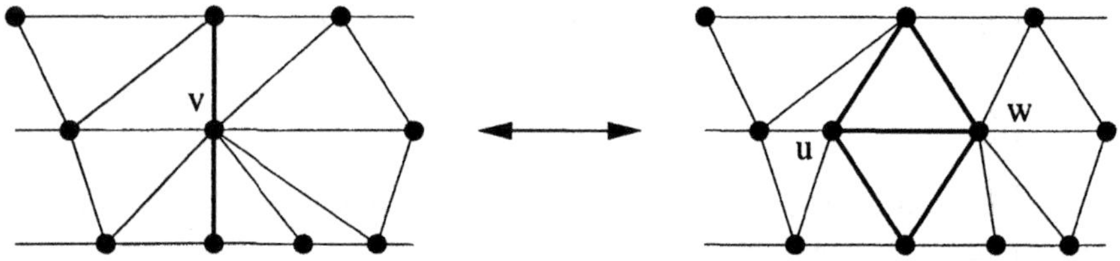

Figure 6. The move used in the Monte Carlo updating of a 2d Lorentzian geometry.

of fixed volume N_2, and repeat the experiment for various values of N_2. This is what one typically does in numerical simulations of field theory. For the case of Euclidean 2d triangulations we are particularly lucky, since the (2,2)-flip move is already by itself ergodic in the space of all triangulations of fixed volume N_2, considerably simplifying the computer simulations. Unfortunately there is no similar result in higher dimensions; we will describe later the method used in this case. The flip move cannot be used in numerical simulations of the Lorentzian model described in Section 2.1, since it is not compatible with the causal structure. One uses instead a version of the point-splitting move (Fig. 6) to update the geometry (see [12] for details).

We conclude this section by describing another type of move which is associated with a large change in matter and geometry (while keeping ΔS small). It is motivated by the baby-universe structure of Euclidean simplicial gravity. Let us consider a two-dimensional triangulated manifold with the topology of a disc. It is characterized by its volume n and boundary length l. For a random geometry these two numbers are not related since for any finite l, n can become arbitrarily large. If we have two such discs, of volume n and $N - n$, but with identical boundary lengths, we can identify their boundaries to create a closed simplicial complex of spherical topology and volume $N_2 = N$. The smaller component is the baby universe and the larger one the parent universe. A typical spherical manifold observed in numerical simulations will contain many structures of this kind, of various sizes and lengths of boundaries, called "necks". The shortest possible neck has length three, and the associated baby universe will be called a minimal baby universe (minbu). We may also have baby universes characterized by longer (but finite and small) neck sizes.

The existence of baby universes opens a completely new possibility for updating the matter sector of the theory. Note that for a finite boundary length l, the interaction between the matter in the baby universe and the parent universe will be of order l. Take the example of Ising spins: even if we flip all spins inside the baby universe, the change in the action will only depend on the interactions across the boundary, and not the size n of the baby universe. This is a completely different situation than on a regular lattice where large domains always have large boundaries. We may use this observation to define a matter update which induces a large change in magnetization and has a large acceptance. The efficiency of such an algorithm will depend on the typical baby universe size. One possible algorithm consists in searching the triangulation for a minimal neck and flip all spins inside the associated minbu with probability W. The same technique can be applied to massless scalar fields. Here we can perform two operations on the fields in the minbu: change all of their signs and/or add the same constant to all of them.

The existence of minbus can be exploited also in the geometric sector. For example, we may use the following three-step algorithm.

- Locate a minimal neck in the triangulation, and cut open the original spherical manifold along the neck. Both of the resulting discs are of the form of a triangulated sphere with a triangle removed.
- Close off both holes by a single triangle, to obtain two spherical triangulations of volumes $n + 1$ and $N - n + 1$ (if the original triangulation had volume N).
- Remove an arbitrary triangle from each of the two manifolds, and glue the resulting discs along their triangular boundaries.

We must compute the correct probability factors for such a move and eventually also include changes in the matter fields. Based on this idea, an extremely efficient "minbu surgery" algorithm has been constructed, shortening the autocorrelation times by three orders of magnitude [16]. The move does not even require a lot of geometric updating if it is accepted, since the connectivities of the manifold are changed only locally. The minbu surgery usually supplements the local moves described earlier, although in 2d one may construct an algorithm that is based exclusively on the cutting and pasting of baby universes, if one permits also longer neck lengths. (A flip move may be viewed as a particular realization of this, if instead of minbus we consider baby universes with boundaries of length $l = 4$.)

5.2. OBSERVABLES IN 2D EUCLIDEAN GRAVITY

The Monte Carlo method described above can be used to generate a set of uncorrelated configurations of volume N_2 which are distributed according to their Boltzmann weights e^{-S}. In order to understand the finite-size effects, one needs to perform the simulations at different values of N_2. The sizes that have been used in numerical simulations to date range from 1000 to 128000 simplices.

In 2d it speeds up the calculations to extend the class of allowed triangulations to include also loops of link length one and two (corresponding to self-energy and tadpole subdiagrams on the dual lattice). Let us denote this extended class by $\mathcal{T}$ and the corresponding class of labelled triangulations by $\tilde{\mathcal{T}}$. Although the complexes constructed in this way are no longer simplicial manifolds, we can still keep the notions of global topology, of local neighbourhoods and of a geodesic distance. Likewise, the relations between the numbers N_k remain unchanged. In some cases (including pure 2d gravity), models based on this set of geometries can be compared directly with the analytic solutions of corresponding matrix models, where the exclusion of tadpole and self-energy subdiagrams corresponds merely to a finite renormalization of the bare coupling constants. On $\mathcal{T}$, the numerical simulations become even simpler. For fixed volume, the flip move is still ergodic and also the minbu surgery moves can be generalized. The advantage of using this class of triangulations is a reduction of the finite-size effects, since it turns out that the local restrictions on the connectivity do not affect the scaling properties of the system.

Our next step will be to describe the measurement of suitable "observables" on the ensemble of configurations generated by the Monte Carlo algorithm. The observables most easily obtained are the critical exponents related to the geometry or the matter fields. In two dimensions, one can sometimes obtain such observables analytically, and use them to test the validity of the numerical results. Recall that in 2d we start from the partition function

$$Z = \sum_{\tilde{\mathcal{T}}} \frac{1}{N_0!} \sum_{\phi_i} e^{-\mu V - S_{matter}(\phi_i)} = \sum_V e^{-\mu V} Z_V, \tag{101}$$

where Z_V is the partition function at fixed volume. If the central charge of matter is $c < 1$, it can be shown analytically that Z_V behaves like

$$Z_V \sim e^{\mu_0 V} V^{\gamma_{str} - 3}(1 + \mathcal{O}(1/V)) \tag{102}$$

for large V and spherical topology. The subleading power contains the critical exponent γ_{str}, which has a known dependence on c (see formula (125) below). The pure-gravity case corresponds to $\gamma_{str} = -1/2$.

There are other statistical systems whose partition function behaves like (102), most notably, various realizations of branched polymers [20, 47]. In these models, γ_{str} is positive and $\leq 1/2$, and V counts the number of vertices.

Measuring the distribution of minbu sizes for a triangulation of fixed volume $N_2 = V$ is an efficient device for determining the critical exponent γ_{str} [48, 49]. The average number $\langle b(n)\rangle_V$ of minbus with volume $0 \ll n < V/2$ is given by

$$\langle b(n)\rangle_V \sim \frac{nZ_n(V-n)Z_{V-n}}{Z_V}, \tag{103}$$

since a minbu of size n can be regarded as a spherical triangulation of volume $n+1$ with one marked triangle. (Since both n and $V-n$ are assumed large we can neglect small corrections to the volume.) Using (102), we obtain

$$\langle b(n)\rangle_V \sim V\,(n(V-n))^{\gamma_{str}-2}. \tag{104}$$

Measuring γ_{str} gives us information about the fractal structure of the theory and provides a simple test of algorithms in cases where its value is known. In all cases with $c < 1$, one finds an excellent numerical agreement with the predicted values. From relation (104) we see that for $-1 < \gamma_{str} < 0$, the average minbu size $\langle n\rangle_V$ remains finite, but that $\langle n^2\rangle_V \sim V^{\gamma_{str}+1}$. This means that with growing γ_{str} we will observe increasingly large fluctuations in minbu size. If $\gamma_{str} > 0$ the average minbu size behaves like $\langle n\rangle_V \sim V^{\gamma_{str}}$.

Another important observable that can be studied by numerical simulations is the volume-volume correlator, which is a particular example of a geometric two-point function. In the continuum theory, it can be defined as

$$G_\Lambda(R) = \int \mathcal{D}[g_{\mu\nu}]\mathcal{D}\phi\, e^{-S[g,\phi]} \iint \mathrm{d}\xi \mathrm{d}\xi' \sqrt{g(\xi)}\sqrt{g(\xi')}\; \delta(d_g(\xi,\xi')-R), \tag{105}$$

where $d_g(\xi,\xi')$ denotes the geodesic distance between two points ξ and ξ', calculated with respect to the metric $g_{\mu\nu}$. $G_\Lambda(R)$ is the partition function for universes with two marked points separated by a geodesic distance R, which we already encountered in Section 3.2. In a discretized theory we can construct the analogous quantity

$$G_\mu(r) = \sum_{\tilde{T}} \sum_{\phi_i} e^{-\mu N_2 - S_{matter}(\phi_i)} \sum_{i,j} \delta_{D(i,j),r}, \tag{106}$$

where $D(i,j)$ is now one of the integer-valued geodesic distances introduced at the end of Section 4.2. The two possibilities will in general differ by a finite scaling factor. For large r, one can show that $G_\mu(r)$ falls off exponentially like

$$G_\mu(r) \sim e^{-m(\mu)r}, \qquad r \gg 1/m(\mu). \tag{107}$$

For small r we expect

$$G_\mu(r) \sim r^{1-\eta}, \tag{108}$$

where η is the anomalous dimension of the two-point function. For $\mu \to \mu_0$ we make the ansatz $m(\mu) \sim (\mu - \mu_0)^{1/d_h}$ where d_h is another critical exponent of the theory. From the functional form of the partition function $G_\mu(r)$, one can derive the estimates $\langle V \rangle_\mu \sim 1/(\mu - \mu_0)$ and $\langle r \rangle_\mu \sim 1/(\mu - \mu_0)^{1/d_h}$. From these one obtains a relation between the average linear extension and the average volume of the configuration, namely,

$$\langle V \rangle_\mu \sim \langle r \rangle_\mu^{d_h}. \tag{109}$$

The exponent d_h is called the cosmological Hausdorff dimension. It is a large-scale property of the average "quantum geometry" of the ensemble, and therefore need not coincide with the dimension of the individual triangulations.

The function $G_\mu(r)$ has been calculated analytically for pure gravity, as explained in Section 3.2, leading to $d_h = 4$. As we have already pointed out, it is convenient in numerical simulations to work with triangulations of fixed total volume. The two-point functions at fixed and unrestricted volumes are related by a discrete Laplace transform. From definition (106) we have

$$G_\mu(r) = \sum_V e^{-\mu V} G_V(r), \tag{110}$$

where $G_V(r)$ is the partition function at fixed volume V. For a particular configuration of the system, $G_V(r)$ is measured as follows. Start with a (dual) vertex i and find all (dual) vertices at (dual) distance 1 from it. By iterating this process, the triangulation is decomposed into shells characterized by their distance r from the initial point i. Note that the shells will in general be disconnected. We can measure the total length of each shell, that is, the number of vertices in the shell.[6] This construction is repeated for all starting points and all configurations in the sample to obtain an averaged distribution. The resulting quantity $\langle n(r) \rangle_V$ is – up to a normalization – a numerical estimate of $G_V(r)$. It is convenient to use the normalization condition

$$\sum_r G_V(r) = V, \tag{111}$$

[6] Note that one could also measure the number of disconnected parts and/or the higher moments of their length distribution.

which leads to an interpretation of $G_V(r)$ as the average volume of a shell with radius r for a triangulation of volume V. For large V one expects that

$$G_V(r) \sim V^{1-1/d_h} F(x), \tag{112}$$

where $x = r/V^{1/d_h}$ is the scaling variable. In terms of x, we have a normalization condition

$$\int dx\ F(x) = 1. \tag{113}$$

From an inverse Laplace transform of relation (107), we can deduce that for large x, $F(x)$ must behave like

$$\log F(x) \sim x^{\frac{d_h}{d_h-1}} \qquad x \gg 1. \tag{114}$$

On the other hand, we expect for small x that

$$F(x) \sim x^{d_H-1}, \tag{115}$$

where d_H is another Hausdorff dimension characterizing the quantum geometry at short distances, and is related to the anomalous dimension η. Measurements of $G_V(r)$ have been performed for systems with a variety of matter types. A simple way to determine the cosmological Hausdorff dimension d_h is to plot $G_V(r)$ against r. Even if the function $F(x)$ is not known explicitly, one can obtain an estimate of d_h by comparing the maxima of these curves for different volumes V. One then plots the rescaled quantities $G_V(r)/V^{1-1/d_h}$ as a function of x, where according to (112) the curves for different volumes V should fall on top of each other. This procedure can be viewed as a *finite-size scaling method*, of the type used in the study of critical phenomena in statistical physics. In the context of quantum gravity, it was introduced in refs. [50, 51] for $d = 2$, and in refs. [52, 53] for $d = 4$.

For pure two-dimensional gravity one can compute $F(x)$ exactly, and finds $d_H = d_h = 4$. Comparing this to the numerical analysis, it was discovered that a scaling form with the correct value of d_h can only be obtained by using a *shifted* scaling variable defined by

$$x_\delta = \frac{r+\delta}{V^{1/4}}, \tag{116}$$

where the constant δ is determined numerically. In the case of pure gravity the shift δ can be calculated analytically, but for more complicated systems this is usually not the case. The agreement between theory and numerical simulations is impressive. In Fig. 7 we show the scaling of $G_V(r)$ for various volumes V *and* the theoretical curve (which can hardly be seen since it coincides perfectly with the numerical data).

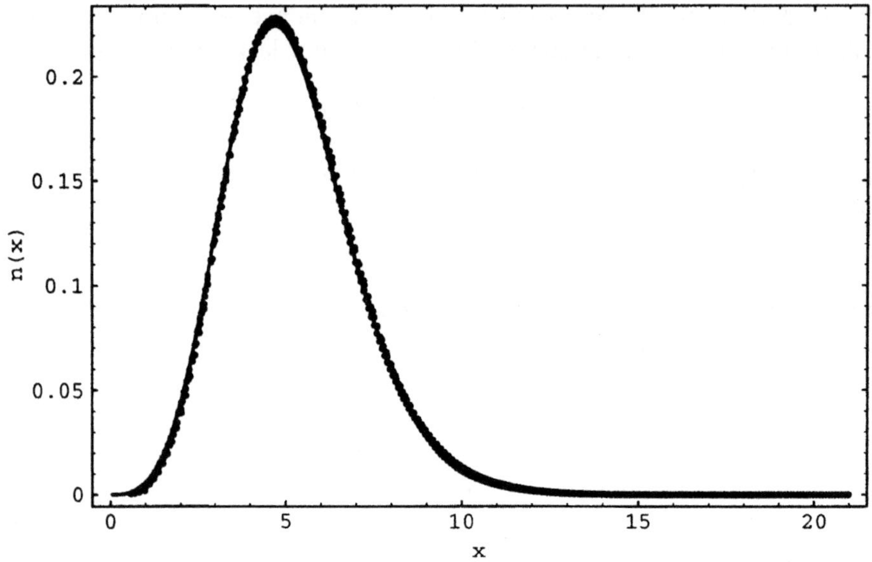

Figure 7. The distributions $n(x) \equiv G_V(x)/V^{1-1/d_h}$ in 2d Euclidean gravity, for volumes $V = 1000$, 2000, 4000, 8000, 16000 and 32000, together with the theoretical distribution $F(x)$.

Such a shift is also necessary in more complicated systems. It was soon realized that for the case of pure gravity this shift is a leading-order finite-size correction to the scaling form (112), which was obtained for $V \to \infty$. This improved scaling could also be used to check the small-x behaviour of $F(x)$ with very good accuracy.

Another important test of our numerical methods is to check whether the scaling limit of the two-point function is indeed independent of the detailed definition of the distance function. Numerically it is easy to measure the two-point function $\bar{G}_V(r)$, where r is now the *link* distance on the triangulation. In this case we have no exact analytical prediction, but we may again fit the numerical results (which look very similar to those for $G_V(r)$) to the scaling relation (112). After a trivial rescaling of the distance, the matching with the theoretical curve $F(x)$ is even better than previously, but again one must include a shift $\bar{\delta}$. Numerically, the shift is much smaller than before. This was to be expected, since a shift in x compensates for small-distance artifacts, which are larger on a lattice with smaller vertex valence. The aspects mentioned above are illustrated in Fig. 8 by data obtained in the study of a $c = -2$ conformal field theory coupled to 2d

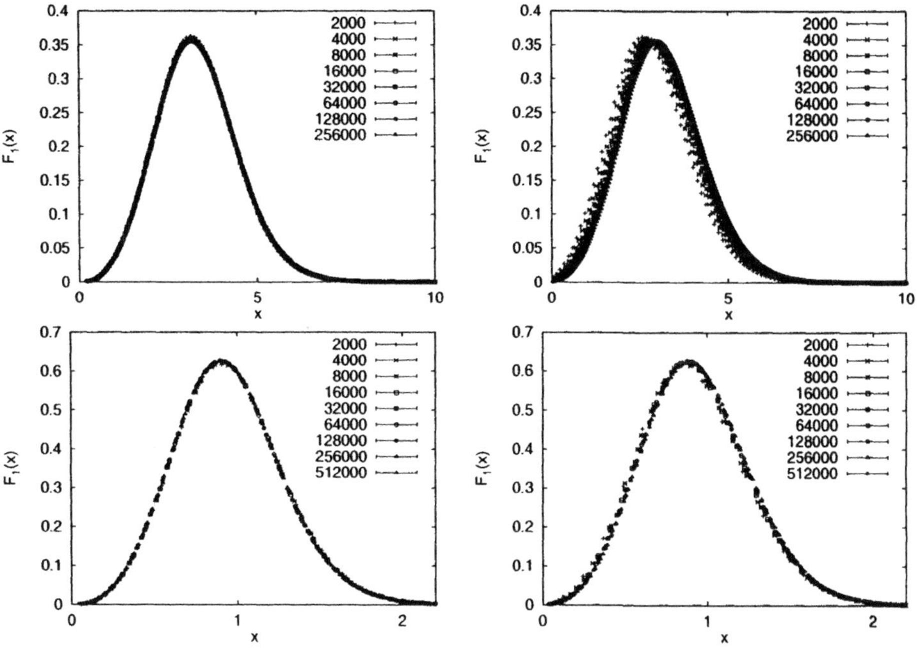

Figure 8. The two upper figures depict the measured distributions $G_V(r)$ of 2d Euclidean gravity coupled to $c = -2$ conformal matter, defined in terms of the dual link distance and rescaled according to eq. (112) to get $F(x)$, with a shift $\delta = 4.5$ (left) and without a shift (right). In the two lower figures, the link distance has been used, again with a shift $\delta = 0.5$ (left) and without(right). The best fit for the Hausdorff dimension d_h extracted from the data is 3.58 ± 0.01, in agreement with theoretical predictions [55].

Euclidean gravity [54].

Analogous results hold for branched-polymer systems, whose two-point functions can be calculated exactly. The distance r is in this case given by the (unique) number of links separating each pair of points. For fixed volume $V \to \infty$, we find a universal scaling of the form (112), but now with an (intrinsic) Hausdorff dimension $d_h = 2$, which also coincides with the value for d_H. Again a shift in x is needed as a leading finite-size correction to obtain optimal agreement with the theoretical results.

To summarize, measuring $G_V(r)$ or $\bar{G}_V(r)$ provides an efficient way for determining the Hausdorff dimension d_h, also in cases where the two-point functions are not known exactly. To take care of short-distance effects, it is important to work with an improved, shifted scaling variable x_δ. In practice one measures $G_V(r)$ for a number of volumes V, and tries to find a best fit to the universal scaling relation (112) by fitting both δ and d_h.

Another geometric observable characterizing a compact metric manifold is its *spectral dimension*, which is related to the diffusion equation and the spectrum of the Laplace operator. If a diffusion process is started with a completely localized initial condition, one can measure the return probability $P(T)$ of a fictitious test particle after "time" T. On a fixed continuum geometry, one obtains after averaging over the initial point that

$$P(T) \sim \frac{1}{T^{d_s/2}} \sum_k a_k T^k, \tag{117}$$

for small T, where the coefficients a_k can be expressed in terms of local geometric invariants. We can measure d_s also in the discretized theory, but some care must be exercised when comparing with a continuum formula like (117), since it is exactly the short-time limit $T \to 0$ that is ill-defined in the discrete case. This can already be seen in the simple case of diffusion on a discrete one-dimensional line, with a discrete Laplacian and a continuous time. The diffusion equation

$$\frac{d\phi_i}{dT} = \frac{\phi_{i+1} + \phi_{i-1} - 2\phi_i}{2}, \tag{118}$$

with the initial condition $\phi_i(0) = \delta_{i0}$ can be solved exactly, yielding

$$P(T) = e^{-T} I_0(T) \sim \frac{1}{T^{1/2}}(1 + \mathcal{O}(1/T)), \tag{119}$$

where $I_0(T)$ is the Bessel function. One therefore rederives the correct spectral dimension ($d_s = 1$) only for *large* T. The short-time behaviour is completely different and dominated by discretization artifacts. For this simple example, one can check numerically the effect of the finite system size, by changing the discrete line to a closed circle with V points. One observes three distinct regions:

- small time T, which is similar to the example above, and dominated by discretization effects;
- intermediate T, where one obtains the correct value of d_s; and
- very large T, where the system approaches a stationary translation-invariant state.

A similar structure is also expected in less trivial cases, such as simplicial quantum gravity in two dimensions and higher. This behaviour is illustrated in Fig. 9, where we show the results of measuring the spectral dimension for 2d Euclidean quantum gravity coupled to matter fields of various central charges c.

The measurement of the spectral dimension in 2d Euclidean quantum gravity is a nice illustration of the fruitful interaction between numerical "experiment" and theory. Since the quantum geometry of 2d Euclidean

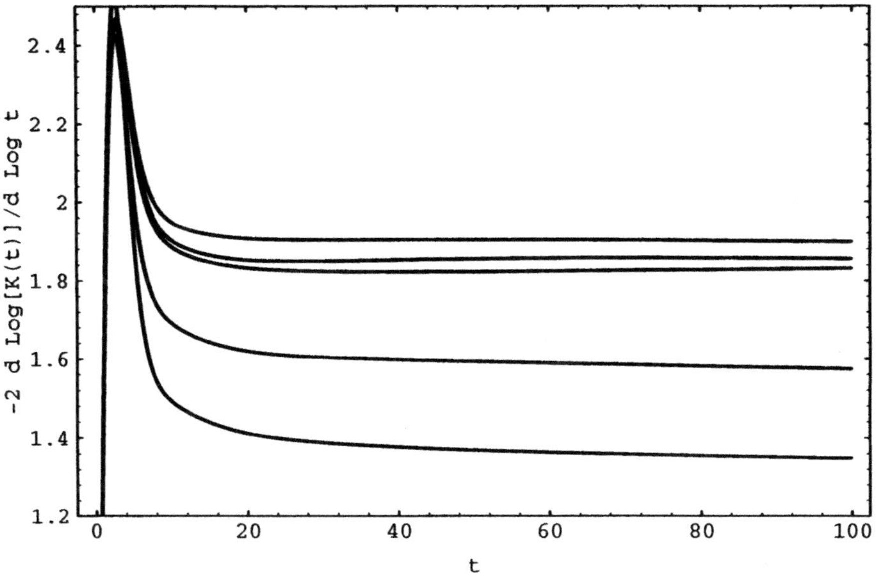

Figure 9. The spectral dimension $d_s \approx -2d\log P(T)/d\log T$ versus T for $c = 0$ (top curve), $c = 1/2$, $c = 1$, $c = 3$ and c=5 (bottom curve) theories coupled to 2d Euclidean quantum gravity. The system size is V=16000 triangles.

gravity has a fractal Hausdorff dimension $d_h = 4$, one could also have expected an anomalous spectral dimension d_s. However, numerical simulations consistently found that $d_s = 2$, no matter which field theory was coupled to gravity, as long as the central charge c remained ≤ 1. This inspired an analytic proof of this value, using continuum conformal Liouville theory and an assumption about finite-size scaling [56, 57]. We should point out that $d_s = 2$ does not imply that one cannot see the fractal structure corresponding to $d_h = 4$ in the diffusion process. It manifests itself in the relation

$$\langle R \rangle_T \sim T^{1/d_h} \tag{120}$$

between the "time" T and the average geodesic distance R diffused in that time. We refer again to [56, 57] for details.

For $c > 1$, the numerical estimate was close to $d_s = 4/3$. Eventually it was proven that generic branched polymers have a spectral dimension $d_s = 4/3$ [58], thus reconciling theory and numerical results. At the same time, it provided independent evidence that for $c > 1$ the space-time degenerates into branched polymers.

We have so far discussed only purely geometric observables. For theories coupled to matter, there will be additional critical exponents characterizing the behaviour of the system close to the critical matter coupling β_c. Near this transition point, we expect to find a divergent correlation length $\xi(\beta) \sim |\beta - \beta_c|^{-\nu}$, for some positive ν. Since we always study systems of finite size, we will never observe a genuine phase transition, but merely a pseudo-critical point $\beta_c^*(V)$, where the linear extension L of the system behaves like $L \sim \xi(\beta)$. For a finite system $L^d = V$. A system with a dynamical geometry may possess two scales: one associated with the cosmological constant when $\mu \to \mu_0$ and another one related to the matter phase transition. Using standard finite-size scaling arguments, we expect that the measured values of the specific heat c_v, the magnetization per unit volume m, and the magnetic susceptibility χ at the pseudo-critical point β_c^* behave like

$$\begin{aligned} c_v(\beta_c^*) &\sim V^{\frac{\alpha}{\nu d}}, \\ m(\beta_c^*) &\sim V^{-\frac{\beta_m}{\nu d}}, \\ \chi(\beta_c^*) &\sim V^{\frac{\gamma_m}{\nu d}}, \end{aligned} \tag{121}$$

where α, β_m and γ_m satisfy the scaling relations

$$\alpha + 2\beta_m + \gamma_m = 2, \qquad 2\beta_m + \gamma_m = \nu d. \tag{122}$$

For magnetic systems corresponding to $0 < c < 1$, one knows from analytical calculations the values of the critical exponents as well as the values of the critical points β_c.

A relation between the scalings in the geometric and matter sectors can be obtained by generalizing the two-point functions $G_\mu(r)$ and $G_V(r)$, to include also correlations between the matter fields at geodesic distance r. One example of a quantity of this kind is

$$G_\mu^{\{\phi,\psi\}}(r) = \sum_{\tilde{T}} \sum_{\phi_i} e^{-\mu N_2 - S_{matter}(\phi_i)} \sum_{i,j} \phi_i \psi_j \delta_{D(i,j),r}, \tag{123}$$

which may be expressed as the discrete Laplace transform of a finite-volume correlator,

$$G_\mu^{\{\phi,\psi\}}(r) = \sum_V e^{-\mu V} G_V^{\{\phi,\psi\}}(r). \tag{124}$$

In this expression, ϕ_i and ψ_j are local functions of the matter or geometric variables at the points i and j. A typical example is the spin-spin correlation function of the Ising system. At the combined critical point of the Ising model and the cosmological constant, (123) yields a reparameterization-invariant definition of a matter correlator in quantum gravity. More precisely, the correlation function is that of a $c = 1/2$ conformal field theory

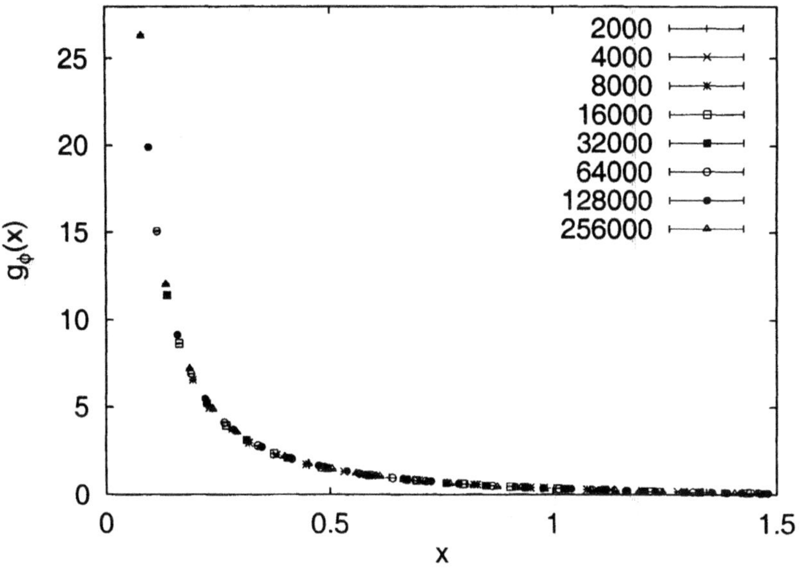

Figure 10. The normalized spin-spin correlation function for the Ising model coupled to 2d Euclidean gravity, plotted as a function of the dimensionless length $x = r/N_T^{1/d_h}$, where r is the geodesic length and N_T the space-time volume. The number of triangles N_T ranges from 2000 to 256000 and the Hausdorff dimension is $d_h = 4.0$. The short-distance behaviour is in accordance with KPZ scaling.

coupled to 2d Euclidean quantum gravity. Note also that the previously defined geometric two-point function (106) is a special case of (123), with the unit operator replacing the spin operators.

This construction has provided us with a new understanding of the KPZ-exponents of conformal field theories coupled to gravity. These exponents may be viewed as the "dressed" scaling exponents of the primary fields in the conformal theory. However, before it was realized that they should be analyzed in terms of the definition (123), it was not clear how to generalize the usual flat-space correlators because of the requirement of reparametrization-invariance (see [59, 60] for further discussion). Fig. 10 shows the result of a numerical simulation of the (suitably normalized) spin-spin correlator.

We should point out a subtlety in the definition of the correlator (123). If the expectation values of the fields entering are not zero, one might be interested in defining the corresponding *connected correlator*. However, there is no straightforward way of defining such an object since (123), in addition to the usual average over field configurations, also includes an average over geometry. A discussion of possible definitions of the connected part of the two-point correlators and their scaling properties can be found in [61].

All of the observables introduced above can also be used in numerical simulations of higher-dimensional gravity-matter systems, and – apart from the ones involving baby-universe counting – also for Lorentzian quantum gravity.

5.3. COMMENTS ON THE 2D RESULTS

Our obvious starting point in 2d Euclidean quantum gravity was the measurement of γ_{str}, in order to compare it with the theoretical prediction for $c < 1$, namely,

$$\gamma_{str} = \frac{c - 1 - \sqrt{(c-1)(c-25)}}{12}. \tag{125}$$

For $c > 1$, this quantity becomes imaginary and its interpretation in terms of Liouville theory breaks down. For $c = 1$, logarithmic corrections appear. For a number of special values of c, we have explicit discrete models, which at their critical points represent conformal field theories with charge c. They are pure 2d gravity ($c = 0$, $\gamma_{str} = -1/2$), the Ising model ($c = 1/2$, $\gamma_{str} = -1/3$), the 3-state Potts model ($c = 4/5$, $\gamma_{str} = -1/5$), the 4-state Potts model ($c = 1$, $\gamma_{str} = 0$), and a single massless scalar field ($c = 1$, $\gamma_{str} = 0$). In all of these cases, formula (125) has been verified with great accuracy. A coupling of several matter fields corresponds to adding their central charges.

Away from their critical points β_c, the finite spin systems approach a pure-gravity behaviour, with $\gamma_{str} \to -1/2$. This observation can be turned into a method for locating β_c, by monitoring the change in γ_{str} as the coupling β is varied (Fig. 11).

For all the spin systems, measurements of the magnetic critical exponents defined in (121) are in perfect agreement with the theoretical predictions[7]. We conclude that for the matter models with $0 \leq c \leq 1$ which have been used in numerical simulations, the scaling limit of the discretized theory corresponds to a unitary matter model coupled to Liouville gravity.

A somewhat surprising result in this range of c is the numerical value of d_h obtained from the scaling of $G_V(r)$. For $0 \leq c < 1$, there are two theoretical predictions for d_h, namely

$$d_h^{(i)} = 2\frac{\sqrt{25-c} + \sqrt{49-c}}{\sqrt{25-c} + \sqrt{1-c}}, \tag{126}$$

[7]This also includes the case $c = 1$, which can be realized either as two Ising spins, as a $q = 4$ Potts model, or as a single Gaussian field coupled to gravity. However, there are logarithmic corrections to formula (102) which have to be included in the fits in order to obtain $\gamma_{str} = 0$ from the data.

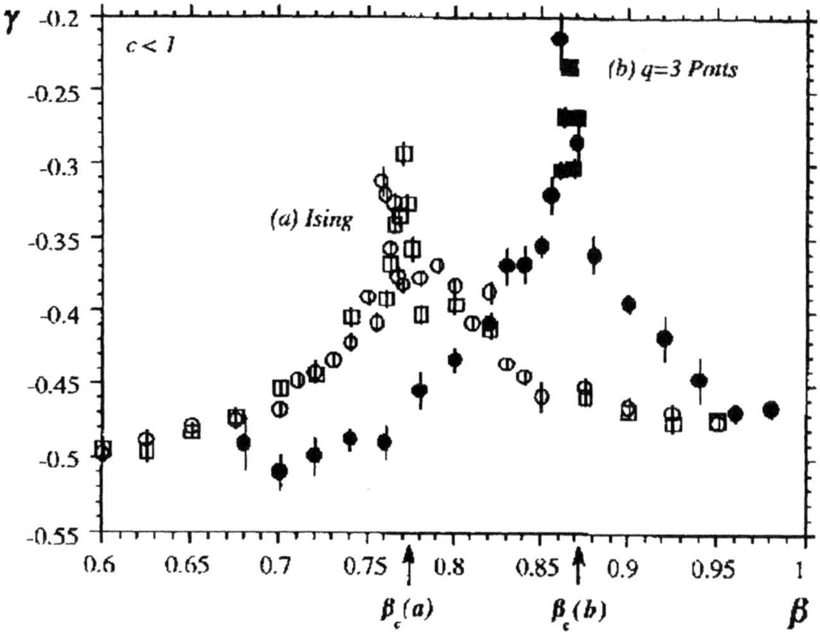

Figure 11. Measurements of γ_{str} as a function of β, for the Ising model (white), and the 3-state Potts model (black), at system sizes $N = 1000$ (circles) and 2000 (squares).

derived by studying diffusion in a fluctuating geometry, using Liouville theory [55], and

$$d_h^{(ii)} = \frac{24}{\sqrt{1-c}\,(\sqrt{1-c}+\sqrt{25-c})}, \tag{127}$$

derived from matrix model considerations [62]. The formulas agree for $c=0$ (where $d_h = 4$), but differ elsewhere. In both derivations there are plausible, but unjustified, assumptions. Neither of them could be confirmed by numerical simulations[8]. Instead, a numerical estimate $d_h \simeq 4$ was obtained for all cases with $0 \leq c \leq 1$ [51, 59] together with $d_H \simeq d_h$ [60, 59] from the small-x behaviour of $F(x)$. The numerical errors seem sufficiently small to support the conjecture that $d_h = 4$ for unitary matter theories coupled to 2d gravity. The spin-spin correlation functions measured for the Ising system ($c = 1/2$) at $\beta = \beta_c$ seem to indicate that the matter degrees of

[8]It is now understood that formula (127) measures the fractal dimension of (generalized) spin clusters of the matter fields, rather than the fractal dimension of the underlying geometry [63].

freedom do not introduce a new scale and that also for these correlations, there is a universal scaling corresponding to $d_h \simeq 4$.

Our numerical information for the case $c < 0$ is only partial. A special case is $c = -2$, where the simulations can be based on a direct generation of diagrams, instead of Monte Carlo methods. As already mentioned in connection with Fig. 8 this led to $d_h = 3.58 \pm 0.02$, which is very close to the value of $d_h = 3.561\ldots$ predicted by (126). Less conclusive numerical data for $c = -5$ also show agreement with (126). Formula (126) also predicts that $d_h = 2$ as $c \to -\infty$, which is the expected flat-space behaviour. It thus seems that eq. (126) could be correct for $c \leq 0$. It is still not understood why the formula fails for $0 < c \leq 1$.

For $c > 1$, the minbu size diverges as $V \to \infty$. Systems with $c > 1$ can easily be studied numerically by coupling several copies of spins or scalar fields. In all cases studied, γ_{str} was found to be positive and to approach $+1/2$ for $c \simeq 4$ and bigger, suggestive of a branched-polymer system. This interpretation would also agree with the scaling analysis of $G_V(r)$ which is consistent with $d_h \to 2$, and the measurement of d_s, consistent with $d_s \to 4/3$. It is not clear to what extent the observed behaviour is universal, since the numerical estimates depend on the details of the regularization. For example, the approach to the branched-polymer phase seems slower when regular triangulations T rather than generalized configurations in $\mathcal{T}$ are used. This could be due to finite-size effects, or alternatively to the presence of other universality classes of branched-polymer systems.

In the 2d Euclidean case, we have learned from both theoretical and numerical investigations that there is only a small "window" of physically sensible theories, by which we mean unitary matter models coupled to gravity. If we move outside this range by adding too many matter fields, the theory breaks down. This indicates that matter and geometry are interacting strongly.

Baby universes play a crucial role in understanding the nature of the spin-gravity interaction, highlighting at the same time the difference between the Euclidean and Lorentzian gravity models. Again the Ising model serves as an ideal illustration. The Ising ground state at zero temperature, for both fixed and fluctuating geometries, is the state where all spins are aligned. Since the energy of a given spin configuration is proportional to the length of the boundary separating spin-up and spin-down regions, the dominant spin configurations at low temperature (equivalently, at large spin coupling) are those with minimal spin boundary lengths.

The lowest spin excitations contributing to the free energy density come from spin clusters with boundaries of minimal length. Unlike on flat, regular lattices, a short spin boundary does not imply that there are few spins inside. On the contrary, the typical situation in Euclidean quantum gravity

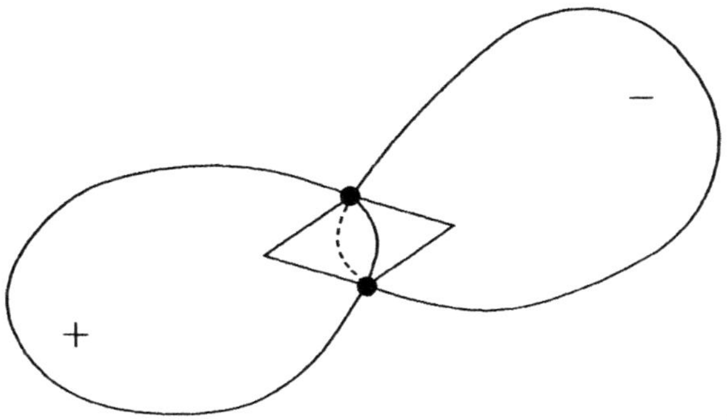

Figure 12. Two spin clusters separated by only two links which form a "bottleneck" on the surface. Two of the four triangles containing one of the two spin-boundary links are also shown.

has an entire baby universe of up-spins, say, on one side of the minimal boundary, and the parent universe with opposite spin orientation on the other, as illustrated in Fig. 12. Since the baby universe can have any size, there is no restriction on the size of spin clusters, even at low temperature. The introduction of a fluctuating geometry with baby universes thus has a strong effect on the matter behaviour. In turn, since short spin boundaries are energetically preferred, the matter has a tendency to "squeeze off" the underlying geometry, generating even more baby universes.

When more than two Ising models are coupled to 2d Euclidean quantum gravity, the matter-geometry interaction becomes so strong that the geometries degenerate into so-called branched polymers, which can be viewed as infinitely branched trees of baby universes of cut-off size (the lattice spacing a). This provides us with an explicit picture of the $c = 1$ barrier of 2d Euclidean quantum gravity. The proliferation of baby universes and its relation to spin clusters are illustrated in Fig. 13.

Since baby universes are absent from Lorentzian gravity, its interaction with matter is much weaker. The Ising model on Lorentzian geometries has not yet been solved analytically, but its critical exponents have been determined both by a high-temperature expansion and Monte Carlo techniques [12]. The Hausdorff dimension of space-time is still $d_h = 2$, and the critical Ising exponents retain the Onsager values found on fixed, regular lattices, in spite of large fluctuations of the geometry. It has also been shown [28] that a particular dimer model[9] coupled to Lorentzian gravity does not change the fractal dimension of the geometry. It is thus tempting to conjecture

[9]It is known that the critical behaviour of dimer models is associated with a $c = -2$ conformal field theory. Since the dimer model considered in [28] imposes certain restrictions on the allowed dimer positions, it strictly speaking has not been proven to lie in the same universality class.

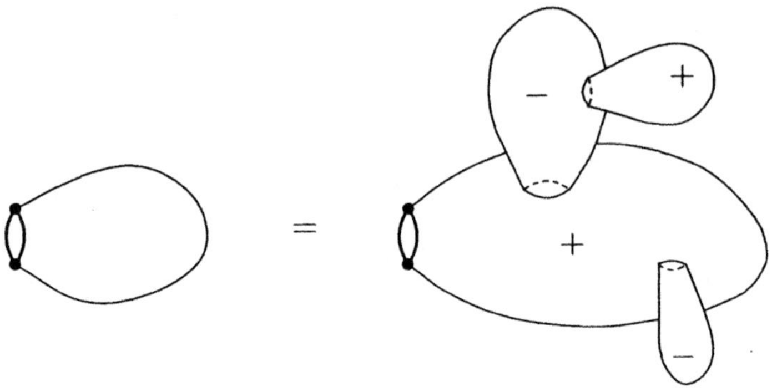

Figure 13. A surface with a minimal boundary and its recursive decomposition into baby universes associated with spin clusters.

that the (fractal) dimension of space-time remains equal to 2 as long as the central change of the matter fields is less than or equal to 1.

It is an interesting question whether there is an analogue of the $c=1$ barrier when sufficiently many matter degrees of freedom (with a sufficiently large central charge c) are added. This is indeed what seems to happen. A phase transition in the geometry has been observed when coupling 8 Ising spin copies (corresponding to $c=4$) to 2d Lorentzian quantum gravity [13]. This effect is illustrated in Fig. 14, where we show typical configurations of the fluctuating Lorentzian geometry in the case of coupling to a single Ising model and to 8 Ising models. The one-Ising triangulation is qualitatively very similar to the fluctuating geometry without any Ising spins, but in the case of 8 Ising spins, the effect of the matter is very pronounced. Since the creation of baby-universe branches is by definition forbidden, the only way for the matter to create short spin-boundaries is by squeezing individual spatial slices to a minimal (cut-off) size, producing the long, stalk-like structure seen in Fig. 14. In the remainder of space-time, the spatial slices attain a macroscopic size. On this extended part of the universe, the scaling of the average spatial length in terms of the cosmological constant is anomalous,

$$\langle L \rangle_\Lambda \sim \frac{1}{\Lambda^{2/3}}, \tag{128}$$

which was not the case for $c \leq 1$. Moreover, the critical Ising exponents, when measured on the extended region, are still given by the Onsager values. We therefore have identified an analogue of the $c = 1$ barrier, at which a geometric phase transition takes place. However, its effect is much milder than in Euclidean gravity. Ising-life continues even beyond the barrier. This underlines yet again the drastic difference between 2d gravity with and without baby universes.

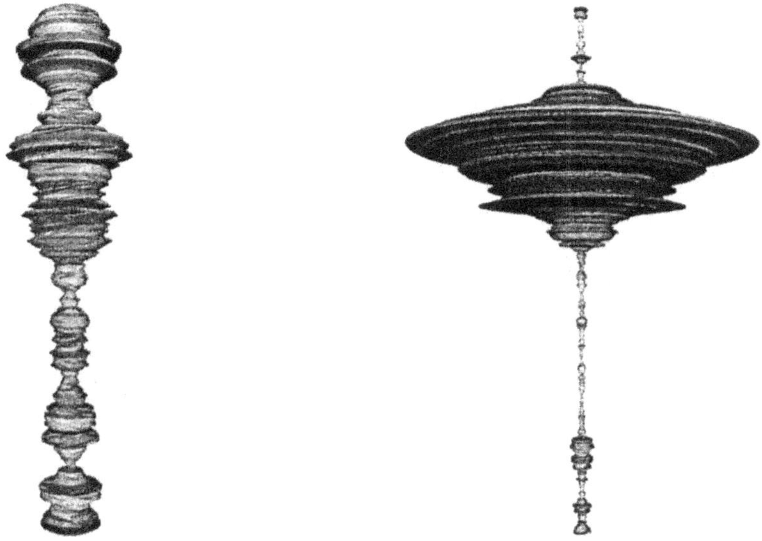

Figure 14. Two typical space-time configurations, resulting from the coupling of one Ising model (left) and eight Ising models (right) to Lorentzian gravity.

6. Dynamically triangulated quantum gravity in $d > 2$

6.1. GENERALIZATION TO HIGHER DIMENSIONS

The general outline of the simplicial regularization of Euclidean quantum gravity presented above remains true in dimension $d > 2$, where the fundamental building blocks are d-simplices with $d+1$ vertices. The partition function for dynamically triangulated pure gravity is given by

$$Z = \sum_{\tilde{T}} \frac{1}{N_0(T)!} e^{-S_{EH}}, \tag{129}$$

where the sum is taken over a class of labelled d-dimensional triangulations with spherical topology. In the following, we will be mainly interested in the cases $d = 3, 4$, where the discretized Einstein-Hilbert action (87) can be written equivalently as

$$S_{EH} = \kappa_d N_d - \kappa_0 N_0, \tag{130}$$

using the Dehn-Sommerville relations. Since for simplicial *manifolds* and fixed volume N_d, the number of vertices has an upper limit $N_0 \propto N_d/d$, the action (130) is bounded for finite volume.

Note that in $d > 2$ it is not possible to realize a locally flat geometry by gluing together equilateral simplices. Only in 2d there is a regular tiling of flat space, in which each vertex is shared by six triangles. This happens because the dihedral angle θ_d of each d-simplex contributing to the parallel transport around a $(d-2)$-simplex satisfies $\theta_d = \arccos 1/d$, which for $d > 2$ is not a rational fraction of 2π. However, this is a short-distance property which should become irrelevant in the scaling limit.

In the simplest case of gravity without matter, there are two couplings: κ_d, which is proportional to the cosmological constant, and κ_0, related to the gravitational constant. Both quantities are dimensionless in the lattice formulation, but will get their dimension back through "dimensional transmutation" in the scaling limit, if it exists.

The gravitational state sum is now

$$Z = \sum_V e^{-\kappa_d V} Z_V(\kappa_0), \tag{131}$$

where $V = N_d$, and $Z_V(\kappa_0)$ is the partition function at fixed volume. In order for the sum (131) to be well defined, $Z_V(\kappa_0)$ must be exponentially bounded for large V,

$$Z_V(\kappa_0) \sim e^{\kappa_d^c(\kappa_0)V} z_V(\kappa_0). \tag{132}$$

The existence of such a bound has been shown [24, 25], but the form of the subleading terms is at present unknown. A power behaviour of z_V leads to a situation similar to (102), but there could in principle also be a logarithmic dependence $\log z_V \sim V^\alpha$, with $\alpha < 1$. For systems with a fixed volume one can replace Z_V by z_V when averaging over configurations. Note that higher-dimensional gravity with the action (130) resembles a spin system in 2d, with the role of the matter coupling β played by κ_0. The cosmological constant will undergo a renormalization according to

$$Z = \sum_V e^{-(\kappa_d - \kappa_d^c)V} z_V(\kappa_0), \tag{133}$$

and $1/(\kappa_d - \kappa_d^c)$ gives an estimate of $\langle V \rangle_{\kappa_d}$.

To understand the critical properties of the system defined by (133), one must investigate the limit $V \to \infty$. For large positive κ_0, configurations with many vertices will be favoured. Comparing with the 2d case, we expect this limit to correspond to a branched-polymer phase. Conversely, for large negative κ_0, the system will move towards lower N_0, leading to "crumpled" configurations. Note that unlike in 2d, N_0 and V are independent, and N_0 can be very small, for example, it is easy to find triangulations such that $N_0 \sim V^a$ with $a < 1$. The fraction $p(\kappa_0) = d \langle N_0 \rangle_V / V$ can be used as an order parameter of the theory, and is related to the average local curvature. It may vary between 0 and 1, corresponding to the two extreme geometric phases. In the limit $p=0$ (κ_0 small), where the geometry is "crumpled", there are only a few vertices and one can move from one to another in only a few steps. The Hausdorff dimension of such geometries is large. In the limit $p=1$ (large κ_0), the number of vertices for a given V is maximal, leading to branched-polymer configurations with Hausdorff dimension two.

Neither of these two limits looks like a promising candidate for a scaling d-dimensional quantum theory. Let us compare this situation with that of a 2d spin system. The two extreme phases correspond to the high- and low-temperature regions of the spin model, where also the spin system has no continuum limit. In the high-temperature limit, the spins fluctuate totally randomly, whereas in the low-temperature limit they hardly fluctuate at all. However, there is a critical temperature T_c where the spin fluctuations become critical and long-ranged. At T_c, one can take a continuum limit and recover a $c=1/2$ conformal field theory. In a similar way one might hope to find a critical gravitational coupling κ_0, where a transition between the two phases of geometries takes place, and where more "reasonable" geometries, exhibiting long-range fluctuations, can be observed. Such a point could serve as a non-perturbative fixed point where a theory of quantum gravity can be defined.

Although the problem of solving higher-dimensional gravity with the simple action (130) is a combinatorial problem which superficially looks quite similar to the 2d case, we still have no hints of how to find an exact solution. There is a mean-field analysis [25, 26] which gives a surprisingly precise description of the phase transition (see also [27]), but most of our understanding of Euclidean dynamical triangulations in $d>2$ comes from performing Monte Carlo simulations, along the lines described in Section 5. Also in these cases a set of ergodic moves is known [46]. Recall that in 2d, one possible set was given by the (p,q)-moves, for integer $p,q>0$ and such that $p+q=d+2$. During a (p,q)-move, a subcomplex of p simplices is replaced by one of q simplices with the same $(d-1)$-dimensional boundary. This set can be generalized to higher dimensions [64, 65], using the observation that by gluing the two subcomplexes from before and after the move, one obtains the boundary of a $(d+1)$-simplex. The boundary can be divided into two connected parts (a part X and its complement $\bar{X}$) in a finite number of ways. Each move corresponds to replacing some $X\to\bar{X}$. In two dimensions, the boundary is that of a tetrahedron, made of four triangles. It can be divided either into two pairs of two adjacent triangles or into a single triangle and a set of three triangles sharing a vertex of order three.

Performing an analogous construction in 3d, one obtains $(1,4)$- and

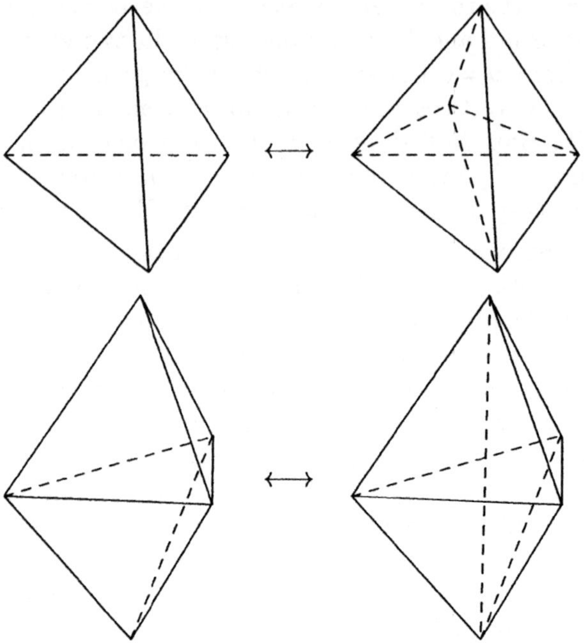

Figure 15. The elementary ergodic moves in $d = 3$. There are four different moves since the moves are different from their inverses.

(2, 3)-moves, together with their inverses (4, 1) and (3, 2). The first move inserts a new vertex of order four at the centre of a tetrahedron, thereby creating four new tetrahedra. A (2, 3)-move takes two tetrahedra which share a triangle, removes this triangle and replaces the two tetrahedra by three tetrahedra, whose common axis (a link of order three) is dual to the removed triangle. Both types of moves are shown in Fig. 15. A similar construction in 4d leads to a set of five (p, q)-moves.

These finite sets of moves are ergodic, that is, by repeating local moves, all configurations in the ensemble can be reached. Unlike in 2d, there are no subsets of moves which are ergodic in the set of triangulations of fixed volume. In 3d, all (p, q)-moves change the volume. If one wants to perform numerical simulations at some fixed V, a modification of the procedure used in the simulations becomes necessary.

A simple way of forcing the volume to lie close to a given V is to use a modified action

$$S'_{EH} = \kappa_d N_d - \kappa_0 N_0 + \epsilon |N_d - V|^\delta \tag{134}$$

in the Monte Carlo simulation, typically with $\delta = 1$ or 2, and a free parameter ϵ. The Monte Carlo process now takes place in the set of all triangulations, with fluctuating volume. If we succeed in tuning the cosmological constant to criticality, $\kappa_d \approx \kappa_d^c(\kappa_0)$, the additional term in the action will force the volume to fluctuate around $N_d = V$. The amplitude of the fluctuations will depend on the free parameter ϵ, which should be sufficiently small to permit big changes in the geometry. We can then simply collect all configurations with $N_d = V$, and use them in measurements, provided they are separated by a number of steps much bigger than the autocorrelation time.

This method seems straightforward, but it leads to the following complication. Since our set of moves is ergodic only in the ensemble of simplicial manifolds of fixed topology and *of arbitrary volume* N_d, restricting the possible volume fluctuations *could* make some parts of the configuration space unreachable. For manifolds which are not *algorithmically recognizable* it can be proven [66] (see also [8]) that the number of steps needed to connect two arbitrary triangulations of identical volume N_d cannot be bounded by a computable function $f(N_d)$ (a function which can itself be computed by a finite algorithm). This situation cannot arise when one has a finite set of moves which *is* ergodic for fixed volume (or even bounded by a computable function), since in this case one can just make a list of all triangulations, which will have a finite, computable length. We conclude that

- *if* a manifold is not algorithmically recognizable there will be "volume barriers" in the sense explained above, effectively separating parts of the configuration space.
- In a dimension d where algorithmically unrecognizable manifolds exist, one cannot have a single, finite set of volume-preserving moves which at the same time is *general*, that is, which works for *any* manifold of the given dimension, and which is ergodic for *all* manifolds.

In $d = 2$ all manifolds are algorithmically recognizable, and the flip move (introduced in Section 5.1) is a volume-preserving move which is ergodic for an arbitrary manifold. It is not known whether there exist three-dimensional manifolds which are not algorithmically recognizable. It has recently been proven that the three-sphere *is* algorithmically recognizable [67], but it is not known whether the same is true for S^4. For $d > 3$ there exist algorithmically unrecognizable manifolds, and for dimension $d > 4$ not even the d-sphere is algorithmically recognizable [68].

In the 4d numerical simulations we use triangulations with S^4-topology, and the reliability of the procedure outlined above could depend on the manifold being algorithmically recognizable. If it is not, one might worry that there are configurations, or even large regions which are separated from the rest of the configuration space by very high "volume barriers". This would imply that in order to reach such a configuration by a successive application of the elementary moves, one would be forced to go through very large intermediate volumes.

Attempts have been made to detect such barriers in numerical simulations on S^4, but without success [69]. Unfortunately, this cannot be taken as an indication that S^4 *is* algorithmically recognizable because even for the five-sphere S^5, which is known to *not* have this property, such barriers have not been seen [70]. There are two ways of interpreting the S^5-result. Either the Monte Carlo simulation explores only a fraction of the configuration space (in some sense it does anyway, since one always deals with a finite sample of "typical" configurations), or the configurations "beyond the barrier" which are hard to reach constitute only a very small fraction of the configuration space. For S^4, there need not be any problems in case it *is* algorithmically recognizable. If it is not, we hopefully are in a situation where the "unreachable" configurations form only a small subset of configurations that becomes negligible in the infinite-volume limit.

Just as in 2d, the local moves can be supplemented by *minbu surgery* moves, using a suitably generalized concept of (minimal) baby universes. They are again defined as parts of the triangulation which are connected to the parent universe by $(d-1)$-dimensional minimal necks, which topologically are boundaries of d-simplices. The baby universes again play an important role, especially in the large-κ_0 phase, and using an improved algorithm has been essential in studying the phase structure of the four-dimensional theory.

The observables described in 2d also exist for gravity in higher dimensions. The minbu distribution can be measured and used to determine γ_{str} (in case the behaviour of $z_V(\kappa_0)$ is power-like), and the scaling properties of the two-point function $G_V(r)$ yield information about the Hausdorff dimension. Matter degrees of freedom can be included without problems, the simplest being spins and massless scalar fields. The coupling of gauge fields to 4d dynamical triangulations will be discussed in the next section. – In all these cases we can employ finite-size scaling and measure the matter correlators.

6.2. NUMERICAL RESULTS IN HIGHER DIMENSIONS

The algorithm outlined in the previous section was first applied to pure 3d gravity [71]. As predicted, both a branched-polymer and a crumpled phase were discovered, but the transition between them turned out to be a very strong first-order phase transition. This manifests itself in a very large hysteresis in the order parameter $p(\kappa_0)$, when changing κ_0 across the phase transition point. The numerical effort spent on the 3d case was relatively small. Since the phase transition is first-order, it cannot give rise to a scaling limit. The phase structure remained unchanged also when spin fields were added [72].

Soon afterwards a numerical algorithm for 4d pure gravity was constructed [73] and its phase structure studied extensively. It looked similar to the one in 3d, with two distinct phases. However, this time the phase transition was much softer, and for a long time was believed to be continuous. This exciting possibility led to many numerical explorations of the properties of the two phases and the nature of the phase transition. It was shown that for large κ_0 one obtains a generic branched-polymer phase with $\gamma_{str} = 1/2$ and $d_h = 2$. Minbu surgery moves were essential in generating independent configurations in this phase. The structure of the crumpled phase seemed rather unusual: the behaviour of $z_V(\kappa_0)$ was no longer dominated by a power of V, and the scaling of $G_V(r)$ suggested that $d_h \to \infty$. (This implies that the average linear extension of the system is almost independent of V.) It was then understood that these pathologies are related to the appearance of exactly two "singular" vertices of very large order $\sim V$. A finite fraction of the simplices of the triangulations shares these two vertices, giving rise to a "volume condensation". The two points are always joined by a link, whose order also grows with the volume [74].

The appearance of these structures is well described both by the mean-field approach mentioned earlier [26], and by a phenomenological, so-called balls-in-boxes model [27]. This model represents the gravitational system as a collection of N_0 boxes, where the i'th box contains q_i balls, analogous to the order $o(i)$ of a vertex i in the simplicial complex. Note that the vertex orders $o(i)$ on a triangulation satisfy

$$\sum_i o(i) = (d+1)N_d. \tag{135}$$

In the simplified model, the q_i are required to satisfy the same constraint. The local probability of finding q balls in a box is given by $p(q) \sim q^{-\beta}$, where β is a free parameter. The model can be solved exactly in the thermodynamic limit, and exhibits two phases: one in which the balls are distributed randomly in all boxes, and a "crumpled" phase at large β where a finite fraction of the balls appears in one box. Note that this model has no concept of neighbouring boxes, and there are no correlations among the q_i (other than the relation (135)).

Although the balls-in-boxes model can hardly claim to reveal much about the geometric structure of quantum gravity, it nevertheless seems to capture some important features of the situation observed in the numerical simulations. In particular, one might wonder whether also in the full simplicial-gravity model there are no correlations between the vertex orders. This conjecture can be tested by investigating the corresponding gravitational two-point correlator for vertices separated by a (geodesic) distance r. Such a correlator is similar to a matter-matter correlator with

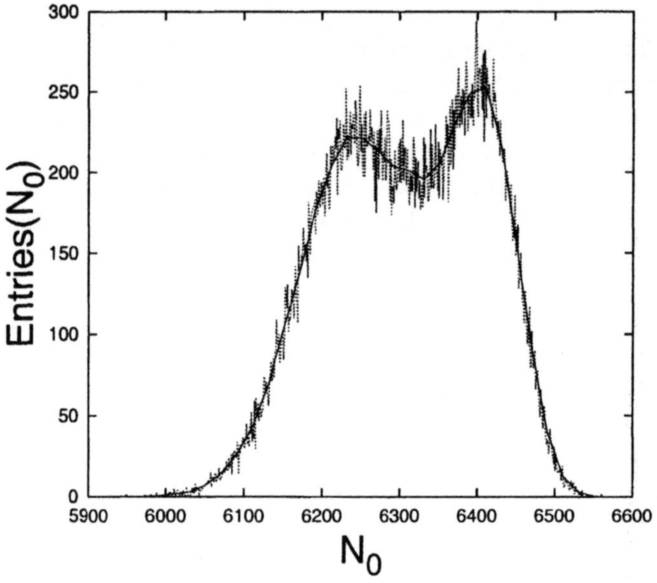

Figure 16. The double peak structure at the phase transion point as a function of the the number of vertices, N_0, for a fixed number of four-simplices, $N_4 = 32000$.

non-vanishing expectation values for the matter fields. (We will not discuss how to define correctly its connected part.) Preliminary measurements indicate that the correlations do not extend beyond a few lattice spacings, that is, they are of the order of the cut-off [77] (see also [78] for an earlier report). Similarly short-ranged correlations have also been observed in the 2d case [61].

There is by now overwhelming numerical evidence that the weak- and strong coupling phases are separated by a weak first-order transition [79]. The determination of the order turned out to be difficult, since no hysteresis was observed for finite systems. The best way of identifying the nature of the phase transition is by measuring the distribution of N_0 in a sample of configurations exactly at the pseudo-critical point. The distribution has two peaks, whose separation grows with the volume. This is shown in Fig. 16. The effect becomes only apparent for sufficiently large systems (of 32000 and 64000 simplices); for smaller volumes the separation cannot be seen.

From the point of view of quantum gravity, this is of course a disappointing result, because there is no new continuum physics associated with the phase transition. The weakness of the transition could be an indication that a continuous phase transition can be reached by enlarging the space of coupling constants, that is, by extending the action S_{EH}. Obvious candidates would be terms containing higher powers of the curvature, which have previously been used in quantum Regge calculus [80].

In both 3d and 4d simplicial gravity, the phases observed on either side of the transition are pathological, in that they do not seem to correspond to an effective quantum geometry of a sensible dimensionality. As already mentioned we find either branched polymers (locally one-dimensional objects) or a condensation of the volume around selected points (yielding an object of arbitrarily large dimension). It is surprising that the rich geometric structure observed in two dimensions has all but disappeared. There are potential cures for this, although there is no guarantee that they will lead to the desired quantum theory.

Within the framework of dynamical triangulations, there are basically two ways of "changing the measure", in order to improve this state of affairs. One is by using a modified action, but leaving the sum over geometries untouched. Gravity with a squared scalar-curvature term was investigated in [81], but the numerical results indicated that the phase structure remained unchanged. Note that also in $d = 2$, the addition of such terms does not change the universality class of pure gravity, as has been shown both numerically (see, for example, [82]) and analytically [83]. The other possibility is to change the sum to a different class of triangulations. A physically motivated example of this kind is the imposition of a causality constraint on Lorentzian geometries. It *does* lead to a change of universality class in $d = 2$, as we have already seen. There are indications that also in $d > 2$ a causality condition acts as an effective regulator for the quantum geometry [84]. Numerical investigations of Lorentzian models in higher dimensions are under way.

Another logical possibility is that only the matter-coupled theory possesses a well-defined scaling limit. Recall that in two dimensions, there is a rather narrow window of physically interesting Euclidean continuum theories, namely, Liouville gravity interacting with conformal fields of charge $0 \leq c \leq 1$. We simply may have been lucky that pure gravity lies within this range.

Matter interactions could in principle suppress the formation of baby universes in higher dimensions, and the inclusion of gauge (vector) fields was suggested as a promising scenario for testing this conjecture [85]. The simplest model of this kind is the non-compact 4d Abelian gauge model formulated in [75]. Like on a regular lattice, the gauge fields $A_{ij} \equiv -A_{ji}$ are associated with oriented lattice links $\{i, j\}$. A discretized version P of the field strength tensor $F_{\mu\nu}$ can be expressed in terms of "plaquettes" around (oriented) triangles $\{ijk\}$,

$$P_{\{ijk\}} = A_{ij} + A_{jk} + A_{ki}. \tag{136}$$

The action was taken to be

$$S_{gauge} = \sum_{\{ijk\}} o(\{ijk\}) P^2_{\{ijk\}}, \tag{137}$$

with $o(\{ijk\})$ denoting the order of the triangle $\{ijk\}$. This factor was included to guarantee that whenever P^2 is constant for all triangles, S_{gauge} is proportional to the volume N_4 of the manifold. The action has a simple gauge symmetry and is relatively easy to use in numerical simulations, since it is Gaussian in the vector fields. The number n_g of gauge fields can be varied, and for $n_g = 3$ a new "crinkled" phase with $\gamma_{str} < 0$ was observed, replacing the branched-polymer phase. Moreover, the phase transition seemed to be continuous rather than first-order.

However, it was soon realized that this new phase can be obtained in a model without gauge fields, by adding a geometric term

$$S_\Delta = \frac{n_g}{2} \sum_{\{ijk\}} \log o(\{ijk\}) \tag{138}$$

to the pure-gravity action [76]. Since S_Δ is an ultra-local measure term, one would not really expect it to lead to new *continuum* physics.

The origin of this term can be traced to a somewhat unfortunate choice of gauge action. At first sight, it seems natural to associate the gauge fields with the links of the triangulation, analogous to what is done on regular, flat lattices. This implies that any charged fields which couple to the gauge fields will be located at the vertices. However, note that scalar fields coupled to gravity in all investigations so far have been placed at the centres of the four-simplexes, to ensure that the number of field degrees of freedom grows proportional to the volume. In the case of fluctuating geometry, this need not be the case if the scalar fields are located at the vertices, since N_0 may grow much slower than N_4.

Thus, if our primary aim is to compare the effects of gauge fields on geometry with those of the previously studied scalar fields, the complex-valued charged fields should be associated with the four-simplexes of the triangulation, and the gauge fields with the links of the *dual* lattice. The pure-gauge action is then a sum over the *dual* plaquettes, which are the natural duals of the triangles in the original triangulation (see [77] for details).

The two models with the gauge potentials A on the links and on the dual links are dual to each other. The theory defined by (137) (on the triangulation) is equivalent to the gauge theory on the dual lattice, if we add by hand a term (138) to the natural gauge action on the dual lattice [77]. Monte Carlo simulations with gauge fields on the dual links without the term (138), as advocated above, have not detected any effect of the gauge fields on the geometry, in line with previous results for scalar fields and Ising spins coupled to gravity.

Putting a more positive spin on this result, the inclusion of the term (138) is the only example of a modification of the action which has led to a genuine change of the geometry. Although it has no obvious physical interpretation and although the crinkled phase is probably not related to interesting continuum physics, it does give us some hope of finding a physically well-motivated term which *does* change the geometry in an interesting way, allowing us to define a continuum theory of gravity.

7. Outlook

In these notes we have described in some detail a non-perturbative path-integral approach to quantum gravity, with and without matter. This "dynamical triangulations" method uses an intermediate regularization, in which the space of all space-time geometries is approximated by a set of simplicial complexes with certain edge length assignments, in the spirit of Regge calculus. For a finite volume (that is, for a complex with a finite number of simplices) and fixed space-time topology, this method yields well-defined, convergent partition functions and propagators. The aim in this approach is to understand the critical behaviour of these statistical systems of fluctuating geometry, and investigate their physical properties in the continuum limit, if it exists.

We believe that this ansatz has a number of virtues that make further research worthwhile. Firstly, it works perfectly in two dimensions, leading to a host of analytical and numerical results. There, it agrees with other discrete or continuum formulations, whenever they can be compared, and in some cases turns out to be even more powerful. This gives us considerable confidence in the validity of the method. Moreover, the two-dimensional construction has served as a useful laboratory for defining diffeomorphism-invariant versions of two-point functions and other field-theoretic observables. Dynamical triangulations as a method can be applied in *any* dimension d, but obviously will lead to very different theories, depending on d. Only in $d = 4$ do we expect to find local, interacting quantum-gravitational degrees of freedom.

Secondly, as has been shown recently, there is also a Lorentzian version of the formulation, where additional causality restrictions are imposed on the space-time geometries, and a well-defined Wick rotation exists. The absence of a natural notion of "analytic continuation" from Euclidean to Lorentzian signature has for a long time stalled progress in discretized path-integral approaches to gravity. Surprisingly, in two space-time dimensions the Euclidean and Lorentzian continuum quantum gravity theories are very different. As a consequence of the causality constraint, the "baby-universe" structure of the Euclidean theory is absent in the Lorentzian sector. This leads to a "smoother" quantum geometry, and well-behaved matter coupling even beyond the $c = 1$ barrier. Such a behaviour is very desirable in dimension $d > 2$, where so far the dominance of "extreme" geometric phases in Euclidean dynamically triangulated models seems to have prevented the existence of an interesting continuum limit.

In higher dimensions, much work remains to be done to establish whether our method leads to a well-defined continuum quantum theory. In the Euclidean sector, a suitable change of the discretized measure or the inclusion of different types of matter may still lead to long-range correlations. In the Lorentzian sector, there is a genuine chance that the causality restrictions will lead to a qualitatively different phase structure, including a second-order phase transition. The investigation of the Lorentzian models for $d > 2$ has only just begun. However, we do not have to start from scratch when trying to solve the theory, but can use the considerable expertise gathered in the Euclidean simulations. Obviously, quantum gravity will not be solved by numerical methods alone, but we have learned from the example of dynamical triangulations in 2d that a mathematically exact analysis and numerical simulations can work hand in hand.

In addition, one can hope for analytical progress in solving the discretized models (which in the case of quantum Regge calculus has always been hampered by the presence of inequalities on the edge lengths, even in $d = 2$). In terms of dynamical triangulations, quantum gravity has been turned into a well-defined combinatorial problem, which has already been solved exactly in two dimensions. It is also likely that – analogous to the Euclidean case – mean-field methods can help to elucidate the Lorentzian phase structure in higher dimensions. This may have interesting ramifications for the general study of spaces of pseudo-Riemannian geometries.

In summary, the dynamical-triangulations approach is a promising tool for both analytical and numerical investigations of quantum gravity. It allows for a natural inclusion of matter, and is explicitly reparametrization-invariant. The path integral can be defined for both Euclidean and Lorentzian geometries. In the Lorentzian case, it involves summing over a class of Wick-rotatable space-times with a natural "time"-slicing, reminiscent of the geometric structures appearing in Hamiltonian quantization schemes. We hope that our construction of a Lorentzian path integral will serve as a bridge between the canonical and covariant formulations, benefitting both and improving our understanding of the nature of quantum gravity.

Acknowledgements

J.A. acknowledges the support of MaPhySto – Center for Mathematical Physics and Stochastics – financed by the National Danish Research Foundation. J.J. was partially supported by the Polish Government KBN grants 2P03B 019 17 and 2P03B 998 14.

References

1. T. Banks, W. Fischler, S.H. Shenker and L. Susskind: M theory as a matrix model: a conjecture, *Phys. Rev. D* **55** (1997) 5112-5128, e-Print Archive: hep-th/9610043.
2. N. Ishibashi, H. Kawai, Y. Kitazawa and A. Tsuchiya: A large-N reduced model as superstring, *Nucl. Phys. B* **498** (1997) 467-491, e-Print Archive: hep-th/9612115.
3. F. David: Planar diagrams, two-dimensional lattice gravity and surface models, *Nucl. Phys. B* **257** (1985) 45-58; A model of random surfaces with nontrivial critical behaviour, *Nucl. Phys. B* **257** (1985) 543-576.
4. J. Ambjørn, B. Durhuus and J. Fröhlich: Diseases of triangulated random surface models, and possible cures, *Nucl. Phys. B* **257** (1985) 433-449;
   J. Ambjørn, B. Durhuus, J. Fröhlich and P. Orland: The appearance of critical dimensions in regulated string theories, *Nucl. Phys. B* **270** (1986) 457-482.
5. V.A. Kazakov, I.K. Kostov and A.A. Migdal: Critical properties of randomly triangulated planar random surfaces, *Phys. Lett. B* **157** (1985) 295-300.
6. J. Ambjørn and J. Jurkiewicz: Four-dimensional simplicial quantum gravity, *Phys. Lett. B* **278** (1992) 42-50.
7. M.E. Agishtein and A.A. Migdal: Simulations of four-dimensional simplicial quantum gravity as dynamical triangulation, *Mod. Phys. Lett. A* **7** (1992) 1039-1061.
8. J. Ambjørn, B. Durhuus and T. Jonsson, *Quantum geometry*, Cambridge Monographs on Mathematical Physics, Cambridge University Press, Cambridge, UK, 1997.
9. A. Krzywicki: Random manifolds and quantum gravity, *preprint* Orsay LPT-ORSAY-99-51, e-Print Archive: hep-lat/9907012.
10. R. Loll: Discrete approaches to quantum gravity in four dimensions, *Living Reviews in Relativity* **13** (1998), http://www.livingreviews.org, e-Print Archive: gr-qc/9805049.
11. J. Ambjørn and R. Loll: Non-perturbative Lorentzian quantum gravity, causality and topology change, *Nucl. Phys. B* **536** (1998) 407-434, e-Print Archive: hep-th/9805108.
12. J. Ambjørn, K.N. Anagnostopoulos and R. Loll: A new perspective on matter coupling in 2d quantum gravity, *Phys. Rev. D* **60** (1999) 104035, e-Print Archive: hep-th/9904012.
13. J. Ambjørn, K.N. Anagnostopoulos and R. Loll: Crossing the c=1 barrier in 2d Lorentzian gravity, to appear in *Phys. Rev. D*, e-Print Archive: hep-lat/9909129.
14. C. Teitelboim: Causality versus gauge invariance in quantum gravity and supergravity, *Phys. Rev. Lett.* **50** (1983) 705-708.
15. L. Bombelli, J. Lee, D. Meyer and R. Sorkin: Space-time as a causal set, *Phys. Rev. Lett.* **59** (1987) 521-524;
    D.P. Rideout and R.D. Sorkin: Classical sequential growth dynamics for causal sets, *Phys. Rev. D* **61** (2000) 024002, e-Print Archive: gr-qc/9904062.
16. F. Markopoulou and L. Smolin: Causal evolution of spin networks, *Nucl. Phys. B* **508** (1997) 409-430, e-Print Archive: gr-qc/9702025; Quantum geometry with intrinsic local causality, *Phys. Rev. D* **58** (1998) 084032, e-Print Archive: gr-qc/9712067.
17. J. Ambjørn, J. Correia, C. Kristjansen and R. Loll: On the relation between Euclidean and Lorentzian 2d quantum gravity, to appear in *Phys. Lett. B*, e-Print

Archive: hep-th/9912267.
18. H. Kawai, N. Kawamoto, T. Mogami and Y. Watabiki: Transfer matrix formalism for two-dimensional quantum gravity and fractal structures of space-time, *Phys. Lett. B* **306** (1993) 19-26, e-Print Archive: hep-th/9302133.
19. J. Ambjørn and Y. Watabiki: Scaling in quantum gravity, *Nucl. Phys. B* **445** (1995) 129-144, e-Print Archive: hep-th/9501049.
20. J. Ambjørn, B. Durhuus and T. Jonsson: A solvable 2-d gravity model with gamma > 0, *Mod. Phys. Lett. A* **9** (1994) 1221-1228, e-Print Archive: hep-th/9401137.
21. J. Ambjørn, K.N. Anagnostopoulos and R. Loll: On the phase diagram of 2d Lorentzian quantum gravity, *preprint* AEI-1999-20, NBI-HE-99-27 e-Print Archive: hep-lat/9908054.
22. J. Ambjørn, K.N. Anagnostopoulos and R. Loll: Making the gravitational path integral more Lorentzian, or Life beyond Liouville gravity, *preprint* AEI-1999-32, NBI-HE-99-43, e-Print Archive: hep-th/9910232.
23. http://www.nbi.dk/~ambjorn/lqg2.
24. M. Carfora and A. Marzuoli: Entropy estimates for simplicial quantum gravity, *J. Geom. Phys.* **16** (1995) 99-119; Holonomy and entropy estimates for dynamically triangulated manifolds, *J. Math. Phys.* **36** (1995) 6353-6376.
25. J. Ambjørn, M. Carfora and A. Marzuoli: *The geometry of dynamical triangulations*, Lecture Notes in Physics, New Series, m50, Springer, Berlin, 1997, e-Print Archive: hep-th/9612069.
26. D. Gabrielli: Polymeric phase of simplicial quantum gravity, *Phys. Lett. B* **421** (1998) 79-85 e-Print Archive: hep-lat/9710055;
J. Ambjørn, M. Carfora, D. Gabrielli and A. Marzuoli: Crumpled triangulations and critical point in 4D simplicial quantum gravity, *Nucl. Phys. B* **542** (1999) 349-394, e-Print Archive: hep-lat/9806035.
27. P. Bialas, L. Bogacz, Z. Burda and D. Johnston: Finite size scaling of the balls in boxes model, *preprint* BI-TP-99-41, e-Print Archive: hep-lat/9910047.
P. Bialas, Z. Burda and D. Johnston: Phase diagram of the mean field model of simplicial gravity, *Nucl. Phys. B* **542** (1999) 413-424, e-Print Archive: gr-qc/9808011; Condensation in the backgammon model, *Nucl. Phys. B* **493** (1997) 505-516.
S. Catterall, R. Renken and J. Kogut: Singular structure in 4-d simplicial gravity, *Phys. Lett. B* **416** (1998) 274-280, e-Print Archive: hep-lat/9709007.
28. P. Di Francesco, E. Guitter and C. Kristjansen: Integrable 2d Lorentzian gravity and random walks, to appear in *Nucl. Phys. B*, e-Print Archive: hep-th/9907084.
29. R.M. Wald: *General relativity*, University of Chicago Press, Chicago, 1984.
30. R. Sorkin: Time-evolution in Regge calculus, *Phys. Rev. D* **12** (1975) 385-396; Err. *ibid.* **23** (1981) 565.
31. R. Nakayama: 2-d quantum gravity in the proper time gauge, *Phys. Lett. B* **325** (1994) 347-353, e-Print Archive: hep-th/9312158.
32. A. Carlini and J. Greensite: Square root actions, metric signature, and the path integral of quantum gravity, *Phys. Rev. D* **52** (1995) 6947-6964, e-Print Archive: gr-qc/9502023.
33. A. Carlini and J. Greensite: Fundamental constants and the problem of time, *Phys. Rev. D* **52** (1995) 936-960, e-Print Archive: gr-qc/9406044.
34. J. Louko and R.D. Sorkin: Complex actions in two-dimensional topology change, *Class. Quant. Grav.* **14** (1997) 179-203, e-Print Archive: gr-qc/9511023.
35. N. Ishibashi and H. Kawai: String field theory of noncritical strings, *Phys. Lett. B* **314** (1993) 190-196, e-Print Archive: hep-th/9307045.
36. Y. Watabiki: Construction of noncritical string field theory by transfer matrix formalism in dynamical triangulation, *Nucl. Phys. B* **441** (1995) 119-166, e-Print Archive: hep-th/9401096.
37. J. Ambjørn, J. Jurkiewicz and Y. Watabiki: Dynamical triangulations, a gateway to quantum gravity?, *J. Math. Phys.* **36** (1995) 6299-6339, e-Print Archive: hep-th/9503108.

38. T. Regge: General relativity without coordinates, *Nuovo Cim. A* **19** (1961) 558-571.
39. E.A. Bender and E.R. Canfield: The asymptotic number of rooted maps on a surface, *J. of Combinatorial Theory A* **43** (1986) 244.
40. H.-C. Ren: Matter fields in lattice gravity, *Nucl. Phys. B* **301** (1988) 661-684.
41. Z. Burda, J. Jurkiewicz and A. Krzywicki, *Phys. Rev. D* **60** (1999) 105029, e-Print Archive: hep-lat/9905015.
42. M.A. Bershadsky, I.D. Vaisburd and A.A. Migdal: Equivalence of the ising model and the fermion model on a riemann surface, *JETP Lett.* **43** (1986) 193-196.
43. G. Thorleifsson: Lattice gravity and random surfaces, *Nucl. Phys. Proc. Suppl.* **73** (1999) 133-145; e-Print Archive: hep-lat/9809131;
M. Bowick: Random surfaces and lattice gravity, *Nucl. Phys. Proc. Suppl.* **63** (1998) 77-88; e-Print Archive: hep-lat/9710005;
D.A. Johnston: Gravity and random surfaces on the lattice: a review, *Nucl. Phys. Proc. Suppl.* **53** (1997) 43-55, e-Print Archive: hep-lat/9607021.
44. S. Bilke, Z. Burda and J. Jurkiewicz: Simplicial quantum gravity on a computer, *Comput. Phys. Commun.* **85** (1995) 278-292; e-Print Archive: hep-lat/9403017.
S. Catterall: Simulations of dynamically triangulated gravity: an algorithm for arbitrary dimension, *Comput. Phys. Commun.* **87** (1995) 409-415, e-Print Archive: hep-lat/9405026.
45. U. Wolff: Critical slowing down, *Nucl. Phys. B (Proc. Suppl.)* **17** (1990) 93-102.
46. J.W. Alexander: The combinatorial theory of complexes, *Ann. Mat.* **31** (1931) 292.
47. J. Ambjørn, B. Durhuus and T. Jónsson: Summing over all genera for d > 1: a toy model, *Phys. Lett. B* **244** (1990) 403-412;
P. Bialas and Z. Burda: Phase transition in fluctuating branched geometry, *Phys. Lett. B* **384** (1996) 75-80, e-Print Archive: hep-lat/9605020;
J.D. Correia, B. Mirza and J.F. Wheater: Branched polymers, complex spins and the freezing transition, *Phys. Lett. B* **415** (1997) 15-23, e-Print Archive: hep-lat/9708027.
48. J. Ambjørn, S. Jain and G. Thorleifsson: Baby universes in 2-d quantum gravity, *Phys. Lett. B* **307** (1993) 34-39, e-Print Archive: hep-th/9303149.
49. J. Ambjørn, S. Jain, J. Jurkiewicz and C.F. Kristjansen: Observing 4d baby universes in quantum gravity, *Phys. Lett. B* **305** (1993) 208-213, e-Print Archive: hep-th/9303041.
50. S. Catterall, G. Thorleifsson, M. Bowick and V. John: Scaling and the fractal geometry of two-dimensional quantum gravity, *Phys. Lett. B* **354** (1995) 58-68, e-Print Archive: hep-lat/9504009.
51. J. Ambjørn, J. Jurkiewicz and Y. Watabiki: On the fractal structure of two-dimensional quantum gravity, *Nucl. Phys. B* **454** (1995) 313-342, e-Print Archive: hep-lat/9507014.
52. B.V. de Bakker and J. Smit: Curvature and scaling in 4-d dynamical triangulation, *Nucl. Phys. B* **439** (1995) 239-258, e-Print Archive: hep-lat/9407014.
53. J. Ambjørn and J. Jurkiewicz: Scaling in four dimensional quantum gravity, *Nucl. Phys. B* **451** (1995) 643-676, e-Print Archive: hep-th/9503006.
54. J. Ambjørn, K. Anagnostopoulos, T. Ichihara, L. Jensen, N. Kawamoto, Y. Watabiki and K. Yotsuji: The quantum space-time of c = -2 gravity, *Nucl. Phys. B* **511** (1998) 673-710. e-Print Archive: hep-lat/9706009; Quantum geometry of topological gravity, *Phys. Lett. B* **397** (1997) 177-184, hep-lat/9611032.
55. Y. Watabiki: Analytic study of fractal structure of quantized surface in two-dimensional quantum gravity, *Prog. Theor. Phys. Suppl.* **114** (1993) 1-17.
56. J. Ambjørn, D. Boulatov, J.L. Nielsen, J. Rolf and Y. Watabiki: The spectral dimension of 2-d quantum gravity, *JHEP* 9802:010 (1998), e-Print Archive: hep-th/9801099.
57. J. Ambjørn, R. Loll, J.L. Nielsen and J. Rolf: Euclidean and Lorentzian quantum gravity: lessons from two-dimensions, *Chaos Solitons Fractals* **10** (1999) 177-195, e-Print Archive: hep-th/9806241.

58. T. Jonsson and J.F. Wheater: The spectral dimension of the branched polymer phase of two-dimensional quantum gravity, *Nucl. Phys. B* **515** (1998) 549-574, e-Print Archive: hep-lat/9710024;
J.D. Correia and J.F. Wheater: The spectral dimension of nongeneric branched polymer ensembles, *Phys. Lett. B* **422** (1998) 76-81, e-Print Archive: hep-th/9712058.
59. J. Ambjørn and K.N. Anagnostopoulos: Quantum geometry of 2-d gravity coupled to unitary matter, *Nucl. Phys. B* **497** (1997) 445-478, e-Print Archive: hep-lat/9701006.
60. J. Ambjørn, K.N. Anagnostopoulos, U. Magnea and G. Thorleifsson: Geometrical interpretation of the KPZ exponents, *Phys. Lett. B* **388** (1996) 713-719, e-Print Archive: hep-lat/9606012.
61. J. Ambjørn, P. Bialas and J. Jurkiewicz: Connected correlators in quantum gravity, *JHEP* 9902:005 (1999), e-Print Archive: hep-lat/9812015.
62. N. Ishibashi and H. Kawai: String field theory of c $\leq$ 1 noncritical strings, *Phys. Lett. B* **322** (1994) 67-78, e-Print Archive: hep-th/9312047.
63. J. Ambjørn, K.N. Anagnostopoulos, J. Jurkiewicz and C.F. Kristjansen: The concept of time in 2-d gravity, *JHEP* 9804:016 (1998), e-Print Archive: hep-th/9802020.
64. U. Pachner: Bistellare Äquivalenz kombinatorischer Mannigfaltigkeiten, *Arch. Math.* **30** (1978) 89; Konstruktionsmethoden und das kombinatorische Homöomorphieproblem für Triangulationen kompakter semilinearer Mannigfaltigkeiten, *Abh. Math. Sem. Univ. Hamburg* **57** (1986) 69-85; P.L. homeomorphic manifolds are equivalent by elementary shellings, *Europ. J. Combinatorics* **12** (1991) 129-145.
65. M. Gross and S. Varsted: Elementary moves and ergodicity in d-dimensional simplicial quantum gravity, *Nucl. Phys. B* **378** (1992) 367-380.
66. A. Nabutovsky and R. Ben-Av: Noncomputability arising in dynamical triangulation model of four-dimensional quantum gravity, *Commun. Math. Phys.* **157** (1993) 93-98.
67. A. Thompson: Thin position and the recognition problem for S^3, *Math. Res. Lett.* **1** (1994) 613.
68. F. Acquistapace, R. Benedetti and F. Broglia: Effectiveness and non-effectiveness in semialgebraic and PL geometry, *Invent. Math.* **102** (1990) 141.
69. J. Ambjørn and J. Jurkiewicz: Computational ergodicity of S^4, *Phys. Lett. B* **345** 435-440, e-Print Archive: hep-lat/9411008.
70. B.V. de Bakker: Absence of barriers in dynamical triangulation, *Phys. Lett. B* **348** 35-38, e-Print Archive: hep-lat/9411070.
71. D.V. Boulatov and A. Krzywicki: On the phase diagram of three-dimensional simplicial quantum gravity, *Mod. Phys. Lett. A* **6** (1991) 3005-3014.
J. Ambjørn and S. Varsted: Three-dimensional simplicial quantum gravity, *Nucl. Phys. B* **373** (1992) 557-580.
J. Ambjørn, D.V. Boulatov, A. Krzywicki and S. Varsted: The vacuum in three-dimensional simplicial quantum gravity, *Phys. Lett. B* **276** (1992) 432-436;
M.E. Agishtein and A.A. Migdal: Three-dimensional quantum gravity as dynamical triangulation, *Mod. Phys. Lett. A* **6** (1991) 1863-1884;
R.L. Renken, S.M. Catterall and J.B. Kogut: Phase structure of dynamical triangulation models in three-dimensions, *Nucl. Phys. B* **523** (1998) 553-568, e-Print Archive: hep-lat/9712011.
72. J. Ambjørn, Z. Burda, J. Jurkiewicz and C.F. Kristjansen: 3-d quantum gravity coupled to matter, *Phys. Lett. B* **297** (1992) 253-260, e-Print Archive: hep-lat/9205021;
R.L. Renken, S.M. Catterall and J.B. Kogut: Three-dimensional quantum gravity coupled to ising matter, *Nucl. Phys. B* **389** (1993) 601-610, e-Print Archive: hep-lat/9205029.
73. J. Ambjørn and J. Jurkiewicz: Four-dimensional simplicial quantum gravity, *Phys. Lett. B* **278** (1992) 42-50; Scaling in four dimensional quantum gravity, *Nucl. Phys. B* **451** (1995) 643-676, e-Print Archive: hep-th/9503006;

M.E. Agishtein and A.A. Migdal: Critical behavior of dynamically triangulated quantum gravity in four-dimensions, *Nucl. Phys. B* **385** (1992) 395-412; e-Print Archive: hep-lat/9204004;
S. Catterall, J. Kogut and R. Renken: Phase structure of four-dimensional simplicial quantum gravity. *Phys. Lett. B* **328** (1994) 277-283; e-Print Archive: hep-lat/9401026.

74. S. Catterall, R. Renken and J. Kogut: Singular structure in 4-d simplicial gravity, *Phys. Lett. B* **416** (1998) 274-280, e-Print Archive: hep-lat/9709007;
S. Catterall, G. Thorleifsson, J. Kogut and R. Renken: Singular vertices and the triangulation space of the d sphere, *Nucl. Phys. B* **468** (1996) 263-276, e-Print Archive: hep-lat/9512012;
T. Hotta, T. Izubuchi and J. Nishimura: Singular vertices in the strong coupling phase of four-dimensional simplicial gravity, *Prog. Theor. Phys.* **94** (1995) 263-270, e-Print Archive: hep-lat/9709073;
P. Bialas, Z. Burda, B. Petersson and J. Tabaczek: Appearance of mother universe and singular vertices in random geometries, *Nucl. Phys. B* **495** (1997) 463-476, e-Print Archive: hep-lat/9608030.
75. S. Bilke, Z. Burda, A. Krzywicki, B. Petersson, J. Tabaczek and G. Thorleifsson: 4-d simplicial quantum gravity interacting with gauge matter fields, *Phys. Lett. B* **418** (1998) 266-272, e-Print Archive: hep-lat/9710077.
76. S. Bilke, Z. Burda, A. Krzywicki, B. Petersson, J. Tabaczek and G. Thorleifsson: 4-d simplicial quantum gravity: matter fields and the corresponding effective action, *Phys. Lett. B* **432** (1998) 279-286, e-Print Archive: hep-lat/9804011.
77. J. Ambjørn, K.N. Anagnostopoulos and J. Jurkiewicz: Abelian gauge fields coupled to simplicial quantum gravity, *JHEP* 9908:016 (1999), e-Print Archive: hep-lat/9907027.
78. B.V. de Bakker and J. Smit: Two point functions in 4-d dynamical triangulation, *Nucl. Phys. B* **454** (1995) 343-356, e-Print Archive: hep-lat/9503004.
79. P. Bialas, Z. Burda, A. Krzywicki and B. Petersson: Focusing on the fixed point of 4-d simplicial gravity, *Nucl. Phys. B* **472** (1996) 293-308, e-Print Archive: hep-lat/9601024;
B.V. de Bakker: Further evidence that the transition of 4-d dynamical triangulation is first order, *Phys. Lett. B* **389** 238-242, e-Print Archive: hep-lat/9603024.
80. H.W. Hamber: Phases of simplicial quantum gravity in four-dimensions: estimates for the critical exponents, *Nucl. Phys. B* **400** (1993) 347-389; On the gravitational scaling dimensions, *preprint* Irvine UCI-99-20, e-Print Archive: hep-th/9912246.
81. J. Ambjørn, J. Jurkiewicz and C.F. Kristjansen: Quantum gravity, dynamical triangulation and higher derivative regularization, *Nucl. Phys. B* **393** (1993) 601-632, e-Print Archive: hep-th/9208032.
82. J. Jurkiewicz and Z. Tabor: Critical properties of the two-dimensional gravity with the R^2 action. *Acta Phys. Polon. B* **25** (1994) 1087-1097.
83. V.A. Kazakov, M. Staudacher and T. Wynter: Exact solution of discrete two-dimensional R^2 gravity, *Nucl. Phys. B* **471** (1996) 309-333, e-Print Archive: hep-th/9601069.
84. J. Ambjørn, J. Jurkiewicz and R. Loll, to be published.
85. I. Antoniadis, P.O. Mazur and E. Mottola: Scaling behavior of quantum four-geometries, *Phys. Lett. B* **323** (1994) 284-291, e-Print Archive: hep-th/9301002; Conformal symmetry and central charges in four-dimensions, *Nucl. Phys. B* **388** (1992) 627-647, e-Print Archive: hep-th/9205015;
I. Antoniadis and E. Mottola: 4-d quantum gravity in the conformal sector, *Phys. Rev. D* **45** (1992) 2013-2025.

INDEX

GPSR Compliance

The European Union's (EU) General Product Safety Regulation (GPSR) is a set of rules that requires consumer products to be safe and our obligations to ensure this.

If you have any concerns about our products, you can contact us on ProductSafety@springernature.com

In case Publisher is established outside the EU, the EU authorized representative is:

Springer Nature Customer Service Center GmbH
Europaplatz 3
69115 Heidelberg, Germany

Batch number: 10755284

Printed by Printforce, the Netherlands